Fundamentals of Tribology and Bridging the Gap Between the Macro- and Micro/Nanoscales

NATO Science Series

A Series presenting the results of scientific meetings supported under the NATO Science Programme.

The Series is published by IOS Press, Amsterdam, and Kluwer Academic Publishers in conjunction with the NATO Scientific Affairs Division

Sub-Series

I. **Life and Behavioural Sciences**	IOS Press
II. **Mathematics, Physics and Chemistry**	Kluwer Academic Publishers
III. **Computer and Systems Science**	IOS Press
IV. **Earth and Environmental Sciences**	Kluwer Academic Publishers

The NATO Science Series continues the series of books published formerly as the NATO ASI Series.

The NATO Science Programme offers support for collaboration in civil science between scientists of countries of the Euro-Atlantic Partnership Council. The types of scientific meeting generally supported are "Advanced Study Institutes" and "Advanced Research Workshops", and the NATO Science Series collects together the results of these meetings. The meetings are co-organized bij scientists from NATO countries and scientists from NATO's Partner countries – countries of the CIS and Central and Eastern Europe.

Advanced Study Institutes are high-level tutorial courses offering in-depth study of latest advances in a field.
Advanced Research Workshops are expert meetings aimed at critical assessment of a field, and identification of directions for future action.

As a consequence of the restructuring of the NATO Science Programme in 1999, the NATO Science Series was re-organized to the four sub-series noted above. Please consult the following web sites for information on previous volumes published in the Series.

http://www.nato.int/science
http://www.wkap.nl
http://www.iospress.nl
http://www.wtv-books.de/nato-pco.htm

Series II: Mathematics, Physics and Chemistry – Vol. 10

Fundamentals of Tribology and Bridging the Gap Between the Macro- and Micro/Nanoscales

edited by

Bharat Bhushan

Ohio Eminent Scholar and The Howard D. Winbigler Professor,
Director, Computer Microtribology and Contamination Laboratory,
Department of Mechanical Engineering,
The Ohio State University, Columbus,
Ohio, U.S.A.

SPRINGER-SCIENCE+BUSINESS MEDIA, B.V.

Proceedings of the NATO Advanced Study Institute on
Fundamentals of Tribology and Bridging the Gap Between the Macro- and
Micro/Nanoscales
Keszthely, Hungary
August 13–25, 2000

A C.I.P. Catalogue record for this book is available from the Library of Congress.

ISBN 978-0-7923-6837-3   ISBN 978-94-010-0736-8 (eBook)
DOI 10.1007/978-94-010-0736-8

Printed on acid-free paper

CONTENTS

vi

3. Wear

4. Lubrication

5. Applications

PREFACE

The word *tribology* was first reported in a landmark report by P. Jost in 1966 (Lubrication (Tribology)--A Report on the Present Position and Industry's Needs, Department of Education and Science, HMSO, London). Tribology is the science and technology of two interacting surfaces in relative motion and of related subjects and practices. The popular equivalent is friction, wear and lubrication. The economic impact of the better understanding of tribology of two interacting surfaces in relative motion is known to be immense. Losses resulting from ignorance of tribology amount in the United States alone to about 6 percent of its GNP or about $200 billion dollars per year (1966), and approximately one-third of the world's energy resources in present use, appear as friction in one form or another. A fundamental understanding of the tribology of the head-medium interface in magnetic recording is crucial to the future growth of the $100 billion per year information storage industry. In the emerging microelectromechanical systems (MEMS) industry, tribology is also recognized as a limiting technology.

The advent of new scanning probe microscopy (SPM) techniques (starting with the invention of the scanning tunneling microscope in 1981) to measure surface topography, adhesion, friction, wear, lubricant-film thickness, mechanical properties all on a micro to nanometer scale, and to image lubricant molecules and the availability of supercomputers to conduct atomic-scale simulations has led to the development of a new field referred to as Microtribology, Nanotribology, or Molecular Tribology (see B. Bhushan, J. N. Israelachvili and U. Landman, "Nanotribology: Friction, Wear and Lubrication at the Atomic Scale," *Nature*, Vol. 374, 1995, pp. 607-616). This field concerns experimental and theoretical investigations of processes, ranging from atomic and molecular scales to microscale, occurring during adhesion, friction, wear and thin-film lubrication at sliding surfaces. Such studies are needed to develop fundamental understanding of interfacial phenomena on a small scale and to study interfacial phenomena in micro- and nanostructures. Friction and wear on micro- and nanoscales have been found to be generally smaller compared to that at macroscales. Therefore, micro/nanotribological studies may identify regimes of ultra-low friction and zero wear.

The field of tribology is truly interdisciplinary. Until recently, it has been dominated by mechanical and chemical engineers who have conducted macro-tests to predict friction and wear lives in machine components and developed new lubricants to minimize friction and wear. Development of the field of micro/nanotribology has attracted many more physicists and chemists who have significantly contributed to the fundamental understanding of friction and wear processes on an atomic scale. Thus, tribology is now studied by both engineers and scientists. The research in micro/nanotribology has grown very rapidly and it has developed as an established field.

Because of the important role tribology plays in the industrial world, it is important that young scientists and engineers be aware of the opportunities for research in this field. Because of the multidisciplinary nature of the research and diversity of the research techniques used, it is difficult for young scientists and engineers to become familiar with the basic theories and applications of this field. The emerging field of micro/nanotribology is not covered in any undergraduate or graduate scientific curriculum.

The first institute focusing on the emerging field of Micro/Nanotribology and Its Applications was organized in Sesimbra, Portugal in June 1996. The major focus at this institute was on studies using atomic force microscopy and related techniques in the areas of roughness, adhesion, friction, wear and lubrication and molecular dynamic simulations. A lot had been accomplished in the meantime and the feeling of some of the researchers was that macro- and micro-researchers may not be communicating enough and not growing together. This was considered to be an opportune time to have an institute where both camps can present, educate, and interact. Researchers from diverse disciplines of science and engineering have been working in micro/nanotribology and this course was expected to bring these people together and help develop the emerging discipline of micro/nanotribology. Furthermore, it helped develop further interactions between macro- and micro/nano-researchers. The scope of the course was broad and was of general interest to all working in the field of tribology.

The primary goal of the proposed ASI was to provide a forum whereby young scientists and engineers can become familiar with the state-of-the art of tribology on macro- and micro/nanoscales. To accomplish this goal, we invited internationally recognized lecturers working on both the fundamental and applied aspects of both macro- and micro/nanotribology. The course started out with the history of macro- and micro/nanotribology, followed by instrumentation, basic theories and latest results on friction, wear, lubrication, and industrial applications. A variety of research tools used in the research on macro- and micro/nanoscales were covered. Both experimental and computer modeling research were covered. Emphasis in the course was on applied aspects. The program for this ASI stressed science and engineering topics that are not typically available to students in any structured university curriculum, special course or a single conference. Participants were invited to give brief oral presentations or to present posters dealing with their current research. We attempted to invite students from both universities and industry in order to provide a strong interaction between academic research scientists and engineers. Further a special effort was made to attract and support students from the NATO partner countries, the Mediterranean Dialogue Countries and such NATO countries as Czech Republic, Poland, Hungary, Turkey, Greece, Spain and Portugal. Special efforts were made to obtain additional funding in order to support a larger number of students.

Response to the institute was overwhelming. There were 15 lecturers, 18 invited speakers and a total of 94 participants that attended the course. Lecturers and participants came from a total of 23 countries (11 NATO countries, 8 NATO Partner countries, 1 Mediterranean Dialogue countries, and 3 Other countries). A vendor exhibition of scientific equipment was held concurrently with the course for three days.

We thank NATO Scientific and Environmental Affairs Division for the financial support. Additional funding to provide travel support to U.S. participants was provided by the National Science Foundation (Dr. J. Larsen-Basse, Manager, Surface Engineering and Tribology Program). Additionally, Directorate for Education and Human Resources of National Science Foundation provided travel support to three U.S. student participants. The Scientific and Technical Research Council of Turkey (TUBITAK) provided travel support to three Turkish participants. The Instituto de Cooperacao Cientifice, e Tecnologica Internacional, Portugal provided travel support to one Portuguese participant.

I would also like to thank members of the International Organizing Committee. Professor Miklos Zrinyi visited Danubius Hotel Helikon in Fall 1999 and selected this resort as a site for the course. He organized an excellent social program. Professors Othmar Marti and Philippe Kapsa made important recommendations to the technical program. Finally, I would like to thank my two students and my secretary. Mr. Wei Peng designed and maintained the web site. Mr. William W. Scott handled all the manuscripts for this proceeding. Ms. Jennifer Pursell handled all communications and mailings. I hope that you find the proceeding productive and enjoyable.

Professor Bharat Bhushan
ASI Director
Columbus, Ohio, U.S.A.
August 25, 2000

Group Photo

Participants list

Mr. Joakim B. ANDERSSON
Uppsala University
The Tribomaterials Group
Angstrom Laboratory
Materials Science, Box 534
SE-751 21
SWEDEN
Tel: 46 18 471 10 86
Fax: 46 18 471 35 72
joakim.andersson@angstrom.uu.se

Dr. Roberto BASSANI
University of Pisa
Department of Mechanical, Nuclear
and Production Engineering
Via Diotisalvi, 2
Pisa 56126
ITALY
Tel: 39-050-585-217
Fax: 39-050-585-265
bassani@ing.unipi.it

Dr. Bharat BHUSHAN
The Ohio State University
Department of Mechanical Engineering
206 West 18th Avenue
Columbus, OH 43210-1107
USA
Tel: 614-292-0651
Fax: 614-292-0325
bhushan.2@osu.edu

Dr. Peter J. BLAU
Oak Ridge National Laboratory
P.O. Box 2008, MS 6063
Oak Ridge, TN 37831-6063
USA
Tel: 865-574-5377
Fax: 865-574-6918
blaupj@ornl.gov

Mr. Renato BUZIO
Universita' di Genova
Dipartimento di Fiscia
Via Dodecaneso 33
Genova 16146
ITALY
Tel: 39-010-353-6356
Fax: 39-010-362-2790
buzio@fisica.unige.it

Dr. Ahmet CAKIR
Dokuz Eylul University
Department of Metallurgical and Materials
Engineering

Bornova/Izmir 35100
TURKEY
Tel: 90-232-388-28-80 / 13
Fax: 90-232-388-78-64
ahmet.cakir@deu.edu.tr

Dr. Sorin N. CIORTAN
Dunarea de Jos University
Department of Mechanical Engineering
47 Domneasca Street
Galati, 6200
ROMANIA
Tel: 40-36-41-48-73
Fax: 40-36-46-13-53
sorin.ciortan@ugal.ro

Dr. Enrico CIULLI
University of Pisa
Department of Mechanical, Nuclear and
Production Engineering
Via Diotisalvi, 2
Pisa 56126
ITALY
Tel: 39-050-585-261
Fax: 39-050-585-265
ciulli@ing.unipi.it

Dr. Hector McI. CLARK
Maros utca 25, I/2
1122 Budapest
HUNGARY
Tel: 36 1 356 2615
Fax: None
bori_clark@compuserve.com

Ms. Tonya Shea COFFEY
North Carolina State University
2808 Isabella Drive
Raleigh, NC 27603
USA
Tel: 919-833-4735
Fax: 919-515-1333
tscoffey@unity.ncsu.edu

Dr. Jaime COLCHERO
Universidad Autonoma de Madrid
Campus Cantoblanco
Departamento de Fisica de la Materia
Condensada C-III-205
E28049, Madrid
SPAIN
Tel: 34-91-39-74-754
Fax: 34-91-39-73-961
jaime.coclhero@uam.es

Mr. Richard S. COWAN
Georgia Institute of Technology
Center for Integrated Diagnostics
801 Ferst Drive
Atlanta, GA 30332-0405
USA
Tel: 404-894-3270
Fax: 404-894-8336
rick.cowan@me.gatech.edu

Dr. Mario D'ACUNTO
via Cecco di Pietro 5
1-56123 Pisa
ITALY
Tel: 39-050-560247
m.dacunto@ing.unipi.it

Dr. Leonid DAIKHIN
Tel Aviv University
School of Chemistry
Ramat Aviv
69978 Tel Aviv
ISRAEL
Tel: 972-3-6408902
Fax: 972-3-6409293
daikhin@post.tau.ac.il

Dr. Jean DENAPE
National Engineering School of Tarbes
BP 1629
65016 TARBES Cedex
FRANCE
Tel: 00-33-562-44-2728
Fax: 00-33-562-44-2708
denape@enit.fr

Dr. Abdallah A. ELSHARKAWY
Kuwait University
Department of Mechanical and
Industrial Engineering
P.O. Box 5969, Safat 13060
KUWAIT
Tel: 965 533 1082
Fax: 965 484 7131
abdallah@kuc01.kuniv.edu.kw

Mr. Michel FAJFROWSKI
MTS Systems
15 av Jean Jaures BP 238
94203 Ivry sur Seine Cedex
FRANCE
Tel: 33-146-70-1180
Fax: 33-146-58-3514
michel.fajfrowski@mts.com

Mr. Jozsef FEHER
Budapest University of Technology
and Economics
Budapest 1521
HUNGARY
Tel: 36-1-463-3229
Fax: 36-1-463-3767
joe.fkt@chem.bme.hu

Ms. Genoveva FILIPCSEI
Budapest University of Technology
and Economics
Budapest 1521
HUNGARY
Tel: 36-1-463-3229
Fax. 36-1-463-3767
geni.fkt@chem.bme.hu

Dr. Friedrich FRANEK
Vienna University of Technology
Institute for Precision Engineering
Floragasse 7
Vienna, A1040
AUSTRIA
Tel: 43-1-58801-35801
Fax: 43-1-58801-35899
friedrich.franek@tuwien.ac.at

Dr. Joost W. M. FRENKEN
Leiden University
Kamerlingh Onnes Laboratory
P. O. Box 9504
2300 RA Leiden
THE NETHERLANDS
Tel: 31-71-5275603
Fax: 31-71-5275404
frenken@phys.leidenuniv.nl

Mr. Vincent FRIDRICI
University Ecole Centrale de Lyon
Ecole Centrale de Lyon
36 Avenue Guy de Collongue BP 163
69130 Ecully
FRANCE
Tel: 33-4-72-18-65-62
Fax: 33-4-78-43-33-83
vincent.fridrici@ec-lyon.fr

Dr. Isaac I. GARBAR
Ben-Gurion University of the Negev
Mechanical Engineering Department
P.O. Box 653
Beer-Sheva, 84105
ISRAEL
Tel: 972-7-6477070
Fax: 972-7-6472813
garbar@menix.bgu.ac.il

Dr. Michael N. GARDOS
Raytheon Electronic Systems
P.O. Box 902, E1/C182
El Segundo, CA 90245
USA
Tel: 310-647-4357
Fax: 310-647-3536
mngardos@west.raytheon.com

Ms. Danya GLUSCHOVE-CORBY
Imperial College of Science, Technology and
Medicine
Tribology Section
Mechanical Engineering Dept.
Exhibition Road
London SW7 2BX, England
UK
Tel: 44-207-594-7236
Fax: 44-207-823-8845
d.gluschove-corby@ic.ac.uk

Mr. Jaroslaw GROBELNY
Univeristy of Lodz
163, Pomorska St. Lodz
90-236
POLAND
Tel: 48-42-635-56-99
Fax: 48-42-678-70-87
jgrobel@krysia.uni.lodz.pl

Dr. Bernard J. HAMROCK
The Ohio State University
Department of Mechanical Engineering
206 West 18th Avenue
Columbus, OH 43210-1107
USA
Tel: 614-292-4930
Fax: 614-292-3163
hamrock.1@osu.edu

Dr. Ude HANGEN
Hysitron/SURFACE
Rhein Str 7
D-41836 Hueckelhoven
GERMANY
Tel: 49-243-397-0305
Fax: 49-243-397-0302
u.hangen@surface-tec.com

Dr. Ken'ichi HIRATSUKA
Chiba Institute of Technology
2-17-1, Tsudanuma, Narashino-shi
Chiba 275-8588
JAPAN
Tel: 81-47-478-0503
Fax: 81-47-478-0529
hiratsuka@pf.it-chiba.ac.jp

Dr. James D. HOLBERY
CSEM Instruments, SA
Jaquet-Droz 1
Neuchatel, CH-2000
SWITZERLAND
Tel: 41-32-720-5847
Fax: 41-32-720-5730
james.holbery@csem.ch

Dr. Stephen M. HSU
National Institute of Standards
and Technology
Room A256, Building 223
Mail Code 8520, NIST
Gaithersburg, MD 20899
USA
Tel: 301-975-6120
Fax: 301-975-5334
stephen.hsu@nist.gov

Dr. Irina HUSSAINOVA
Tallinn Technical University
Ehitatate Tee 5
Tallinn 19086
ESTONIA
Tel: 372-620-3303
Fax: 372-620-2020
irina@meo.ttu.ee

Mr. Tae-Yeon HWANG
Samsung Advanced Institute
of Technology
95 West Plumeria Drive
San Jose, CA 95134
USA
Tel: 408-544-5696
Fax: 408-544-5665
tyhwang@sisa.samsung.com

Dr. Jacob ISRAELACHVILI
University of California at
Santa Barbara
Department of Chemical Engineering
Engineering II, Room 3357
Santa Barbara, CA 93106
USA
Tel: 805-893-8407
Fax: 805-893-7870
jacob@engineering.ucsb.edu

Dr. Yutaka IWAMOTO
Sony Corporation
Communication and Network Co.
2-1-1 Shinsakuragaoka, Hodogaya-ku
Yokohama-shi, Kanagawa-ken,
240-0036
JAPAN
Tel: 81-45-353-6840
Fax: 81-45-353-6907
yutaka.iwamoto@jp.sony.com

xviii

Dr. Czeslaw K. KAJDAS
Warsaw University of Technology
Department of Chemistry
17 Lukasiewicza Street
09-400 Plock
POLAND
Tel: 48-24-367-2191
Fax: 48-24-262-7494
ckajdas@zto.pw.plock.pl

Dr. Philippe KAPSA
Laboratory of Tribology and Systems Dynamics
UMCR CNRS 5513
Ecole Centrale de Lyon, BP 163
69131 Ecully Cedex
FRANCE
Tel: 33-472-186274
Fax: 33-478-433383
philippe.kapsa@ec-lyon.fr

Dr. Rainer KASSING
University of Kassel
Institute for Technology Physics
D-34109 Kassel
GERMANY
Tel: 0561-804-4532
Fax: 0561-804-4136
kassing@physik.uni-kassel.de

Dr. Alexey I. KHARLAMOV
Institute for Problems of Materials
Science
3, Krjijanovski str
03142 Kiev
UKRAINE
Tel: 038-044-4440256
Fax: 038-044-4442131
dep73@ipms.kiev.ua

Dr. Jacob KLEIN
Wiezmann Institute of Science
P.O. Box 26
Rehovot 76100
ISRAEL
Tel: 972-8-934-3823
Fax: 972-8-934-4138
jacob.klein@weizmann.ac.il

Dr. Eugeniusz KLUGMANN
Technical University of Gdansk
ETI Faculty
Narutowicza 11-12
PL-80-952 Gdansk
POLAND
Tel: 48-58-347-17-34
Fax: 48-58-41-61-32
eklug@pg.gda.pl

Mr. Lior KOGUT
University - Technion - Israel Institute
of Technology
Faculty of Mechanical Engineering - Technion
Haifa 32000
ISRAEL
Tel: 972-04-8292090
Fax: 972-04-8292065
mekogut@tx.technion.ac.il

Dr. Mihaly KOZMA
Budapest University of Technology
and Economics
1521, Budapest, P.O. Box 91
Budapest 1521
HUNGARY
Tel: 36-1-463-2363
Fax: 36-1-463-3510
kozma@xenia.gee.bme.hu

Mr. Andrzej J. KULIK
Ecole Polytechnique Federale de Lausanne
EPFL-DP/IGA
CH-1015 Lausanne
SWITZERLAND
Tel: 41-21-693-3359
Fax: 41-21-693-4470
root@igahpse.epfl.ch

Dr. Uzi LANDMAN
Georgia Institute of Technology
School of Physics
837 State Street, NW
Atlanta, GA 30332-0430
USA
Tel: 404-894-3368
Fax: 404-894-7747
uzi.landman@physics.gatech.edu

Mr. Karsten LANDWEHR
Hysitron/SURFACE
Rhein Str 7
D-41836 Hueckelhoven
GERMANY
Tel: 49-243-397-0305
Fax: 49-243-397-0302
k.landwehr@surface-tec.com

Dr. Jorn LARSEN-BASSE
National Science Foundation
4201 Wilson Boulevard
Room 545
Arlington, VA 22230
USA
Tel: 703-306-1361
Fax: 703-306-0291
jlarsenb@nsf.gov

Dr. Thomas LIEW
Data Storage Institute
National University of Singapore
5 Engineering Drive 1
Singapore 117608
SINGAPORE
Tel: 65-874-8519
Fax: 65-777-2406
tomliew@mail.dsi.nus.edu.sg

Dr. Jean-Luc LOUBET
CNRS
Ecole Centrale de Lyon
LTDS - Bat H10 - BP 163
69131 Ecully Cedex
FRANCE
Tel: (33) 472 186281
Fax: (33) 478 433383
jean-luc.loubet@ec-lyon.fr

Dr. Kenneth C. LUDEMA
University of Michigan
Mechanical Engineering and Applied Mechanics
2250 GG Brown
Ann Arbor, MI 48109
USA
Tel: 734-764-3364
Fax: 734-647-3170
kenlud@umich.edu

Dr. Gustavo S. LUENGO
L'OREAL
Applied Physics Department
90 rue General Roguet
Clichy, 92583 Cedex
FRANCE
Tel: 33-1-47567326
Fax: 33-1-47567962
gluengo@recherche.loreal.com

Dr. Othmar MARTI
University of Ulm
Department of Experimental Physics
Albert-Einstein-Allee 11
89069 Ulm
GERMANY
Tel: 49 731 502 3011
Fax: 49 731 502 3036
othmar.marti@physik.uni-ulm.de

Dr. Ernst MEYER
Institute of Physics
University of Basel
Klingelbergstr. 82
4056 Basel
SWITZERLAND
Tel: 41 61 267 37 24 or 67
Fax: 41 61 267 37 95
ernst.meyer@unibas.ch

Dr. Seizo MORITA
Osaka University
Department of Electronic Engineering
Osaka University, 2-1 Yamado-Oka
Suita, Osaka, 565-0871
JAPAN
Tel: 81-6-6879-7761
Fax: 81-6-6879-7764
smorita@ele.eng.osaka-u.ac.jp

Dr. Martin H. MUESER
Johannes Gutenberg-Universitat
Institut fur Physik, WA 331
55099, Mainz
GERMANY
Tel: 49-6131-392-3646
Fax: 49-6131-392-5441
martin.mueser@uni-mainz.de

Mr. Matthias MUELLER
Universitaet Karlsruhe
D-76128 Kalsruhe
GERMANY
Tel: 49-721-608-3423
Fax: 49-721-607-593
matthias.mueller@physik.uni-karlsruhe.de

Dr. Nikolai K. MYSHKIN
Metal-Polymer Research Institute of Belarus
National Academy of Sciences
32 a Kirov Street
Gomel 246050
BELARUS
Tel: 375-232-526273
Fax: 375-323-526273
nkmyshkin@mail.ru

Dr. Masayuki NAKAYAMA
Dai-ichi Unviersity
1-10-2, Chuo, Kokubu-shi
Kagoshima-ken
JAPAN
Tel: 81-995-45-0640, ext. 3221
Fax: 81-337-00-2234
nakamasa@ca2.so-net.ne.jp

Mr. Michael O'HERN
MTS - Nano Instruments Innovation Center
1001 Larson Drive
Oak Ridge, TN 37830
USA
Tel: 865-481-8451
Fax: 865-481-8455
mike.ohern@mts.com

Dr. Yilmaz OZMEN
Pamukkale University
Denizli Meslek Yuksekokulu, Ulucarsi
20100 Denizli

TURKEY
Tel: 90-258-263-3794
Fax: 90-258-263-3794
yozmen@hotmail.com

Dr. Liviu C. PALAGHIAN
Dunarea de Jos University
Department of Mechanical Engineering
47, Domneasca Street
Galati, 6200
ROMANIA
Tel: 40-36-41-48-73
Fax: 40-36-46-13-53
liviu.palaghian@ugal.ro

Mr. Wei PENG
The Ohio State University
Department of Mechanical Engineering
206 West 18th Avenue
Columbus, OH 43210-1107
USA
Tel: 614-292-4825
Fax: 614-292-3163
peng.44@osu.edu

Ms. Ulrika B. PETTERSSON
Uppsala University
Tribomaterials Group
The Angstrom Laboratory
P.O. Box 534
S-751 21
SWEDEN
Tel: 48-18-471 31 14
Fax: 46-18-471 35 72
ulrika.pettersson@material.uu.se

Dr. Ireneusz PIWONSKI
University of Lodz
163, Pomorska St. Lodz
90-236
POLAND
Tel: 48-42-635-58-33
Fax: 48-42-678-70-87
irek@krysia.uni.lodz.pl

Dr. Adam POLAK
Cracow University of Technology M4
ul. Warszawski 24
31155 Krakow
POLAND
Tel: 48 12 648 0555, ext. 3543
Fax: 48 12 648 8131
aspolak@mech.pk.edu.pl

Dr. Andreas A. POLYCARPOU
University of Illinois at
Urbana-Champaign
Department of Mechanical and
Industrial Engineering
1206 West Green Street

332DMEB
MC-244
Urbana, IL 61801
USA
Tel: 217-244-1970
Fax: 217-244-6534
polycarp@uiuc.edu

Dr. Stanislaw J. PYTKO
The Technical University of Mining
and Metallurgy
Al.Mickiewicza 30
30-059 Krakow
POLAND
Tel: 48-12-617-3065
Fax: 48-12-411-6295
s_pytko@uci.agh.edu.pl

Ms. Minodora I. RIPA
Universitatea of Dunarea de Jos
Str. Domneasca 47
6200 Galati
ROMANIA
Tel: 40-36-46-13-53
Fax: 40-36-46-13-53
minodora.ripa@ugal.ro

Dr. Zygmunt RYMUZA
Warsaw University of Technology
Chodkiewicza 8
02-525 Warsaw
POLAND
Tel: 48-22-660-8602
Fax: 48-22-660-8601
kup_ryz@mech.pw.edu.pl

Dr. Salim SAHIN
Celal Bayar University
Faculty of Engineering
Department of Mechanical Engineering
Manisa/45140
TURKEY
Tel: 90-236-241-2144 ext. 274
Fax: 90-236-241-2143
salimsahin@hotmail.com

Dr. Miquel SALMERON
Lawrence Berkeley National Laboratory
Materials Sciences Division
Mail Stop 66
Berkeley, CA 94720
USA
Tel: 510-486-6230
Fax: 510-486-4995
salmeron@stm.lbl.gov

Dr. Steven R. SCHMID
University of Notre Dame
377 Fitzpatrick Hall
Notre Dame, IN 46556

USA
Tel: 219-631-9489
Fax: 219-631-8341
schmid.2@nd.edu

Dr. Thomas SCHIMMEL
Universitaet Karlsruhe
D-76128 Kalsruhe
GERMANY
Tel: 49 721 608 3570
Fax: 49 721 607 593
thomas.schimmel@physik.uni-karlsruhe.de

Dr. Udo D. SCHWARZ
Institute of Applied Physics
University of Hamburg
Jungiusstr, 11
Hamburg, D-22087
GERMANY
Tel: 49-40-42838-6297
Fax: 49-40-42838-5311
schwarzu@physnet.uni-hamburg.de

Mr. William W. SCOTT
The Ohio State University
Department of Mechanical Engineering
206 West 18th Avenue
Columbus, OH 43210-1107
USA
Tel: 614-292-4825
Fax: 614-292-3163
scott.445@osu.edu

Mr. Tamas SEBESTYEN
Budapest University of Technology and
Economics, BAYATI
Bertalan L. u. 2. 608.
H-1111
HUNGARY
Tel: 361-463-1694
Fax: 361-463-3467
sebi@bzaka.hu

Dr. Valentinas SNITKA
Kaunas University of Technology
Research Center for Microsystems
and Nanotechnology
Kaunas 3031
LITHUANIA
Tel: 37 07 45 15 88
Fax: 37 07 45 15 93
vsnitka@microsys.ktu.lt

Dr. Andres SOOM
University at Buffalo
SEAS 412 Bonner Hall
Buffalo, NY 14260
USA

Tel: 716-645-2772, ext. 1105
Fax: 716-645-2495
soom@eng.buffalo.edu

Dr. Hugh A. SPIKES
Imperial College
Department of Mechanical Engineering
London SW7 2BX
UK
Tel: 44-207-594-7064
Fax: 44-207-823-8845
h.spikes@ic.ac.uk

Mr. Thomas STIFTER
University of Ulm
Department of Experimental Physics
D-89069 Ulm
GERMANY
Tel: 69 737 5023078
Fax: 69 737 5023036
thomas.stifter@physik.uni-ulm.de

Dr. Jeffrey L. STREATOR
Georgia Institute of Technology
Room 4206, MRDC Building
801 Ferst Street
Atlanta, GA 30332-0405
USA
Tel: 404-894-2742
Fax: 404-894-8336
jeffrey.streator@me.gatech.edu

Dr. Andras Z. SZERI
University of Delaware
126 Spencer Lab
Newark, DE 19716
USA
Tel: 302-831-8017
Fax: 302-831-6751
szeri@me.udel.edu

Dr. Matthew TIRRELL
University of California at
Santa Barbara
Office of the Dean
College of Engineering
Santa Barbara, CA 93106-5130
USA
Tel: 805-893-3141
Fax: 805-893-8124
tirrell@engineering.ucsb.edu

Dr. Mustafa TOPARLI
Dokuz Eylul University
Faculty of Engineering
Department of Metallurgical and
Materials Engineering
Bornova/Izmir 35100

TURKEY
Tel: 90-232-388-2880-17
Fax: 90-232-388-7864
mustafa.toparli@deu.edu.tr

Mrs. Ksenija TOPOLOVEC MIKLOZIC
Imperial College of Science Technology and
Medicine
Tribology Section
Mechanical Engineering Dept.
Exhibition Road
London SW7 2BX, England
UK
Tel: 44-207-594-7236
Fax: 44-207-823-8845
k.topolovec@ic.ac.uk

Dr. Waldemar TUSZYNSKI
Institute for Terotechnology
ul. K. Pulaskiego 6/10
Radom 26-600
POLAND
Tel: (0048 48) 44241 ext. 209
Fax: (0048 48) 44760 or 44765
waldemar.tuszynski@itee.radom.pl

Dr. Mahomed Hanif USSMAN
University of Beira Interior
Department of Textile Engineering
Rua Marques Avila e Bolama
6200 Covilha
PORTUGAL
Tel: 351 275 319700
Fax: 351 275 319888
ussamn@ciunix.ubi.pt

Dr. Brian L. WEICK
University of the Pacific
3601 Pacific Avenue
Stockton, CA 95211
USA
Tel: 209-946-3084
Fax: 209-946-3102
bweick@uop.edu

Mr. Ludger WEISSER
Digital Instruments GmbH - Veeco
Janderstrasse G
D-68199 Mannheim
GERMANY
Tel: 49-621-842100
Fax: 49-621-842-1022
weisser@digmbh.de

Dr. Bodo WOLF
Tu Dresden, Physik, JKFP
D-01062 Dresden
GERMANY
Tel: 49-351-463-5522
Fax: 49-351-463-7048
wolf@physik.phy.tu-dresden.de

Mr. Thomas WYROBEK
Hysitron, Inc.
5251 West 73rd Street
Minneapolis, MN 55439
USA
Tel: 612-835-6166
Fax: 612-835-6166
wyrobek@hysitron.com

Ms. Xiaoyin XU
Imperial College of Science Technology and
Medicine
Tribology Section
Mechanical Engineering Dept.
South Kensington
London SW7 2BX, England
UK
Tel: 44-207-594-7236
Fax: 44-207-832-8845
xiaoyin.xu@ic.ac.uk
Dr. Anatoly L. ZHARIN
Belarussian State Research & Production
Powder Metallurgy Concern
Deputy Director General
Head of Tribology Laboratory
41 Platonov Street
Minsk, 220071
BELARUS
Tel: 375 (17) 232-83-62
Fax: 375 (17) 210-09-77
zharin@hotmail.com

Dr. Miklos ZRINYI
Budapest University of Technology
and Economics
Budapest 1521
HUNGARY
Tel: 36-1-463-3229
Fax: 36-1-463-3767
zrinyi.fkt@chem.bme.hu

HISTORY OF TRIBOLOGY
AND ITS INDUSTRIAL SIGNIFICANCE

Ken Ludema, Professor Emeritus
Department of Mechanical Engineering and Applied Mechanics
The University of Michigan,
Ann Arbor, MI, USA, 48109-2125

1. Introduction

THE definitive and complete early history of tribology has been written by Duncan Dowson (1979). He wisely focused on authors and events that have withstood the test of scientific and technological relevance. Some examples in his writing display the ingenuity of man in such items as support-pivots for the gates of ancient cities, the axles of chariots and wagons, and various articles of war. These items may seem mundane to us but they were vital for survival at the time. Friction was a major concern of that day because motive power was supplied largely by people or animals.

Since we know of no formal body of knowledge that guided the design of these early items we assume that they were empirically developed, that is, designed from the experience of artisans. Slowly, the scientists and mathematicians began to formalize knowledge in tribology, including the oft quoted daVinci, Amontons and Coulomb. Empiricism in tribology surely continues to this day, however, and seems to yield very slowly to scientific methods. Many, if not most, of the simpler consumer products we can buy today are still designed and made by skilled people who spend little time on tribological aspects of a product. They select available materials and lubricants, not particularly from thorough knowledge, and certainly without themselves realizing that there are several sciences "behind" what they choose. The more complicated products are developed with the aid of extensive testing, often inadequately directed or focused on the real tribological issues involved.

Historical events will be mentioned further, but this paper is mostly about the industrial significance of tribology today. The fact that this topic is on the agenda of this conference belies a suspicion, or even a conviction that industry does not see our contributions as particularly relevant. We should know why since we see very important discoveries emerging from our work. There are two concerns about our relevance:

a. Have we made significant contributions within the industries of our nations? Surely "we" have but mostly in the days in which products were simpler and the contribution of any of the specialized skills in tribology produced major advances in the products.

b. From the industrial point of view, are our tribological contributions seen as significant to industry? They clearly were in times past, but this view is changing for the worse. The likely reason is that products are moving toward optimization and higher sophistication so that single disciplines no longer suffice and and "we" have not kept pace with cross disciplinary solutions to problems.

Since we will be using the term "tribology" often it would be well to define the term.

1

B. Bhushan (ed.),
Fundamentals of Tribology and Bridging the Gap between the Macro- and Micro/Nanoscales, 1–11.
© 2001 *Kluwer Academic Publishers.*

2. Tribology and Tribogists

Here, from The American Heritage Dictionary:
  Tribology:
"The science of the (basic) mechanisms of friction, lubrication and wear of interacting mechanisms (ie, mechanical 'things') that are in relative motion".

   The "science"?   Not the technology?

  Science:
a. "The observation, identification, description, experimental investigation and theoretical explanation of natural phenomena". [a human endeavor]
b. "An activity that *appears* to require study and method".
c. "Knowledge gained through experience". [a body of knowledge]
        (A scientist is one who is actively engaged in the observation ...... of natural phenomena.)

  Technology:
a. "The application of science, esp, to industrial or commercial objectives",
b. "The entire body of methods and materials used to achieve such objectives".

   Figure 1 shows several possible scopes of the term "tribology," encompassing the sciences and technology.

3. The Transition of Tribology from Art Toward a Science

   The railroads and ship building industries may have propelled tribology as much as any industries. Ships and railroad trains were the first large machines that had to survive though far from their maintenance bases. By the mid 19th century, people knew that steel on steel is a vulnerable combination in sliding bearings but is adequate for rolling contact. They therefore were quite willing to use the lead based alloy of Isaac Babbitt in those bearings. It remained then to find good lubricants. For some service, as in the journal / bearing pair of car (carriage) axles, pork fat was adequate for a few decades whereas refined oils were best for engine mechanisms.
   Toward the end of the 19th century the electrical power business began to grow and that brought attention to bearing performance in both steam turbines and in electrical generators. By this time the use of bearings in which the journal or shaft "floated" on oil was well known and studies were under way on two important issues, that is, minimizing the hazard of start-stop sequences and minimizing energy loss in the bearings. Many mechanical engineers turned their attention to bearing studies and the technical literature was rich in those days on their progress. Rolling element bearings were developing nicely for small applications but were not ready for service in turbines.
   By 1890 fluid film lubrication caught the attention of the fluid mechanics community, which began to formalize knowledge in this area. Reynolds and others

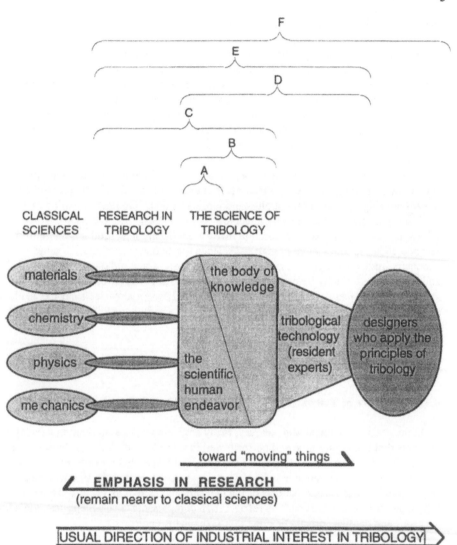

Figure 1. Various views on the breadth of the field of tribology.

explored the characteristics of the oil lubricated bearing and thus hydrodynamics was born, not in a vacuum or without any previous work, but as a stroke of insight into all the work that had gone on before. Hydrodynamics grew into elastohydrodynamics in about 1937 when the equations of hydrodynamics were found inadequate for high contact pressures. The equations of hydrodynamics predicted fluid films that were much thinner than seemed to exist: the missing quantities were the elastic deflection in the contact region and the increase in oil viscosity due to high pressure between the journal and bearing. At about the same time lubricant chemistry was found to strongly influence bearing life, particularly due to the sulfur compounds in oil that were not extracted in the refining processes.

About 50 years ago, active minds began to go beyond empiricism in tribology. Instead of searching for existing materials to satisfy some need, efforts were applied toward developing materials, coatings and lubricants that were intended to be superior to existing products. The science of tribology began to anticipate the needs, which followed the course of basic research in many areas. Many new products came from recently acquired knowledge in material science (particularly coatings), organic chemistry, molecular dynamics and such. Furthermore, some old questions on adhesion and friction were finally being explored in terms of the behavior of atoms and sub atomic particles, just in time to meet the needs of MEMS (micro-electrical-mechanical-systems). Tribology moves on, at least in the minds of tribologists if not in the minds of the captains of industry.

4. One Measure of Tribological Progress, and a Bit of Regression

In the great mechanical 20th century, many engineers were employed in developing products that last a long time and perform well in that lifetime. Some of the progress is seen in the evolution of the automobile from the 1916 Maxwell (luxury touring car) to the 1996 Ford Taurus (midsize sedan). This may be seen, indirectly, in the lubrication requirements in the owner manuals of the two cars and shown in Table 1. There are very many more examples such as refrigeration compressors, aircraft engines, tools for surgeons and dentists, prosthetic bone joints: the list is endless.

Though it is possible to trace the advances in tribology in most mechanical products, it is not so obvious that designated "tribologists" participated in many of these advances. Rather, the advances were more likely the work of diligent engineers and scientists of all kinds. This works well where specialists in one area, such as oil chemists work closely with hydrodynamicists, for example. Such close working is not common however, since chemists are likely to be under a different vice president than are hydrodynamicists, and likely housed in different buildings as well. With such separation the vital questions about a tribological issue do not often come to mind. The result is a great amount of lost work, expensive but fruitless testing, and products that are not really ready for the market. This appears finally in unsatisfactory performance of many products, such as squealing brakes, noisy chairs sliding on marble floors, railroad trains that require great braking distances, squeaking door hinges, short lived prosthetic hip joints, engines that need frequent servicing: this list is also endless.

On occasion the separation of disciplines has rather unexpected consequences. Recently the US Air Force explored the development of a new jet engine for which the ball bearings would be exposed to higher temperatures and higher loads than previously. The call went out for an improved bearing steel (with higher fatigue life and greater resistance to corrosion) and a more thermally stable lubricant. The new steel and the new lubricant became available in due course, only the combination does not work well. There had apparently been no thought given to the mechanism of bearing lubrication in the previous versions of steel and lubricant. This mechanism involves a chemical reaction between sliding steel surfaces and some active constituents in the lubricant. Since the new steel and new lubricant were both made to be more chemically inert than previously, the necessary chemical reaction between the new steel and lubricant does not occur. The steel people did their best and the lubricant people did their best but there was no tribologist present who knew of these diverging developments. Likely this same omission produced hundreds of "dead ends" in many other products. But good minds made progress anyway, and we now have a technology that would boggle the mind of 18th century man.

5. The Early Industrial View of Tribology and Tribologists

Tribologists were not often so designated before 1966 when the term was coined by the Oxford Dictionary. Small industries rarely have specialists in any topic such as tribology, but the bigger industries do. There were to be found in the "lubricants lab" or the "bearing group" or the "mechanics group", usually well separated as mentioned before.

The reasons are not clear but tribologists of all types enjoyed increasing esteem from the 1930's onward in industry. Perhaps the reason was that all technical topics prospered in that era, or perhaps some of the most visible progress in new products was made in product wear life. In any case, the industrial people were prominent in the formation of several technical societies in tribology. The mechanical engineers formed the Lubrication Division (now called the Tribology Division) within the American Society of Mechanical Engineers (ASME). From this group came the independent American Society of Lubrication Engineers (now called the Society of Tribologists and Lubrication Engineers, the STLE) and several committees within the American Society for Testing and Materials (ASTM). The lubricants specialists were active in the American Chemical Society (ACS) and several committees in the ASTM. The materials specialists (metallurgists before polymers and ceramics became widely used) were active in the American Society for Metals (ASM) and again in the ASTM committees. All of these societies prospered from the financial support of industry, but also because industrial employees were given time to keep the societies and committees moving. Industries even supported their employees in travels to conferences in foreign nations. The several conferences on tribology held high interest and the bigger ones attracted 300 to 500 participants.

A measure of the topical interest in early tribology conferences can be seen from the programs of the biannual and week long Gordon Research Conferences in the 1960's and 1970's. Three days were devoted to the mechanics of contact and lubrication, the fourth day was usually on the chemical aspects of lubrication and the fifth day was on the material aspects of friction and wear. Participants from academia and government labs were in small minority.

6. Parallel Trends in Tribology

Over time since the 1940's the three US military branches had formed tribology research groups, and sponsored a great amount of research in academia and in industry. NASA (National Aviation and Space Agency) had a large group of tribologists to help design the space vehicles. The NSF (National Science Foundation) had a Lubrication Program (renamed a separate Tribology Program) which was distinct from its many other programs. Later the DOE (Department of Energy) was formed and sponsored large programs in the conservation of energy and materials, both by in-house research and by sponsorship of research elsewhere. The 1970's were the heyday of Tribology.

Since then there has been a slow erosion of tribology. Any remaining Tribology Program in the branches of the military, in NASA and in the NSF are now found as small sub section under other programs, though NSF has done fairly well in sponsoring young investigators and in establishing focused research centers. The DOE continues also with a reduced program.

7. The Modern Trends in Industry

Most industries in the massive consumer product sector have "downsized" their tribology groups as well since the 1970's, some entirely and some to a small fraction of the former size. Tribology-intensive products (bearings, cutting tools, diesel engines, e.g.) and capital products (jet engines, e.g.) industries have not downsized as much since most such products gain market advantage by extending the reliable life of their products. The oil companies have downsized their tribology groups very severely because the profit margins in lubricants and fuels does not warrant much product improvement since product improvement is likely to reduce the use of their products!

The downsizing in the consumer product industry is probably not connected with the downsizing in the government agencies. The government downsizing is most likely a matter of less emphasis on military needs and more emphasis on social issues. Industry, however, responds to the economics of production of consumer products. In industry, therefore, tribology may have "run its course" in one way or other. Perhaps better tribological performance of products has less value to consumers than such frills as the color schemes on I-Macs or small creature comfort appointments in automobiles.

Or perhaps tribology has fallen from prominence because of tribologists themselves!! In some product lines industry would rather pay warranty costs than fund research to improve those products. Perhaps industry despairs of finding and managing interdisciplinary groups to do the research, but how can we expect them to if we ourselves had not displayed cross disciplinary interests either!. This will be explored in the next sections.

The downsizing of tribology in industry has had the great effect of diminishing the industrial presence in professional societies and publications. Not only are there fewer tribologists in industry but there are now greater restrictions on what may be published for proprietary reasons. The privilege of attending conferences has also declined considerable. The great majority of participants in conferences are now from academia and government groups so that conferences are very much more basic science oriented. These same groups also the dominant the technical journals which almost no product designer reads.

The fact of the diminishing of tribology industry is cause enough for concern, but it does seem that computer intensive groups, the analytical instrument groups, mechanics groups and a few others thrive (survive) while tribology groups disappear. Tribologists had tried to demonstrate our value to an economy by showing how "good tribology" could save 5 or so percent of the the GNP (gross national product) of a nation by conserving energy and materials. It apparently was not a convincing argument. It was probably a case of that 5 percent not appearing in the annual corporate profits. Doubtless the other specialties had also been busy justifying their own existence all along too.

We might be tempted to attribute our plight to the claim that tribology is very complicated, which is, of course true, but it is no more complicated than any other field. More likely we have simply not prepared ourselves to meet the broad needs of the field. Practical tribology problems encompass issues in chemistry, physical chemistry, material science (and engineering), physics, mechanics and several more disciplines. The necessary confluence of these disciplines in common mechanical systems did not seem obvious when many problems could be solved by one discipline. An example is the contributions made by the field of mechanics, solid and fluid. A century ago the methods of mechanics were found to add much understanding to tribology problems. Most other aspects of tribology were dealt with in the background while mechanics papers filled the literature. But, all the while the chemical and materials aspects of problems went only partly solved. More recently the more highly developed mechanics areas have diminished in prominence while material science (and engineering) came forward. A specific example of this transition is frictional vibration (stick-slip?). Quite clearly, to us at least, the vibration of sliding systems originates in the interactions between the sliding surfaces. In the heyday of mechanics, frictional vibration was studied by specialists in mechanical dynamics where vibrations were influenced by the proper amount of damping in the proper places in a mechanism. The very preponderance of such papers strongly suggested that frictional vibrations originate in the system surrounding the sliding surfaces, not the sliding surfaces themselves. Tribologists had no alternative solutions, and still do not, except to use an oil containing "friction modifier". The tendency for sliding pairs to vibrate upon sliding, though a frictional issue, is rarely mentioned in tables of values of coefficient of friction.

Given the lack of interdisciplinary perspective, tribologists cannot often explain the behavior of tribological systems briefly, coherently or with great confidence. We do well in our sub-specialties, but these topics rarely seem germane to product designers. We offer little help to product designers, and in fact, about the most useless information in handbooks are the tables of coefficient of friction. No one who does even the simplest tests (e.g.., using a tilting plane) will believe published results, even the results of modern researchers! Few designers even look for information on wear resistance of materials because of the exaggerated claims made for so many "new" materials over the years. No one seems to comprehend just how lubricant chemistry works, even the chemists who formulate lubricants. Would tribologists place much confidence in any other field with this record of performance?

This is not to say that we have zero value to the practical designer. We can claim many triumphs and rightly so, but many failures too. Recall that I attribute this state of affairs to the great complexity of the field.

8. Improving the General Knowledge in Tribology

Having implied that tribological design is mostly done by people with little knowledge of tribology, a logical suggestion would be to teach tribology in the undergraduate mechanical design courses in colleges and universities. Very little takes place in the US at this time. Of the 150+ curricula in mechanical design in universities in North America, fewer than 30 offer much teaching in tribology and most of these are short presentations in contact mechanics and hydrodynamics, usually in design classes. Fewer than 10 offer intensive courses in any tribological topic at all and fewer still offer broad and interdisciplinary courses.

Why is so little being taught? Few design faculty know why, and fewer faculty seem inclined to introduce the topic at this point in time. There appear to be four reasons:
a. There is little authoritative information available on which to base analytical exercises for students,
b. Student laboratories in solid mechanics and fluid mechanics are less costly and more quickly done than are wear tests,
c. Mechanical design courses most often emphasize the overall functioning and making of mechanical devices with little concern for the inner details,
d. The research interests of many design faculty focuses on "design optimization", which requires models that can be manipulated in computers: there are no(?) credible models in friction, wear or the chemical mechanisms of lubrication to manipulate.

Graduate programs in the field would of necessity include some chemistry, some material science, some mechanics, etc., enough to function well in tribology. This diversity of topics will be difficult enough given the great aversion most students have to taking courses outside their undergraduate specialty. Graduate curricula will also require interdisciplinary research to develop the mind of tribologists, which will require the presence of faculty from the several represented disciplines. This presents the major problem that traditional academic departments, in the US at least, are reluctant to add faculty for interdisciplinary programs. And the glue that holds an interdisciplinary group together is interdisciplinary funding, usually with "seed" money from the government and supporting funds from industry. But the success of such funding is strongly influenced by the usual dichotomy between the specialized interest of scientists in the research groups and the interests of the industrial liaison members from industry, few of whom are well versed in tribology. More common is the example of the lone faculty member, supported by industrial or government funds, who sponsor students in some topic. Whereas the students find their research work to be interesting, only a few find employment in tribology.

9. What Can We Do To Improve the Image of Tribology?

Few of the reasons for the present status of tribology can be reversed in a short time. Perhaps the expansion of the teaching of tribology to students in mechanical design is the easiest, but that will require organized efforts. Perhaps the professional societies could take this as their responsibilities. The development of interdisciplinary graduate curricula in tribology must await the increased credibility of tribology and tribologists. This would be a profitable exercise whether it leads to graduate curricula in tribology or not.

Increasing the credibility of tribology and tribologists will require at least two efforts:

a. By cooperative effort of tribologists in all disciplines, define the academic diversity of practical tribology and show how to integrate all facets of the field into one coherent continuum. This would serve the field in at least three ways, namely to:

1. Establish a template for graduate college curricula,

2. Help employers determine what tribological skills they need and identify people with the desired skills,

3. Help industry support necessary research in the most appropriate ways.

b. Begin a concerted effort to model (write equations for) friction, wear , and the chemical aspects of lubrication, to name a few. A most useful goal would be to produce models that mechanical designers can use. In some topics this will require first identifying the underlying mechanisms, but some modeling action should begin nonetheless. We could begin by publishing a reliable table of values of the coefficient of friction. Full scale modeling can not be done rigorously at first because of the great complexity of tribology, but the scientific mind surely can do more than has been done to date.

The exercise of modeling by itself will have the great benefit of formalizing the field. Tribology in the minds of many observers is a chaotic field and this chaos must be reduced quickly.

Scientists may feel that practical modeling is for others to do. The fact is that "the others" have not done it: a new approach is needed. Our industrial significance depends on it.

Reference
Dowson, D., (1979) *History of Tribology*, Longman, London

Table 1. A Comparison of the Lubrication Requirements of Two Automobiles,
1916 - 1996, as Written in the Owner Manuals.

The 1916 Deluxe Maxwell
  A. Lubrication:
1. Every day or every 100 miles:
a. Check oil level in the engine, wet clutch, transmission and differential gear housing,
b . Turn grease cup caps on the 8 spring bolts, one turn ($\approx$ 0.05 cu.in.),
c. Apply a few drops of engine oil to steering knuckles, tie rod clevises and fan hub.
2. Each week or 500 miles,
a. Apply a few drops of engine oil to the: spark and throttle cross-shaft brackets, the
starter shift and switch rods, starter motor front bearing,
        the steering column oiler, the speedometer parts, all brake clevises, oilers and
cross-shaft brackets, at least 12 locations,
b. Force a "grease gun full" (half cup) of grease into the universal joint,
c. Turn the grease cup on the generator drive shaft bearing, one turn,
d. Pack the ball joints of the steering mechanism with grease ($\approx$ 1/4 cup).
3. Each month or 1500 miles,
a. Force a "grease gun full" of grease into the engine timing gear, and into the steering
gear case,
b. Apply a few drops of 3 in 1 oil to the magneto bearings,
c. Pack the wheel hubs with grease ($\approx$ 1/4 cup each),
d. Turn the grease cup on the rear axle spring seat, 2 turns.
4. Each 2000 miles,
a. Drain crank case, the wet clutch, the transmission and the rear axle; flush each with
kerosene, and refill (several quarts),
b. Jack up car by the frame, pry spring leaves apart and insert graphite grease between
the leaves.

  B. Other Maintenance for perspective,
_____Every two weeks:
a. Check engine valve action,
b. Listen for crankshaft bearing noises,
c. Check battery fluid level and color,
d. Adjust carburetor mixtures,
e. Check fan belt tension,
f. Drain water from carburetor bowl,
g. Tighten body and fender bolts,
h. Check strength of magneto spark,
i. Check for spark knock (to determine when thick layers carbon should be cleaned
(chipped off) from "inside" the head of the engine).

  On a regular basis:
a. Check engine compression,
b. Inspect ignition wiring,
c. Clean and regap spark plugs,
d. Inspect cooling system for leaks,
e. Clean gasoline strainer,

f. Inspect steering parts,

g. Inspect springs,

h. Check effectiveness of brakes (two wheel brakes),

i. Examine tires for cuts or bruises. (Those tires needed to be pumped up often, leaks needed to be sealed every few hundreds miles and they lasted less than 10,000 miles (albeit on bad roads) whereas today tires last over 60,000 miles, and at much higher speeds and loads, with very little attention.)

j. Adjust alcohol/water ratio in radiator (ethylene glycol was not yet available).

The 1996 Taurus (lubrication only)

1. Change engine oil and filter every 7500 miles or 6 months,

2. Lubricate steering linkages every 15,000 miles,

3. Change trans-axle fluid *every* 120,000 miles   (240,000, 360,000?)
     ----- nothing else!

The consumption of oil, grease and kerosene for the Maxwell works out to about 9 or 10 quarts (plus any oil that was burned) per 1000 miles, and little was recycled. With the Taurus it comes to about 1.5 quarts per 1000 miles and most of that is recycled or used in some non polluting way. Where did all that grease and oil go? Most of it dripped onto the roads (some roads had wet streaks (of oil) between tire tracks, all of them had layers of oxidized grease) and soaked into and upon the dirt floors in the garages, several inches thick.

Of course, the auto industry did not rise by itself, but rather was elevated by several attending technologies. These included:

1. Fuels and lubricants,

2. Bearing materials,

3. Manufacturing precision,

4. Shaft and face seals,

5. Coatings and surface processes.

6. Rubber technology.

We hardly recognize the magnitude of these achievements.

FRICTION, WEAR, LUBRICATION, AND MATERIALS CHARACTERIZATION USING SCANNING PROBE MICROSCOPY

BHARAT BHUSHAN
Ohio Eminent Scholar and The Howard D. Winbigler Professor
Dept. of Mechanical Engineering
The Ohio State University
206 W 18th Avenue
Columbus, OH 43210-1107 U.S.A.

1. Introduction

The atomistic mechanisms and dynamics of the interactions of two materials during relative motion, need to be understood in order to develop fundamental understanding of adhesion, friction, wear, indentation and lubrication processes. At most solid-solid interfaces of technological relevance, contact occurs at many asperities. Consequently the importance of investigating single asperity contacts in studies of the fundamental micromechanical and tribological properties of surfaces and interfaces has long been recognized. The recent emergence and proliferation of proximal probes, in particular scanning probe microscopies (the scanning tunneling microscope and the atomic force microscope) and the surface force apparatus, and of computational techniques for simulating tip-surface interactions and interfacial properties, has allowed systematic investigations of interfacial problems with high resolution as well as ways and means for modifying and manipulating nanoscale structures. These advances have led to the appearance of the new field of micro/nanotribology, which pertains to experimental and theoretical investigations of interfacial processes on scales ranging from the atomic- and molecular- to the microscale, occurring during adhesion, friction, scratching, wear, nanoindentation, and thin-film lubrication at sliding surfaces (Singer and Pollock, 1992; Persson and Tosatti, 1996; Bhushan, 1995, 1997, 1998a, b, 1999a, b, c, 2001; Bhushan et al., 1995a; Guntherodt et al., 1995).

The micro/nanotribological studies are needed to develop fundamental understanding of interfacial phenomena on a small scale and to study interfacial phenomena in micro– and nanostructures used in magnetic storage systems, microelectromechanical systems (MEMS) and other applications (Bhushan, 1996, 1997, 1998a). Friction and wear of lightly loaded micro/nanocomponents are highly dependent on the surface interactions (few atomic layers). These structures are generally lubricated with molecularly thin films. Micro/nanotribological studies are also valuable in fundamental understanding of interfacial phenomena in macrostructures to provide a bridge between science and engineering.

The surface force apparatus (SFA), the scanning tunneling microscopes (STM), atomic force and friction force microscopes (AFM and FFM) are widely used in micro/nanotribological studies. Typical operating parameters are compared in Table 1. The SFA was developed in 1968 and is commonly employed to study both static and dynamic properties of molecularly thin films sandwiched between two molecularly smooth surfaces. The STM, developed in 1981, allows imaging of electrically

B. Bhushan (ed.),
Fundamentals of Tribology and Bridging the Gap between the Macro- and Micro/Nanoscales, 13–39.
© 2001 *Kluwer Academic Publishers.*

Table 1. Comparison of typical operating parameters in SFA, STM and AFM/FFM used for micro/nanotribological Studies

Operating Parameter	SFA	STM*	AFM/FFM
Radius of mating surface/tip	~ 10 mm	5 - 100 nm	5 - 100 nm
Radius of contact area	10 - 40 μm	N/A	0.05 – 0.5 nm
Normal load	10 – 100 mN	N/A	<0.1 nN – 500 nN
Sliding velocity	0.001 – 100 μm/s	0.02 – 2 μm/s (scan size ~1 nm x 1 nm to 125 μm x 125 μm; scan rate <1-122 Hz)	0.02 – 2 μm/s (scan size ~1 nm x 1 nm to 125 μm x 125 μm; scan rate <1-122 Hz)
Sample limitations	Typically atomically–smooth, optically transparent mica; opaque ceramic, smooth surfaces can also be used	Electrically-conducting samples	None

*Can only be used for atomic-scale imaging

conducting surfaces with atomic resolution, and has been used for imaging of clean surfaces as well as of lubricant molecules. The introduction of the atomic force microscope in 1985 provided a method for measuring ultra-small forces between a probe tip and an engineering (electrically conducting or insulating) surface, and has been used for topographical measurements of surfaces on the nanoscale, as well as for adhesion and electrostatic force measurements. Subsequent modifications of the AFM led to the development of the friction force microscope (FFM), designed for atomic-scale and microscale studies of friction. This instrument measures forces transverse to the surface. The AFM is also being used for investigations of scratching wear, indentation, detection of transfer of material, boundary lubrication, and fabrication and machining. Meanwhile, significant progress in understanding the fundamental nature of bonding and interactions in materials, combined with advances in computer-based modeling and simulation methods, has allowed theoretical studies of complex interfacial phenomena with high resolution in space and time. Such simulations provide insights into atomic-scale energetics, structure, dynamics, thermodynamics, transport and rheological aspects of tribological processes.

The nature of interactions between two surfaces brought close together, and those between two surfaces in contact as they are separated, have been studied experimentally with the surface force apparatus. This has led to a basic understanding of the normal forces between surfaces, and the way in which these are modified by the presence of a thin liquid or a polymer film. The frictional properties of such systems have been studied by moving the surfaces laterally, and such experiments have provided insights into the molecular-scale operation of lubricants such as thin liquid or polymer films. Complementary to these studies are those in which the AFM or FFM is used to provide a model asperity in contact with a solid or lubricated surface. These experiments have demonstrated that the relationship between friction and surface roughness is not always simple or obvious. AFM studies have also revealed much about the nanoscale nature of intimate contact during wear and indentation.

In this chapter, we present a review of significant aspects of micro/nanotribological studies conducted using AFM/FFM.

2. Friction and Adhesion

2.1. ATOMIC-SCALE FRICTION

To study friction mechanisms on an atomic scale, a well-characterized freshly cleaved surface of highly oriented pyrolytic graphite (HOPG) has been studied by Mate et al. (1984) and Ruan and Bhushan (1994b). [For the friction calibration technique, see Ruan and Bhushan (1994a) or Bhushan (1999a).] The atomic-scale friction force of HOPG exhibited the same periodicity as that of the corresponding topography (Figure 1a), but the peaks in friction and those in topography were displaced relative to each other (Figure 1b). A Fourier expansion of the interatomic potential was used to calculate the conservative interatomic forces between atoms of the FFM tip and those of the graphite surface. Maxima in the interatomic forces in the normal and lateral directions do not occur at the same location, which explains the observed shift between the peaks in the lateral force and those in the corresponding topography. Furthermore, the observed local variations in friction force were explained by variation in the intrinsic lateral force between the sample and the FFM tip (Ruan and Bhushan, 1994b) and these variations may not necessarily occur as a result of a atomic-scale stick-slip process (Mate et al., 1984), but can be due to variation in the intrinsic lateral force between the sample and the FFM tip.

2.2. MICROSCALE FRICTION

Local variations in the microscale friction of cleaved graphite are observed. These arise from structural changes that occur during the cleaving process (Ruan and Bhushan, 1994c). The cleaved HOPG surface is largely atomically smooth but exhibits

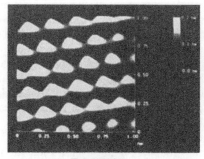

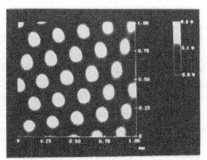

Topography   Friction

(a)

Sliding direction

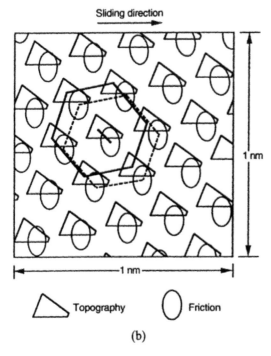

1 nm

|← 1 nm →|

△ Topography   ○ Friction

(b)

Figure 1. (a) Gray-scale plots of surface topography and friction force maps of a 1 nm x 1 nm area of freshly cleaved HOPG, showing the atomic-scale variation of topography and friction, and (b) schematic of superimposed topography and friction maps from (a); the symbols correspond to maxima. Note the spatial shift between the two plots (Ruan and Bhushan, 1994b).

line-shaped regions in which the coefficient of friction is more than an order of magnitude larger, Figure 2. Transmission electron microscopy indicates that the line-shaped regions consist of graphite planes of different orientation, as well as of amorphous carbon. Differences in friction have also been observed for multi-phase ceramic materials (Koinkar and Bhushan, 1996c). Figure 3 shows the surface roughness and friction force maps of Al_2O_3-TiC (70 – 30 wt %). TiC grains have a Knoop hardness of about 2800 kg/mm^2; therefore, they do not polish as much and result in a slightly higher elevation (about 2 - 3 nm higher than that of Al_2O_3 grains). TiC grains exhibit higher friction force than Al_2O_3 grains. The coefficients of friction of TiC and Al_2O_3 grains are 0.034 and 0.026, respectively, and the coefficient of friction of Al_2O_3- TiC composite is 0.03. Local variation in friction force also arises from the scratches present on the Al_2O_3- TiC surface. Meyer et al. (1992) also used FFM to measure structural variations of organic mono- and multi-layer films. All of these measurements suggest that the FFM can be used for structural mapping of the surfaces. FFM measurements can be used to map chemical variations, as indicated by the use of the FFM with a modified probe tip to map the spatial arrangement of chemical functional groups in mixed organic monolayer films (Frisbie et al., 1994). Here, sample regions that had stronger interactions with the functionalized probe tip exhibited larger friction.

Local variations in the microscale friction of nominally rough, homogeneous surfaces can be significant, and are seen to depend on the local surface slope rather than the surface height distribution. This dependence was first reported by Bhushan and Ruan (1994) and Bhushan et al. (1994) and later discussed in more detail by Koinkar and Bhushan (1997b) and Sundararajan and Bhushan (2000b). In order to show elegantly any correlation between local values of friction and surface roughness, surface roughness and friction force maps of a silicon grid with square pits were obtained, Figure 4 (Sundararajan and Bhushan (2000b). Figure 4 shows the surface roughness map, the slopes of the roughness map taken along the sliding direction

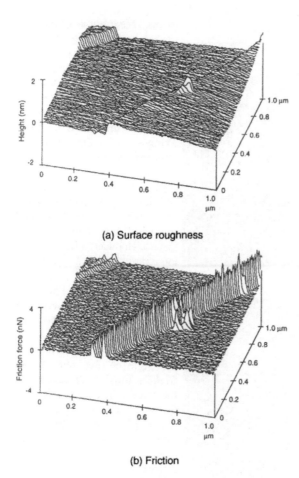

(a) Surface roughness

(b) Friction

Figure 2. (a) Surface roughness and (b) friction force maps at a normal load of 42 nN of freshly cleaved HOPG surface against a Si_3N_4 FFM tip.

Friction in the line-shaped region is over an order of magnitude larger than the smooth areas (Ruan and Bhushan, 1994b).

(surface slope map) and the friction force map for various samples. There is a strong correlation between the surface slopes and friction forces. For example, in Figure 4, friction force is high locally at the edge of the pits with a positive slope and is low at the edges with a negative slope.

We now examine the mechanism of microscale friction, which may explain the resemblance between the slope of surface roughness maps and the corresponding friction force maps (Ruan and Bhushan, 1994 b, c; Bhushan and Ruan, 1994; Bhushan et al., 1994; Bhushan, 1999a, b; Sundararjan and Bhushan, 2000b). A ratchet mechanism is believed to be the dominant mechanism for the local variations in the friction force map, which suggests a direct correspondence between the coefficient of friction and the surface slope. With the tip sliding over the leading (ascending) edge of an asperity, the surface slope is positive; it is negative during sliding over the trailing (descending) edge of an asperity. Thus, measured friction is high at the leading edge of asperities and low at the trailing edge. In addition to the slope effect, the collision of tip when encountering an asperity with a positive slope produces additional torsion of the cantilever beam leading to higher measured friction force. When encountering an asperity with a same negative slope, however, there is no collision effect and hence no effect on torsion. This effect also contributes to the difference in friction forces when the tip scans up and down on the same topography feature. The ratchet mechanism and the collision effects thus semi-quantitatively explain the correlation between the slopes of the roughness maps and friction maps observed in Figure 4. Next, we study the directionality effect on friction. During friction measurements, the friction force data from both the forward (trace) and backward (retrace) scans is useful in understanding the origins of the observed friction

18

Surface Topography

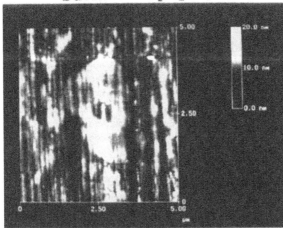

Friction Force

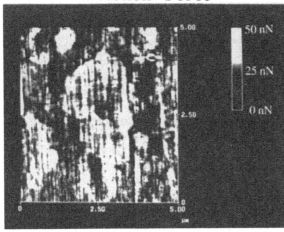

Figure 3. Gray scale surface roughness ($\sigma = 0.80$ nm) and friction force maps (mean = 7.0 nN, σ = 0.90 nN) for Al_2O_3-TiC (70-30 wt %) at a normal load of 138 nN (Koinkar and Bhushan, 1996c).

forces. Magnitudes of material - induced effects are independent of the scan direction whereas topography-induced effects are different between forward and backward scanning directions. Since the sign of the friction force changes as the scanning direction is reversed (because of the reversal of torque applied to the end of the tip), addition of the friction force data of the forward and backward scan eliminates the material-induced effects while topography induced effects still remain. Subtraction of the data between forward and backward scans does not eliminate either effects, Figure 5 (Sundararajan and Bhushan, 2000b).

Due to the reversal of the sign of the retrace (R) friction force with respect to the trace (T) data; the friction force variations due to topography are in the same direction (peaks in trace corresponds to peaks in retrace). However, the magnitudes of the peaks in trace and retrace at a given location are different. An increase in the friction force experienced by the tip when scanning up a sharp change in topography is more than the decrease in the friction force experienced when scanning down the same topography change, partly because of collision effects discussed earlier. Asperities on engineering surfaces are asymmetrical which also affect the magnitude of friction force in the two directions. Asymmetry in tip shape may also have an effect on the directionality effect of friction. We notice that since magnitude of surface slopes are virtually identical; the tip shape asymmetry should not have much effect. Because of the differences in the magnitude of friction forces in the two directions, subtracting the two friction data yields a residual peak.

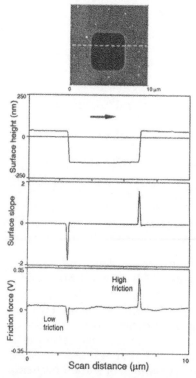

Figure 4. Surface roughness map, surface slope map taken in the sample sliding direction (the horizontal axis), and friction force map for a silicon grid (with 5 μm square pits of depth 180 nm and a pitch of 10 μm) (Sundararajan and Bhushan, 2000b).

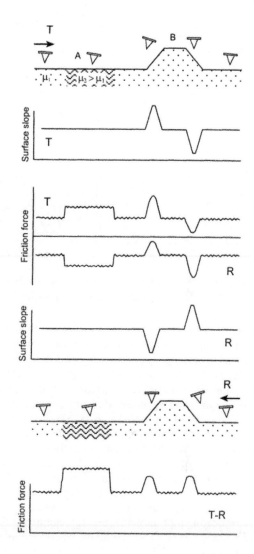

Figure 5. Schematic illustration showing the effect of an asperity (making an angle θ with the horizontal plane) on the surface in contact with the tip on local friction in the presence of adhesive friction mechanism. W and F are the normal and friction forces, respectively, and S and N are the force components along and perpendicular to the local surface of the sample at the contact point, respectively.

In order to facilitate comparison of directionality effect on friction, it is important to take into account the sign change of the surface slope and friction force in the trace and retrace directions. Figure 6 shows topography, slope and friction force data for the silicon grid in the trace and retrace directions. The correlation between surface slope and friction forces is clear. The third column in the figures shows retrace slope and friction data with an inverted sign (-retrace). Now we can compare trace data with -retrace data. It is clear the friction experienced by the tip is dependent upon the

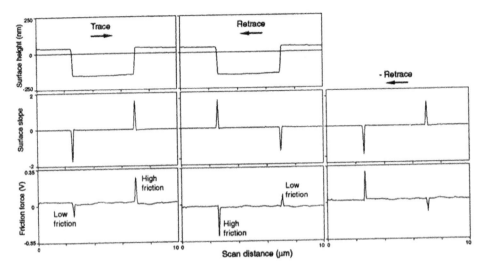

Figure 6. Two-dimensional profiles of surface heights, surface slopes and friction forces for scans across the silicon grid pit (Sundararajan and Bhushan, 2000b).

scanning direction because of surface topography. In addition to the effect of topographical changes discussed earlier, during surface finishing processes, material can be transferred preferentially onto one side of the asperities, which also causes asymmetry and direction dependence. Reduction in local variations and in directionality of friction properties requires careful optimization of surface roughness distributions and of surface finishing processes.

The directionality as a result of surface asperities effect will be also manifested in macroscopic friction data, that is, the coefficient of friction may be different in one sliding direction than that in the other direction. Asymmetrical asperities accentuate this effect. The frictional directionality can also exist in materials with particles having a preferred orientation. The directionality effect in friction on a macroscale is observed in some magnetic tapes. In a macroscale test, a 12.7-mm wide polymeric magnetic tape was wrapped over an aluminum drum and slid in a reciprocating motion with a normal load of 0.5 N and a sliding speed of about 60 mm/s (Bhushan, 1997). The coefficient of friction as a function of sliding distance in either direction is shown in Figure 7. We note that the coefficient of friction on a macroscale for this tape is different in different directions. Directionality in friction is sometimes observed on the macroscale; on the microscale this is the norm (Bhushan 1996, 1999a). On the macroscale, the effect of surface asperities normally is averaged out over a large number of contacting asperities.

AFM/FFM experiments are generally conducted at relative velocities as high as few tens of μm/s. To simulate applications, it is of interest to conduct friction experiments at higher velocities. High velocities can be achieved by mounting either the sample or the cantilever beam on a shear wave transducer (ultrasonic transducer) to produce surface oscillations at MHz frequencies (Scherer et al., 1998, 1999). The velocities on the order of few mm/s can thus be achieved. The effect of in-plane and out-of-plane sample vibration amplitude on the coefficient of friction is shown in

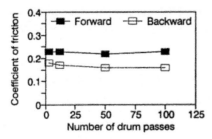

Figure 7. Coefficient of macroscale friction as a function of drum passes for a polymeric magnetic tape sliding over an aluminum drum in a reciprocating mode in both directions. Normal load = 0.5 N over 12.7-mm wide tape, sliding speed = 60 mm/s (Bhushan, 1995).

Figure 8. Vibration of a sample at ultrasonic frequencies (> 20 kHz) can substantially reduce the coefficient of friction, known as ultrasonic lubrication or sonolubrication. When the surface is vibrated in-plane, classical hydrodynamic lubrication develops hydrodynamic pressure, which supports the tip and reduces friction. When the surface is vibrated out-of-plane, a lift-off caused by the squeeze-film lubrication (a form of hydrodynamic lubrication), reduces friction.

2.3. COMPARISON OF MICROSCALE AND MACROSCALE FRICTION DATA

Table 2 shows the coefficient of friction measured for single-crystal silicon surfaces on micro- and macroscales. To study the effect of sample material, data for virgin and ion-implanted silicon are presented. The values on the microscale are much lower than those on the macroscale. There can be the following four and possibly more differences in the operating conditions responsible for these differences. First, the contact stresses at AFM conditions generally do not exceed the sample hardness which minimizes plastic deformation. Second, when measured for the small contact areas and very low loads used in microscale studies, indentation hardness and modulus of elasticity are

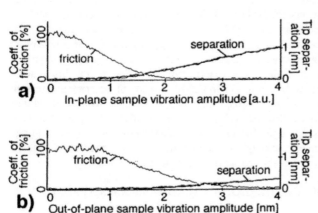

Figure 8. Reduction of coefficient of friction, measured at a normal load of 100 nN and average tip separation as a function of surface amplitude on a single-crystal silicon subjected to a) in-plane and b) out-of-plane vibrations at about 1 MHz against a silicon nitride tip (Scherer et al., 1998).

higher than at the macroscale, as will be discussed later (Bhushan et al., 1995a, 1996). Lack of plastic deformation and improved mechanical properties reduce the degree of wear and friction. Next, the small apparent areas of contact reduce the number of particles trapped at the interface, and thus minimize the plowing contribution to the friction force (Bhushan et al., 1995a). As a fourth and final difference, we will note in the next section that coefficient of friction increases with an increase in the AFM tip radius. AFM data presented so far were taken with a sharp tip whereas asperities

Table 2.   Surface roughness (standard deviation of surface heights or σ) and coefficients of friction on micro- and macroscales of single-crystal silicon samples in air.

Material	σ (nm)	Microscale coefficient of friction versus Si_3N_4 tip[1]	Macroscale coefficient of friction versus alumina ball[2]
Si (111)	0.11	0.03	0.18
C^+-implanted Si (111)	0.33	0.02	0.18

[1]Tip radius of about 50 nm in the load range of 10-150 nN (2.5 – 6.1 GPa), a scanning speed of 5 μm/s and scan area of 1 μm x 1 μm.

[2]Ball radius of 3 mm at a normal load of 0.1 N (0.3 GPa) and average sliding speed of 0.8 mm/s.

coming in contact in macroscale tests range from nanoasperities too much larger asperities, which may be responsible for larger values of friction force on macroscale.

To demonstrate the load dependence on the coefficient on friction, stiff cantilevers were used to conduct friction experiments at high loads, Figure 9 (Bhushan and Kulkarni, 1996). At higher loads (with contact stresses exceeding the hardness of the softer material), as anticipated, the coefficient of friction for microscale measurements increases towards values comparable with those obtained from macroscale measurements, and surface damage also increases. Thus Amontons' law of friction, which states that the coefficient of friction is independent of apparent contact area and normal load, does not hold for microscale measurements. These findings suggest that microcomponents sliding under lightly loaded conditions should experience ultra low friction and near-zero wear (Bhushan et al., 1995a).

2.4. EFFECT OF TIP RADII AND HUMIDITY ON ADHESION AND FRICTION

The tip radius and relative humidity affect adhesion and friction for dry and lubricated surfaces (Bhushan and Sundararajan, 1998; Bhushan and Dandavate, 2000). Figure 10 shows the variation of single point adhesive force measurements as a function of tip radius on a Si(100) sample for a given humidity for several humidities. The adhesive force data are also plotted as a function of relative humidity for a given tip radius for several tip radii. The general trend at humidities up to the ambient is that a 50-nm radius Si_3N_4 tip exhibits a slightly lower adhesive force as compared to the other microtips of larger radii; in the latter case, values are similar. Thus for the microtips there is no appreciable variation in adhesive force with tip radius at a given humidity up to the ambient. The adhesive force increases as relative humidity increases for all tips. The trend in adhesive forces as a function of tip radii and relative humidity can be explained by the presence of meniscus forces, which arise from capillary condensation of water vapor from the environment forming meniscus bridges. If enough liquid is present to form a meniscus bridge, the meniscus force should increase with an increase in tip radius (proportional to tip radius for a spherical tip) and should be independent of

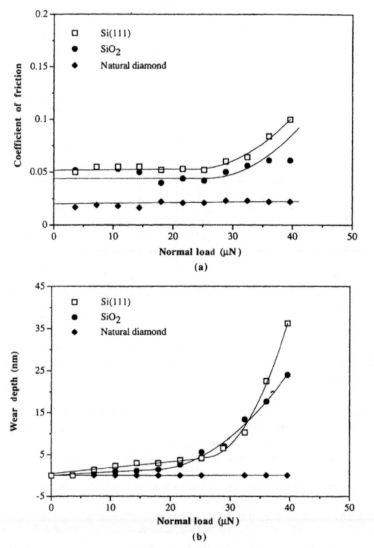

Figure 9. (a) Coefficient of friction as a function of normal load and (b) corresponding wear depth as a function of normal load for silicon, SiO_2 coating and natural diamond. Inflections in the curves for silicon and SiO_2 correspond to the contact stresses equal to the hardnesses of these materials (Bhushan and Kulkarni, 1996).

the relative humidity or water film thickness. In addition, an increase in tip radius in a dry environment results in increased contact area leading to higher values of van der Waals forces. However, if nanoasperities on the tip and the sample are considered then the number of contacting and near- contacting asperities forming meniscus bridges increases with an increase of humidity leading to an increase in meniscus forces. This

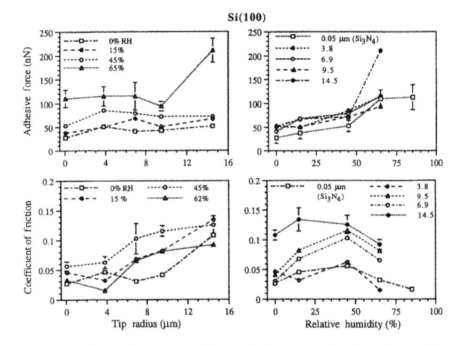

Figure 10. Adhesive force and coefficient of friction as a function of tip radius at several humidities and as a function of relative humidity at several tip radii on Si(100) (Bhushan and Sundararajan, 1998).

explains the trends observed in Figure 10. From the data, the tip radius has little effect on the adhesive forces at low humidities but increases with tip radius at high humidity. Adhesive force also increases with an increase in humidity for all tips. This observation suggests that thickness of the liquid film at low humidities is insufficient to form continuous meniscus bridges to affect adhesive forces in the case of all tips.

Figure 10 also shows the variation in coefficient of friction as a function of tip radius at a given humidity, and as a function of relative humidity for a given tip radius for Si(100). It can be seen that for 0% RH, the coefficient of friction is about the same for the tip radii except for the largest tip, which shows a higher value. At all other humidities, the trend consistently shows that the coefficient of friction increases with tip radius. An increase in friction with tip radius at low to moderate humidities arises from increased contact area (higher van der Waals forces) and higher values of shear forces required for larger contact area. At high humidities, similar to adhesive force data, an increase with tip radius occurs because of both contact area and meniscus effects. Although AFM/FFM measurements are able to measure the combined effect of the contribution of van der Waals and meniscus forces towards friction force or adhesive force, it is difficult to measure their individual contributions separately. It can be seen that for all tips, the coefficient of friction increases with humidity to about ambient, beyond which it starts to decrease. The initial increase in the coefficient of friction with humidity arises from the fact that the thickness of the water film increases with an increase in the humidity, which results in a larger number of nanoasperities forming meniscus bridges and leads to higher friction (larger shear force). The same trend is expected with the microtips beyond 65% RH. This is attributed to the fact that at higher humidities, the adsorbed water film on the surface acts as a lubricant between the two surfaces. Thus the interface is changed at higher humidities, resulting in lower shear strength and hence lower friction force and coefficient of friction.

3. Scratching, Wear and Fabrication/Machining

3.1. MICROSCALE SCRATCHING

The AFM can be used to investigate how surface materials can be moved or removed on micro- to nanoscales, for example, in scratching and wear (Bhushan, 1999a) (where these things are undesirable), and nanofabrication/ nanomachining (where they are desirable). Scratching can be performed at random loads to determine the scratch resistance of materials and coatings. The coefficient of friction is measured during scratching and the load at which the coefficient of friction increases rapidly is known as the "critical load" which is a measure of scratch resistance. In addition, the post scratch imaging can be performed in-situ with the AFM in tapping mode to study failure mechanisms. Figure 11 shows data from a scratch test on Si(100) with a scratch length of 25 μm and the scratching velocity of 0.5 μm/s. At the beginning of the scratch, the coefficient of friction is 0.04 which is a typical value for silicon. At about 35 μN (indicated by the arrow in the figure), there is a sharp increase in the coefficient of friction, which is the critical load. Beyond the critical load, the coefficient of friction continues to increase steadily. In the post-scratch image, we note that at the critical load, a clear groove starts to form. This implies that Si(100) was damaged by plowing at the critical load, associated with the plastic flow of the material. At and after the critical load, small and uniform debris is observed and the amount of debris increases with increasing normal load. Sundararajan and Bhushan (2000a) have also used this technique to measure scratch resistance of diamondlike carbon coatings ranging in thickness from 3.5 to 20 nm.

3.2. MICROSCALE WEAR

By scanning the sample in two dimensions with the AFM, wear scars are generated on the surface. Figure 12 shows the effect of normal load on the wear depth. We note that wear rate is very small below 20 μN of normal load (Koinkar and Bhushan, 1997c; Zhao and Bhushan, 1998). A normal load of 20 μN corresponds to contact stresses comparable to the hardness of the silicon. Primarily, elastic deformation at loads below 20 μN is responsible for low wear (Bhushan and Kulkarni, 1996).

To understand wear mechanisms, evolution of wear can be studied using AFM. Figure 13 shows evolution of wear marks of a DLC coated disk sample. The data illustrate how the microwear profile for a load of 20 μN develops as a function of the number of scanning cycles (Bhushan et al., 1994). Wear is not uniform, but is initiated at the nanoscratches. Surface defects (with high surface energy) present at nanoscratches act as initiation sites for wear. Coating deposition also may not be uniform on and near nanoscratches, which may lead to coating delamination. Thus, scratch-free surfaces will be relatively resistant to wear.

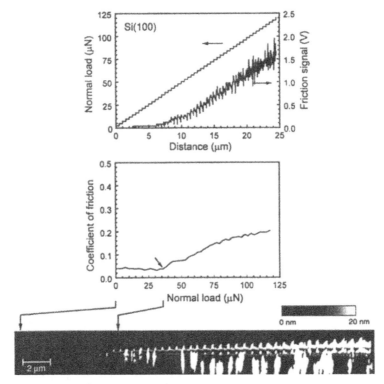

Figure 11. (a) Applied normal load and friction signal measured during the microscratch experiment on Si(100) as a function of scratch distance, (b) friction data plotted in the form of coefficient of friction as a function of normal load, and (c) AFM surface height image of scratch obtained in tapping mode (Sundararajan and Bhushan, 2000a).

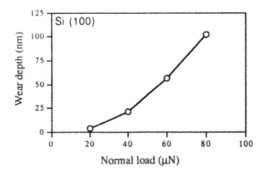

Figure 12. Wear depth as a function of normal load for Si(100) after one cycle (Zhao and Bhushan, 1998).

Wear precursors (precursors to measurable wear) can be studied by making surface potential measurements (DeVecchio and Bhushan, 1998; Bhushan and Goldade, 2000a, 2000b). The contact potential difference or simply surface potential between two surfaces depends on a variety of parameters such as electronic work function, adsorption and oxide layers. The surface potential map of an

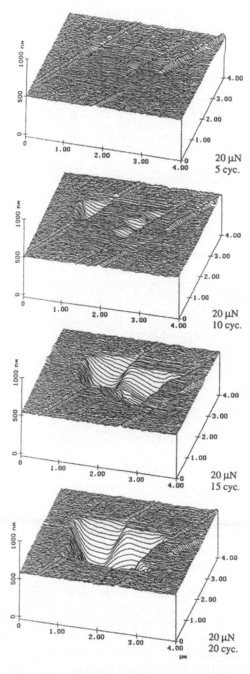

Figure 13. Surface plots of diamond-like carbon-coated thin-film disk showing the worn region; the normal load and number of test cycles are indicated (Bhushan et al., 1994).

interface gives a measure of changes in the work function, which is sensitive to both physical and chemical conditions of the surfaces including structural and chemical changes. Before material is actually removed in a wear process, the surface experiences stresses that result in surface and subsurface changes of structure and/or chemistry. These can cause changes in the measured potential of a surface. An AFM tip allows mapping of surface potential with nanoscale resolution. Surface height and change in surface potential maps of a polished single-crystal aluminum (100) sample abraded using a diamond tip at two loads of approximately 1 μN and 9 μN, are shown in Figure 14a [Note that sign of the change in surface potential is reversed here from that in DeVecchio and Bhushan (1998)]. It is evident that both abraded regions show a large potential contrast (~0.17 Volts), with respect to the non-abraded area. The black region in the lower right hand part of the topography scan shows a step that was created during the polishing phase. There is no potential contrast between the high region and the low region of the sample indicating that the technique is independent of surface height. Figure 14b shows a close up scan of the upper (low load) wear region in Figure 14a. Notice that while there is no detectable change in the surface topography, there is nonetheless a large change in the potential of the surface in the worn region. Indeed the wear mark of Figure 14b might not be visible at all

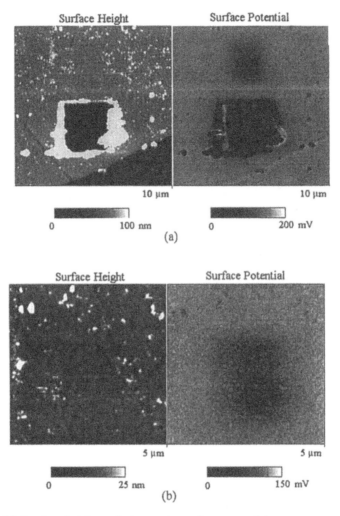

Figure 14. (a) Surface height and change in surface potential maps of wear regions generated at 1 μN (top) and 9 μN (bottom) on a single crystal aluminum sample showing bright contrast in the surface potential map on the worn regions. (b) Close up of upper (low load) wear region (DeVecchio and Bhushan, 1998).

was it not for the noted absence of wear debris generated nearby and then swept off during the low load scan. Thus, even in the case of zero wear (no measurable deformation of the surface using AFM), there can be a significant change in the surface potential inside the wear mark which is useful for study of wear precursors. It is believed that the removal of thin contaminant layer including natural oxide layer gives rise to the initial change in surface potential. The structural changes, which precede generation of wear debris and/or measurable wear scars, occur under ultralow loads in the top few nanometers of the sample, and are primarily responsible for the subsequent changes in surface potential.

3.3. IN-SITU CHARACTERIZATION OF LOCAL DEFORMATION

In-situ surface characterization of local deformation of materials and thin coatings can be carried out using an AFM. Failure mechanisms of polymeric thin coatings were studied by Bobji and Bhushan (2000, 2001). They studied nucleation and growth of microcracks developed in multi-layered polymeric/metallic magnetic tapes. Figure 15 shows topographical images of the three magnetic tapes and the PET substrate after being strained to 3.75 percent, which is well beyond the elastic limit of the substrate. They reported that cracking of the coatings started at about 1 percent strain for all tapes much before the substrate starts to yield at about 2 percent strain. Metal-particle (MP) tape develops short and numerous cracks perpendicular to the direction of loading. In tapes with metallic coating, the cracks extend throughout the tape width. In metal evaporated (ME) tape with DLC coating, there is a bulge in the coating around the primary cracks that are initiated when the substrate is still elastic, like crack 'A' in the figure. The white band on the right side of the figure is the bulge of another crack. The secondary cracks like 'B' and 'C' are generated at higher strains and are straighter compared to the primary cracks. In ME tape which has a Co-O film on PET substrate, with a thickness ratio of 0.03, both with and without DLC coating, no difference is observed in the rate of growth between primary and secondary cracks. The failure is cohesive with no bulging of the coating. This seems to suggest that the DLC coating has residual stresses that relax when the coating cracks, causing delamination. Since the stresses are already relaxed, the secondary crack does not delaminate. The presence of the residual stress is confirmed by the fact that a free standing ME tape curls up (in a cylindrical form with its axis perpendicular to the tape length) with a radius of curvature of about 6 mm and the ME tape without DLC does not curl. The front side of PET substrate is much smoother at smaller scan lengths. However, in 20 μm scans, it has lot of bulging outs, which appear as white spots in the figure. These spots change shape even while scanning the samples in tapping mode at very low contact forces.

4. Indentation

Mechanical properties, such as hardness and Young's modulus of elasticity can be determined on micro- to picoscales using the AFM (Bhushan and Ruan, 1994; Bhushan et al., 1994; Bhushan and Koinkar, 1994a, b) and a depth-sensing indentation system used in conjunction with an AFM (Bhushan et al., 1996; Kulkarni and Bhushan, 1996a, b, 1997).

4.1. NANOSCALE INDENTATION

To make accurate measurements of hardness at shallow depths, a depth-sensing indentation system is used. Figure 16 shows the load-displacement curves at different peak loads for Si(100). Load-displacement data at residual depths as low as about 1 nm can be obtained. Loading/unloading curves are not smooth, but exhibit sharp discontinuities particularly at high loads. Any discontinuities in the loading part of the curve probably result from slip of the tip. The sharp discontinuities in the unloading part of the curves are believed to be due to formation of lateral cracks, which form at the base of the median crack, which results in the surface of the specimen being thrust upward. The indentation hardness of surface films with an indentation depth of as small as about 1 nm has been measured for Si(100) (Bhushan et al., 1996; Kulkarni and

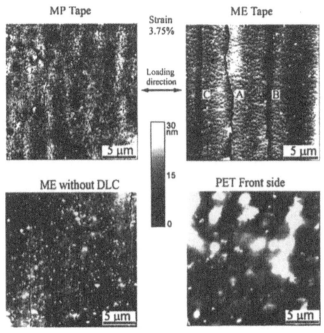

MP Tape

Strain 3.75%

Loading direction

ME Tape

C A B

30 nm

15

0

5 µm

5 µm

ME without DLC

5 µm

PET Front side

5 µm

Figure 15. Comparison of the crack morphologies at 3.75 percent strain in three magnetic tapes and PET substrate. Cracks B and C, nucleated at higher strains are more linear than crack A (Bobji and Bhushan, 2001).

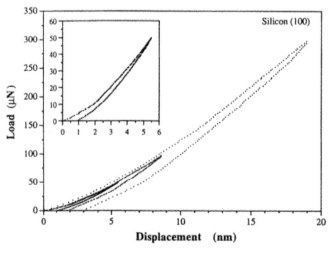

Figure 16. Load-displacement curves at various peak loads for Si(100) (Bhushan et al., 1996).

Bhushan, 1996a, b, 1997). The hardness of silicon on a nanoscale is found to be higher than on a microscale, Figure 17. This decrease in hardness with an increase in indentation depth can be rationalized on the basis that as the volume of deformed material increases, there is a higher probability of encountering material defects. Thus mechanical properties show size effect.

Bhushan and Koinkar (1994a) have used AFM measurements to show that ion implantation of silicon surfaces increases their hardness and thus their wear resistance. Formation of surface alloy films with improved mechanical properties by ion implantation is of growing technological importance as a means of improving the mechanical properties of materials. Hardness of 20 nm thick diamond like carbon films has been measured by Kulkarni and Bhushan (1997).

The creep and strain-rate effects (viscoelastic effects) of ceramics can be studied using a depth-sensing indentation system. Bhushan et al. (1996) and Kulkarni et al. (1996a, b, 1997) have reported that ceramics exhibit significant plasticity and creep on a nanoscale.

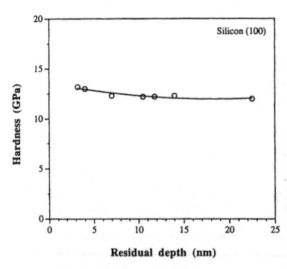

Figure 18 shows the load-displacement curves for single crystal silicon at various peak loads held at 180s. To demonstrate the creep effects, the load-displacement curves for a 500 µN peak load held at 0 and 30 s are also shown as an inset. Note that significant creep occurs at room temperature. The mechanism of dislocation glide plasticity is believed to dominate the indentation creep process on the macroscale.

The Young's modulus of elasticity is calculated from the slope of the indentation curve during unloading.

Figure 17. Indentation hardness as a function of residual indentation depth for Si(100) (Bhushan et al., 1996).

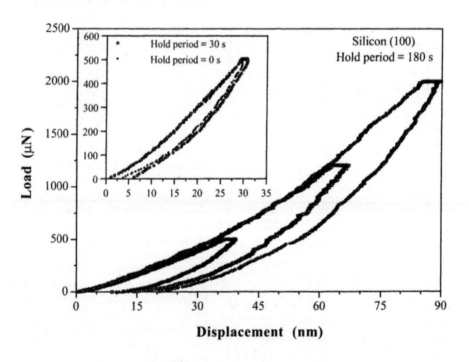

Figure 18. Creep behavior of Si(100) (Bhushan et al., 1996).

5. Boundary Lubrication

The classical approach to lubrication uses freely supported multimolecular layers of liquid lubricants (Bowden and Tabor, 1950; Bhushan, 1996). The liquid lubricants are sometimes chemically bonded to improve their wear resistance (Bhushan, 1996). To study depletion of boundary layers, the micro-scale friction measurements are made as a function of number of cycles. For an example of experiments with virgin Si(100) surfaces and silicon surfaces lubricated with Z-15 and Z-Dol PFPE lubricants, see Figure 19 (Koinkar and Bhushan, 1996a, b). Z-Dol is PFPE lubricant with hydroxyl end groups. Its film was thermally bonded at 150^0 C for 30 minutes (BUW-bonded, unwashed) and, in some cases the unbonded fraction was washed off with a solvent to provide a chemically bonded layer of the lubricant (BW) film. In Figure 19a, the unlubricated silicon sample shows a slight increase in friction force followed by a drop to a lower steady state value after 20 cycles. Depletion of native oxide and possible roughening of the silicon sample are believed to be responsible for the decrease in friction force after 20 cycles. The initial friction force for the Z-15 lubricated sample is lower than that of unlubricated silicon and increases gradually to a friction force value comparable to that of the silicon after 20 cycles. This suggests the depletion of the Z-15 lubricant in the wear track. In the case of the Z-Dol coated silicon sample, the friction force starts out very low and remains low during the 100 cycles test. This suggests that Z-Dol does not get displaced/depleted as readily as Z-15. The nanowear results for BW and BUW/Z-Dol samples with different film thicknesses are shown in Figure 19b. The BW with thickness of 2.3 nm exhibits an initial decrease in the friction force in the first few cycles and then remains steady for more than 100 cycles. The decrease in friction force possible arises from the alignment of any free liquid lubricant present over the bonded lubricant layer. BUW with a thickness of 4.0 nm exhibits behavior similar to BW (2.3 nm). Lubricated BW and BUW samples with thinner films exhibit a higher value of coefficient of friction. Among the BW and BUW samples, BUW samples show the lower friction because of extra-unbonded fraction of the lubricant.

The effect of the operating environment on coefficient of friction of unlubricated and lubricated

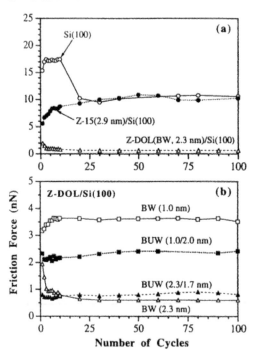

Figure 19. Friction force as a function of number of cycles using Si_3N_4 tip at a normal load of 300 nN for (a) unlubricated Si(100), Z-15 and bonded washed (BW) Z-Dol and (b) bonded Z-Dol before washing (BUW) and after washing (BW) with different film thicknesses (Koinkar and Bhushan, 1996a).

samples is shown in Figure 20. Silicon (100) samples were lubricated with 2.9 nm thick Z-15 and 2.3 nm thick Z-Dol bonded and washed (BW) lubricants. The coefficient of friction in a dry environment is lower than that in a high humidity environment. We believe that in the humid environment, the condensed water from the humid environment competes with the liquid film present on the sample surface and interaction of the liquid film (water for the unlubricated sample and polymer lubricant for the lubricated sample) to the substrate is weakened and a boundary layer of the liquid forms puddles. This dewetting results in poorer lubrication performance resulting in high friction. Since Z-Dol is a bonded lubricant with superior frictional properties, the dewetting effect in a humid environment for Z-Dol is more pronounced than for Z-15.

The effect of scanning speed on the coefficient of friction of unlubricated and lubricated samples is shown in Figure 21. The coefficient of friction for an unlubricated silicon sample and a lubricated sample with Z-15 decreases with an increase in scanning velocity in the ambient environment. These samples are insensitive to scanning velocity in dry environments. Samples lubricated with Z-Dol do not show any effect of scanning velocity on the friction. Alignment of liquid molecules (shear thinning) is believed to be responsible for the drop in friction with an increase in scanning velocity for samples with mobile films and exposed to ambient environment.

For lubrication of microdevices, a more effective approach involves the deposition of organized, dense molecular layers of long-chain molecules on the surface contact. Such monolayers and thin films are commonly produced by Langmuir-Blodgett (L-B) deposition and by chemical grafting of molecules into self-assembled monolayers (SAMs) (Bhushan et al., 1995b; Bhushan, 2001; Bhushan and Liu, 2001).

Liquid film thickness measurement of thin lubricant films (on the order of 10 nm or thicker) with nanometer lateral resolution can be made with the AFM (Bhushan and Blackman, 1991; Bhushan, 1999a). The lubricant thickness is obtained by

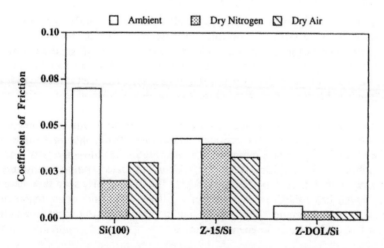

Figure 20. Coefficient of friction for unlubricated and lubricated Si(100) samples in ambient (~50% RH), dry nitrogen (~5% RH) and dry air (~5 RH) (Koinkar and Bhushan, 1996b).

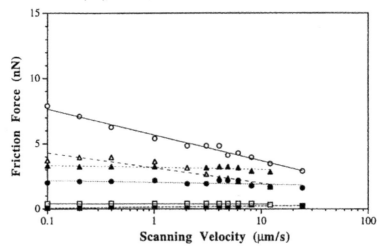

Figure 21. Coefficient of friction as a function of scanning velocity for unlubricated and lubricated Si(100) samples in ambient (~50% RH), dry nitrogen (~5% RH) and dry air (~5% RH) (Koinkar and Bhushan, 1996b).

measuring the force on the tip as it approaches, contacts and pushes through the liquid film and ultimately contacts the substrate. The distance between the sharp snap-in (owing to the formation of a liquid meniscus between the film and the tip) at the liquid surface and the hard repulsion at the substrate surface is a measure of the liquid film thickness.

Lubricant film thickness mapping of ultra-thin films (on the order of couple of 2 nm) can be obtained using friction force microscopy (Koinkar and Bhushan, 1996a) and adhesive force mapping (Bhushan and Dandavate, 2000). Figure 22 shows gray scale plots of the surface topography and friction force obtained simultaneously for unbonded Demnum type PFPE lubricant film on silicon. The friction force plot shows well distinguished low and high friction regions roughly corresponding to high and low regions in surface topography (thick and thin lubricant regions). A uniformly lubricated sample does not show such a variation in the friction. Friction force imaging can thus be used to measure the lubricant uniformity on the sample surface, which cannot be identified by surface topography alone. Figure 23 shows the gray scale plots of the adhesive force distribution for silicon samples coated uniformly and nonuniformly with Z-DOL type PFPE lubricant. It can be clearly seen that there exists a region, which has adhesive force distinctly different from the other region for the nonuniformly coated sample. This implies that the liquid film thickness is nonuniform giving rise to a difference in the meniscus forces.

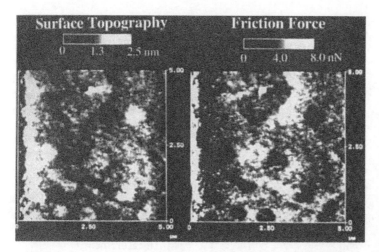

6. Closure

At most solid-solid interfaces of technological relevance, contact occurs at many asperities. A sharp AFM/FFM tip sliding on a surface simulates just one such contact. However, asperities come in all shapes and sizes. The effect of radius of a single asperity (tip) on the friction/adhesion

Figure 22. Gray scale plots of the surface topography and friction force obtained simultaneously for unbonded perfluoropolyether lubricant film on silicon (Koinkar and Bhushan, 1996a).

performance can be studied using tips of different radii. AFM/FFM are used to study various tribological phenomena, which include surface roughness, adhesion, friction, scratching, wear, indentation, detection of material transfer, and boundary lubrication. Measurement of atomic-scale friction of a freshly–cleaved highly–oriented pyrolytic graphite exhibits the same periodicity as that of the corresponding topography. However, the peaks in friction and those in the corresponding topography are displaced relative to each other. Variations in atomic-scale friction and the observed displacement can be explained by the variation in interatomic forces in the normal and lateral directions. Local variations in microscale friction occur and are found to correspond to the local slopes, suggesting that a ratchet mechanism and collision effects are responsible for this variation. Directionality in the friction is observed on both micro- and macroscales, which results from the surface roughness and surface preparation. Anisotropy in surface roughness accentuates this effect. Microscale friction is generally found to be smaller than the macrofriction as there is less plowing contribution in microscale measurements. Microscale friction is load dependent and friction values increase with an increase in the normal load, approaching to the macrofriction at contact stresses higher than the hardness of the softer material. The tip radius also has an effect on the adhesion and friction.

Mechanism of material removal on the microscale is studied. Wear precursors can be detected at early stages of wear using a surface potential measurement. Wear rate for single-crystal silicon is negligible below 20 μN and is much higher and remains approximately constant at higher loads. Elastic deformation at low loads is responsible for negligible wear. Most of the wear debris is loose. Evolution of wear has also been studied using AFM. Wear is found to be initiated at nanoscratches. For a sliding interface requiring near-zero friction and wear, contact stresses should be below the hardness of the softer material to minimize plastic deformation and surfaces should be free of nanoscratches. Further, wear precursors can be studied by making surface potential measurements. It is found that even in the case of zero wear (no measurable

Adhesive force

3.5 nm uniform Z-DOL/Si (100)

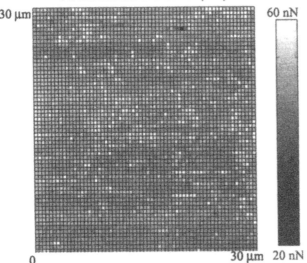

3-10 nm non-uniform Z-DOL/Si (100)

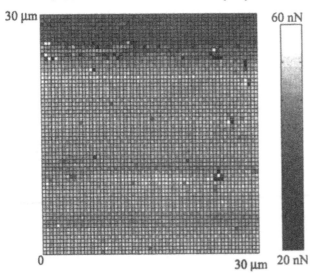

Figure 23. Gray scale plots of the adhesive force distribution of a uniformly-coated, 3.5-nm thick unbonded perfluoropolyether lubricant film on silicon and 3- to 10-nm thick unbonded perfluoropolyether lubricant film on silicon that was deliberately coated non-uniformly by vibrating the sample during the coating process (Bhushan and Dandavate, 2000).

deformation of the surface using AFM); there can be a significant change in the surface potential inside the wear mark which is useful for study of wear precursors. In-situ surface characterization of local deformation of materials is studied to better understand failure mechanisms.

Modified AFM can be used to obtain load-displacement curves and for measurement of nanoindentation hardness and Young's modulus of elasticity, with depth of indentation as low as 1 nm. Hardness of ceramics on nanoscales is found to be higher than that on the microscale. Ceramics exhibit significant plasticity and creep on a nanoscale. Scratching and indentation on nanoscales are powerful ways to screen for adhesion and resistance to deformation of ultrathin films. Detection of material transfer on a nanoscale is possible with AFM.

Boundary lubrication studies and measurement of lubricant-film thickness with a lateral resolution on a nanoscale can be conducted using AFM. Chemically bonded lubricant films with a mobile fraction are superior in wear resistance.

Investigations of wear, scratching and indentation on nanoscales using the AFM can provide insights into failure mechanisms of materials.

Coefficients of friction, wear rates and mechanical properties such as hardness have been found to be different on the nanoscale than on the macroscale; generally, coefficients of friction and wear rates on micro- and nanoscales are smaller, whereas hardness is greater. Therefore, micro/nanotribological studies may help define the regimes for ultra-low friction and near zero wear.

7. References

Bhushan, B. (1995), "Micro/Nanotribology and its Applications to Magnetic Storage Devices and MEMS," *Tribol. Int.* **28**, 85-95.

Bhushan, B. (1996), *Tribology and Mechanics of Magnetic Storage Devices*, Second ed., Springer-Verlag, New York.

Bhushan, B. (1997), *Micro/Nanotribology and its Applications*, Vol. E330, Kluwer Academic Pub., Dordrecht, Netherlands.

Bhushan, B. (1998a), *Tribology Issues and Opportunities in MEMS*, Kluwer Academic Pub., Dordrecht, Netherlands.

Bhushan, B. (1998b), "Micro/nanotribology Using Atomic Force/Friction Force Microscopy: State of the Art," *Proc. Inst. Mech. Engrs. Part J: J. Eng. Tribol.* **212**, 1-18.

Bhushan, B. (1999a), *Handbook of Micro/Nanotribology*, Second ed., CRC Press, Boca Raton, FL.

Bhushan, B. (1999b), "Nanoscale Tribophysics and Tribomechanics," *Wear* **225-229**, 465-492.

Bhushan, B. (1999c), "Wear and Mechanical Characterisation on Micro- to Picoscales Using AFM," *Int. Mat. Rev.* **44**, 105-117.

Bhushan, B. (1999d), *Principles and Applications of Tribology*, Wiley, NY.

Bhushan, B. (2001), *Modern Tribology Handbook*, CRC Press, Boca Raton, Florida.

Bhushan, B. and Blackman, G.S. (1991), "Atomic Force Microscopy of Magnetic Rigid Disks and Sliders and its Applications to Tribology," *ASME J. Tribol.* **113**, 452-458.

Bhushan, B. and Dandavate, C. (2000), "Thin-film Friction and Adhesion Studies Using Atomic Force Microscopy," *J. Appl. Phys.* **87**, 1201-1210.

Bhushan, B. and Goldade, A. V. (2000a), "Measurements and Analysis of Surface Potential Change During Wear of Single Crystal Silicon (100) at Ultralow Loads Using Kelvin Probe Microscopy", *Appl. Surf.* **157**, 373-381.

Bhushan, B. and Goldade, A. V. (2000b), "Kelvin Probe Microscopy Measurements of Surface Potential Change Under Wear at Low Loads," *Wear* (in press).

Bhushan, B. and Koinkar, V. N. (1994a), "Tribological Studies of Silicon for Magnetic Recording Applications", *J. Appl. Phys.* **75**, 5741-5746.

Bhushan, B. and Koinkar V. N. (1994b), "Nanoindentation Hardness Measurements Using Atomic Force Microscopy", *Appl. Phys. Lett.* **64**, 1653-1655.

Bhushan, B. and Kulkarni, A. V. (1996), "Effect of Normal Load on Microscale Friction Measurements", *Thin Solid Films* **278**, 49-56; **293**, 333.

Bhushan, B. and Liu, H. (2001), "Nanotribological Properties and Mechanisms of Alkylthiol and Biphenyl Thiol Self-Assembled Monolayers Studied by AFM," *J. Phys. Chem.*, submitted for publication.

Bhushan, B. and Ruan, J. (1994), "Atomic-scale Friction Measurements Using Friction Force Microscopy: Part II - Application to Magnetic Media", *ASME J. Trib.* **116**, 389-396.

Bhushan, B. and Sundararajan, S. (1998), "Micro/nanoscale Friction and Wear Mechanisms of Thin Films Using Atomic Force and Friction Force Microscopy", *Acta Mater.* **46**, 3793-3804.

Bhushan, B., Koinkar V. N. and Ruan, J. (1994), "Microtribology of Magnetic Media", *Proc. Inst. Mech. Eng., Part J: J. Eng. Tribol.* **208**, 17-29.

Bhushan, B., Israelachvili, J. N. and Landman, U. (1995a), "Nanotribology: Friction, Wear and Lubrication at the Atomic Scale", *Nature* **374**, 607-616.

Bhushan, B., Kulkarni, A. V., Koinkar, V. N., Boehm, M., Odoni, L., Martelet, C. and Belin, M. (1995b), "Microtribological Characterization of Self-assembled and Langmuir-Blodgett Monolayers by Atomic and Friction Force Microscopy," *Langmuir* **11**, 3189-3198.

Bhushan, B., Kulkarni, A. V., Bonin, W. and Wyrobek, J. T. (1996), "Nano/Picoindentation Measurement Using a Capacitance Transducer System in Atomic Force Microscopy," *Philos. Mag.* **74**, 1117-1128.

Bobji, M. S. and Bhushan, B. (2000), "Atomic Force Microscopic Study of the Micro-Cracking of Magnetic Thin Films Under Tension," *Scripta Mater.* (in press).

Bobji, M. S. and Bhushan, B. (2001), "In-Situ Microscopic Surface Characterization Studies of Polymeric Thin Films During Tensile Deformation Using Atomic Force Microscopy," *J. Mater. Res.*, submitted for publication.

Bowden, F. P. and Tabor, D. (1950), *The Friction and Lubrication of Solids*, Part 1, Clarendon Press, Oxford, U.K.

DeVecchio, D. and Bhushan, B. (1998), "Use of a Nanoscale Kelvin Probe for Detecting Wear Precursors," *Rev. Sci. Instrumen.* **69**, 3618-3624.

Frisbie, C.D., Rozsnyai, L.F., Noy, A., Wrighton M.S. and Lieber, C.M. (1994), "Functional Group Imaging by Chemical Force Microscopy," *Science* **265**, 2071-2074.

Guntherodt, H.J., Anselmetti, D. and Meyer, E. (1995), *Forces in Scanning Probe Methods*, Vol. E286, Kluwer Academic Pub., Dordrecht, Netherlands.

Koinkar, V.N. and Bhushan, B. (1996a), "Micro/nanoscale Studies of Boundary Layers of Liquid Lubricants for Magnetic Disks," *J. Appl. Phys.* **79**, 8071-8075.

Koinkar, V.N. and Bhushan, B. (1996b), "Microtribological Studies of Unlubricated and Lubricated Surfaces Using Atomic Force/Friction Force Microscopy," *J. Vac. Sci. Technol. A* **14**, 2378-2391.

Koinkar, V.N. and Bhushan, B. (1996c), "Microtribological Studies of Al_2O_3-TiC, Polycrystalline and Single-Crystal Mn-Zn Ferrite and SiC Head Slider Materials," *Wear* **202**, 110-122.

Koinkar, V.N. and Bhushan, B. (1997a), "Microtribological Properties of Hard Amorphous Carbon Protective Coatings for Thin Film Magnetic Disks and Heads," *Proc. Inst. Mech. Eng. Part J: J. Eng. Tribol.* **211**, 365-372.

Koinkar, V.N. and Bhushan, B. (1997b), "Effect of Scan Size and Surface Roughness on Microscale Friction Measurements," *J. Appl. Phys.* **81**, 2472-2479.

Koinkar, V.N. and Bhushan, B. (1997c), "Scanning and Transmission Electron Microscopies of Single-crystal Silicon Microworn/machined Using Atomic Force Microscopy," *J. Mater. Res.* **12**, 3219-3224.

Kulkarni, A.V. and Bhushan, B. (1996a), "Nanoscale Mechanical Property Measurements Using Modified Atomic Force Microscopy," *Thin Solid Films* **290-291**, 206-210.

Kulkarni, A.V. and Bhushan, B. (1996b), "Nano/picoindentation Measurements on Single-crystal Aluminum Using Modified Atomic Force Microscopy," *Materials Letters* **29**, 221-227.

Kulkarni, A.V. and Bhushan, B. (1997), "Nanoindentation Measurement of Amorphous Carbon Coatings," *J. Mater. Res.* **12**, 2707-2714.

Mate, C.M., McClelland, G.M., Erlandsson R. and Chiang, S. (1987), "Atomic-scale Friction of a Tungsten Tip on a Graphite Surface," *Phys. Rev. Lett.* **59**, 1942-1945.

Meyer, E., Overney, R., Luthi, R., Brodbeck, D., Howald, L., Frommer, J., Guntherodt, H.J., Wolter, O., Fujihira, M., Takano, T. and Gotoh, Y. (1992), "Friction Force Microscopy of Mixed Langmuir-Blodgett Films," *Thin Solid Films* **220**, 132-137.

Persson, B.N.J. and Tosatti, E. (1996), *Physics of Sliding Friction*, Vol. E311, Kluwer Academic Pub., Dordrecht, Netherlands.

Ruan, J. and Bhushan, B. (1994a), "Atomic-scale Friction Measurements Using Friction Force Microscopy: Part I General Principles and New Measurement Techniques," *ASME J. Tribol.* **116**, 378-388.

Ruan, J. and Bhushan, B. (1994b), "Atomic-scale and Microscale Friction of Graphite and Diamond Using Friction Force Microscopy," *J. Appl. Phys.* **76**, 5022-5035.

Ruan, J. and Bhushan, B. (1994c), "Frictional Behavior of Highly Oriented Pyrolytic Graphite," *J. Appl. Phys.* **76**, 8117-8120.

Scherer, V., Arnold W. and Bhushan, B. (1998), "Active Friction Control Using Ultrasonic Vibration," in *Tribology Issues and Opportunities in MEMS* (B. Bhushan, ed.), pp. 463-469, Kluwer Academic Pub., Dordrecht, Netherlands.

Scherer, V., Arnold, W. and Bhushan, B. (1999), "Lateral Force Microscopy Using Acoustic Friction Force Microscopy," *Surface and Interface Anal.* **27**, 578-587.

Singer, I.L. and Pollock, H.M. (1992), *Fundamentals of Friction: Macroscopic and Microscopic Processes,* Vol. E220, Kluwer Academic Pub., Dordrecht, Netherlands.

Sundararajan, S. and Bhushan, B. (2000a), "Development of a Continuous Microscratch Technique in an Atomic Force Microscope and its Application to Study Scratch Resistance of Ultra-Thin Hard Amorphous Carbon Coatings," J. Mater. Res., (in press).

Sundararajan, S. and Bhushan, B. (2000b), "Topography-Induced Contributions to Friction Forces Measured Using an Atomic Force/Friction Force Microscope," *J. Appl. Phys.* **88** (in press).

Zhao, X. and Bhushan, B. (1998), "Material Removal Mechanism of Single-crystal Silicon on Nanoscale and at Ultralow Loads," *Wear* **223**, 66-78.

ATOMIC SCALE ORIGIN OF ADHESION AND FRICTION
Viscoelastic Effects in Model Lubricant Monolayers

MIQUEL SALMERON AND SUSANNE KOPTA
Materials Sciences Division, Lawrence Berkeley National Laboratory, University of California, Berkeley, California 94720, USA

ESTHER BARRENA AND CARMEN OCAL
Departamento de Intercaras y Crecimiento, Instituto de Ciencia de Materiales de Madrid, Consejo Superior de Investigaciones Científicas, 28049 Madrid, Spain

Abstract

We discuss the origin of energy dissipation in friction with model lubricant monolayers of alkyl-silanes and thiols. Atomic Force Microscopy (AFM), Surface Forces Apparatus (SFA) and Sum Frequency Generation vibrational spectrocopy (SFG) were used to study the structure of the films, (ordering, thickness, molecular conformation), as a function of applied load. We show that energy dissipation can be described in terms of excitation of specific molecular rearrangements. This include generation of terminal gauche defects at loads of 20 to 80 MPa and collective tilts of the alkyl chains at certain threshold loads in the GPa regime. The rigid molecular tilts are accompanied by stepwise increases in friction. Only a discrete number of tilt angles and corresponding film heights are observed. The results are explained by a simple close packing and chain interlocking model driven by the van der Waals attractive energy. Friction energy is spent in overcoming an activation energy barrier for tilting. The activation consists in a slight increase in the separation (~0.2 Å) between chains to allow the interlocked alkyl chains to slide past each other. The increase in separation causes a loss of cohesive energy of the film.

1. Introduction

Tribology is the study of the phenomena that occur in the contact between two materials. Its main topics include adhesion, friction and wear (Bowden and Tabor, 1967). A large part of the accumulated knowledge in the field is of a phenomenological nature with little understanding of the processes occurring at the atomic and molecular scale in the buried contact interface. Classical continuum mechanics has provided a basis for rationalizing the observations of the variation of contact area as a function of applied load. The

B. Bhushan (ed.),
Fundamentals of Tribology and Bridging the Gap between the Macro- and Micro/Nanoscales, 41–52.
© 2001 *Kluwer Academic Publishers.*

connection with friction is established through the assumption that this force is proportional to the contact area. This assumption and the validity of the continuum mechanics approach has been shown to be valid down to the nanometer scale, in contacts that contain a number of atoms ranging from tens to several hundreds (Carpick et al., 1996a; Carpick et al., 1996b).

In macroscopic contacts, the interface consists of many micrometer and submicrometer size asperities where the real contact occurs. The real area of contact is generally a very small fraction of the apparent contact area. In addition, during sliding the contact points shift and deform. For that reason, and because of the inaccessibility of the buried interface to spectroscopic probes (except in the case of transparent materials), fundamental studies of tribology are always difficult. Today, thanks to the development of proximal probes, such as the atomic force microscope (AFM), and the surface forces apparatus (SFA), which are illustrated in Figure 1, it is possible to perform experiments on single asperities. In addition, it is possible to combine the SFA with modern surface specific vibrational spectroscopies, such as Sum Frequency Generation (SFG), to interrogate the interface. This lecture will be devoted to tribological studies in single asperities using AFM, SFA and SFG.

The aim of our studies is to determine the atomic scale processes responsible for the energy loss in friction. Ideally these processes should be described in terms of elementary excitations that require a discrete amount of energy. For example, excitation of phonons and electronic transitions (Dayo et al., 1998; Persson, 1998) or the rupture or modification of specific chemical bonds. Each process has an associated characteristic lifetime. This is the time it takes for the excitation to decay back to the original or ground state. Electronic excitations, for example, have characteristic

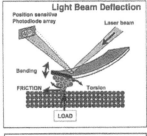

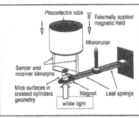

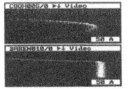

Figure 1. The tools of modern nanoscale tribology. Top: Schematic representation of an atomic force microscope (AFM), showing the microfabricated cantilever bending and twisting in response to vertical and parallel forces applied to the tip apex. The motion of the lever is detected by the light reflected from the back side of the lever and impinging on a position-sensitive photodiode arrangement. Middle: The surface forces apparatus (SFA), where two crossed cylindrical lenses are used to confine material interposed. The lenses support mica sheets and can be displaced normally to applied load, and horizontally for shear force measurements. The separation is measured by multiple light interference between the half-silvered backside surfaces of the mica sheets. Bottom: interference fringes obtained after dispersing the light transmitted through a slit by the two mica mirrors. The shape of the fringes is a direct measure of the geometry of the contact. The top fringe is parabollic and its position reflects the separation of the mica sheets. The bottom fringe has a flattened apex, indicating contact and allowing a direct measurement of the contact diameter.

times on the order of femtoseconds (1 fs = 10^{-15} seconds). Electronic excitations will take place in metals and semiconductors where the overlap of atomic orbitals at the interface and their sliding motion create electron hole pairs that are dragged along the direction of motion. On insulators, electronic excitations require very large energies, on the order of the band gap (several eVs), and thus will not be important. Energy is lost by de-excitation through collisions with impurities and surface defects as in an electrical resistance. Next we will consider excitation of phonon modes. The atoms are plucked away from their equilibrium position and are then released during atomic scale stick-slip, similar to the plucking of a violin string. Their vibration couples to other vibrational modes of lower frequency and propagates into the bulk, removing energy from the contact region. The typical lifetime of vibrational modes is measured in picoseconds (1 ps = 10^{-12} second). Lifetime values can be determined from the width of the peaks in electronic and vibrational spectroscopies.

Other excitations to consider are soft phonon modes (*i.e.*, of very low frequency), such as molecular displacements, rotations and librations, which should be particularly important in polymer materials and on films of organic molecules (Salmeron, 1998; Carpick and Salmeron, 1997; Xiao et al., 1996). Other excitations give rise to geometrical changes of the molecules, such as conformation changes through the creation of gauche defects. These various excitation modes determine the viscoelastic behavior of materials. The lifetime of these modes is much larger than the previous times and is due to the entanglement of the usually long molecules with the neighboring ones. Typical lifetime values can be on a few orders of magnitude around 1 second, depending strongly on temperature. Because this lifetime can be comparable to sliding times in friction experiments, viscoelastic excitations can give rise to velocity dependence effects in friction and to hysteresis in adhesion that is observed in the formation and rupture of contacts.

To complete the description of energy loss mechanisms, we should include point defects (ad-atoms, vacancies, intestitials, *etc.*), which can be produced during sliding at or near the interface. Defects that are produced below the surface of the contacting bodies can accumulate into dislocation and slippage planes (Salmeron, 1992; Kelchner et al., 1998; Kiely and Houston, 1998). Production of such defects is the basic mechanism of wear. The lifetime of these defects is practically infinite, and the surface of the materials might not recover its original state unless heated to high temperature to anneal off the defects.

The present lecture will focus on the excitation of viscoelastic defects in monolayer films of organic molecules. These are used as model lubricants and are formed by self-assembly and Langmuir-Blodgett techniques (SAM and LB, respectively). Before presenting details of experiments performed in my laboratory to study these defects, we will illustrate with one example the progressive excitation of new energy loss mechanisms as the load increases.

2. Load-dependent excitation of different energy dissipation modes

When two materials are pressed against each other with an increasing load (L), the area of the contact increases in the manner predicted by continuum mechanic models. These are basically the Hertz model (Bowden and Tabor, 1967) and its modified versions, which include short- and long-range adhesive interactions (JKR, DMT and intermediate modes) (Johnson et al., 1971; Derjaguin et al., 1975; Carpick et al., 1999). In the Hertz model, the contact area of spherical bodies increases as $L^{2/3}$ due to elastic deformation. The pressure or force per atom, however, increases much more slowly, as $L^{1/3}$. We expect the friction to follow a $L^{2/3}$ dependence (or JKR, *etc.*) on load as long as the dissipation processes do not change. At some point, however, the load per atom will reach a value where new elementary excitations of higher energy may come into play. At these threshold loads, the friction will depart from the smooth continuum mechanics curve. The example of Figure 2 illustrates this incremental turning on of new energy dissipation modes. The graph shows how the friction between a mica substrate covered by SAM of octadecylsiloxane (OTS) and the silicon nitride tip of an AFM evolves as the load is changed over a large range. At low load, the tip (with a radius of ~500 Å) is riding on top of the silane molecules and the friction increases monotonously (not visible on the scale of the figure). Studies using this and other molecules containing long alkane chains (thiols on gold and siloxanes on mica) have shown that, in this region, the friction curve follows a JKR behavior. Between 50 and 130 nN load, however, a rapid departure from this regime occurs. This is due to the successive turning on of new energy loss mechanisms that are related to the deformation of alkyl chains. The rest of this lecture is devoted to analyzing this load regime and to determine the nature of these deformations. Above 130 nN, the tip has displaced the OTS molecules completely and the friction is characteristic of bare mica. At even higher loads, approximately 270 nN in this case, the load per atom is large enough to initiate yet another energy dissipation mechanism: the creation of point defects leading to wear of the mica substrate (Hu et al., 1995; Kopta and Salmeron, 2000). The values of the loads where these changes occur depend strongly on the tip radius.

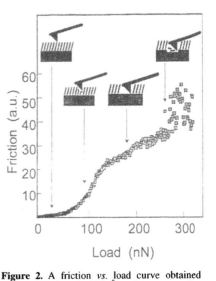

Figure 2. A friction *vs.* load curve obtained with a SiN tip of ~500 Å radius on a mica surface coated with a monolayer of octadecylsiloxane (OTS). At low load, the tip is fully supported by the tightly packed upright molecules and friction is low. Although the scale is too small to show it clearly, in that region the curve can be fit with a JKR formula for the contact area. Friction energy is mostly dissipated in phonon modes. As the load increases, first the tip starts to tilt and displace the OTS molecules until it establishes direct contact with the mica. The friction energy is spent in overcoming barriers to molecular tilts and other deformation modes. At the highest loads, point defects start being produced that end up wearing the mica.

3. Viscoelastic deformations in alkylsilane and alkylthiol monolayers

Since we will deal with alkyl-thiols and alkyl-silanes, we will use the short notation Cn for the chain length, where n represents the number of C atoms. The number of possibilities to deform a long alkane chain molecule is very large. A 120° rotation along the axis joining any consecutive pair of C atoms produces a gauche deformation of the molecule. In principle, a free molecule should exhibit $(n-2)^3$ gauche distortion modes. In SAM and LB films, however, geometrical constrictions put severe limitations on the type of deformations that can occur. The close-packed arrangement of molecules effectively hinders the distortions because they require space, which can only be created by the local displacement of molecules in the film, an energetically costly process. In spite of the not too large value of the activation energy for gauche defect production in isolated molecules (~0.2 eV), we expect these distortions to be produced only at very high loads in well-packed monolayers. One exception to this argument applies to the terminal CH_3 group. A gauche rotation around the next CC bond axis requires less than 1 Å of additional space, which can be easily accommodated by small displacements of the neighboring molecules. The schematic drawings of Figure 3 illustrate the geometry of terminal and internal gauche distortions.

Another possible deformation of the molecules in SAM and LB films consists of the rigid tilt of several molecules, as illustrated at the bottom of Figure 3. This collective tilt requires also extra space, *i.e.*, the neighboring molecules have to displace in order to accommodate the tilt. Because of the concerted tilt of many molecules, however, the space requirements are less severe than in the case of internal gauche distortions where each individual molecule needs extra space. One might thus expect that collective tilts should be produced more easily than internal gauche distortions.

In the following sections, we present evidence of this intuitively simple deformation sequence as a function of load: terminal gauche rotations, rigid tilts and internal gauche defects. On poorly-packed layers, of course, this sequence will not be followed and many modes can be excited at any time. This explains the increase in friction force observed for short chain lengths, which is due to their decreased packing density.

a) Terminal gauche deformations :

b) Internal gauche deformations :

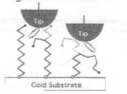

c) Rigid chain tilts :

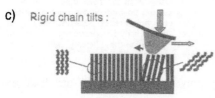

Figure 3. Schematic diagram illustrating several molecular deformation modes of long chain alkanes.

It also explains that the "stiffness" of a film, as measured by the load needed to deform or displace the molecules, depends on preparation conditions and on the time elapsed since preparation. During this time, the molecules diffuse and move about so that the film becomes more compact and ordered (Barrena et al., 1999).

4. Sum Frequency Experiments in the Surface Forces Apparatus

To test the formation of gauche defects, we performed experiments using a SFA consisting of a spherical lens of a few cm radius pressing against an optically flat quartz plate coated with SAM and LB of various molecules with alkyl chains (Du et al., 1995). The example of Figure 4 shows the results obtained with OTS. The top panel shows the SFG spectrum in the region of the C-H stretch region. Only a strong resonance peak is observed at 2875 cm^{-1}, which corresponds to the symmetric CH$_3$ modes of the terminal group. The other large peak is a combination mode of the symmetric stretch and a bending mode. The lack of CH$_2$ stretch peaks indicates that the alkyl chains are straight and well-packed. Next the lens is pressed against the flat until the contact area has grown to be about 1 mm. This occurred at an applied pressure of 50 MPa. Under these conditions, the visible and infrared laser spots could be focused entirely inside the contact area, as indicated schematically in the drawing at the center of Figure 4. The bottom panel shows the variation of the intensity of the 2875 cm^{-1} mode as a function of the position of the laser spot by scanning the sample across it. The oscillations observed outside are due to interference effects of the increasing gap between the sphere and the flat. Inside, however, a decrease of intensity by a factor of 15 is observed. Further increase of the pressure to 80 MPa reduced the intensity of the peak by more than 100 times, below the noise level of

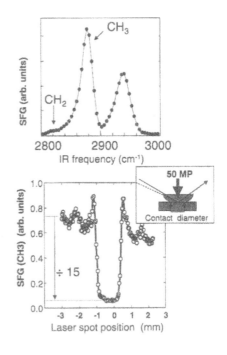

Figure 4. Top: Sum Frequency Generation spectrum in the C-H stretch region of an OTS monolayer on a flat quartz. Notice the low intensity of the CH$_2$ stretch mode, which is not allowed by symmetry in the strait all-trans configuration. Only the CH$_3$ stretch mode is non-centro-symmetric and thus allowed. Center: diagram of the SFA set up. A spherical lens of about 5 cm radius is pressed against the quartz plate until a flat contact of about 1 mm diameter is formed by elastic deformation. The visible and infrared laser beams are focused inside. Bottom: Intensity of the 2875 cm^{-1} mode as a function of laser position across the lens-quartz contact at a pressure of 50 MPa. Inside the contact, the intensity is reduced by a factor of ~15. At 80 MPa, the intensity decreases to below the noise level.

our measurement. The signal recovered completely upon removal of the load and separation of the surfaces.

These results were interpreted as indicative of the formation of terminal gauche distortions in the alkyl end groups, as in the schematic of Figure 3a. This deformation reorients the dynamic dipole moment of the symmetric CH_3 mode to being nearly parallel to the surface (in vertical molecules). In addition, and more importantly, the dipole orientations are randomized under compression. This would cause the extinction of the second order polarizability and explain the loss of intensity of the peak. Notice that the pressure of 80 MPa is at the very low end of the pressures typically achieved during AFM imaging in contact mode, where due to the sharp tips (a few hundred Å radius) at loads of ~1 nN, pressures of 0.1 GPa and higher are easily produced. This implies that, in most AFM images, the end group of the alkyl chain molecules in SAM monolayers of thiols and silanes is distorted. This does not prevent, in the case of thiols, the existence of order with $\sqrt{3} \times \sqrt{3}$ R30 periodicity, due to the close packing of chains.

5. Tilting of Alkyl Chains

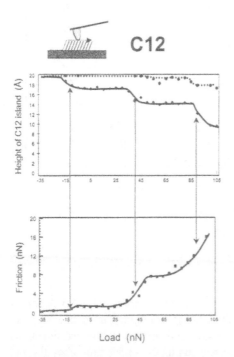

Figure 5. Height (top) and friction (bottom) of a dodecylsilane island (~1 μm diameter) on mica as a function of applied load. Negative loads mean that the lever is being pulled out of the surface to partially compensate the adhesive force (~-35 nN, in this case). Notice the stepwise decrease of the island height and the simultaneous increases in friction force. The points at the top in the broken line represent the island height after returning the load to its lowest value.

Using AFM, we studied the response of islands of SAM of C12, C16 and C18 siloxanes on mica and of C16 thiols on Au(111) films to applied pressure (Barrena et al., 1999; Barrena et al., 2000). By using islands instead of complete monolayers, the height of the film could be accurately determined. The preparation of these islands has been described in detail in recent publications. Friction was also measured at the same time as the film height. The results are summarized in Figures 5 and 6 for C12 silanes and C16 thiols, respectively. The most striking result is the appearance of steps in both film height and friction force as a function of applied load. Notice that the load starts at negative values, which correspond to the tip being pulled out of the surface to compensate for the adhesive forces. The lowest value is close to the pull-off point.

The following summarizes observations from many experiments. For a given Cn alkyl chain, a finite number of discrete heights is observed, always with the same values regardless of tip, island size and preparation conditions. The load at the threshold that separates the various plateaus depends on preparation conditions. In general, if the islands are allowed to grow large, the threshold moves to higher loads. As a function of time to allow the diffusion and annealing of defects, the threshold loads change as well. Ripe islands are harder to tilt than freshly prepared ones (Du et al., 1995). When returning to low load, the original film height is recovered (except at the highest load values that disrupt the island), indicating that the molecules tilt back to their most upright position when the load is removed.

The appearance of well-defined heights of the film suggest a model where molecules tilt under the pressure exerted by the tip to adopt specific configurations of maximum stability. The simplest model is one that implies interlocking of neighboring chains such that maximum space filling is achieved. This maximizes the van der Waals cohesive energy of the film. With this model, which is illustrated in Figure 7, it is possible to predict all the observed film heights. For silanes on mica, excellent agreement is obtained, starting with an upright configuration of the molecules and considering only a one-dimensional tilt (Barrena et al., 1999). The tilt angles must obey the relation:

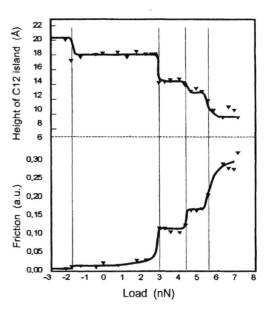

C16 on Au(111)

Figure 6. Same as in Figure 5 for an hexadecylthiol island on Au(111).

$$\tan\theta = na / d \ (n = 0, 1, ...)$$

for maximum space filling, which generates all the observed heights using the values of $a = 2.5$ Å, for the C-C-C distance projected along the chain length, and $d = 4.5$ Å for the chain diameter. For thiols on gold, where the starting configuration is one with the molecules already tilted by ~35°, one has to consider tilts in two orthogonal directions, as illustrated at the bottom of Figure 7. The corresponding tilt angles now obey the two relationships:

$$\tan\theta_x = na / d_x \ (n = 0, 0.5, 1, 1.5, ...)$$
$$\tan\theta_y = ma / d_y \ (m/2 = 0, 0.5, 1, 1.5, ...)$$

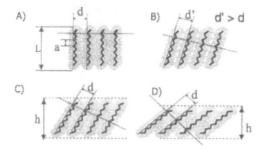

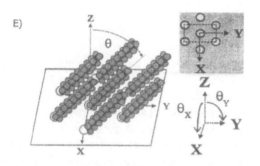

Figure 7. Top: Schematic illustration of the space filling and chain interlock model used to explain the stepwise tilting of molecules in the experiments described in Figures 5 and 6. In the transition state, shown in B, the separation chain between adjacent chains d' increases by about 0.2 Å, over that in the ground states A and C. Bottom: For thiols on Au(111), one has to consider tilts in two directions, along the first nearest neighbors (X) and the second nearest neighbors (Y).

for tilts along the x- and y-axes (Barrena et al., 2000). The heights of the plateaus in Figure 6 correspond to the values of $(n,m) = (0,2)$ for $\theta = 35°$, h = 20 Å; $(1,2)$ for $\theta = 43°$, h = 18 Å; $(2,2)$ for $\theta = 55°$, h = 14 Å; and $(2.5, 2)$ for $\theta = 59°$, h = 12 Å.

6. Tilting of Chains as a Mechanism of Energy Dissipation

The success of the geometrical space filling and chain interlock model in explaining all the observed heights of thiols and silanes provides a firm basis for an attempt to explain the friction increases. The model we adopt is one where molecular packing is dictated exclusively by van der Waals attraction between molecular chains. We can compute the cohesive energy by assuming additivity, as in the Hamaker method, and sum energies terms between methylene groups of the form a/r^6 (Israelachvili, 1992). The geometry is illustrated in Figure 8. The parameter a in the van der Waals energy can be determined by equating the cohesive energy of long chains (n > 12) to the sublimation energy of the corresponding alkane. The details of the calculation are described in another publication (Salmeron, 2000) and here we will give only the main results. As the drawings at the top of Figure 7 illustrate, to tilt the molecules one must first introduce a small separation between chains, on the order of the van der Waals corrugation of methylene units along the chain, to allow "disengagement" and sliding. The

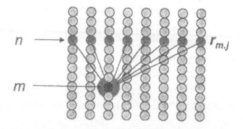

Figure 8. Model used to calculate the van der Waals cohesive energy of films made of straight alkane chains. The energy per CH_2 group at position n (each group is represented by a circle) is computed by adding $1/r^6$ terms from all the CH_2 groups in a plane at position m. This simple model allows us to calculate the relative stability of films of different number of CH_2 groups (i.e., the chain length), and also the change in cohesive energy due to changes in chain-chain separation during tilting of the molecules.

separation between chains at the transition point d′ (Figure 7b) is larger than that in equilibrium by about 0.2 Å, as estimated from a rigid ball model of the methylene units with their corresponding van der Waals radii. With this assumption, one can repeat the calculation of the van der Waals sums to obtain the cohesive energy loss in going from a to b in Figure 7. The result is 0.2 eV for a C12 chain. The link to friction force is obtained by multiplying the number of molecules swept by the tip per unit length by 0.2 eV and equating this to the friction force. Since the tip radius and contact area are approximately known (we use Hertzian mechanics to calculate the contact diameter), we obtain a friction force on the order of 2 nN, which is close to the observed values (see Figure 6) within a factor of 2 to 3.

7. Conclusions

Using atomic force microscopy and the surface forces apparatus combined with sum frequency generation vibrational spectroscopy, we have studied structural and friction properties of model organic monolayers under compressive loads. The effect of the compressive load on structure is the distortion of the alkyl chain. First, terminal gauche defects involving mainly the end CH_3 group take place at pressures below 0.1 GPa. Collective tilts are next produced in well-packed monolayers. Collective tilts are the less demanding distortions in terms of spatial requirements. Loosely packed films, as those that can be formed freshly after preparation (*i.e.*, not well-annealed films), tilt more easily and probably allow deep chain distortions. The same occurs in films formed by short chains where the van der Waals cohesive energy between chains is reduced. We have shown that the energy dissipated during friction can be understood on the basis of a model involving the excitation of specific conformation and geometrical changes of the molecules. Using islands, we have seen that this generates a series of plateaus with discrete height values. A simple geometrical model involving space filling and chain interlocking to maximize van der Waals attraction is capable of explaining all the observations. The model also explains that the energy dissipated by the tilting process is used in overcoming the barrier for molecular disentanglement and sliding along the axis.

Acknowledgments

This work was supported by the Director, Office of Science, Office of Basic Energy Sciences, of the U.S. Department of Energy under Contract No. DE-AC03-76SF00098.

References

Barrena, E., Ocal, C. and Salmeron, M. (2000), "Molecular packing changes of alkanethiols monolayers on Au(111) under applied pressure", *J. Chem. Phys.* **113**, 2413-2418.

Barrena, E., Ocal, C. and Salmeron, M. (1999a), "Evolution of the structure and mechanical stability of self-assembled alkane-thiol islands on Au(111) due to diffusion and ripening", *J. Chem. Phys.* **111**, 9797-9802.

Barrena, E., Kopta, S., Ogletree, D.F., Charych, D.H. and Salmeron, M. (1999b), "The relationship between friction and molecular structure: Alkylsilane lubricant films under pressure", *Phys. Rev. Lett.* **82**, 2880-2883.

Bowden, F.P. and Tabor D. (1967), *Friction and Lubrication,* Methuen, London, England.

Carpick, R.W., Ogletree, D.F. and Salmeron, M. (1999), "A general equation for fitting single asperity contact area and friction measurements", *J. Colloid Interface Sci.* **211**, 395-400.

Carpick, R.W. and Salmeron, M. (1997), "Scratching the surface: Fundamental investigations of tribology with atomic force microscopy", *Chem. Rev.* **97**, 1163-1194.

Carpick, R.W., Agrait, N., Ogletree, D.F. and Salmeron, M. (1996a), "Measurement of interfacial shear (friction) with an ultrahigh vacuum atomic force microscope", *J. Vac. Sci. Technol. B* **14**, 1289-1295.

Carpick, R.W., Agrait, N., Ogletree, D.F. and Salmeron, M. (1996b), "Variation of the interfacial shear strength and adhesion of a nanometer-sized contact", *Langmuir* **12**, 3334-3340.

Dayo, A., Alnasrallah, W. and Krim, J. (1998), "Superconductivity-dependent sliding friction", *Phys. Rev. Lett.* **80**, 1690-1693.

Derjaguin, B.V., Müller, V.M. and Toporov, Yu.P. (1975), "Effect of contact deformations on the adhesion of particles", *J. Colloid Interface Sci.* **53**, 314-326.

Du, Q., Xiao, X.-d., Charych, D., Wolf, F., Frantz, P., Ogletree, D.F., Shen, Y.R. and Salmeron, M. (1995), "Nonlinear optical studies of monomolecular films under pressure", *Phys. Rev. B* **51**, 7456-7463.

Hu, J., Xiao, X.-d., Ogletree, D.F. and Salmeron, M. (1995), "Atomic scale friction and wear of mica", *Surf. Sci.* **327**, 358-370.

Israelachvili, J. (1992), *Intermolecular and surface forces,* Academic Press, London, England.

Johnson, K.L., Kendall, K. and Roberts, A.D. (1971), "Surface energy and the contact of elastic solids", *Proc. Roy. Soc. London A* **324**, 301-313.

Kelchner, C.L., Plimpton, S.J. and Hamilton, J.C. (1998), "Dislocation nucleation and defect structure during surface indentation", *Phys. Rev. B* **58**, 11085-11088.

Kiely, J.D. and Houston, J.E. (1998), "Nanomechanical properties of Au (111), (001), and (110) surfaces", *Phys. Rev. B* **57**, 12588-12594 (1998).

Kopta, S. and Salmeron, M. (2000), *J. Chem. Phys.,* in press.

Persson, B.N.J. (1998), *Sliding Friction: Physical Principles and Applications,* Springer, Heidelberg, Germany.

Salmeron, M. (2000), *Trib. Lett.,* in press.

52

Salmeron, M. (1998), "Observing friction at work", *Chemtech* **28**, 17-23.

Salmeron, M., Folch, A., Neubauer, G., Tomitori, M., Ogletree, D.F. and Kolbe, W. (1992), "Nanometer scale mechanical properties of Au(111) thin films", *Langmuir* **8**, 2832-2842.

Xiao, X.-D., Hu, J., Charych, D.H. and Salmeron, M. (1996), "Chain length dependence of the frictional properties of alkylsilane molecules self-assembled on mica by AFM", *Langmuir* **12**, 235-237.

ATOMIC-SCALE STICK SLIP

R. Bennewitz, E. Meyer, M. Bammerlin, T. Gyalog
Institute of Physics, University of Basel, Klingelbergstrasse 82,
CH-4056 Basel, Switzerland
E. Gnecco
Institute of Physics, University of Genova, via Dodecaneso, 33,
I-16146 Genova, Italy

ABSTRACT. Atomic-scale stick-slip is one of the fundamental friction processes. It has been observed on layered materials, such as graphite, or ionic crystals, such as NaCl(001). Recently, wearless friction was also observed on clean metallic surfaces, such as Cu(111).

The friction force vs. lateral position traces show stick slip with the periodicity of the atomic lattice. The probing tip sticks at certain positions, builts up elastic deformation until a threshold value is reached. Then the tip jumps one unit cell to the next sticking site. Friction force loops show that the energy which is released during one slip is typically 1eV.

The velocity dependence of atomic-scale stick-slip was investigated. A logarithmic dependence of friction as a function of velocity is found. The results are discussed in terms of a Tomlinson model, which takes into account thermal activation. At low velocities, the tip may slip at lower lateral forces because of thermal activation. At higher velocities the probability is lower to overcome the barrier by thermal activation.

1. Introduction

Friction force loops [See list of recent reviews] show hysteresis between the back- and forward scan, which is associated with the dissipation of energy (non-conservative forces). At low loads the hysterisis is barely visible, but increases with increasing normal force. A sawtooth pattern becomes visible, which varies on the atomic scale. By acquiring images where the tip is scanned in x- and y-direction, Mate et al. determined that the periodicity of friction is that of the atomic lattice of graphite. A surprising observation was that even at loads of 10^{-5}N, where continuum models suggest contact diameters of 100nm, atomic-scale stick slip is visible. Pethica suggested that eventually graphite flakes were broken of the surface and adhered to the tip. Thus, friction between commensurate surfaces is observed. An alternative explanation was given by McClelland that the tip and surface make contact at only a few nm-scale asperities, so that the corrugation is not entirely averaged out. More recent measurements on non-layered materials in ultrahigh

B. Bhushan (ed.),
Fundamentals of Tribology and Bridging the Gap between the Macro- and Micro/Nanoscales, 53–66.
© 2001 *Kluwer Academic Publishers.*

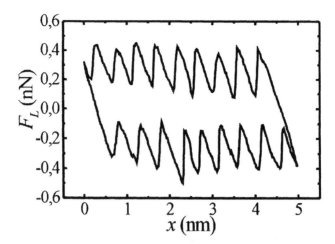

Figure 1: Frictional forces between a silicon tip and a NaCl(001)-surface in ultrahigh vacuum as a function of sample position. Atomic stick slip is observed, where the spacing is given by the atomic lattice.

vacuum have shown that atomic-scale stick slip is limited to a rather low load regime for sharp tips [Lüthi (1996)]. A higher loads, plastic deformations of the sample or tip are observed. Thus, the observation of Mate et al. of atomic-scale stick slip at high loads might be restricted to layered materials or the presence of some lubricating contamination films. However, one has to emphasize that atomic-scale stick slip at low loads is observed on practically all materials and is probably a major source of dissipation. The mechanisms of this atomic-scale slip will be discussed below.

2. Loading dependence

Mate et al. presented the first normal force dependence of a tungsten tip on graphite . They found a rather linear dependence with a friction coefficient of about 0.01. Although, linear dependences of friction vs. normal force are common in macroscopic experiments, this observation is not expected in microscopic, single asperity experiments. The linear dependence in macroscopic contacts arises from the increase of contacting asperities with increasing load, which has been explained by statistical arguments. For a single-asperity contact a non-linear dependence is expected. A simple Hertzian contact would result in a loading dependent contact area:

$$A = \pi \left(\frac{3RF_N}{4E*} \right)^{2/3} \tag{1}$$

$$E^* = \left(\frac{1 - \nu_1^2}{E_1} + \frac{1 - \nu_2^2}{E_2} \right)^{-1} \tag{2}$$

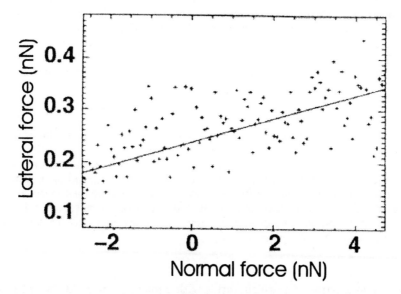

Figure 2: Frictional forces in ultrahigh vacuum between a silicon tip and a KBr(001) surface as a function of normal force. All data were extracted from atomic-scale stick slip friction loops. A linear dependence is found with a friction coefficient of 0.02.

where E_1 and E_2 are the Young's moduli of the sample and probing tip and ν_1 and ν_2 are the Poisson ratios. Assuming that the Bowden-Tabor adhesion model is valid, a loading dependence of friction of the form

$$F_F = \tau A = \tau\pi(\frac{3RF_N}{4E_*})^{2/3} \qquad (3)$$

is expected, where τ is the shear strength. Thus, the proportionality observed by Mate et al. might result from a multiasperity contact. The tip may have nm-scale roughness and the contacting asperities may increase with normal load as discussed in macroscopic experiments. Some evidence has been found by the Kaneko group [Putman], who found that the linear dependence of a silicon nitride tip on mica, observed in dry conditions, can change to non-linear dependence at high humidity. The explanation given by Putman et al. was that capillary condensation of water on the rough probing tip leads to a smoothening of the tip. With this smoothened tip, single-asperity behaviour is observed. Alternatively, the experiment of Mate et al. may also reflect a pressure-dependence of the shear strength τ. Briscoe pointed out that the shear strength is given in first approximation by $\tau = \tau_0 + \alpha p$, where $p = F_N/A$ is the pressure and α the proportionality constant. Thus, in the framework of Hertzian deformation, a loading dependence of the form:

$$F_F = \tau A = \tau_0\pi(\frac{3RF_N}{4E_*})^{2/3} + \alpha F_N \qquad (4)$$

Thus, for a large α-parameter a linear behaviour is observed. Whereas, small α-values lead to the non-linear dependence. This explanation found some support by measurements from Schwarz et al., where both linear and non-linear behaviour were observed with the same tip on heterogeneous surfaces, consisting of islands of C_{60} on GeS in air. Linear behaviour was found on GeS and non-linear behaviour on C_{60}. Schwarz et al. interpreted their data with the Fogden-White model, which can be interpreted as an extended Hertz-model, where long-range capillary forces cause an additional attractive force, which shifts the normal force scale:

$$F_F = \tau A = \tau_0 \pi \left(\frac{3R(F_N + F_0)}{4E_*} \right)^{2/3} + \alpha(F_N + F_0) \tag{5}$$

where $F_0 = 4\pi R\gamma(1-D_0/2r_k)$, γ is the surface tension of the liquid-vapor interface, D_0 is the separation of the surfaces and r_k represents the radius of the meniscus. In the case of Schwarz et al., capillary condensation of water may occur in air. The experimental pull-off force of 6.7nN was consistent with a tip radius of $R \approx 7.8$nm. With this model, they found different α-values and τ_0-values for the two materials. However, Schwarz et al. also pointed out that deviations of the spherical geometry of the tip may explain the observed behaviour. They suggested that the contact area may follow a more general law:

$$A = CF_n^m \tag{6}$$

where m is a parameter which varies with geometry: $m=2/3$ for sphere-plane geometry and $m = 1/2$ for pyramidal-shaped tip. With the fit parameters m and σ_0 (α =0) they found good agreement with the experiment. Thus, one has to make sure that the tip shape is spherical, in order to determine α and σ_0 accurately.

The effect of tip shape on the loading dependence of friction was studied by Carpick et al. (1996) . The sliding of Pt-coated tips with different geometries on mica was investigated. The shape of the probing tip was characterized by imaging a stepped $SrTiO_3(305)$ surface. At the step sites the image is a convolution with the probing tip and allows one to determine the profile of the tip apex. The profiles were fitted with polynoms of the form $z \approx r^n$. Best fits for blunted tips were between r^4 to r^6. Carpick et al. analyse the friction vs. normal force curves with an extended JKR-model, where the tip profile was included. Best agreement is found with the profiles r^6, which is in agreement with the independently measured tip profile. Thus, Carpick et al. find that the friction is proportional to contact area and that the shear strength τ is constant in first approximation for the case of Pt on mica in UHV. It is important to take into account deviations from the spherical geometry. Also, the results were applied to rather large radii of curvature ($\approx$140nm).

3. 2d-histogram technique

The 2d-histogram technique is a method to measure the loading dependence of friction as a function of normal force [Meyer (1996), Lüthi (1995)]. In contrast to conventional friction vs. normal force curves, the method is based upon the acquisition of images, so called friction force maps. During the acquisition of these data, the loading is increased or decreased. The probing tip can be either scanned on the same line or can change the position in y-direction after the acquisition of each friction loop. Forward- and backward scan images are then subtracted from each other and divided by two. For compensation of thermal and piezoelectric creep, the images can be shifted horizontally (along x-axis). Finally, the data are used to

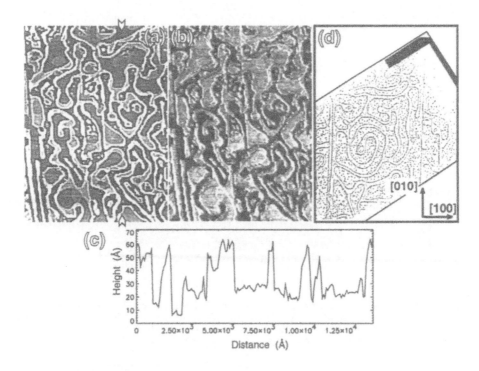

Figure 3: AgBr(001)-islands grown on NaCl(001). (a) Topography image. (b) Corresponding friction force map. High friction is observed on the AgBr(001) areas. (c) Profile as indicated in (a). The islands are 1-5 nm high. (d) Comparison with an scanning electron image of a replica, produced by metal decoration. The step structure is in good agreement with the structure observed by FFM.

compute a 2d-histogram. The method has the advantage that good statistics can be achieved. It also does not presume any functional dependence between normal force and frictional forces. Each (F_F, F_N)-data point gives a contribution to the 2d-histogram. For weak or absent correlation, the data are randomly distributed in the (F_F, F_N)-plane. For strong correlation, the data are piling up in distinct regions of the (F_F, F_N)-plane, reflecting the functional or multifunctional dependence $F_F^i(F_N)$, where the index i represents different materials or inequivalent sites.

Thin films of AgBr(001) are grown epitaxially on NaCl(001). As shown in Fig. 3 islands of about 1-5nm, corresponding to 2-10 unit cell heights (a_0=5.77Å), can be observed. The 2d-histogram shows three different regimes, as can be recognized from the corresponding lateral force image: 1) wear-less friction. 2) wave-like structure 3) droplet-like structure. Regimes 2) and 3) show drastic changes of the morphology of the film during scanning and are therefore accompanied with wear processes. It is quite remarkable, that only the 2d-histogram technique allows us to distinguish clearly between wear-less friction and friction with wear. Spikes at normal forces above 13nN are therefore related to the transport of AgBr-material by the action of the probing tip. On the average, a rather linear behaviour is

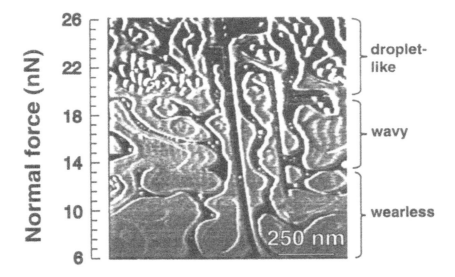

Figure 4: Friction force map acquired with variable normal force. Different regimes can be distinguished. The lower part (6-13nN) corresponds to wear-less friction. Between 13-20nN wave-like structure are visible, showing the first stage of wear. Above 20nN we observe droplet-like feature corresponding to strong deformation of the AgBr-islands.

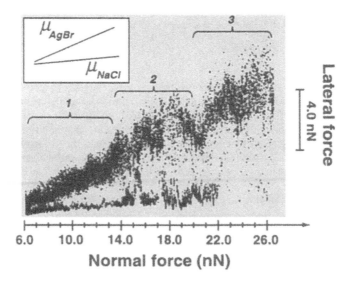

Figure 5: 2d-Histogram of the data shown in Fig. 4. The three regimes are indicated in the figure. Note that only regime 1 corresponds to the case of wear-less friction. The spikes on the histogram originate from the transport of AgBr-material.

observed. The increase of the friction forces on AgBr compared to NaCl by a factor of 30 is quite drastic [Lüthi (1995)]. For comparison, the reduction of friction by an excellent boundary lubricant, such as Cd-arachidate, compared to a silicate substrate, is only a factor of ten. In literature, it is not well established that NaCl(001) is a reasonable lubricant under ultrahigh vacuum conditions. However, it is known from the field of extreme pressure lubricants (E.P.) that chlorides are excellent lubricants of steel at temperatures above 300°C where water can be excluded [Bowden].

Although, both AgBr and NaCl are ionic crystals of similar structure, there seem to be fundamental differences. Several mechanisms may lead to this drastic difference in friction: 1) The Youngs modulus of AgBr is reduced (E_{NaCl}=3.61· 10^{10}N/m^2, E_{AgBr}=2.53· 10^{10}N/m^2) leading to an increased contact area for the same normal force. 2) AgBr is known for its high surface energy (γ_{NaCl}=0.18 J / m^2, γ_{AgBr} =0.29J / m^2) which leads to increased adhesive force, increased net normal force and therefore also enlarges the contact area. Both contributions increase the contact area by a factor of $(\frac{\gamma_{AgBr}(1-\nu_{AgBr}^2)E_{NaCl}}{\gamma_{NaCl}(1-\nu_{NaCl}^2)E_{AgBr}})^{2/3}$=1.63. Thus, these contributions are not dominant.

Another possibility is surface diffusion: The surface diffusion coefficient of Ag$^+$-ions and Br$^-$-ions is very high. Performing local scratch experiments with the AFM and observing the time evolution of the created structure (refilling of holes by surface diffusion) a diffusion coefficient of $9·10^{-14}$cm^2/s could be determined [Meyer (1992)]. This increased surface diffusion might also be related to the contrast formation of friction. E.g., Ag-ions or Br-ions could be moved into interstitial positions. Even, a small ploughing term might be existent on the AgBr-surface, leading to the larger lateral forces. The small scratches may not be observable for small loadings, because they are refilled immediately after the passage of the tip. From the observation of wear processes above 13nN it becomes obvious that the mobility of AgBr is very high and that the suggested contrast formation is quite reasonable.

4. Lateral contact stiffness measurements: Ways to determine the contact area in FFM

The FFM is a useful tool to study friction, adhesion, lubrication and wear. Under certain conditions (low load, smooth tip shape, unreactive surfaces...) a single asperity contact is formed and wearless friction is observed. In this regime, several groups have observed that friction is proportional to the contact area. Thus, the Bowden-Tabor adhesion model seems to be valid even at scales of 1-100nm. Accordingly, the friction force, F_F is given by

$$F_F = \tau A = \tau \pi a^2 \tag{7}$$

where τ is the shear strength and A is the contact area. One major problem for a quantitative analysis of friction data is the knowledge of the contact area. Unfortunately, the contact area is not directly measured in FFM. This is in contrast to the surface force apparatus, where the contact area is optically visible. Thus, the contact area has to be calculated with models, such as continuum elasticity models (e.g., Hertz, Fogden-White, Johnson-Kendall-Roberts or Maugis-Dugdale).

The particular choice of the model depends on the range of adhesive interactions, the load and tip shape. In addition, the shear strength depends also on the load (usually treated as a second order effect) and velocity (FFM is rather slow and can be treated as the quasi-static case). The relative motion and the lateral force may also affect the contact area, which is neglected in continuum elasticity models.

An independent way to determine the contact area has been introduced by the Welland group in Cambridge [Lantz (1997a/1997c)] and the Salmeron group in Berkeley [Carpick (1997)]. Instead of measuring the normal stiffness, where normal spring constants are commonly rather small, these authors suggested to measure the lateral contact stiffness, where typical spring constants are around 50-200N/m. Thus, the accuracy of the measurement is better than in normal contact stiffness measurements (at least for common cantilever geometries).

For a sphere-plane geometry, the lateral stiffness of the contact is given by:

$$k^x_{contact} = 8aG^* \tag{8}$$

where a is contact radius and

$$G^* = (\frac{2 - \nu_1^2}{G_1} + \frac{2 - \nu_2^2}{G_2})^{-1} \tag{9}$$

and G_1 and G_2 are the shear moduli of t he sample and probing tip and ν_1 and ν_2 are the Poisson ratios [Johnson,Colchero (1996a/b)]. The equation is valid for various continuum elasticity models and does not depend on the interaction forces [Johnson,Carpick (1997)]. This is in contrast to the analogous equation of the normal stiffness, which is only valid for the Hertzian case. For other models, the equation of the normal stiffness has to be modified.

A simple explanation has been given by Carpick et al. (1997): The lateral force $dF_{lateral}$ corresponds to a lateral stress $d\sigma$ which is distributed across the contact area A. This stress produces a strain $d\varepsilon \approx dx/a$, since a is the length scale of the stress distribution. Taking into account Hooke's law $d\sigma = G \cdot \varepsilon$, one can conclude that $dF_{lateral}/dx \approx G/a$. It has been assumed, that the lateral displacement does not change the contact area, which is reasonable for small displacements. However, in the case of the normal stiffness, the contact area changes with normal displacement, which explains that the normal stiffness is not generally proportional to the contact radius.

In close analogy to the treatment of normal stiffness, the elastic response in the experimental set-up is described by a series of springs. A lateral displacement of the sample Δz is distributed between three springs:

$$\Delta x = \Delta x_{contact} + \Delta x_{tip} + \Delta x_{cantilever} \tag{10}$$

where $\Delta x_{contact}$ is the elastic deformation of the contact, Δx_{tip} is the elastic deformation of the tip and $\Delta x_{cantilever}$ the elastic deformation of the cantilever. The lateral force acting on each of the springs is equal and the effective spring constant is given by:

$$F_{lateral} = k^x_{eff}\Delta x \tag{11}$$

where

$$k^x_{eff} = (\frac{1}{k^x_{contact}} + \frac{1}{k^x_{tip}} + \frac{1}{c_x})^{-1} \tag{12}$$

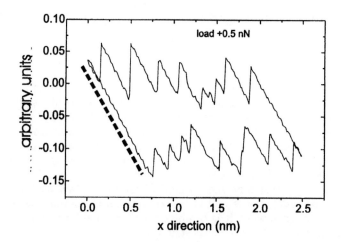

Figure 6: a) How to determine the lateral contact stiffness: Determine the slope dF/dx of the sticking part of a friction loop (lateral force vs. sample position x).

where $k^x_{contact}$ is the lateral stiffness of the contact, k^x_{tip} the lateral stiffness of the tip and c_x the lateral spring constant of tip (for most commercial cantilevers, the torsional spring constant is relevant!).

The experimental procedure to determine k^x_{eff} is as follows: Typical lateral force vs. lateral position loops, also called friction force loops consist of a sticking part, where the tip essentially stays at the same position, and a sliding part, where the tip starts to move and atomic-stick slip is observed. The slope of the sticking part corresponds to the effective spring constant:

$$dF_{lateral}/dx = k^x_{eff} \qquad (13)$$

If the piezoelectric scanner is calibrated, the lateral position can be measured accurately. Second, the lateral force scale depends on an accurate knowledge of the cantilever spring constant, c_x. Calibration procedures are described in the appendix. Having determined the effective spring constant, k^x_{eff}, the lateral stiffness of the contact can be determined according to equation 12. It was Lantz et al. (1997a) , who pointed out that the lateral stiffness of the tip, k_{tip} can be important. Scanning transmission electron microscopy was used to determine the tip shape accurately. Finite element analysis was applied to determine the displacement at a given force. The lateral stiffness of commercial probing tips was found to be comparable to the lateral spring constant. For silicon levers, Lantz et al. found even a smaller lateral stiffness for the tip than for the cantilever $(k_{tip} < c_x)$ [Lantz (1997b)].

Carpick et al. investigated the dependence of the lateral stiffness of the contact $k^x_{contact}$ as a function of loading for mica in humid air. They found good agreement with an extended Hertz model, also called Fogden and White model . According to this model the loading axis has to be shifted by the pull-off force F_P. Thus, the contact stiffness is given by

$$k_{contact}^x = 8G^* (\frac{3R}{4E^*}(F_N + F_P))^{1/3} \tag{14}$$

Lantz et al. applied the same method to NbSe$_2$ in ultrahigh vacuum conditions. They found that the contact radii calculated by the lateral contact stiffness measurement and independently by the Maugis-Dugdale theory (using a radius of curvature, determined from electron microscopy) were in good agreement (typical contact radius of 1-2nm). The loading dependence of the lateral contact stiffness and the corresponding contact radius was also found in good agreement with the Maugis-Dugdale theory [Lantz (1997c)].

One has to emphasize that the method of lateral contact stiffness measurements is a real extension to the more common friction measurements, which gives access to shear strength, when it is combined with normal friction measurements. Combining equations 8, 9, 7 the shear strength τ is given by:

$$\tau = \frac{64G^2 F_F}{\pi (k_{contact}^x)^2} \tag{15}$$

Thus, the lateral contact stiffness $k_{contact}^x$, combined with a friction force measurement F_F gives quantitative values of the shear strength, which depend less on the used continuum elasticity model than the normal stiffness measurement. The tip radius is not needed for the calculation. However, the tip shape has been assumed to be spherical or parabolic, which can be tested by scanning high aspect ratio features on surfaces or observed with an electron microscope. If the tip shape deviates, the prefactor in equation 8 has to be changed, accordingly [Johnson].

5. Velocity dependence of atomic-scale friction

The classical Coulomb law (sometimes also called the third da Vinci-Amontons law) states that friction is independent of velocity. However, this law is only approximately valid for unlubricated contacts of rather hard metals, such as titanium or steel. For soft materials, such as lead or indium, an increase of friction at low velocities (10^{-13} to 10^{-8} m/s) is observed [Rabinowicz]. At high velocities (above 10^{-4} m/s) a decrease of friction is found again. The low velocity dependence is assumed to be related to creep. The high velocity dependence is related to the velocity dependence of adhesion: A large contact needs some time to form a commensurate contact. This process is also called contact ageing. One may say that a clear understanding of these macroscopic observations is lacking.

The dependence of friction on the microscopic scale has been studied recently. Zwörner et al. found that the friction forces between silicon tips and different carbon compounds are constant over a wide range of velocities [Zwörner]. Bouhacina et al. report a logarithmic increase in friction with velocity between a tip and polymers grafted on silicon surfaces. Gnecco et al. have reported the dependence of atomic scale friction between a silicon tip and a NaCl(001) as a function of velocity in ultra-high vacuum. The velocity was varied between 1nm/s up to 1μm/s. No changes in topography were observed after several scans. Therefore, one can assume that the case of wearless friction is studied. With normal loads of 0.44nN

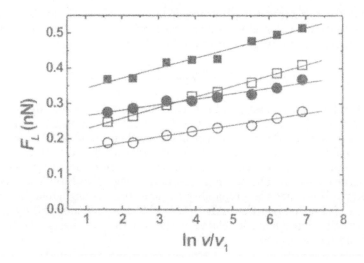

Figure 7: Friction as a function of scanning velocity at F_N 0.44nN (circles) and F_N=0.65nN (squares) loads. Open and solid sympbols refer to $\langle F_L \rangle$ (mean of the absolute value of the lateral force maps) and $\langle F_{Lmax} \rangle$ (mean absolute value of the peaks in the friction loops).

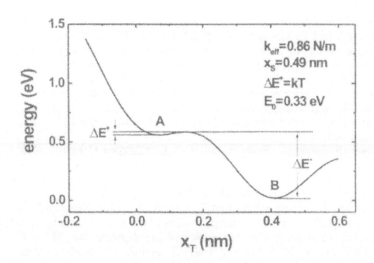

Figure 8: Potential $V tot$ used for the Tomlinson model including thermal activation.

and 0.65nN a logarithmic dependence of atomic-scale stick slip is observed. A low velocities the friction loops are narrower, whereas at higher velocities a large average friction is observed. A fit of the logarithmic depencence:

$$F_L = F_{L0} + \frac{k_B T}{\lambda} ln\frac{v}{v_1} \tag{16}$$

yields values of λ=1.56eV/nN for F_N=0.44nN and λ=0.88eV/nN for F_N=0.65nN, where k_B is the Boltzmann constant, T is the temperature and v_1=1nm/s. The theoretical analysis of these measurements is based on a modified Tomlinson model, which includes thermal activation.If the tip slides at low velocity the probability is higher that the jump occurs at lateral forces, where the energy barrier ΔE^+ is comparable to k_BT. The potential is shown in fig. 8 with typical experimental parameters. Note that the propability to jump back is negligible, because ΔE^- (≈ 0.33eV) is much larger than ΔE^+ ($\approx k_B T$). The influence of the normal force on the velocity dependence is related to the fact that the barrier heights depend strongly on the normal force.

6. Acknowledgement

This work was supported by the Swiss National Science Foundation, the Kommission für Technologie und Innovation and the Swiss Priority Program MINAST.

7. References

For recent reviews see e.g.:
R. Overney, and E. Meyer (1993) *MRS Bulletin*, **18**, 26-35.
I.L. Singer (1993) in *Dissipative Process in Tribology*, Edts. Dowson, D., Taylor, C.M., Childs, T.H.C., Gopdet, M. and Dalmaz, G., Proceedings of the 20th Leed-Lyon Symposium on Tribology, Villeurbanne, 7-10 Sept.
E. Meyer, R. Overney, and J. Frommer (1994) in *Handbook of Micro/Nanotribology*, Edt. B. Bhushan, CRC Press Inc.
O. Marti (1993) Nanotribology: Friction on a Nanometer Scale, *Physica Scripta* **T49**, 599-604.
E. Meyer, R. Lüthi, L. Howald and H.-J. Güntherodt (1995) p. 285 in *Forces in Scanning Probe Methods*, Eds. H.-J. Güntherodt, D. Anselmetti and E. Meyer, NATO ASI Series E: Applied Sciences Vol. 286, Kluwer Academic publishers.
J. Krim (1995) *Comments Condens. Mater. Phys.*, **17** 263-270.
B. Bhushan, J.N. Israelachvili and U. Landman (1995) *Nature* **374**, 607-610.
C.M. Mate (1995) Force microscopy studies of the molecular origins of friction and lubrication, *IBM Journal of Research and Development*, **39**, 617-627.
R.W. Carpick and M. Salmeron (1997) Scratching the Surface: Fundamental Investigations of Tribology with Atomic Force Microscopy, *Chemical Reviews* **97**, 1163-1194.
E. Meyer, R.M. Overney, K. Dransfeld, and T. Gyalog (1998) *Nanoscience: Friction and Rheology on the Nanometer Scale* World Scientific Publishing, Singapore.

T. Bouhacina, J.P. Aime, S. Gauthier, D. Michel, and V. Heroguez (1997) *Phys.*

Rev. **B 56**, 7694-7700.

F.P. Bowden and D. Tabor (1967) *Friction and Lubrication* , London, Methuen, revised edition.

B. Briscoe and D.C.B. Evans (1982) The shear properties of Langmuir-Blodgett layers, *Proc. R. Soc. London A* **380**, 389-407.

B.J. Briscoe and A.C. Smith, The interfacial shear strength of molybdenum disulfide and graphite films, *ASLE Transactions* **25**, 349-354.

R.W. Carpick, N. Agrait, D.F. Ogletree adn M. Salmeron (1996) Measurement of interfacial shear (friction) with an ultrahigh vacumm force microscope, *J. Vac. Sci. Technol. B* **14**, 1289-1295.

R.W. Carpick, D.F. Ogletree and M. Salmeron (1997) Lateral stiffness: A new nanomechanical measurement for the determination of shear strengths with friction force microscopy, *Appl. Phys. Lett.* **70**, 1548-1550.

J. Colchero, M. Luna and A.M. Baro (1996a) Lock-in technique for measuring friction on a nanometer scale, *Appl. Phys. Lett.*, **68**, 2896-2898.

J. Colchero, M. Luna and A.M. Baro (1996b) Energy dissipation in scanning force microscopy - friction on an atomic scale, *Tribology Letters*, **2**, 327-343.

A. Fogden and L.R. White (1990) *J. Colloid Interface Sci.* **138**, 414-418.

E. Gnecco, R. Bennewitz, T. Gyalog, Ch. Loppacher, M. Bammerlin, E. Meyer and H.-J. Güntherodt (2000), *Phys. Rev. Lett.* **84** 1172-1174.

K.L. Johnson (1985) *Contact Mechanics*, Cambridge University Press, Cambridge, United Kingdom.

M.A. Lantz, S.J. O'Shea, A.C.F. Hoole and M.E. Welland (1997a) Lateral stiffness of the tip and tip-sample contact in frictional force microscopy, *Appl. Phys. Lett.*, **70**, 970-972.

Lantz et al. (1997b) found for a silicon tip: c_n=1.1N/m ; c_x=110N/m ; k_{tip}=84N/m and for a Si_3N_4-tip: c_n=0.6N/m ; c_x=8.2N/m ; k_{tip}=39N/m [Data from Lantz (1997a)]

M.A. Lantz, S.J. O'Shea, M.E. Welland and K.L. Johnson (1997c) Atomic force microscope study of contact area and friction on $NbSe_2$, *Phys. Rev. B*, **55**, 10776-10780.

R. Lüthi, E. Meyer, H. Haefke, L. Howald, W. Gutmannsbauer, M. Guggisberg, M. Bammerlin and H.-J. Güntherodt (1995) *Surf. Sci.* **338**, 247-251.

R. Lüthi, E. Meyer, M. Bammerlin, L. Howald, H. Haefke, T. Lehmann, C. Loppacher, H.-J. Güntherodt, T. Gyalog and H. Thomas (1996) Friction on the atomic scale: An ultrahigh vacuum atomic force microscopy study on ionic crystals, *J. Vac. Sci. Technol. B* **14**, 1280-1284.

C.M. Mate, G.M. McClelland, R. Erlandsson, and S. Chiang (1987) Atomic-Scale Friction of a Tungsten Tip on a Graphite Surface, *Phys. Rev. Lett.* **59**, 1942-1945.

E. Meyer, L. Howald, R. Overney, D. Brodbeck, R. Lüthi, H. Haefke, J. Frommer and H.-J. Güntherodt (1992) *Ultramicroscopy* **42-44**, 274-278.

E. Meyer et al. (1996) in *Physics of Sliding Friction*, edited by B.N.J. Persson and E. Tosatti, Series E: Applied Sciences, Vol. 311, Kluwer Academic Publishers, 349-356.

J.B. Pethica (1986) Comment on Interatomic Forces in Scanning Tuneling Microscopy: Giant Corrugations of the Graphite Surface, *Phys. Rev. Lett.* **57**, 3235.

C.A.J. Putmann, M. Igarshi and R. Kaneko (1995) Single-asperity friction in friction force microscopy: The composite-tip model, *Appl. Phys. Lett.* **66**, 3221-3223.

E. Rabinowicz (1965) *Friction and Wear of Materials*, John Wiley&Sons.

U.D. Schwarz, W. Allers, G. Gensterblum and R. Wiesendanger (1995) Low-load friction behaviour of epitaxial C_{60} monolayers under Hertzian contact, *Phys. Rev. B* **52**, 14976-14984.

O. Zwörner, H. Hölscher, U.D. Schwarz, and R. Wiesendanger (1998) *Appl. Phys. Lett.* **66**, S263-267.

DISSIPATION MECHANISMS STUDIED BY DYNAMIC FORCE MICROSCOPIES

E. Meyer, R. Bennewitz, O. Pfeiffer, V. Barwich, M. Guggisberg,
S. Schär, M. Bammerlin, Ch. Loppacher, U. Gysin,
Ch. Wattinger and A. Baratoff
Institut für Physik, Universität Basel, Klingelbergstrasse 82,
4056 Basel, Switzerland

ABSTRACT. The dissipation mechanisms of contact force microscopy on solid surfaces are related to the fast motion during the slip process. Different degrees of freedom can be excited, such as phonons or electronic excitations. The dissipation mechanisms of dynamic force microscopy (DFM) were recently investigated due to the improvement in large amplitude DFM, also called dissipation force microscopy. Experimental methods to determine damping with DFM will be discussed. When an electrical field is applied between probing tip and sample, damping is observed, which depends on voltage. This type of damping is related to mirror charges, which move in the sample and/or tip because of the motion of the cantilever. When the contact potential is compensated, this long-range part is minimized. Under these conditions, only short-range damping can be measured, which appears at distances of about 1nm and increases exponentially with closer separation. Recent models of this type of damping show, that there might be a relationship to the local phonon density.

1. Introduction

Dynamic force microscopy [Mc Clelland, Martin, Nonnenmacher] has made excellent progress: True atomic resolution has been achieved on insulators [Bammerlin] , semiconductors [Giessibl (1995),Sugawara] and metals [Loppacher (1998/1999)]. Quantitative understanding of frequency shifts has been achieved by the use of first order perturbation theory. The frequency shifts were related to conservative forces, where some analytical formulae were given by Giessibl et al. [Giessibl (1997)] for power laws and exponential laws. Dürig has shown that an inversion procedure can be applied to derive the force and energy vs. distance curves from any frequency shift vs. distance curves [Dürig (1999b/2000a)]. In this article, dissipative forces are studied, which are characterized by an energy loss during each oscillation cycle, ΔE. A common way to describe dissipation is to indicate the Q-factor, which is defined by:

B. Bhushan (ed.),
Fundamentals of Tribology and Bridging the Gap between the Macro- and Micro/Nanoscales, 67–81.
© 2001 *Kluwer Academic Publishers.*

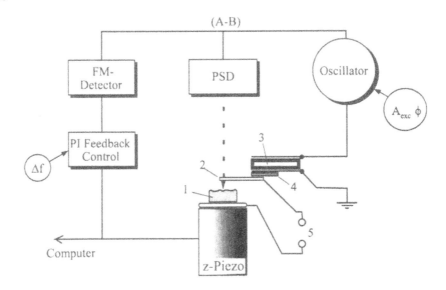

Figure 1: Scheme of dynamic force microscopy. 1: sample, 2: cantilever with integrated probing tip, 3: piezoelectric actuator, 4: isolation between cantilever and piezo, 5: bias voltage. In the case of the phase locked loop detection (PLL), both the FM-detection and the oscillation are integrated in the PLL-circuit [Loppacher (1998/1999)] .

$$Q = 2\pi \frac{E}{\Delta E} \tag{1}$$

where $E = 1/2kA^2$ is the stored energy of the oscillator with the spring constant k and the amplitude of the oscillation A. Experimentally, several possibilities are available to determine the Q-factor of the cantilever, such as the frequency spectrum, the ring down method, the phase variation experiment or the determination of the driving force in constant amplitude mode. These methods will be described in more detail below. Usually, the experimentalist starts to determine the Q-factor of the free cantilever (far away from the surface). An approach of the probing tip to the surface will cause a small change of the Q-factor, which corresponds to a small energy loss during each cycle:

$$\Delta E = -2\pi E \frac{\Delta Q}{Q^2} . \tag{2}$$

The high Q-factors of microfabricated silicon cantilevers (10'000-100'000) and the $1/Q^2$-dependence give the opportunity to measure extremely small energy losses of the order of milli-electron volts per cycle.

Alternatively, the dissipated power P can be calculated by:

$$P = \frac{\Delta E}{\Delta t} = 2\pi f \frac{1/2kA^2}{Q} \tag{3}$$

where typical power losses of 10^{-14} W are observed. For comparison, the power loss in contact force microscopy is given by the average frictional force

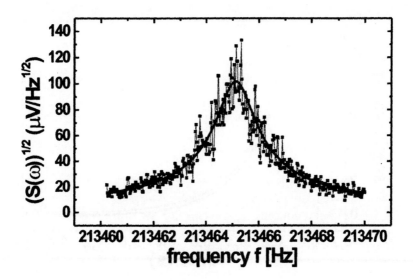

Figure 2: Example of a thermal spectrum, which has been fitted by the equation 5 and Q=142'914 has been found.

$\overline{F_F} \approx$ nN and the velocity ($v \approx$ 50nm/s): $\overline{P} = \overline{F_F}v \approx 10^{-16}$W. Eventhough, the characteristic energy loss per slip in contact force microscopy ($\approx eV$) is larger, we find that the dissipated power is small compared to the case of dynamic force microscopy. The reason might be related to the larger effective velocity, which is given by the resonance frequency and the amplitude:

$$v = \omega A \sin \omega t \tag{4}$$

with A=10nm and $\omega = 2\pi \cdot$ 150kHz, maximum velocities of about 10^{-2}m/s are found, which is 6 orders of magnitude larger than the scanning speed in contact force microscopy.

2. Methods to measure Q-factors by dynamic force microscopy (DFM)

Frequency spectra

The properties of a cantilever can be measured with frequency sweep spectra where the frequency is swept with constant excitation A_{exc} and the amplitude A and phase φ of the cantilever response are determined. Alternatively, no external excitation (A_{exc}=0) is applied, but thermal noise is used for excitation. Then, the thermal spectrum is given by:

$$S(\omega) = \frac{2k_B T \omega_0^3}{Qk((\omega^2 - \omega_0^2)^2 + \frac{\omega_0^2 \omega^2}{Q^2})} \tag{5}$$

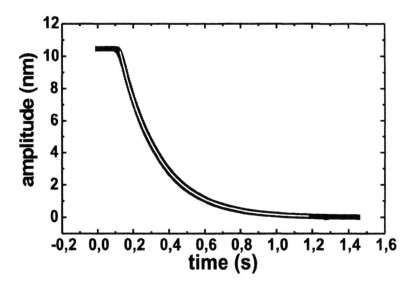

Figure 3: Oscillation of the cantilever as a function of time. This method is called ring down method. The curve has been fitted by the equation 6 and $Q=142'280$ has been found, which is in good agreement with the value from the thermal spectrum. See Fig. 2.

where $S(\omega)$ is the spectral amplitude density, $\omega = 2\pi f$, the radial frequency and k the spring constant.

Ring down method

An accurate method for cantilevers with high Q-factor is the ring down method. The amplitude A is measured as a function of time t. An example is shown in fig. 3. The curve is analyzed with the equation:

$$A(t) = A(0)e^{-\pi(f_0/Q)t} \tag{6}$$

Phase variation method

At close separation between probing tip and sample the amplitude A has to be regulated in order not crash the tip. Therefore, the phase φ is varied at constant amplitude A. The phase locked loop yields the resulting frequency $f = \omega/(2\pi)$ or the frequency shift $\Delta f = f - f_0$ relative to the resonance frequency of the free cantilever $f_0 = \omega_0/(2\pi)$. The equation of motion is given by

$$\ddot{z} + \frac{\omega_0}{Q}\dot{z} + (\omega_0)^2(z - z_{exc}) = 0. \tag{7}$$

With the ansatz:

$$z(t) = z_0 + A\cos(\omega t) \tag{8}$$

$$z_{exc} = A_{exc}\cos(\omega t + \varphi) \tag{9}$$

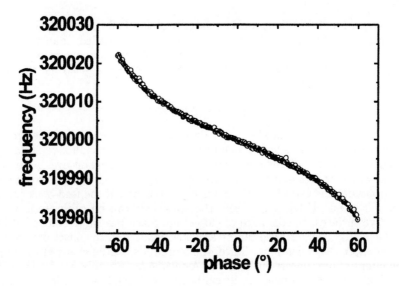

Figure 4: Phase variation experiment, where equation 12 is used to determine the Q-factor.

the excitation amplitude A_{exc} and the phase φ are related to the frequency by the equations:

$$A_{exc} = \sqrt{(f^2 - f_0^2)^2 + \frac{f_0^2 f^2}{Q^2} \frac{A_0}{f_0}} \tag{10}$$

$$\varphi = \arctan\left(\frac{f^2 - f_0^2}{f_0 f} Q\right) \tag{11}$$

which can be approximated for small frequency shifts $(\Delta f / f_0 \ll 1)$ by

$$f = f_0 \left(1 - \frac{1}{2Q} \tan(\varphi)\right). \tag{12}$$

Phase variation experiments, as shown in figure 4, are in good agreement with frequency spectra and ring down experiments and can be extended to close tip sample separations [Loppacher (1998/1999)].

A_{exc}-measurements

The most common way to determine the dissipative forces during imaging is to measure the output signal of amplitude feedback, A_{exc}, which keeps the amplitude A constant. $F_0 = kA_{exc}$ is also called the driving force, which compensates the dissipative forces exactly. Under the assumption that the phase is optimally adjusted $(\varphi = 90°\,,\ f = f_0)$ the output signal A_{exc} is given by

$$A_{exc} = A/Q \tag{13}$$

(cf. equation 10). Thus, local variations of the Q-factor or local dissipative forces can be determined. This mode is also simply called "damping" mode, because A_{exc} is proportional the local changes of the damping rate $f\Delta E/E = 2\pi f A_{exc}/A$. For small changes of A_{exc}, compared to $A_{exc0} = A/Q_0$, the changes of the Q-factor are given by

$$\Delta A_{exc} = -\Delta Q \cdot A/Q^2. \tag{14}$$

Some comments should also be made about the calibration of the amplitude A, which is essential to acquire quantitative results. On conductive surfaces, the procedure is simple: (1) The microscope is operated at constant average tunneling current, which ensures that the distance of closest approach remains constant. (2) The amplitude A is varied and the change of z-position of the feed-back controller Δz_0 is measured. Using $\Delta z_0 = \Delta A$ the amplitude can be calibrated. On non-conductive surfaces, the distance of closest approach has to be controlled by the force itself. E.g., the distance can be adjusted relative to the onset of short-range forces or short-range damping. Since, the amplitude can be changed by some nm to some tens of nm, uncertainties in the Å-regime are not relevant for the calibration. Finally, the A_{exc}-signal can be calibrated with the free cantilever, where the Q-factor is measured by the ring down method or frequency spectrum. Alternatively, the phase variation method can be used.

3. Determination of power dissipation

A simple way to interpret the local measurements of the excitation signal A_{exc} is to calculate the power dissipation. As suggested by Cleveland et al. and Gotsmann et al. the power, which is dissipated by the interaction between tip and sample P_{tip} is given by the difference between the power which is delivered by the piezo actuator to the cantilever base P_{in} and the power which is used by the intrinsic damping of the cantilever (background dissipation) P_0:

$$P_{tip} = P_{in} - P_0 \tag{15}$$

The power fed into the cantilever-tip system can be calculated by

$$P_{in} = k(z - z_{exc})\dot{z}_{exc}. \tag{16}$$

Assuming that the motion of the cantilever is harmonic,

$$z = A\cos(\omega t + \varphi), \tag{17}$$

which is valid for non-contact operation, and that the excitation signal of the base is given by

$$z_{exc} = A_{exc}\cos(\omega t) \tag{18}$$

The time averaged input power is given by

$$\overline{P_{in}} = \frac{\omega k A A_{exc}\sin(\varphi)}{2} \tag{19}$$

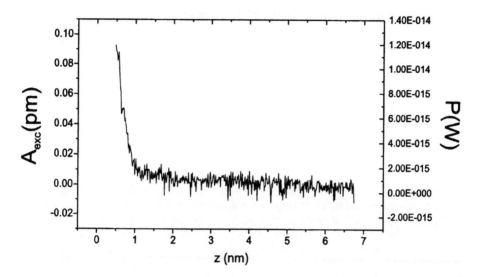

Figure 5: Measurement of A_{exc} and power P as a function of distance of a silicon tip on a Si(111)7x7 surface. Parameters were f_0=160'269Hz, Q_0=15700, A=9.2nm and k=28.5N/m. The power was derived according to equation 22.

and the internal damping of the cantilever is

$$\overline{P_0} = \gamma \dot{z}^2 = \frac{k\omega^2 A^2}{2\omega_0 Q_0} \qquad (20)$$

where $\gamma = k/(Q_0\omega_0)$ is the damping rate of the cantilever, ω is the resonance frequency at the separation of interest, ω_0 is the resonance frequency of the free cantilever, Q_0 is the Q-factor of the unperturbated cantilever and k is the spring constant. Finally, the power, which is dissipated by the tip, is given by

$$\overline{P_{tip}} = \frac{1}{2}\frac{kA^2\omega}{Q_0}[\frac{Q_0 A_{exc} \sin\varphi}{A} - \frac{\omega}{\omega_0}] \qquad (21)$$

Since frequency shifts of the cantilever are relatively small, the term ω/ω_0 can be approximated by 1. Furthermore, the phase shift is given by φ=90° for nc-AFM, which leads us to the equation

$$\overline{P_{tip}} = \frac{1}{2}\frac{kA^2\omega}{Q_0}[\frac{Q_0 A_{exc}}{A} - 1] \qquad (22)$$

A typical measurement is shown in Fig. 5 with values of the order of 10^{-14}W. The curves do depend on the applied voltage, which is associated to dissipation due to moving charges in the tip and/or sample. At close separations of about 1nm another dissipation channel is opened which increases rapidly with distance. This short-range damping will be discussed below.

4. Determination of dissipative forces

Dissipation is related to hysteresis of the force vs. distance curve. The energy loss per cycle is given by the integral:

$$\Delta E = \oint \vec{F}(\vec{r}) \cdot d\vec{r} \tag{23}$$

A convenient way to describe dissipation is to introduce a velocity dependent friction force

$$F_{diss} = \gamma(z)\dot{z} \tag{24}$$

where γ is a "friction coefficient", which depends on distance, and $\dot{z}$ is the velocity (as described in 4). It is important to take into account that $F_{diss}(z-z_0) = -F_{diss}((-1)(z - z_0))$ is asymmetric with respect to path inversion. In the case of viscous forces, such as Stokes damping in air or liquids, γ is the distance-independent viscosity in units of Ns/m=Poise. Apart from this simple case, the definition 24 can be used to describe any dissipation mechanism. In this generic form $\gamma(z)$ is a function of distance. The advantage of definition 24 is that $\gamma(z)$ is an even function with respect to path inversion and can be easily plotted. The aim of dissipation force microscopy is to determine $F_{diss}(z)$ or $\gamma(z)$. In the static mode (non-vibrating cantilever), the procedure is rather simple. The force vs. distance curve is acquired during approach F_a and during retraction F_r. In the case of adhesion hysteresis a difference is found between back and forward scan. The conservative part is given by $F_{cons} = (F_a + F_r)/2$ and the dissipative part is $F_{diss} = (F_a - F_r)/2$. In the dynamic mode, the task is more complex. Dürig has shown that it is possible to recover even a complex adhesion hysteresis force law by the measurement of higher harmonics. At present, this method has not been implemented in experiments. Therefore, the reader is referred to the original paper of Dürig (2000b) . In the case of a genuine viscous force that is strictly proportional to the velocity, the analysis is simpler. It is sufficient to detect only the first harmonic. The excitation signal $A_{exc}(z)$ at constant amplitude, which is equivalent to the driving force $F_0(z) = kA_{exc}(z)$, can then be used to calculate the dissipative force F_{diss} as a function of distance. The exact relation is:

$$F_0(z) = -\frac{2\omega}{\pi} \int_z^{z+2A} \gamma(x) \left(1 - \left(\frac{x-z}{A} - 1\right)^2\right)^{1/2} dx \tag{25}$$

where $z = z_0 - A$ is the distance of closest approach $x = z_0 + A\cos(\omega t)$ is the physical distance between tip and sample. Under the assumption of large amplitudes A compared to the range of interaction, the convolution integral 25 can be simplified to

$$F_0(z) = -\frac{2^{3/2}\omega}{A^{1/2}\pi} \int_z^\infty \gamma(x)(x-z)^{1/2} dx \tag{26}$$

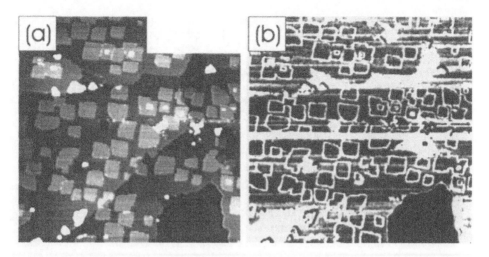

Figure 6: (a) Constant frequency shift image of a NaCl(100) film on Cu(111). (b) A_{exc}-map.

which can be inverted by Laplace transformation:

$$\gamma(x) = -\frac{\sqrt{A}}{\sqrt{2}\omega} \int_x^\infty \frac{\frac{d^2}{dz^2}F_0(z)}{\sqrt{(z-x)}}dz. \qquad (27)$$

As discussed in the original paper by Dürig (2000b), the exact dissipation function $\gamma(x)$ can be recovered by an iterative method, where equations 25 and 27 are used subsequently.

5. Imaging with Dissipation Force Microscopy and Influence of Tip Structure

Imaging of heterogeneous surfaces is interesting for the understanding of the dissipation processes and also offers a novel type of operation mode to achieve material-specific contrast. In nc-AFM, the microscope is operated in constant frequency shift mode and the amplitude A is also kept constant. The topography map is a landscape of constant frequency shift, which is essentially a constant conservative force image. Small NaCl(100) islands on top of continous NaCl(001) film are observed (cf. Fig 6). Only in the lower right an uncovered Cu(111) area is visible. The excitation signal A_{exc} map shows large differences between the NaCl-covered areas and the copper area. Therefore, dissipation force microscopy yields material-specific contrast. The increased excitation signal at the step sites of the NaCl-islands also shows that the coordination site plays an essential role in contrast formation. Fig. 7 represents a high resolution area where step and kink sites are atomically resolved. The strongest damping contrast is observed on the kink sites. Increased contrast is found on the step sites and the smallest contrast is visible on the terrace site. In order to exclude a convolution with the topography and the long-range depedence of damping, one has to perform

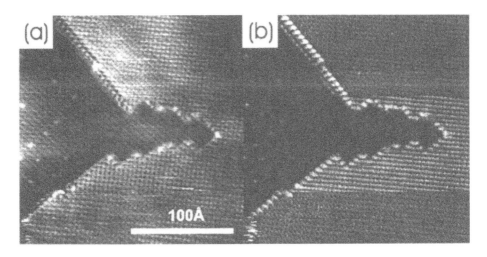

Figure 7: (a) Constant frequency shift image of a NaCl(100) film on Cu(111). (b) A_{exc}-map. Parameters were f_0=158'300Hz, Δf=-185Hz, A=0.7nm and k=26N/m.

experiments at constant height. Loppacher et al. have shown that the atomic-scale contrast of damping is visible also in the constant height mode. A small tip change is visible in the middle of the image 7. This tip change causes strong variations of A_{exc} but shows only minor changes of the topography. Measurements on Cu(111) by Loppacher et al. have shown that some tip changes are not visible at all in the topography but can be detected in the excitation signal [Loppacher (2000a)]. The determination of the short-range forces as a function of distance by Loppacher et al. (2000a/b) has shown that the frequency shift and excitation signal decay exponentially with distance. The decay length of the conservative force (frequency shift) varies from 0.33 to 0.45nm, where the decay length of the dissipative force (excitation or damping) varies between 0.21 to 0.28nm. In all cases the ratio between the two decay lengths is about 1.5. In the same study, it was observed that the decay length of the tunneling current was twice as large as the conservative force. Therefore, we conclude that the short-range dissipative forces decay more rapidly than the conservative forces, which explains the sensitivity of the damping to tip changes. In agreement with the prediction of Gauthier (1999), Loppacher et al. (2000b) found that the power is proportional to the square root of the amplitude $P \propto \sqrt{A}$. For future applications of dissipation force microscopy it will be essential to prepare tips in a well-defined way. Characterization by techniques, such fiel ion microscopy, appears adequate. Chemically more inert tips, such as carbon nanotubes, is of interest. Fig. 8 shows silicon tip with an attached multi-wall nanotube tip [Barwich]. The tube has been glued to the silicon tip and shortened in order to reduce thermal vibrations to the sub-Å-level. Fig. 9 shows frequency vs. distance and damping vs. distance curves, which indicate that the nanotube tip reduces long-range conservative forces as well as long-range dissipative forces.

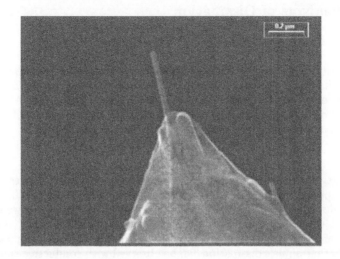

Figure 8: Example of a nanotube tip. The length is 200 nm. The radius of curvature is $R = 15$ nm. The thermal vibration is $\overline{x_{th}} = 0.02$ nm.

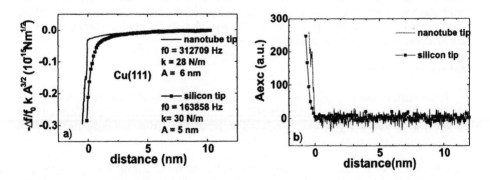

Figure 9: (Left) Constant frequency shift vs. distance curve, (right) A_{exc} vs. distance curve with a nanotube tip and a silicon tip. The long-range forces are reduced. The damping signal of the nanotube tip decays more strongly.

6. Mechanisms of Dissipation

Joule's dissipation

Localized charges on the sample surface and/or tip influence the contrast formation of dissipation force microscopy. These charges induce mirror charges on the counter electrodes, which are moving during the oscillation cycle. These moving charges are associated with currents I. Due to the resistance of the probing tip and/or sample, the currents are related to energy dissipation:

$$P = R_s I^2 \qquad (28)$$

where R_s is the spreading resistance of the sample/tip electrodes. Denk and Pohl have applied this method on GaAs heterostructures. A direct relationship between the doping level and the spreading resistance was found. Stowe et al. have shown that this Joule's damping depends on the applied voltage, dopant concentration and tip-sample distance on semiconducting surfaces. Loppacher et al. have shown that Joule's dissipation can be observed with conductive silicon tips on metallic surfaces, such as $Cu(111)$. The minimum dissipation is observed with compensated contact potential. The voltage dependence is of the form $P \propto (V - V_c)^2$. The distance dependence could be fitted with $P = c/(d - d_0)^2$.

Magnetic hysteresis

The oscillating tip of a magnetic force microscope induces a local oscillating magnetic field at the sample. Therefore, magnetic loops are probed during approach and retraction. This magnetic hysteresis can be measured with the excitation signal A_{exc} which is directly related to the loss of energy. Grütter et al. have observed that the main contrast in dissipation is observed at domain walls, which can be moved by the action of the tip. Experimentally, energy losses of about $\Delta E = 20 \text{meV}$ per cycle were observed.

Adhesion hysteresis

From surface force apparatus experiments by Israelachvili it has been concluded that adhesion hysteresis is intimately related to lateral friction forces [Israelachvili]. In general, adhesion hysteresis is related to bond formation between probing tip and sample. Therefore, adhesion hysteresis does depend on the chemical state of the tip. In analogy to friction force microscopy, material-specific contrast can be observed, which may be related to the different bond formation. An example has been given by Dürig (2000b) where a grating of chromium and glass could be imaged with an oxidized silicon tip. Large adhesion hysteresis is found on quartz compared to the chromium covered parts of the sample. This observation is in accordance with the principle that similar materials adhere more strongly. Another example of possible adhesion hysteresis is shown in Fig. 6, where NaCl(001) films on Cu(111) were imaged.

Creation of phonons

From the performance of music instruments, such as drums or violins, we know that knocking on a surface can create phonons in an efficient way. Dürig (1999a) has investigated the formation of a small repulsive contact, which deforms the surface in a reversible manner. Based on continuum mechanics ($E=10^{11}N/m^2$, $c=1000m/s$) Dürig estimates energy losses of $\Delta E \approx 10^{-8}eV$ for a repulsive force of 1nN and a resonance frequency of 100kHz. Therefore, he concludes that this type of dissipation does not play an important role for hard materials. However, the mechanism might be more important for soft materials, such as polymers. Gauthier and Tsukada (1999) calculated dissipation with the help of the fluctuation-dissipation theorem. They could relate the dissipative force to the local phonon density. Furthermore, they showed that this type of dissipation force should be proportional the square of the conservative force gradient and square root of the oscillation amplitude. Unfortunately, the predicted dissipative forces are orders of magnitudes lower than the experimental observations. Gauthier and Tsukada also created the term "dissipation force microscopy" for this type of force microscopy.

Non-linear effects

Gauthier et al. (2000) proposed that the experimental results of dissipation force microscopy might be influenced by non-linear effects. A non-linear force law can cause a frequency spectrum with manifolds, where only one branch is accessible for the experiment. These asymmetric frequency spectra might mimick an increase of the Q-factor. The authors propose to examine frequency spectra in more detail to disentangle this effect from real dissipation effects. Experiments by Erlandsson support the suggestion of Gauthier, because some asymmetry of the amplitude vs. frequency curves was found. Phase variations experiments by Loppacher et al. (1998/1999) are in accordance with the harmonic approximation.

The role of higher oscillation modes

Another possibility to explain the changes of the driving force at the resonance frequency is to involve higher oscillation modes. The short-range forces contribute only during a fraction of the oscillation cycle. Therefore, the force vs. time resembles a pulse-like wave-form. It is reasonable to expect that higher oscillation modes are excited. However, Pfeiffer et al. have shown that the higher bending modes are negligible with good accuracy during constant amplitude operation. Higher modes are not excited within the experimental error. Higher harmonics $n \cdot f_0$ may play a more important role. However, these harmonics are difficult to measured due to the artificial excitation of higher harmonics due to electronic effects.

Tomlinson effect in vertical direction

Sasaki et al. proposed that instabilities on the tip or on the sample may be important for the understanding of dissipation force microscopy. In close analogy to the Tomlinson model in friction force microscopy, a second minimum is formed during the approach, where a tip or sample atom jumps out of its equilibrium position. During retraction the atom jumps back to the original position. This type of mechanism may explain the rather large energy losses without involving permanent changes of the tip sample geometry.

80

7. Acknowledgement

We thank Ch. Gerber for stimulating discussions. This work was supported by the Swiss National Science Foundation and the Bundesamt für Bildung und Wissenschaft (BBW).

8. References

Albrecht, T.R., Grütter, P., Horne, D., and Rugar, D. (1991) "Frequency modulation detection using high-Q cantilevers for enhanced force microscope sensitivity" *J. Appl. Phys.* **69**, 668-674.

Bammerlin, M. et al. (1998) "True Atomic Resolution on the Surface of an Insulator via Ultrahigh Vacuum Dynamic Force Microscopy", *Probe Microscopy* **1**, 3-7.

Barwich, V., Bammerlin, M., Bennewitz, R., Guggisberg, M., Loppacher, C., Pfeiffer, O., Meyer, E., Güntherodt, H.-J., Salvetat, J.P., Bonard, J.M., and Forro, L. (2000) "Carbon nanotubes as tips in non-contact SFM", *Appl. Surf. Sci.* **157**, 269-273.

Cleveland, J., Anczykowski, B., Schmid, A., and Elings, V. (1998) "Energy dissipation in tapping-mode atomic force microscopy", *Appl. Phys. Lett.* **72**, 2613-2615.

Denk, W., and Pohl, D.W. (1991) "Local electrical dissipation imaged by scanning force microscopy", *Appl. Phys. Lett.* **59**, 2171-2174.

Dürig, U. (1999a) "Conservative and Dissipative Interactions in Dynamic Force Microscopy" *Surf. Interface Anal.* **27**, 467-473.

Dürig, U. (1999b) "Relations between interaction force and frequency shift in large-amplitude dynamic force microscopy" *Appl. Phys. Lett.* **75**, 433-435.

Dürig, U. (2000a) "Extracting interaction forces and complementary observables in dynamic probe microscopy" *Appl. Phys. Lett.* **76**, 1203-1205.

Dürig, U. (2000b) "Interaction sensing in dynamic force microscopy", *New Journal of Physics* **2**, 5.1-5.12.

Erlandsson, R., Olsson, L., and Martensson, P. (1996) "Inequivalent atoms and imaging mechanisms in ac-mode atomic-force microscopy of Si(111)7x7", *Phys. Rev. B* **54**, R8309-R8312.

Gauthier, M., and Tsukada, M. (1999) "Theory of noncontact dissipation force microscopy", *Phys. Rev. B* **60**, 11716-11722.

Gauthier, M., et al. (2000) 3rd Workshop of Non-contact AFM, Hamburg, to appear in Appl. Phys. A.

Giessibl, F.J. (1995) "Atomic Resolution of Silicon(111)7x7 by Atomic Force Microscopy Through Repulsive and Attractive Forces", *Science* **267**, 68-72.

Giessibl, F.J. (1997) "Forces and frequency shifts in atomic-resolution dynamic-force microscopy", *Phys. Rev. B* **56**, 16010-16015.

Gotsmann, B., Seidel, C., Anczykowski, B., and Fuchs, H. (1999) "Conservative and dissipative tip-sample interaction forces probed with dynamic AFM", *Phys. Rev. B* **60**, 11051-11061.

Israelachvili, J.N. (1985) *Intermolecular and Surface Forces*, Academic Press, London.

Grütter, P., Liu, Y., LeBlanc, P. and Dürig, U. (1997) "Magnetic dissipation force microscopy", *Appl. Phys. Lett.* **71**, 5279-5282.

Loppacher, Ch., Bammerlin, M., Battiston, F.M., Guggisberg, M., Müller, D., Hidber, H.R., Lüthi, R., Meyer, E., Güntherodt, H.-J. (1998) "Fast Digital Electronics for Application in Dynamic Force Microscopy Using High-Q Cantilevers", *Appl. Phys. A* **66**, 215-220.

Loppacher, Ch., Bammerlin, M., Guggisberg, M., Battiston, F.M., Bennewitz, R., Rast, S., Baratoff, A., Meyer, E., Güntherodt, H.-J. (1999) "Phase Variation Experiments in Non-Contact Dynamic Force Microscopy Using Phase Locked Loop Techniques", *Appl. Surf. Sci.* **140**, 287-291.

Loppacher, Ch., Bammerlin, M., Guggisberg, M., Schär, S., Bennewitz, R., Baratoff, A., Meyer, E., Güntherodt, H.-J. (2000a) "Dynamic force microscopy of copper surfaces- Atomic resolution and distance dependence of tip-sample interaction and tunneling current", submitted to *Phys. Rev. B*.

Loppacher, C., Bennewitz, R., Pfeiffer, O., Guggisberg, M., Bammerlin, M., Schär, S., Barwich, V., Baratoff, A and Meyer, E. (2000b) "Experimental Aspects of Dissipation Force Microscopy", to appear in *Phys. Rev. B*.

Martin, Y., Williams, C.C., and Wickramasinghe, H.K. (1989) "Atomic force microscope-force mapping and profiling on a sub 100-Å scale", *J. Appl. Phys.* **61**, 4723.

McClelland, G.M., Erlandsson, R., and Chiang, S. (1987) in *Review of Progress in Quantitative Non-Destructrive Evaluation*, edited by D.O. Thompson and D. E. Chimenti (Plenum, New York), Vol. 6B, p. 1307-1312.

Nonnenmacher, M., Greschner, J., Wolter, O., and Kassing, R. (1991) "Scannning force microscopy with micromachined silicon devices", *J. Vac. Sci. Technol. B* **9** 1358-1362.

Pfeiffer, O., Loppacher, C., Wattinger, C., Bammerlin, M., Gysin, U., Guggisberg, M., Rast, S., Bennewitz, R., Meyer, E., and Güntherodt, H.-J. (2000) "Using higher flexural modes in non-contact force microscopy", *Appl. Surf. Sci.* **157**, 337-342.

Sasaki, N. et al. (2000), 3rd Workshop of Non-contact AFM, Hamburg, to appear in Appl. Phys. A.

Sugawara, Y., Ohta, M., Ueyama, H. and Morita, S. (1995) "Defect motion on an InP(110) surface observed with non-contact force microscopy" *Science* **270**, 1646.

Stowe, T., Kenny, T., Thomson, D., Rugar, D. (1999) "Silicon dopant imaging by dissipation force microscopy" *Appl. Phys. Lett.* **75**, 2785.

FRICTIONAL-FORCE IMAGING AND FRICTION MECHANISMS WITH A LATTICE PERIODICITY

S.MORITA[1], Y.SUGAWARA[1], K.YOKOYAMA[1] AND S.FUJISAWA[2]
[1] Department of Electronic Engineering,
Graduate School of Engineering, Osaka University,
2-1-1 Yamada-Oka, Suita, Osaka 565-0871, JAPAN
[2] Tribology Division, Mechanical Engineering Laboratory,
1-2 Namiki, Tsukuba city 305-8564, Japan

ABSTRACT: The atomic force / lateral force microscope (AFM/LFM) using the optical lever method works as a two-dimensional frictional-force microscope (2D-FFM) in case of the atomically flat sample surface. On an atomic scale, the sliding body cannot be approximated to a uniform one but becomes a periodic one. As a result, the friction motion shows a lattice periodicity and becomes two-dimensional, because of the symmetry breaking of the stick-point distribution along the sliding (scanning) direction. Using a 2D-FFM, we can investigate the elementary process of the friction under the elastic deformation region, i.e., the two-dimensional nature of the atomic-scale friction with a lattice periodicity. Here, we show how to analyze the experimentally obtained 2D-FFM image patterns and how to deduce the load dependence of the effective adhesive region and also the stick-point distributions.

1. Introduction

Friction is a familiar process, which always occurs at the interface in sliding contact. At the macroscopic contact, numerous asperities are in contact, where all the adhesions between asperities contribute to the friction (Bowden and Tabor, 1954). However, the mechanism of friction is based on the atomistic motion at the sliding interface. So, to understand the mechanism of friction further, the corresponding atomic-scale processes at the microscopic contact, for instance a single asperity and an atomically flat surface, should be investigated experimentally.

Rapid advances in technology such as the surface force apparatus (SFA) (Israelachivili, 1972; 1985), the atomic force microscope (Binnig et al., 1986) and the molecular dynamics (MD) simulation (Glosi et al., 1993) brought us to the point where we can investigate atomic-scale processes at sliding interfaces. Experimentally, the SFA affords atomic resolution only in the vertical direction. Recently, the frictional-force microscope (FFM), which is one of the modified AFMs, enabled us to investigate the friction of a single asperity, resolved on an atomic scale not only in the vertical direction but also in the lateral direction (Mate et al., 1987). On the other hand,

83

B. Bhushan (ed.),
Fundamentals of Tribology and Bridging the Gap between the Macro- and Micro/Nanoscales, 83–101.
© 2001 Kluwer Academic Publishers.

84

in the early FFM studies only one component of the frictional force vector was studied, although the frictional force vector is two-dimensional on an atomically flat surface (Fujisawa et al., 1993a). Thus to understand the atomic-scale friction in more detail, we have to investigate the two-dimensional nature of the frictional-force vector.

The atomic force / lateral force microscope (AFM/LFM) using the optical lever method works as a two-dimensional frictional-force microscope (2D-FFM) in case of the atomically flat sample surface (Fujisawa et al., 1993b). On an atomic scale, the sliding body cannot be approximated to a uniform one but becomes a periodic one composed of atoms with a lattice periodicity. So, the distribution of the stick-points has the lattice periodicity. The friction process hence shows a lattice periodicity (Mate et al., 1987; Erlandsson et al., 1988), and becomes two dimensional (Fujisawa et al., 1994a), because of the symmetry breaking of the stick-point distribution along the sliding/scanning direction (Morita et al., 1996). Using a 2D-FFM, we can investigate the two-dimensional nature of the atomic-scale friction with a lattice periodicity. In our previous publications, we already reported on the two-dimensional atomic scale friction between a single asperity and atomically flat cleaved surfaces of mica (Fujisawa et al., 1993a), MoS_2 (Fujisawa et al., 1994a; 1995a; 1995b), NaF (Fujisawa et al., 1994b; 1996) and graphite (Fujisawa et al., 1995c; 1996b, 1998).

The present review article deals with the experimental results on the elementary process of the friction under the elastic deformation region, i.e., the two-dimensional nature of the atomic-scale friction with a lattice periodicity. We mainly investigate the load dependence of the two-dimensional periodic friction of graphite. Here, we show how to analyze the experimentally obtained 2D-FFM image and how to deduce the load dependence of the effective adhesive region and also the stick-point distributions.

2. Two-dimensional Frictional-force Microscope (2D-FFM)

First, we comment on the force components of the normal reacting force and the frictional force measured with an AFM/LFM using an optical lever deflection method with a quadrant position sensitive detector. The AFM/LFM simultaneously detects the angle changes of the cantilever due to the deflection and the torsion by the AFM and the LFM functions, respectively. Under the contact mode, both the normal reacting and the frictional forces act to the tip apex of the AFM cantilever. On an atomically flat sample surface, the normal reactive force becomes normal to the X-Y plane and is given by $N=(0,0,N_Z)$ as shown in Fig.1, while the frictional force becomes parallel to the X-Y plane and is given by $f=$ ($f_X,f_Y,0$) as shown in Fig.1. The normal reactive force N_Z induces the deflection of the cantilever. The Y component f_Y of the frictional force also induces the deflection of the cantilever. On the contrary, the X component f_X of the frictional force induces the

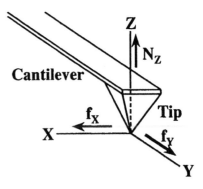

Figure 1. The X, Y, Z directions are defined as the directions across, along, and normal to the cantilever, respectively.

torsion of the cantilever. Therefore, the AFM/LFM using an optical lever technique detects both the N_Z and the f_Y by the deflection, and the f_X by the torsion. As a result, the deflection output voltage V_{def} is given by

$$V_{def}=aN_Z+bf_Y, \tag{1}$$

while the torsion output voltage V_{tor} is given by

$$V_{tor}=cf_X. \tag{2}$$

Here, a, b and c are force sensitivity constants which depend on the efficiency of the optical system. It should be noted that the AFM function measures the change $\triangle N_Z=k_Z\triangle Z$ of the normal reacting force N_Z due to the surface corrugation $\triangle Z$ and is independent of the normal load N_Z itself, although the f_Y generally depends on the normal load N_Z (Fujisawa et al., 1995b). Here, the k_Z is the spring constant for the Z force component. As a result, the deflection output voltage V_{def} should be modified as

$$V_{def}=ak_Z\triangle Z+bf_Y. \tag{3}$$

Consequently, at rather high normal load where $|ak_Z\triangle Z|\ll|bf_Y|$ is satisfied, the change of the deflection output voltage V_{def} is approximated (Fujisawa et al., 1994c) as

$$V_{def}\fallingdotseq bf_Y. \tag{4}$$

Thus the optical lever AFM/LFM works as a two-dimensional frictional-force microscope (2D-FFM) to measure the change of the two-dimensional frictional-force vector f with the components $(f_X,f_Y,0)$ (Fujisawa et al., 1993b). Lateral displacements $\triangle X$ and $\triangle Y$ of the canti-lever due to the f_X and the f_Y can be defined as

$$\triangle X= f_X/k_X \tag{5}$$

and

$$\triangle Y= f_Y/k_Y, \tag{6}$$

using the spring constants k_X and k_Y for the X and the Y force components.

To calibrate the lateral displacements, the concept of the lateral force curve shown in Fig.2 (b) is very useful. As shown in Fig.2 (a), the friction occurs between two bodies in sliding contact. At the beginning of the sliding motion (scanning), the two bodies are sticking and move together due to the static friction. This means that the maximum displacement D of the

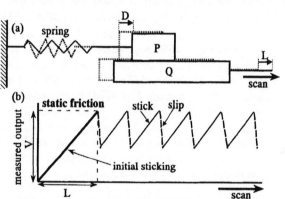

Figure 2. (a) The scheme of the simple one-dimensional frictional-force measurement system. Q is pulled by the spring, so that between P and Q the friction occurs. During the static friction, displacement D of the spring is equal to the scanned distance L. (b) Output signal due to the spring displacement vs. the scanned distance. Initial sticking appears as the linear increase of the output signal up to V with the scanned distance up to L.

spring is equal to the scanned distance L. Therefore, measured output V is given by V=dkL. Here, d and k are the force sensitivity constant given by the efficiency of the used optical system with electronic circuits and the spring constant, respectively. As a result, the coefficient dk is given by dk=V/L. Using such a calibration method, we can obtain the lateral displacements $\triangle X$ and $\triangle Y$ as

$$\triangle X = f_X/k_X = V_{tor}/ck_X = V_{tor}(L_X/V_X) \qquad (7)$$

and

$$\triangle Y = f_Y/k_Y \doteqdot V_{def}/bk_Y = V_{def}(L_Y/V_Y), \qquad (8)$$

respectively. V_X, L_X, V_Y and L_Y are the measured output V and the scanned distance L corresponding to the initial sticking in the lateral force curves for the X and the Y scanning direction, i.e., V_{tor} vs. X scan and V_{def} vs. Y scan, respectively.

3. Frictional-force Imaging of Graphite with a Lattice Periodicity

3.1. EXPERIMENTAL PROCEDURES

We used a weak feedback control of Z direction in order to compensate slow Z direction drift. A cleaved (0001) surface of graphite (PGCCL) was used as an atomically flat sample surface, because it consists of only carbon atoms and hence simple material. The lattice structure exhibits the threefold symmetry with a lattice constant of 0.246nm. As the AFM cantilever, we used a rectangular microcantilever which is made of Si_3N_4. The radius of curvature on the sharp tip apex is ~20nm. The tip height is 2.9 micrometer (μ m). The length, the width and the thickness of the cantilever are 100 μ m, 40 μ m and 0.8 μ m, respectively. Assumed elastic modulus and poisson ratio for the cantilever are 1.46×10^{11} N/m and 0.25, respectively. The calculated spring constants of the deflection and the torsion of the cantilever are $k_Z = 0.75$ N/m and $k_X = 550$ N/m, respectively.

To investigate the load dependence of the frictional force, we increased the load step by step from a 22 nN to ~ 327 nN, and at each load we measured the two-dimensional frictional-force microscope (2D-FFM) image with the raster scan. The raster scan size is 1.3 nm $\times$ 1.3 nm. The raster scan rate was set at ~21 nm/s for the fast line-scan and ~51 pm/s for the slow-scan. The raster scan consists of 256 fast line-scans. One of the lattice direction of the graphite surface was set to be parallel to the Y direction, i.e., along the cantilever. Soon after the cleavage of the graphite surface, measurements were performed in air at room temperature.

3.2. ATOMIC-SCALE FRICTION OF A GRAPHITE (0001) SURFACE

Figures 3 (a) and (b) show the f_X/k_X image obtained from V_{tor} and the f_Y/k_Y image obtained from V_{def}, respectively. The fast line-scan direction is across the cantilever, i.e., X direction. These images were simultaneously obtained by stacking the fast line-scan data step by step toward the slow-scan direction, i.e., Y direction. Here, the magnitude of the f_X/k_X and the f_Y/k_Y was indicated by the brightness.

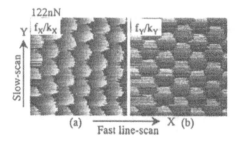

Fig.3 Simultaneously obtained 2D-FFM images of graphite at the load of 122 nN. (a) f_X/k_X vs. X scan and (b) f_Y/k_Y vs. X scan.

Fast line-scan and slow scan were performed from left to right and from bottom to top of the images, respectively. So the changes of the fast line-scan data with the slow scan create the contrast of these images. If we ignore fine structures or small differences of the unit cell, they all seem to have a structure of the three-fold symmetry with the periodicity of 0.25 ± 0.03nm. This agrees well with the lattice structure of the cleaved graphite surface, which implies that the change of the two-dimensional frictional-force vector has the lattice periodicity.

3.3. TWO-DIMENSIONAL STICK-SLIP MODEL

To investigate the atomic nature of the frictional force, we investigated every fast line-scan data. Figures 4 (a) and (b) show simultaneously obtained typical data of the f_X/k_X vs. X scan and the f_Y/k_Y vs. X scan, respectively. The f_X/k_X vs. X scan shows the sawtooth behavior similar to Fig.2 (b), which is the typical frictional signal and is called the stick-slip motion. The vertical axis of the f_X/k_X is calibrated using the intial sticking due to the static friction, i.e., using eq.(7). On the other hand, the f_Y/k_Y vs. X scan shows the square-wave behavior made of sharp step-like rise and fall with a constant amplitude, which rise and fall are synchronized with the sharp fall of the slip signal of the f_X/k_X vs. X scan. The vertical axis of the f_Y/k_Y is calibrated by measuring the sawtooth behavior of the f_Y/k_Y vs. Y scan, i.e., using eq.(8). The averaged periodicity and the amplitude of the sawtooth behavior are 0.21 ± 0.04nm and 0.20 ± 0.03nm, respectively,

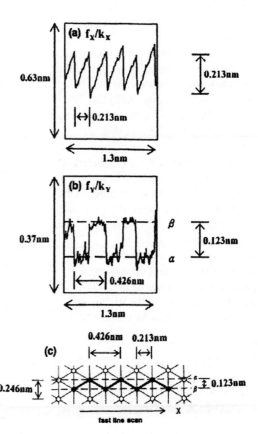

Fig.4 Simultaneously obtained typical data of (a) f_X/k_X vs. X scan and (b) f_Y/k_Y vs. X scan due to single fast line-scan along the X direction. The vertical axes are experimentally calibrated using the lateral force curve. (c) The two-dimensional stick-slip model on the graphite surface. Open and closed circles show stick-points with the lattice periodicity of the graphite surface. The closed circles represent the trace of the tip apex which was deduced from the frictional behaviors of Figs.4 (a) and (b).

while the averaged periodicity and the amplitude of the square-wave behavior are 0.44 ± 0.03nm and 0.11 ± 0.02nm, respectively. The sawtooth behavior of Fig.4 (a) indicates that the stick and the slip occur every ca.0.21nm scan along the X direction. On the other hand, the square-wave behavior of Fig.4 (b) indicates that the tip apex alternately slips ca.0.11nm to the Y or the −Y directions for every ca. 0.44nm scan along the X direction. From these values, we can determine the zigzag trajectory of the tip-apex as shown by the closed circles in Fig.4 (c). Thus, the distribution of the stick-points has the lattice periodicity and hence the frictional motion becomes two-dimensional on an atomically flat surface. The two-dimensional stick-slip model of Fig.4 (c) implies that the origin of the stick-slip phenomena with the lattice periodicity is quite different from that of the macroscopic scale, where the stick and slip are induced due to the difference between the static and the kinetic frictions (Bowden and Tabor, 1954). Furthermore, the two-dimensional stick-slip model can quantitatively predict the averaged periodicity, the averaged amplitude and the wave-form of the frictional-force vector of the f_X/k_X and the f_Y/k_Y.

4. Load Dependence of the Slip Places

4.1. LOAD DEPENDENCE OF 2D-FFM IMAGES

Figures 5 (a)-(c) show typical images of the f_Y/k_Y (left) and the f_X/k_X (right) at the loads of 44nN, 122nN and 327nN, respectively, for the fast line scan direction along the Y axis. The lattice periodicity of the graphite surface

Fig.6 Line-scan data of frictional-force vector f_Y/k_Y and f_X/k_X. (a) Fast line-scan data obtained at the α of Fig.5 (a), and (b) the slow scan data obtained at the β of Fig.5 (a)

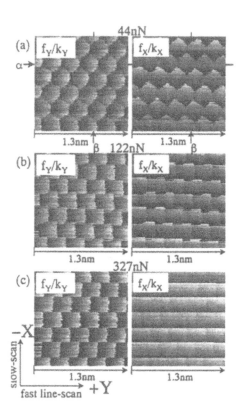

Fig.5 2D-FFM images, i.e., f_Y/k_Y vs. Y scan (left images) and f_X/k_X vs. Y scan (right images), of graphite obtained at the loads of (a) 44nN, (b) 122 nN and (c) 327nN.

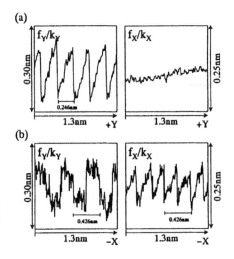

appears in all images. Besides, by increasing the load, the frictional-force pattern of the unit cell with the lattice periodicity changes drastically.

4.2. SLIP-PLACE IMAGES

The two-dimensional stick-slip model implies that the pattern of the 2D-FFM image is composed of the slip place. To analyze these patterns, we therefore deduce the slip place from each fast line-scan data such as Figs.6 (a) and (b). Figure 6 (a) shows the fast line-scan data of the f_Y/k_Y and the f_X/k_X

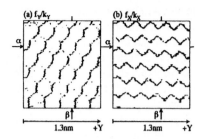

Fig.7 Images of slip places obtained from (a) f_Y/k_Y line-scan data and (b) f_X/k_X line-scan data such as Fig.6.

obtained at the α of Fig.5 (a), while Fig.6 (b) shows the slow-scan data of the f_Y/k_Y and the f_X/k_X obtained at the β of Fig.5 (a). The wave forms along to the scan direction, i.e., the f_Y/k_Y vs. +Y fast line-scan and the f_X/k_X vs. −X slow scan, clearly show the sawtooth behavior. On the other hand, the wave forms across the scan direction, i.e., the f_Y/k_Y vs. -X fast line-scan and the f_X/k_X vs. +Y slow scan, show the square wave behavior and a monotonic behavior, respectively. These behaviors agree well with the predictions by the two-dimensional stick-slip model. For an example, in case of slip places observed in a fast line-scan data such as Figs.6 (a), X positions are fixed to the α position while Y positions were deduced from slip positions of the f_Y/k_Y and the f_X/k_X. Thus, X and Y positions of slip places were obtained as shown in Figs.7 (a) and (b). In general, the f_Y/k_Y image shows slip places across the Y direction, while the f_X/k_X image shows slip places across the X direction as Figs.7 (a) and (b) show. Therefore, by putting slip places of the f_X/k_X upon those of the f_Y/k_Y, we can

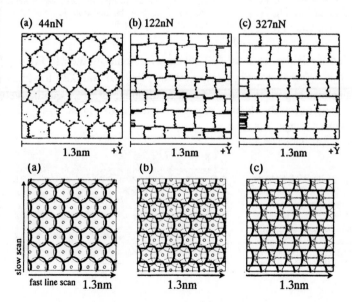

Fig.8 Typical slip place images of frictional force obtained at the loads of (a) 44nN, (b) 122nN and (c) 327 nN for the +Y fast line-scan.

Fig.9 Calculated typical slip place images of frictional-force for three different effective adhesive radius r.

deduce a more complete pattern of slip places for the frictional force as Fig.8 (a) obtained at the load of 44nN shows. Figures 8 (a)-(c) show typical slip place images of the frictional-force obtained at the loads of 44nN, 122nN and 327nN in Figs.5 (a)-(c), respectively. These pattern change of slip place images can be explained by the change of the effective adhesive radius (Fujisawa et at., 1995b ; Sasaki et al., 1996; Kerssemakers and De Hosson, 1996; Hölscher et al., 1997). Figures 9 (a)-(c) show calculated typical slip place images of the frictional force for three different effective adhesive radii r, which is r=0.14nm for Fig.9(a), r=0.17nm for Fig.9(b) and r=0.22nm for Fig.9 (c), respectively (see Appendix).

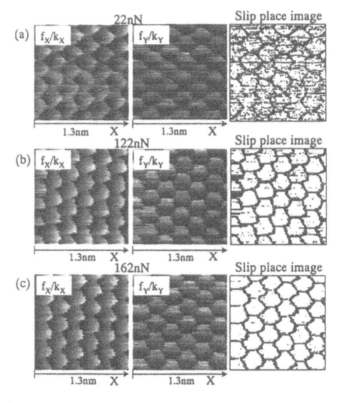

Fig.10 2D-FFM images, i.e., f_X/k_X vs. X scan (left images) and f_Y/k_Y vs. X scan (center images), and corresponding slip place images of graphite obtained at the loads of (a) 22nN, (b) 122nN and (c) 162nN, respectively.

Using the lattice constant ℓ=0.246nm of graphite written in Fig.4 (c), the ratio of r/ℓ used is r/ℓ=0.57 for Fig.9 (a), 0.69 for Fig.9 (b) and 0.89 for Fig.9 (c), respectively. Calculated slip place patterns for three different ratios r/ℓ qualitatively agree well with the experimentally obtained slip place patterns of Figs.8 (a)-(c). We also deduced the slip place images for the X fast line-scan

Fig.11 2D-FFM images, i.e., f_Y/k_Y vs. Y scan (left images) and f_X/k_X vs. Y scan (center images), and corresponding slip place image of graphite obtained at the load of 22nN.

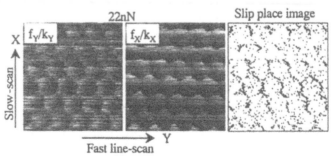

as shown in Figs.10 (a)-(c). Here, 2D-FFM images of Fig.10 (b) at the load of 122nN is the same data with Fig.3. Slip place in Fig.10 (a) obtained at the rather low load of 22nN is scattering, which seems to be induced due to the weak sticking compared with the spring constant (Sasaki et al., 1996). For the Y fast line-scan, we also found the similar scattering phenomenon of the slip place at the same low load of 22nN as shown in Fig.11.

5. Load Dependence of Effective Adhesive Region

Slip place image at the load of 122nN of Fig.8 (b) obtained for the Y fast line-scan direction shows a different pattern from that of Fig.10 (b) obtained at the same load but for the X fast line-scan direction. This result can be explained that the effective adhesive radius r for the Y direction is different from that for the X direction. In short, the shape of the effective adhesive region is not circle. Therefore, to investigate the shape of the effective adhesive region and its load dependence seems to be very important to make clear the imaging mechanism of 2D-FFM in more detail.

As the calculated typical slip place images of Figs.9 (a)-(c) indicate, only a part of the unit cell of the slip place pattern is determined by the effective adhesive region, while the other slip places distributing along the fast line-scan direction is determined by the lattice periodicity (Sasaki et al., 1996, 1997, 1998a, 1998b; Kerssemakers and De Hosson, 1996; Hölscher et al., 1997; Fujisawa et at., 1998) Therefore, we deduced a part of effective adhesive region such as Fig.12 (a) from slip place images such as Fig.8 (a) by comparing with calculated images such as Fig.9 (a). Next, we summed up all the slip places by assuming the distribution of stick points with the lattice periodicity as shown in Fig.4 (c). As a result, we obtained the shape of the effective adhesive region along the +Y direction as Fig.12 (b) shows. We therefore repeated the same procedure for 2D-FFM images obtained for the −Y, the +X and the −X fast line-scan directions. At last, by summing up

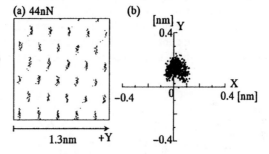

Fig.12 (a) Effective adhesive region deduced from the slip place image of Fig.8 (a). (b) Effective adhesive region summed up from Fig.12 (a) by assuming the distribution of stick points with the lattice periodicity.

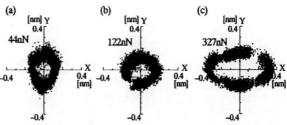

Fig.13 Experimentally obtained two-dimensional shape of the effective adhesive region of graphite surface at the loads of (a) 44nN, (b) 122nN and (c) 327nN.

all the parts of the effective adhesive region, we obtained the whole shape of the effective adhesive region and its load dependence as Figs.13 (a)-(c) show. Experimentally obtained two-dimensional effective adhesive region of graphite surface clearly shows the load dependence of its shape, which averaged effective adhesive radius increases at the high load by increasing the load (Fujisawa et al., 1995b). As Figs.13 (a)-(c) have the two-fold symmetry rather than the three fold symmetry, these shapes may reflect the shape of the cantilever in Fig.1 rather than the three-fold symmetry of the graphite surface shown in Fig.4 (c). We furthermore investigated the reproducibility of these shapes, and confirmed that the shape of the effective adhesive region has the two-fold symmetry rather than the three-fold symmetry. However, the shape itself does not show the reproducibility. Hence the shape seems to depend on the contact at the sliding interface. The scattering width of the effective adhesive region does not show the load dependence, so that the noise seems to be origin of the scattering. These procedures to deduce the whole shape of the two-dimensional effective adhesive region will be useful to investigate the friction mechanism on an atomic scale.

6. Load Dependence of Stick-Point Distributions with a Definite Extent

Figures 14 (a), (b), (c) and (d) show typical square-wave signals of the f_X/k_X vs. $+Y$ fast line-scan at the loads of 22nN, 44nN, 91nN and 327nN partly shown in Fig.5, respectively. The amplitude of the square-wave signals increases by increasing the load. Furthermore at the load of 22nN, the slope of the slip signals is not so sharp as those at 44nN, 91nN and 327nN, so that the wave-form seems to be rather rounded.

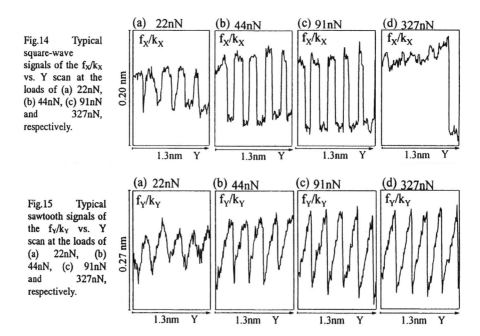

Fig.14   Typical square-wave signals of the f_X/k_X vs. Y scan at the loads of (a) 22nN, (b) 44nN, (c) 91nN and 327nN, respectively.

Fig.15   Typical sawtooth signals of the f_Y/k_Y vs. Y scan at the loads of (a) 22nN, (b) 44nN, (c) 91nN and 327nN, respectively.

Figures 15 (a), (b), (c) and (d) also show typical sawtooth signals of the f_Y/k_Y vs. $+Y$ fast line-scan at the loads of 22nN, 44nN, 91nN and 327nN partly shown in Fig.5, respectively. All the data show sawtooth signals with the lattice periodicity. The amplitude of the slip signals increases by increasing the load. Furthermore, at the load of 22nN, the slope of the slip signals is not so sharp as those at 44nN, 91nN and 327nN so that the wave-form seems to be rather rounded. On the other hand, at 22nN and 44nN the slope of the sticking signal is not so steep as those at 91nN and 327nN. The amplitude of the slip signal corresponds to the slip distance of the tip between the stick-points, according to the two-dimensional stick-slip model. So the measured slip distance should be smaller than the lattice constant of 0.246 nm, at least at the load smaller than 91nN. This is not consistent with the assumption in the simple two-dimensional stick-slip model that the stick-point is a point and the slip distance is just the lattice constant. Therefore, we conjecture that the assumed stick-point should not be a point but a region with a definite extent. Hereafter, we refer this region as "sticking-domain". Further, from the increase of the amplitude of the slip signal by increasing the load, we expect that the extent of the sticking-domain should decrease by increasing the load as predicted by Sasaki et al. (1996).

The sticking-domain can be estimated by a spatial distribution of the tip position by the raster scan. The spatial distribution can be reconstructed from experimental data by converting the frictional-force signals to the displacement of the tip, based on the method introduced by Kawakatsu et al. (1996). Using the raster scan position (X_0, Y_0) and the calibrated lateral displacements $(\triangle X, \triangle Y)$ due to the frictional force, the X and the Y positions of the tip apex where the tip apex is stuck are given by

$$X = X_0 + \triangle X, \tag{9}$$

and

$$Y = Y_0 + \triangle Y. \tag{10}$$

Here, $\triangle X$ and $\triangle Y$ can be deduced using eqs. (7) and (8). We calibrated the lateral displacements by using the lateral force curve obtained at the rather high load where the amplitude of the sawtooth signal becomes constant and the extent of the sticking-domain seems to be enough small.

Figures 16 (a), (b), (c) and (d) show the spatial distributions of the tip position at the loads of 22nN, 44nN, 91nN and 327nN, respectively. These maps are obtained from corresponding images shown in Fig.5 and Fig.11 except for that of 91nN. The tip position is indicated by the dot, where the existence frequency of the tip is indicated by the darkness. At the load of 22nN, the distribution of the tip position is rather continuous and extends to the whole surface, although it seems to be a little

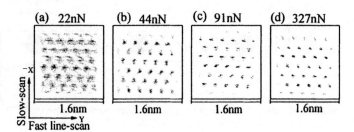

Fig.16 Tip position maps at the loads of (a) 22nN, (b) 44nN, (c) 91nN and (d) 327nN. The frame corresponds to 1.6nm $\times$ 1.6nm.

(a) 22nN   (b) 44nN   (c) 91nN   (d) 327nN

Slow-scan   1.6nm   1.6nm   1.6nm   1.6nm

Fast line-scan

concentrated with the lattice periodicity. This result suggests that the tip moves rather smoothly without stick-slip motion, although the tip motion has two-dimensionality with the lattice periodicity. By increasing the load up to 44nN, the distribution of tip position becomes rather discrete with a clear lattice periodicity, which suggests the two-dimensionally discrete friction with the lattice periodicity. This discretely concentrated area of tip position corresponds to the sticking-domain. By increasing the load more than 44nN, the extent of the sticking-domain decreases as we expected. On the other hand, even at the load of 327 nN, the tip does not stick completely but moves gradually in the sticking-domain. Further, the rather smooth motion at the load of 22nN could be interpreted that the sticking-domain extends to the whole area of the surface. These behaviors of the sticking-domain as a function of the load are similar to those obtained as a function of the spring constant predicted by Sasaki et al. (1986). This result indicates that the extent of the sticking-domain is determined by the competition between the load and the spring constant of the cantilever.

By comparing the typical fast line-scan data with the obtained tip position map, we conjecture that the round-shape signals in both square-wave and sawtooth signals at the 22nN shown in Figs. 14 (a) and 15 (a) correspond to the smooth motion of the tip (Sasaki et al., 1996). These results agree with the scattering of slip place at the 22nN shown in Fig.10 (a) and Fig.11. Furthermore, the gentle slope of the sticking signal in the sawtooth signal at 44nN shown in Fig.15 (b) seems to correspond to the gradual motion of the tip in the sticking-domain.

There is possibility that a part of the observed motion of the tip is due to the lateral deformation of tip-sample contact or lateral bending of the tip. However, the contribution of the lateral deformation or lateral bending is smaller than the actual tip motion. This is suggested from the result in Fig.16 that the observed sticking-domain becomes smaller, while the lateral deformation and bending should become larger, with the increase of the frictional force by increasing the load.

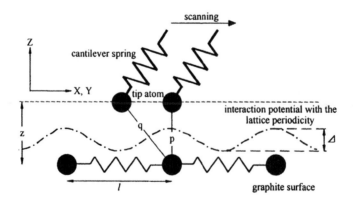

Fig.17 A simple one-dimensional model of a frictional-force microscope with a single-atom tip and atomically flat surface. Closed circles represent the center position of the atoms. The sinelike wave shown by a dotted broken line represents the amplitude Δ of the interaction potential between the tip and surface. The dashed line shows the trajectory of the tip over the surface by the scanning.

7. Origin of the frictional force with a lattice periodicity

We qualitatively explain the load dependence of effective adhesive radius using an interaction potential between a single atom tip and a monolayer surface corresponding to a simple one-dimensional model of frictional force microscope, as shown in Fig.17. Firstly, we explain that effective adhesive radius depends on the amplitude of the interaction potential by using energy minimization in a total potential model (Sasaki et al., 1996; Fujisawa et al., 1998). Then, the load dependence of the potential amplitude is explained by the Lennard-Jones potential between the tip and surface atoms.

Since the single atom tip is used, the discussion by using this model is limited for the behavior of a single atom friction or the practical single atom friction such as two-dimensionally coherent or commensurate friction between the layered material surface and its flake stuck to the tip, where we assumes that the tip would move in similar way to the single atom tip (Morita et al., 1996).

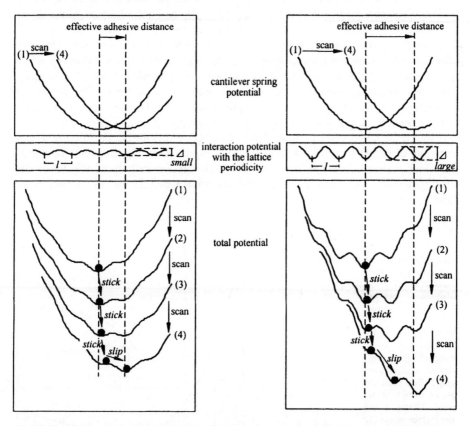

Fig.18 The relation of the effective adhesive distance and amplitude of the interaction potential. The left and right figures show two cases that the amplitude of the interaction potential $\triangle$ is small and large, respectively. The bottom part shows the summation of the cantilever spring potential (top part) and the interaction potential between the tip and the surface (middle part). (1) ~ (4) represent the evolution of the total potential by the scanning from (1) to (4).

7.1. FRICTION MECHANISM WITH A LATTICE PERIODICITY

Figure 18 shows the total potential model of Fig.17. In this one-dimensional model, effective adhesive radius corresponds to the scanning distance required to cause the tip slip from the center of the sticking domain. We refer this distance as an effective adhesive distance. The effective adhesive distance is estimated by considering one stick-slip process of the tip with the evolution of the total potential by the scanning. The total potential is given by the summation of the cantilever spring potential and the interaction potential with the lattice periodicity. In the total potential model, the tip position corresponds to one of local minima, which is represented by the closed circle.

The state (1) in Fig.18 represents the state that the scan-point locates at the center of a sticking domain. The scan-point, which is represented by the vertical dashed line, indicates the basal position of the tip atom at the center of the parabola representing the cantilever spring potential. By the scanning, the scan-point moves to the right, i.e., the relative position of the cantilever spring and the surface changes from (1) to (4) gradually. As the total potential evolves to (2) and (3), the local minimum where the tip stuck becomes shallower. At (4), the local minimum where the tip stuck from (1) to (3) disappears. It makes the tip jump or slip to the nearest local minimum. Thus, the effective adhesive distance corresponds to the scanning distance from (1) to (4), which is represented by the distance between the two vertical dashed lines (Sasaki et al., 1996). By comparing two cases with the small and large amplitudes of the interaction potential, as shown in the left and the right hands in Fig.18, the effective adhesive distance becomes larger as the amplitude of the interaction potential increases.

It should be noted that by the evolution from (1) to (4) during the stick, the local minimum of the tip position moves gradually along the scanning direction. This small and slow motion of the tip corresponds to the gentle motion of the tip in the sticking-domain, and gives a sticking-domain with a definite extent (Sasaki et al., 1996).

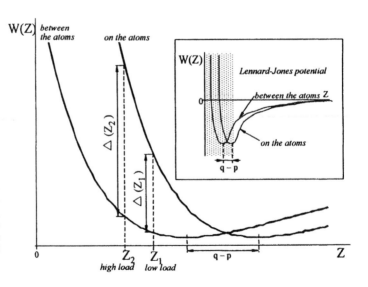

Fig.19 Lennard-Jones potential $W(z)$ as a function of the distance Z between the tip and the sample surface in repulsive force region, which corresponds to enlarged figure of the shaded area of inset graph. Two cases are shown whether the tip atom locates above the surface atoms (on atoms) or above the gaps between the surface atoms (off atoms).

7.2. LOAD DEPENDENCE OF THE INTERACTION POTENTIAL

The amplitude, i.e., the periodic change of the interaction potential with the lattice periodicity is qualitatively explained by the difference of the distance between the tip atom and surface atom. As shown in Fig.17, the tip surface distance of p (=z) where the tip atom is above the surface atom is shorter than q where the tip atom is above the gap between the surface atoms. So, based on the Lennard-Jones potential, the potential value between the tip and surface atoms changes depending on the tip position relative to the surface atoms, as shown in Fig.19. This difference of $\triangle$ gives the potential amplitude with the lattice periodicity. The $\triangle$ increases from $\triangle(Z_1)$ to $\triangle(Z_2)$, by decreasing the tip surface distance from Z_1 to Z_2 due to the increase of the load, as shown in Fig.19. Thus, the potential amplitude increases by increasing the load.

We mention that in the two-dimensional stick-slip model a sticking force or attractive force works to the tip. The force between the tip and the surface is, however, repulsive, since the tip and the surface are in contact. This contradiction is explained by that the weak place in the repulsive force distribution such as the site of q (=z) with the longest tip-surface-distance appears to be a pseudo attractive place with the lattice periodicity.

8. Conclusion

The atomic force / lateral force microscope (AFM/LFM) using the optical lever method works as a two-dimensional frictional-force microscope (2D-FFM) in case of the atomically flat sample surface. On an atomic scale, the sliding body cannot be approximated to a uniform one but becomes a periodic one composed of atoms with a lattice periodicity. As a result, the friction motion shows a lattice periodicity and becomes two-dimensional. Using a 2D-FFM, we can investigate the elementary process of friction under the elastic deformation region, i.e., the two-dimensional nature of atomic-scale friction with a lattice periodicity.

We investigated the load dependence of an atomic-scale friction between a single asperity tip and the atomically flat sample surface of graphite experimentally, by using the 2D-FFM. As a result, we found that the slip signal, which constructs the force boundary of the lattice periodicity image, changes drastically by increasing the load.

From the slip signals, we deduced the load dependence of the slip place images. Then we showed the method to obtain the two-dimensional shape of the effective adhesive region from the slip place images, and deduced its load dependence. We found that the shape of the experimentally obtained effective adhesive region is not circle but has a two-fold symmetry and that the averaged radius increases at the high load by increasing the load.

From the load dependence of the slip signal, we found that the tip gradually moves while the tip sticks to the sticking-domain. The extent of the sticking-domain decreases by increasing the load. At low load region the sticking-domain seems to cover whole the surface. Origin of load dependence of the effective adhesive region is attributed to the load dependence of the surface potential amplitude. These results are explained by the simple one-dimensional potential model between the tip and the surface atoms.

98

References

Binnig, G., Quate, C.F. and Gerber, Ch. (1986), "Atomic Force Microscope", Phys.Rev.Lett. **56**, 930-933.

Bowden, F.P. and Tabor, D. (1954), *The Friction and Lubrication of Solids*, Clarendon, Oxford.

Erlandsson, R., Hadziloannou, G., Mate, C.M., McClelland, G.M. and Chiang, S. (1988), "Atomic-scale friction between the muscovite mica cleavage plane and a tungsten tip", J.Chem.Phys. **89**, 5190-5193.

Fujisawa, S., Sugawara, Y., Ito, S., Mishima, S., Okada, T. and Morita, S. (1993a), "The two-dimensional stick-slip phenomenon with atomic resolution", Nanotechnology, **4**, 138-142.

Fujisawa, S., Sugawara, Y. and Morita, S. (1993b), "Origins of Forces Measured by Atomic Force/Lateral Force Microscope", Microbeam Analysis, **2**, 311-316.

Fujisawa, S., Kishi, E., Sugawara, Y. and Morita, S. (1994a), "Fluctuation in Two-Dimensional Stick-Slip Phenomenon Observed with Two-Dimensional Frictional Force Microscope", Jpn.J.Appl.Phys., **33**, 3752-3755.

Fujisawa, S., Kishi, E., Sugawara, Y. and Morita, S. (1994b), "Two-dimensionally discrete friction on the NaF(100) surface with the lattice periodicity", Nanotechnology, **5**, 8-11.

Fujisawa, S., Ohta, M., Konishi, T., Sugawara, Y. and Morita, S. (1994c), "Difference between the forces measured by an optical lever deflection and by an optical interferometer in an atomic force microscope", Rev.Sci.Instrum., **65**, 644-647.

Fujisawa, S., Kishi, E., Sugawara, Y. and Morita, S. (1995a), "Atomic-scale friction observed with a two-dimensional frictional-force microscope", Phys.Rev.B, **51**, 7849-7857.

Fujisawa, S., Kishi, E., Sugawara, Y. and Morita, S. (1995b), "Load dependence of two-dimensional atomic-scale friction", Phys.Rev.B, **52**, 5302-5305.

Fujisawa, S., Kishi, E., Sugawara, Y. and Morita, S. (1995c), "Two-dimensionally quantized friction observed with two-dimensional frictional microscope", Tribology Letters, **1**, 121-127.

Fujisawa, S., Sugawara, Y. and Morita, S. (1996a), "Load dependence of the periodicity in friction force images on the NaF(100) surface", Philosophical Magazine A, **74**, 1329-1337.

Fujisawa, S., Sugawara, Y. and Morita, S. (1996b), "Localized Fluctuation od a Two-Dimensional Atomic-Scale Friction", Jpn.J.Appl.Phys., **35**, 5909-5913.

Fujisawa, S., Yokoyama, K., Sugawara, Y. and Morita, S. (1998), "Analysis of experimental load dependence of two-dimensional atomic-scale friction", Phys.Rev.B, **58**, 4909-4916.

Glosli, J.N. and McClelland, G.M. (1993), "Molecular Dynamics Study of Sliding Friction of Ordered Organic Monolayers", Phys.Rev.Lett. **70**, 1960-1963.

Hölscher, H., Schwarz, U.D., Zwörner, O. and Wiesendanger, R. (1997), "Stick-slip movement of a scanned tip on a graphite surface in scanning force microscopy", Z.Phys.B, **104**, 295-297.

Israelachvili, J.N. and Tabor, D. (1972), "The Measurement of van der Waals Dispersion Foreces in the Range of 1.5 to 130 nm", Proc.R.Soc.London, **A331**, 19-38.

Israelachvili, J.N. (1985), *Intermolecular and Surface Forces*, Academic Press, London.

Kawakatsu, H. and Saito, T. (1996), "Scanning force microscopy with two optical levers for detection of deformations of the cantilever", J.Vac.Sci.Technol.B, **14**, 872-876.

Kerssemakers, J., and De Hosson, J.Th.M. (1996), "Influence of spring stiffness and anisotropy on stick-slip atomic force microscopy imaging", J.Appl.Phys., **80**, 623-632.

Mate, C.M., McClelland, G.M., Erlandsson, R. and Chiang, S. (1987), "Atomic-Scale Friction of a Tungsten Tip on a Graphite Surface", Phys.Rev.Lett. **59**, 1942-1945.

Morita, S., Fujisawa, S. and Sugawara, Y. (1996), "Spatially quantized friction with a lattice periodicity", Surface Science Reports, **23**, 1-42.

Sasaki, N., Kobayashi, K. and Tsukada, M. (1996), "Atomic-scale friction of graphite in atomic-force microscopy", Phys.Rev.B, **54**, 2138-2149.

Sasaki, N., Tsukada, M., Fujisawa, S., Sugawara, Y., Morita, S. and Kobayashi, K. (1997), "Analysis of frictional-force image patterns of a graphite surface", J.Vac.Sci.Technol.B, **15**, 1479-1482.

Sasaki, N., Tsukada, M., Fujisawa, S., Sugawara, Y. and Morita, S. (1998a), "Theoretical analysis of atomic-scale friction in frictional-force microscopy", Tribology Letters, **4**, 125-128.

Sasaki, N., Tsukada, M., Fujisawa, S., Sugawara, Y., Morita, S. and Kobayashi, K. (1998b), "Load dependence of the frictional-force microscopy image pattern of the graphite surface", Phys.Rev.B, **57**, 3785-3786.

Appendix: Relaxation Model of Scan Point

Figures 20 (a) and (b) show two kinds of two-dimensional stick-slip model. Here, small open circles and also A, B, C and D show the distribution of the stick-points with the three-fold symmetry of the lattice. Large dotted or broken open circles indicate the effective adhesive region determined by the effective adhesive radius r where the thick broken open circle corresponds to the effective adhesive region of the stick-point A. Hexagons show corresponding unit cell belonging to each stick-point. Under the condition that the tip atom of Fig.17 is sticking to the stick-point A, if the scan point along the fast scan-line shown by the dotted line reaches to the point S_1 (a slip place) on the boundary of the effective adhesive region of the stick-point A, the tip atom slips to the nearest stick-point B as shown in the slip model of the tip atom in Fig.20 (a), where the relaxation of the spring elongation or the attractive force due to the next stick-point becomes maximum. Here, the slip distance of the tip atom from the stick-point A to the B agrees with the lattice constant ℓ. The thick solid arc on the boundary of the effective adhesive region of the stick-point A indicates the part where the stick-point B is the nearest stick-point and the tip atom slips from the stick-point A to the B. Such two-dimensional stick-slip model of Fig.20 (a) shows the tip atom motion by the scanning and the elongation/relaxation of the spring due to the slip of the tip atom, so that the slip model of Fig.20 (a) can be used to explain the wave-form of the line-scan data as shown in Figs.4 (a), (b) and Fig.6 (a) (Morita et al., 1996).

On the other hand, the relaxation model of Figure 20 (b) (Kerssemakers et al., 1996) can explain the pattern of the slip-place. In this model, when the scan point reaches to the point S_1 (a slip place) on the boundary of the effective adhesive region of the stick-point A, the scan point instead of the tip atom will relax from the point S_1 to the point S_2 with the lattice constant ℓ where the relaxation vector S_1S_2 is the opposite one of the slip vector **AB**. Because of the lattice periodicity, the stick-point B is equivalent

(a)   (b)

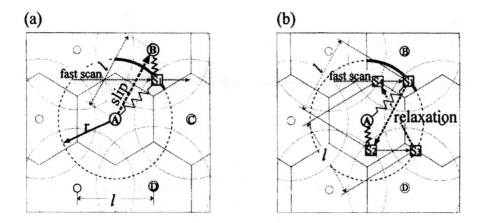

Figure 20. Two-dimensional stick-slip models. (a) Slip model of the tip atom and (b) relaxation model of the scan point.

to the stick-point A so that the point S_1 measured from the stick-point B, i.e., the vector BS_1, is equivalent to the point S_2 measured from the stick-point A, i.e., the vector AS_2. By scanning further, the scan point reaches to another slip place S_3 and then relaxes to the next scan point S_4. Finally, the scan point returns to the point S_1. Thus, the motion of the scan point shows the two-dimensional closed route of $(S_1-S_2-S_3-S_4-)$. These two types of stick-slip models of Figs.20 (a) and (b) can be interpreted as an extended stick-slip model and a reduced stick-slip model on the analogy of the extended zone scheme and the reduced zone scheme of the electron energy diagram, respectively.

Figure 21 shows a typical unit cell of the two-dimensional slip place pattern corresponding to Fig.9 (b). Thick lines and arcs indicate the slip place pattern of the stick-point A. Arcs b, c and d correspond to the slip places from the stick-point A to the stick-points B, C and D, respectively. Arcs b', c' and d' correspond to the relaxation places of the scan points from arcs b, c and d, respectively. Arcs b', c' and d' also correspond to the slip places from the stick-points B', C' and D' to the stick-point A, respectively. Two-dimensional closed routes of the scan point shown by 1 and 2 correspond to the zigzag slip motion due to the fast line-scan. Arcs b, d, b' and d' indicate the slip places of the zigzag slip motion. On the other hand, one-dimensional closed routes of the scan point shown by 3 and 4 correspond to the straight slip motion due to the fast line-scan. Arcs c and c' indicate the slip places of the straight slip motion. The region surrounded by the arcs c and c', and the scan lines 3 and 4 corresponds to the straight slip motion. The boundary of the straight slip motion is determined by the cross-points between the effective adhesive region of the stick-point A (C') and the hexagon boundaries, i.e., bisectors of B-C and C-D (A-D' and A-B') (Fujisawa et al., 1998). The region surrounded by the arcs b, d', and the scan lines 1 and 2, (and also surrounded by the arcs d and b', and the scan lines 1 and

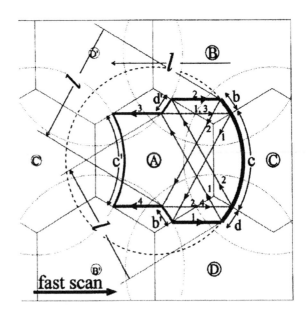

Figure 21. Calculated typical slip place pattern which corresponds to Fig.9 (b).

2) is determined by the cross-point between the effective adhesive region of the stick-points D' (B') and the hexagon boundary, i.e., bisector of A-B (A-D) (Fujisawa et al., 1998). As a result, the region due to the straight slip motion becomes larger, but that due to the zigzag slip motion becomes smaller by increasing the effective adhesive radius r as Figs.9 (a), (b) and (c) show.

ATOMIC SCALE ORIGINS OF FORCE INTERACTION

S.MORITA[1], Y.SUGAWARA[1], K.YOKOYAMA[1] AND
T.UCHIHASHI[2]
[1] Department of Electronic Engineering,
Graduate School of Engineering, Osaka University,
2-1-1 Yamada-Oka, Suita, Osaka 565-0871, JAPAN
[2] Department of Electronics, Faculty of Engineering,
Himeji Institute of Technology, 2167 Shyosya, Himeji 671-2201, JAPAN

ABSTRACT: The noncontact atomic force microscope (NC-AFM) using the frequency modulation (FM) detection method works as a three-dimensional force-mapping tool to investigate atomic scale origins of force interaction. For examples, we will show two different methods to map atomic force three-dimensionally; one by measuring site-dependent frequency-shift curves on an atomic scale as a function of the tip-sample distance between a clean Si(111)7x7 surface and a clean Si tip, and the other by measuring the tip-sample distance dependence of atomically resolved NC-AFM image on Si(111)$\sqrt{3}$x$\sqrt{3}$-Ag sample surface. Further, by placing a suitable atom on the tip apex, we investigated the possibility to control the interaction force between the tip and sample atoms on an atomic scale.

1. Introduction

The atomic force microscope (AFM) invented by Binnig et al. (1986) is a unique microscope based on a mechanical method which has the following characteristics: (1) it has true atomic resolution, (2) it can measure atomic force (so-called atomic force spectroscopy), (3) it can observe even insulators, and (4) it can measure mechanical responses such as elastic deformation. Therefore, the AFM has a potential to investigate atomic scale origins of force interaction which is very important to clarify atomic-scale friction mechanism.

However, usual contact mode imaging seems to be rather destructive (Sugawara et al., 1994) and cannot routinely achieve observation of an atomic defect (Ohta et al., 1993), although the periodic lattice structure can be imaged (Ohta et al., 1995). On the other hand, noncontact mode imaging with the noncontact atomic force microscope (NC-AFM) utilizing a frequency modulation (FM) detection method achieved true atomic resolution in 1995 for Si(111)7x7 (Giessibl, 1995; Kitamura et at., 1995) and InP(110) (Sugawara et al., 1995). Until now, various kinds of sample surfaces such as Si(100)2x1 (Uchihashi et al., 1999a), GaAs(110) (Sugawara et al., 1999), NaCl(100)

B. Bhushan (ed.),
Fundamentals of Tribology and Bridging the Gap between the Macro- and Micro/Nanoscales, 103–120.
© 2001 Kluwer Academic Publishers.

(Bammerlin et al., 1997), $TiO_2(110)$ (Fukui et al., 1999), $TiO_2(100)$ (Raza et al., 1999), Si(111)$\sqrt{3}$x$\sqrt{3}$-Ag (Minobe et al., 1999), Ag(111) (Orisaka et al., 1999), Cu(111) (Loppacher et al., 1999), graphite(0001) (Allers et al., 1999), TGS (Eng et al., 1999), NiO(100) (Hosoi et al., 2000), C_{60} (Kobayashi et al., 1999), Si(100)2x1:H (Yokoyama et al., 2000), $HCOO^-/TiO_2(110)$ (Fukui et al., 1999), and self-assembled monolayer of adenine base (Uchihashi et al., 1999b) have been observed with atomic resolution. Therefore, the NC-AFM may be used to investigate atomic scale origins of force interaction between various kinds of sample surface and a tip.

The present review article deals with the experimental demonstrations on the three-dimensional force-mapping to investigate atomic scale origins of force interaction. For examples, we will show two different methods to map atomic force three-dimensionally; one by measuring site-dependent frequency-shift curves on an atomic scale as a function of the tip-sample distance, and the other by measuring the tip-sample distance dependence of atomically resolved NC-AFM image. Further, by placing a suitable atom on the tip apex, we will show the possibility to control the interaction force between the tip and sample atoms on an atomic scale.

2. Noncontact Atomic force Microscope (NC-AFM)

2.1. CONTACT AFM

Contact mode AFM can obtain the periodic lattice image even under a normal load more than 10^{-6}N (Ishizaka et al., 1990). Under the approximation that the single bond strength between adjacent atoms is of the order of 10^{-9}N, this result suggests that the contact area will be roughly made of more than ~1,000 single bonds, i.e., ~1,000 atoms (Morita et al., 1996). Therefore, as shown in Fig.1, usual contact mode imaging seems to be rather destructive and cannot routinely achieve observation of an atomic defect (Ohta et al., 1993).

2.2. NONCONTACT AFM

NC-AFM using the frequency modulation (FM) detection

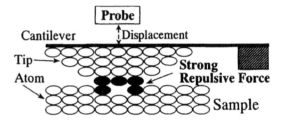

Fig.1 Schematic model of contact mode AFM. This figure shows that the contact mode AFM is destructive and has a large contact area because of the strong repulsive force.

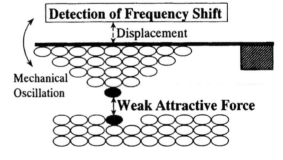

Fig.2 Schematic model of NC-AFM. This figure shows that the NC-AFM is nondestructive and can observe even an atomic point defect if weak attractive force can be detected as a frequency shift of the mechanical oscillation of the cantilever.

method (Albrecht et al., 1991) succeeded to obtain true atomic resolution in 1995 for Si(111)7x7 (Giessibl, 1995; Kitamura et at., 1995), and to observe atomic point defects (Ueyama et al., 1995) and even defect motion on InP(110) surface (Sugawara et al., 1995). Therefore, the NC-AFM with the FM detection method can sensitively detect the frequency shift of the mechanical oscillation of the cantilever due to the force interaction between the tip and the sample surface.

2.3. FREQUENCY MODULATION (FM) DETECTION METHOD

The displacement of the cantilever is detected with a fiber-optic interferometer or optical beam deflection. Figure.3 shows a schematic diagram of the NC-AFM. The force interaction acting on the probe tip was detected as a frequency shift of the mechanical oscillation of the cantilever utilizing an FM detection method. The piezoelectric scanner was also used to oscillate the cantilever at its mechanical resonant frequency. The positive feedback system with an automatic gain control (AGC) circuit was used to maintain the oscillation amplitude constant (switch 1 is "ON") in the constant-oscillation mode, while the excitation voltage supplied to the piezoelectric tube scanner was maintained constant in the constant-excitation mode (switch 2 is "ON") (Ueyama et al., 1998). In the present experiment, we mainly used the latter mode to weaken the damage due to the sudden contact between the tip and the sample surface. The frequency shift of the cantilever shown in Fig.4 was detected by a tunable analog FM demodulator in

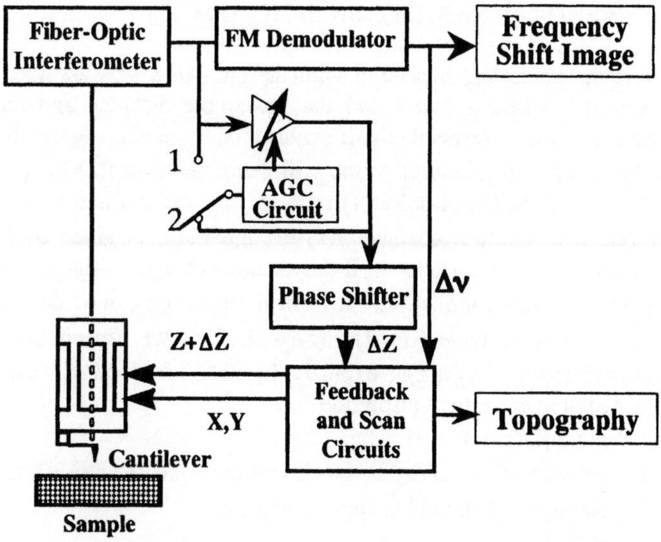

Fig.3 Schematic diagram of the NC-AFM using the FM detection method.

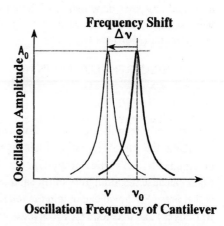

Fig.4 Schematic model of the frequency shift due to the force interaction between the tip and the sample surface.

Fig.3. Under the assumptions of very small oscillation amplitude and weak interaction, the force interaction acting on the probe tip will be approximately proportional to the force derivative, although such assumptions are not achieved in the NC-AFM measurements (Giessibl, 1997).

3. Guidelines for the achievement of true atomic resolution with NC-AFM

3.1. VERTICAL RESOLUTION IN NC-AFM

Here the value $f(z)$ measured with the NC-AFM was assumed to be proportional to $\exp(-z/L)$, where z and L are the tip-sample distance and the decay length of the measured value, respectively. It means that, by increasing the tip-sample distance from z to $z+\delta z$, the measured value will decrease from $f(z)$ to $f(z+\delta z)=f(z)\exp(-\delta z/L)$. Then the signal(S)-to-noise(N) ratio of $f(z)$ was defined as $k=S/N$. It means that the measurable smallest change $\delta f(z)$ of signal $f(z)$ is given by $\delta f(z)= f(z)/k$, which is equivalent to the noise level. If the smallest change of tip-sample distance controllable by the feedback loop is δz, the vertical resolution can be defined as δz. It also means that $\delta f(z)$ is equivalent to $f(z)-f(z+\delta z)$. Thus we can obtain the relation of $\delta f(z)= f(z)/k=f(z)-f(z+\delta z)=f(z)[1-\exp(-\delta z/L)]$. Then from $1/k=[1-\exp(-\delta z/L)]$, we obtained the equation of vertical resolution

$$\delta z=L\ln[k/(k-1)] \tag{1}$$

as a function of the decay length L and the signal-to-noise ratio k. By assuming $\delta z/L \ll 1$, that is, $k \gg 1$, Eq.(1) is approximated as

$$\delta z \doteqdot L/k. \tag{2}$$

Eq.(2) suggests that the scanning probe microscope (SPM) as well as the NC-AFM can obtain better vertical resolution by decreasing the decay length, i.e., by making tip-sample distance dependence of the measured value stronger, and by increasing the signal-to-noise ratio k larger (Morita et al., 1999).

3.2. LATERAL RESOLUTION IN NC-AFM

3.2.1 *Case of a large tip*

By assuming that the radius R of tip curvature is much larger than the tip-sample distance z, i.e., $R \gg z$, the lateral resolution δx is approximately determined by the vertical resolution δz as shown in Fig.5, because the ambiguity of tip-sample distance is δz. From Fig.5, we can obtain the relation $\delta z= R(1-\cos\theta)$ and $\delta x=R\sin\theta$. From these equations, we can derive the equation of lateral resolution

$$\delta x \doteqdot [\delta z(2R-\delta z)]^{1/2}. \tag{3}$$

By assuming $R \gg \delta z$, we can rewrite Eq.(3) as

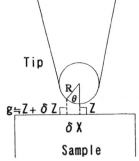

Fig.5 Schematic simple one-dimensional model of tip-sample configuration in case of $R \gg z$ (Morita et al., 1999).

$$\delta x \doteqdot [2R \, \delta z]^{1/2}. \qquad (4)$$

Eq.(4) suggests that we can obtain better lateral resolution by decreasing the radius R of tip curvature, and by improving the vertical resolution δz, i.e., by decreasing the decay length L and by increasing the signal-to-noise ratio k (Morita et al., 1999).

3.2.2. Case of a Single Atom Tip

By assuming that the radius of tip curvature is much smaller than the tip-sample distance z, i.e., $R \ll z$, the lateral resolution δx is approximately determined by the vertical resolution δz as shown in Fig.6, because the ambiguity of tip-sample distance is δz. From Fig.6, we can obtain the relation $(z + \delta z)^2 = z^2 + \delta x^2$. From this relation, we can derive the equation of lateral resolution

$$\delta x \doteqdot [\delta z (2z + \delta z)]^{1/2}. \qquad (5)$$

By assuming $z \gg \delta z$, we can rewrite Eq.(5) as

$$\delta x \doteqdot [2z \, \delta z]^{1/2}. \qquad (6)$$

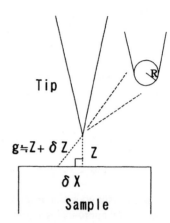

Fig.6 Schematic simple one-dimensional model of tip-sample configuration in case of $R \ll z$ (Morita et al., 1999).

Eq.(6) suggests that we can obtain better lateral resolution by decreasing the tip-sample distance z, and by improving the vertical resolution δz, i.e., by decreasing the decay length L and by increasing the signal-to-noise ratio k (Morita et al., 1999).

3.2.3. Decay Length of Frequency Shift

Now we will investigate the possible decay length of frequency shift $\Delta v = f(z)$. At first, we will assume tip-sample distance dependence of frequency shift as $f(z) = A/z^n$ (n;integer, A;constant). Next we will approximate tip-sample distance dependence of frequency shift as $f(z_0) = B \exp(-z_0/L)$ around the distance z_0 where we will measure the frequency shift. By increasing the tip-sample distance from z_0 to $z_0 + \delta z$, $f(z_0) = A/z^n$ decreases to $f(z_0 + \delta z) = A/(z_0 + \delta z)^n$ while $f(z_0) = B \exp(-z_0/L)$ decreases to $f(z_0 + \delta z) = f(z_0) \exp(- \delta z/L)$. Using $f(z_0 + \delta z)/f(z_0) = \exp(- \delta z/L) = [z_0/(z_0 + \delta z)]^n$, we can obtain $1 - \delta z/L \doteqdot 1 - n \delta z/z_0$ under the assumption of $\delta z \ll z_0$, and can derive the decay length

$$L \doteqdot z_0/n. \qquad (7)$$

Thus, for a smaller tip-sample distance z_0 and for a force interaction with a larger integer n, we can obtain a smaller decay length L (Morita et al., 1999).

3.2.4. Guidelines for the Achievement of True Atomic Resolution

From Eqs.(2), (4), (6) and (7), guidelines for the achievement of true atomic resolution with the NC-AFM will be given by (1) good signal-to-noise ratio (large k), (2) small decay length (small L), (3) small radius of tip curvature (small R), (4) small tip-sample distance (small z) and (5) large n (Morita et al., 1999). It should be noted that the experimental decay length L becomes small by decreasing the effective radius R of tip curvature because of the decrease of the volume integration effect around the tip apex, by removing the contamination on the tip apex which mechanically induces the wide-area contact, and by decreasing the mechanical oscillation amplitude of the cantilever which introduces the averaging effect of the force interaction. In the present experiment, to obtain the clean Si tip, both contamination and native oxide on the virgin Si tip (the spring constant; 27-41N/m, the mechanical resonant frequency; 151-172kHz, the nominal radius of the tip apex; 5-10nm, and the Q factor of the cantilever; 38,000 in UHV) were removed by in situ Ar-ion sputtering.

4. Tip-Sample Distance for Obtaining True Atomic Resolution

4.1. CONTACT POINT AND NONCONTACT REGION

Figure.7 shows simultaneously measured approaching curves of the frequency shift and the oscillation amplitude under the constant-excitation mode. From the oscillation amplitude curve in Fig.7, we determined the contact point Z=0 nm where the oscillation amplitude begins to decrease due to the rather strong repulsive force by decreasing the tip-sample distance. Then we assigned the contact and noncontact regions to the regions below and above the contact point as shown in Fig.7 (Ueyama et al, 1998; Morita et al., 2000a).

It should be noted that the absolute value of the frequency

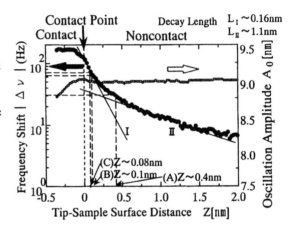

Fig.7 Simultaneously measured frequency shift and oscillation amplitude curves. Sample is Zn-doped p-GaAs(110) cleaved surface (Morita et al., 2000a).

shift in the noncontact region at first increases slowly then quickly by decreasing the tip-sample distance. The decay lengths obtained in the corresponding regions II and I were 1.1nm and 0.16nm, respectively. In general, the decay length takes the smallest value just before the contact point as shown in Fig.7. This phenomenon also suggests that the rather strong repulsive force works below the contact point and suppresses the increase of the absolute value of the frequency shift by decreasing the tip-sample distance.

4.2. TIP-SAMPLE DISTANCE FOR OBTAINING TRUE ATOMIC RESOLUTION

We then obtained the NC-AFM images at three different tip-sample distances: (A) z~0.4nm, (B) z~0.1nm and (C) z~0.08nm before the contact point as shown in Fig.7. Figures 8 (A), (B) and (C) show measured NC-AFM images and line profiles along the white lines in the corresponding NC-AFM images. The variable frequency shift mode which measures the frequency shift image (see Fig.3) was used under weak feedback condition to suppress the thermal drift. From Fig.8 (A) measured at z~0.4nm, we found that the NC-AFM image close to region II shows only a large-scale contrast, perhaps, due to defects, but no atomic-scale contrast. On the other hand, from Fig.8 (B) measured at z~0.1nm, we found that the NC-

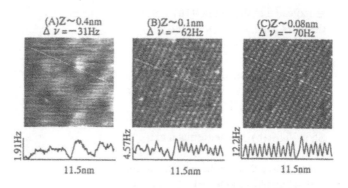

Fig.8 Distance dependence of the frequency shift image and cross-sectional line profile along white lines indicated in NC-AFM images measured at (A) z~0.4nm, (B) z~0.1nm and (C) z~0.08nm. Measurements were done under the constant excitation-voltage mode. Scan area is 10nm × 10nm (Morita et al., 2000a).

AFM image in region I shows atomic-scale point defects as well as the periodic lattice structures, although the image seems a little vague. Furthermore, from Fig.8 (C) measured at z~0.08nm, we found that the NC-AFM image just before contact becomes very clear. Thus the distance dependence of the NC-AFM image is very strong. This result qualitatively agrees with the expectation predicted by the tip-sample distance dependence of the lateral resolution given by Eq.(6) and of the decay length given by Eq.(7). On the other hand, in the contact region shown in Fig.7, we could not obtain the atomic point-defect stably because of the strong contact between the tip and sample surfaces. Therefore, we conjectured that the true atomic resolution can be experimentally achievable just before the contact point.

5. Mapping and Control of Atomic Force on Si(111)7x7 Surface

5.1. SITE-DEPENDENT FREQUENCY-SHIFT CURVE MEASURED WITH Si TIP

To map atomic force three dimensionally on an atomic scale, we measured the frequency-shift curve between the clean Si tip and a Si(111)7x7 sample by gating the feedback loop during a raster scan. As a result, from several times of NC-AFM topography measurements on Si(111)7x7 sample surfaces such as Fig.9 (a), we obtained more than 1000 site-dependent frequency-shift curves (Morita et al., 2000b).

Then, we investigated the site dependence of the frequency-shift curve on a Si(111)7x7 sample surface measured by the clean Si tip. Here, we specified each site, above which site-dependent frequency-shift curve was obtained, from each point where the feedback was gated by investigating the corresponding point of the NC-AFM topography. As a result, frequency-shift curves measured at the site above Si adatoms (on adatom) such as indicated by Sa in Fig.9 (a) showed a discontinuity such as indicated by Sa in Fig.9 (b). On the other hand, frequency- shift curves measured at the site above gaps between adjacent Si adatoms (off adatom) as indicated by Sh in Fig.9 (a) did not show a discontinuity similar to Sa in Fig.9 (b), but showed only a continuous smooth curve as indicated by Sh in Fig.9 (b). These results make clear that the force interaction mechanism acting above Si adatoms is different from that acting above gaps between adjacent Si adatoms (Morita et al., 2000).

The clean Si tip and a Si(111)7x7 sample surface have active dangling bonds. Each dangling bond directed towards the vertex of a tetrahedron determined by sp^3 hybrid orbitals. In case of the Si(111)7x7 surface structure, dangling bonds of Si adatoms direct towards the normal line. On the other hand, the direction of the dangling bond of the clean Si tip apex can not be specified because of the uncertainty of the crystal structure at the tip apex. However, if we will assume that the dangling bond of the clean Si tip apex directs right beneath, there are roughly two cases whether the dangling bond of the clean Si tip apex directs towards Si adatoms (Sa) (on adatom) or gaps between adjacent Si adatoms (Sh) (off adatom), respectively. In case of on adatom, the onset of the strong covalent bond between dangling bonds of the clean Si tip apex and a Si adatom will suddenly induce strong attractive

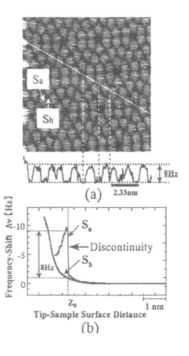

Fig.9 (a) The NC-AFM image of a Si(111)7x7 surface and a cross-sectional line profile along the white line in the NC-AFM image obtained with the clean Si tip. The averaged frequency shift is –5Hz. (b) Simultaneously obtained site-dependent frequency-shift curves as a function of tip-sample distance (Morita et al., 2000b).

force by approaching the tip towards a Si(111)7x7 surface. On the other hand, in case of off adatom, the covalent bond and, hence, strong attractive force will not occur because of a mismatch between directed bonds. Therefore, we conjectured that the observed discontinuity of the frequency-shift curve above Si adatoms indicated by Sa in Fig.9 (b) was caused by the sudden onset of strong attractive force due to the onset of the strong covalent bond between dangling bonds of the atom on the clean Si tip apex and a Si adatom. In this case, there is the possibility that the Si atom on the tip apex and a Si adatom will microscopically jump together to form the sp^3 bond state. On the other hand, we concluded that the force interaction mechanism of the frequency-shift curves above gaps between adjacent Si adatoms is the van der Waals force and/or the Coulomb force.

We obtained a variable frequency-shift image of a Si(111)7x7 surface as shown in Fig.9 (a) using the NC-AFM at the distance Z_0 indicated in Fig.9 (b) where the discontinuity occurred. It should be noted that the change of the frequency-shift due to the discontinuity nearly agrees with the maximum change of the frequency-shift in the cross-sectional line profile along the white solid line in the Si(111)7x7 NC-AFM image of Fig.9 (a). Therefore, the NC-AFM image of Fig.9 (a) may be a kind of the dangling bond image of Si adatoms on the Si(111)7x7 surface.

Thus, we succeeded to obtain site-dependent frequency-shift curves on an atomic scale between a Si(111)7x7 surface and the clean active Si tip with a dangling bond by using the NC-AFM. These results suggest that the NC-AFM can develop into a kind of spectroscopic tool, i.e., atomic force spectroscopy, which can measure the three-dimensional force-related map with true atomic resolution.

5.2. CONTROL OF ATOMIC FORCE WITH AN OXIDIZED Si TIP

Then, we investigated frequency-shift curves between a Si(111)7x7 surface and the oxidized Si tip, and also a corresponding NC-AFM image of a Si(111)7x7 surface measured by the oxidized Si tip. Here, to obtain the inactive Si tip without a dangling bond, the virgin Si tip with both contamination and native oxide was used without in situ Ar-ion sputtering (Uchihashi et al., 1997). As shown in Fig.10 (a), frequency-shift curve measured by the oxidized Si tip did not show a discontinuity but showed only a continuous smooth curve. On the other hand, frequency- shift curves measured by the clean Si tip clearly showed a discontinuous jump above Si adatoms as shown in Fig.10 (c). Further, the NC-AFM images measured by the oxidized Si tip became unclear as shown in Fig. 10 (b) compared with those measured by the clean Si tip as shown in Fig. 10(d). The corresponding cross-sectional line profile of Fig.10 (b) measured by the oxidized Si tip along the white line in the NC-AFM image of Fig.10 (b) showed about 3 Hz frequency-change which is one-fifth of 15Hz shown in Fig.10 (d) (Uchihashi et al., 1997).

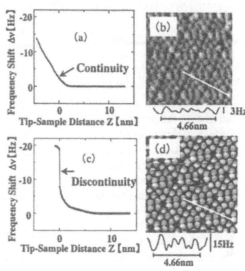

Fig.10 (a) Typical frequency-shift curves between a Si(111)7x7 sample and the oxidized Si tip. (b) Corresponding NC-AFM image and a cross-sectional line profiles along the white line in the NC-AFM image. The averaged frequency shift was −1.1Hz. (c) Typical frequency-shift curves between a Si(111)7x7 sample and the clean Si tip. (d) Corresponding NC-AFM image and a cross-sectional line profiles along the white line in the NC-AFM image. The averaged frequency shift was −13Hz. (Uchihashi et al., 1997).

Therefore, we concluded that the strong interaction force, i.e., the discontinuity in the frequency-shift curve, due to the onset of the covalent bond between dangling bonds of the clean active Si tip and a Si(111)7x7 sample surface was suppressed by replacing the clean active Si tip apex with the oxidized inactive Si tip apex. We also concluded that the force interaction mechanism between a Si(111)7x7 sample surface and the oxidized Si tip was mainly the van der Waals force and/or the Coulomb force. Present result suggests that we can control interaction force between the tip and sample atoms on an atomic scale by placing a suitable atom on the tip apex.

It should be noted that present experimental results qualitatively agree with the calculated results by Pérez et al. (1997) and Sasaki and Tsukada (1999), and also agree with the recent experimental results by Lantz et al. (2000) who carefully investigated the tip-sample distance dependence of line sections at the low temperature.

6. Mapping and Control of Atomic Force on Si(111)$\sqrt{3}$x$\sqrt{3}$-Ag Surface

6.1. HONEYCOMB-CHAINED TRIMER (HCT) MODEL OF Si(111)$\sqrt{3}$x$\sqrt{3}$-Ag

The honeycomb-chained trimer (HCT) model (Katayama et al., 1991) has been accepted as the appropriated model for the Si(111)$\sqrt{3}$x$\sqrt{3}$-Ag surface. In this model, the surface contains Ag trimer at the first layer 0.075nm above the Si trimer at the second layer as shown in Fig.11. The Ag atom forms the covalent bond with the Si atom 0.075nm below Ag atom. The distances between the nearest neighbor Ag atoms forming the Ag trimers and between those Si atoms forming the Si trimers are

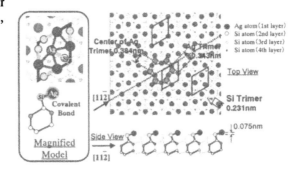

Fig.11 Top view and side view of a Si(111)$\sqrt{3}$x$\sqrt{3}$-Ag sample surface. This surface contains the Ag trimer at the first layer 0.075nm above the Si trimer at the second layer (Katayama et al., 1991).

0.343nm and 0.231nm, respectively. Also, the distance between the centers of Ag trimers forming the honeycomb arrangement is 0.384nm. There are three Ag atoms and three Si atoms per $\sqrt{3}$x$\sqrt{3}$ unit cell as shown by rhombuses in the top view of Fig.11, while the number of centers of Ag trimer are two per $\sqrt{3}$x$\sqrt{3}$ unit cell. From first-principle calculations for the HCT model, it was proved that highly unoccupied electronic surface states are distributed in the center of the Ag trimer (Watanabe et al., 1991).

6.2. TIP-SAMPLE DISTANCE DEPENDENCE OF NC-AFM IMAGE MEASURED WITH Si TIP

On the Si(111)$\sqrt{3}$x$\sqrt{3}$-Ag region, we investigated the tip-sample distance dependence of the NC-AFM image between the clean Si tip and Si(111)$\sqrt{3}$x$\sqrt{3}$-Ag sample surface as shown in Figs.12 (a)-(c) (Minobe et al., 1999). Here, Figs.12 (a)-(c) were NC-AFM images measured under the constant frequency-shift mode at the frequency-shift of (a) $\Delta v =$ -37 Hz, (b) $\Delta v =$ -43 Hz and (c) $\Delta v =$ -51 Hz, respectively. Here, z=0 of tip-sample distance z is the contact point. Tip-sample distance z from the contact point was deduced from the simultaneously measured frequency-shift curve. In Fig.12, the pattern of the NC-AFM image changed remarkably when the tip-sample distance z was decreased. At the rather far distance of z= 0.2-0.3nm, bright spots where attractive force is stronger constituted hexagonal ring forming a honeycomb structure. There are two bright spots in the $\sqrt{3}$x$\sqrt{3}$ unit cell as shown by a rhombus in Fig.12 (a). At the intermediate distance of z=0.05nm, dark lines where attractive force is weaker constituted trefoil pattern consisting of three dark lines as shown in Fig.12 (b). Just before the contact point, i.e., at the distance of z= 0.03nm, bright spots constituted triangular pattern consisting of three bright spots. There are three bright spots in the $\sqrt{3}$x$\sqrt{3}$ unit cell as shown by a rhombus in Fig.12 (c). These results indicate that the imaging site and imaging atom as well as the imaging mechanism depend on the Si tip-sample distance on Si(111)$\sqrt{3}$x$\sqrt{3}$-Ag sample surface (Minobe et al., 1999). Therefore, Figs.12 (a)-(c) are the direct evidence that we can achieve the three-dimensional force mapping with true atomic resolution by measuring the tip-sample distance dependence of the NC-AFM image instead of site-dependent frequency-

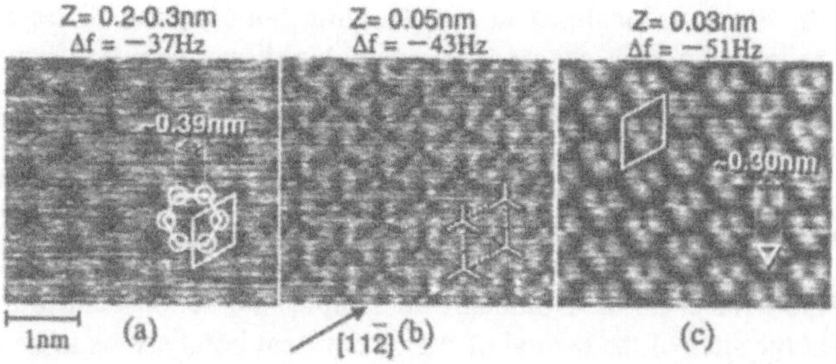

Fig.12 Tip-sample distance dependence of the NC-AFM image obtained with the clean Si tip. (a) Image at the rather far distance of Z= 0.2-0.3nm, (b) image at the intermediate distance of Z= 0.05nm and (c) image at just before the contact point of Z= 0.03nm (Minobe et al., 1999).

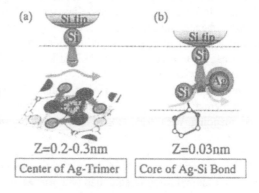

Fig.13 Schematic model of the NC-AFM imaging mechanism on the Si(111)$\sqrt{3}$x $\sqrt{3}$-Ag surface obtained with the clean Si tip (a) at the far distance of Z= 0.2-0.3nm and (b) near contact of Z= 0.03nm.

shift curves. Recently, Lantz et al. (2000) achieved a similar experiment where distance dependence of line sections were investigated using a Si(111)7x7 sample.

By comparing the bright spot pattern of hexagonal ring and ca.0.39nm spacing of bright spots in Fig.12 (a) with the HCT model of Fig.11, we concluded that the center of Ag trimers was imaged as bright spots at the rather far distance. Dominant tip-sample interaction contributing to the attractive force at the center of Ag trimer seems to be physical force interaction, i.e., either van der Waals or Coulomb force. There is a possibility that van der Waals attractive force acts on the AFM tip most strongly at the center of Ag trimer composed of three Ag atoms at the topmost layer. On the other hand, there is another possibility that the Coulomb attractive force acts on the electron localized at the dangling bond of the Si tip most strongly at this site, because positive charge induced by highly unoccupied electronic surface states seems to exist most densely at the center of Ag trimer (Minobe et al., 1999) as shown in Fig.13 (a).

By comparing the triangular pattern consisting of three bright spots and ca.0.30nm spacing of bright spots in Fig.12 (c) with the HCT model of Fig.11, we concluded that the sites of the orbital of Ag-Si covalent bond were imaged as bright spots just before the contact point. Dominant tip-sample interaction contributing to the attractive force at the sites of the orbital of Ag-Si covalent bond seems to be chemical force interaction, i.e., attractive force due to the onset of hybridization between the dangling bond of the Si tip apex and the orbital of the Ag-Si covalent bond. Namely, when the Si tip apex with the dangling bond approaches toward the covalent bond between Ag and Si atoms, the orbital of the dangling bond of the tip apex Si atom will begin to hybridize with the orbital of the covalent bond between Ag and Si atoms, and then a strong attractive tip-sample interaction force will be induced at the sites of the orbital of the Ag-Si covalent bond (Minobe et al., 1999) as shown in Fig.13 (b).

The trefoil pattern at the intermediate distance in Fig.12 (b) can be explained by the overlap of hexagonal bright spots in Fig.12 (a) with triangular bright spots in Fig.12 (c). We concluded that the physical and chemical force interactions equivalently contribute to the imaging mechanism of the NC-AFM at this distance, because the trefoil pattern appeared at the intermediate distance.

Thus, using Si(111)$\sqrt{3}$x$\sqrt{3}$-Ag sample with two atom species of Ag and Si atoms, we succeeded in demonstrating alternative method to acieve the three-dimensional force mapping with true atomic resolution by measuring the tip-sample distance dependence of the NC-AFM image instead of site-dependent frequency-shift curves. Present result suggests that the NC-AFM can be developed into a kind of spectroscopic tool, i.e., atomically resolved atomic force spectroscopy, which can measure the three-dimensional force-related map with true atomic resolution.

6.3. ATOMIC FORCE CONTROL BY PLACING AN Ag ATOM ON THE TIP APEX

Then, we transferred an Ag atom to the Si tip apex from the Si(111)$\sqrt{3}$x$\sqrt{3}$-Ag sample surface by the contact of the Si tip with the Si(111)$\sqrt{3}$x$\sqrt{3}$-Ag. After that, other type of the NC-AFM image was reproducibly obtained on the Si(111)$\sqrt{3}$x$\sqrt{3}$-Ag sample surface, and then the tip-sample distance dependence of the NC-AFM image pattern disappeared as shown in Figs.14 (a)-(c) (Yokoyama et al., 1999). Before the contact, the topmost atom of the tip apex was a Si atom, because the Si tip was cleaned by *in situ* Ar ion sputtering before the NC-AFM imaging and maintained under the UHV better than 2×10^{-10} Torr. On the other hand, there are only two atom species (Si and Ag) on the Si(111)$\sqrt{3}\times\sqrt{3}$-Ag sample surface. If an Ag atom was picked up on the Si tip apex, the NC-AFM image may change drastically. However, if a Si atom was picked up on the Si tip apex or dropped off to the Si(111)$\sqrt{3}\times\sqrt{3}$-Ag sample surface, the NC-AFM image will not change. Thus, we concluded that an Ag atom was picked up on the Si tip apex.

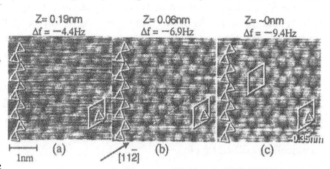

Figures 14 (a)-(c) show the typical atomic-resolution images obtained after the Ag atom picking up from the Si(111)$\sqrt{3}\times\sqrt{3}$-Ag sample surface.

Fig.14 Tip-sample distance dependence of the NC-AFM image obtained with the Ag adsorbed tip. (a) Image at the rather far distance of z= 0.19nm, (b) image at the intermediate distance of z= 0.06nm and (c) image at near contact point of z= ~0nm (Yokoyama et al., 1999).

Here, these NC-AFM images were obtained at the frequency-shift of (a) Δv = -4.4 Hz, (b) Δv = -6.9 Hz and (c) Δv = -9.4 Hz, respectively. Corresponding tip-sample distances are roughly estimated to be z= 0.19nm, 0.06nm and ~0nm, respectively. As shown in Fig.14, when the tip approached the surface, the NC-AFM image pattern did not show distance dependence, and only the image contrast improved. We can observe zigzag triangular pattern as shown in Fig.14. Here, each triangle was composed of three bright spots located in the $\sqrt{3}\times\sqrt{3}$ unit cell. The distance between the bright spots is 0.35 ±0.02 nm Directions of the apexes of all triangles are tilted a little around the [11$\bar{2}$] direction.

Now, by comparing Figs.14 (a)-(c) measured by the Ag adsorbed tip with the HCT model, we will consider which site on the Si(111)$\sqrt{3}\times\sqrt{3}$-Ag sample surface were observed as the triangular pattern in the NC-AFM images. There are two types of triangle structures on the surface. One is the Ag trimer at the first layer, and the other is the Si trimer located at the second layer 0.075nm below the first layer. The measured distance between the bright spots of 0.35 ±0.02nm agrees with the Ag atom spacing 0.343nm composing of the Ag trimer within the experimental error, while it does not agree with 0.231nm composing of the Si trimer. Further, the measured direction of the apexes of the triangle patterns tilted around the [11$\bar{2}$] direction agrees with that for the Ag trimer, while it does not agree with the [1$\bar{1}$2] direction for the Si trimer. So, we concluded that the most appropriate site for the triangle pattern observed using the Ag adsorbed tip is the site of individual Ag atoms forming the Ag trimer (Yokoyama et al., 1999).

Then, we will consider the imaging mechanism of individual Ag atoms on the

Si(111)$\sqrt{3}$x$\sqrt{3}$-Ag sample surface using the Ag adsorbed tip. Due to the adsorption of the Ag atom onto the Si tip apex, it is expected that a dangling bond out from a Si tip apex was terminated by the Ag atom. In this case, the tip-sample force contributing to the NC-AFM imaging seems to originate from the van der Waals force and/or onset of the overlap of orbitals between the Ag atom on the tip and the Ag atom on the surface. As a result, the top site of the individual Ag atoms (or nearly true atomic topography) will appear as a bright spot. Hence, NC-AFM image pattern will not show the tip-sample distance dependence. In order to prove the proposed imaging mechanism of the NC-AFM on the Si(111)$\sqrt{3}$x$\sqrt{3}$-Ag sample surface, the detailed theoretical investigations are expected.

Present results on the Si(111)$\sqrt{3}$x$\sqrt{3}$-Ag sample surface with two atom species of Ag and Si atoms suggest that we can control the interaction force between the tip and the surface on an atomic scale by placing a suitable atom on the tip apex such as the clean Si tip, the oxidized Si tip, the Ag adsorbed Si tip, the H terminated tip and the chemically modified tip etc.

7. NC-AFM Image of Si(100)2x1 Compared with Si(100)2x1:H Surface

In order to apply the NC-AFM as a science tool in a variety of fields such as surface science, it is very important to understand the imaging mechanisms on various surfaces. Here we investigated the NC-AFM imaging mechanisms of a Si(100)2x1 surface and a Si(100)2x1:H. Si(100)2x1 surface has tilted dangling bonds and hence reactive, while Si(100)2x1:H surface is terminated by hydrogen atoms and hence inactive. Further, these surfaces have nearly the same structure. Therefore, the tilted dangling bond effect on the NC-AFM imaging of Si(100)2x1 surface and the hydrogen termination effect on the NC-AFM imaging of Si(100)2x1:H surface can be clarified by comparing both NC-AFM images with the same Si tip.

7.1. NC-AFM IMAGES OF Si(100)2x1 AND Si(100)2x1:H

Figure 15 shows the atomic resolution image of Si(100)-2x1 surface measured at $\langle\Delta v\rangle$ = -30 Hz (Yokoyama et al., 2000). The scan area was 6.9nm×4.6nm. The paired bright spots (imaged dimer) constituting rows with a 2x1 symmetry was clearly observed. Further, the distance between paired bright spots is 0.32 ± 0.01nm. As shown in Fig.15, the change of the frequency shift along a white solid line in the NC-AFM image of Fig.15 was estimated about 9 Hz.

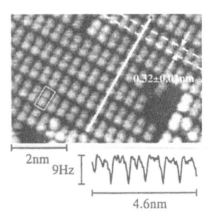

Fig.15 NC-AFM image of Si(100)2x1 reconstructed surface. White dotted lines shows the dimer arrangement. The cross-sectional line profile along a white solid line. Averaging frequency shift was set to be $\langle\Delta v\rangle$ = -30 Hz (Yokoyama et al., 2000).

Figure.16 shows the atomic resolution image of hydrogen terminated silicon [Si(100)2x1:H] surface measured at <Δv> = -11 Hz. The scan area was 6.9nm×4.6nm. The paired bright spots (imaged dimer) constituting rows with a 2×1 symmetry was observed. Further, the distance between paired bright spots is 0.35 ± 0.01nm, which approximately agrees with the distance between hydrogen atoms on monohydride surface, i.e., 0.352nm. As shown in Fig.16, the averaged change of the frequency shift along a white solid line in the NC-AFM image of Fig.16 was estimated ca.3 Hz.

We will discuss the origin of the imaged dimer in NC-AFM image on Si(100)-2x1 surface (Fig.15) and Si(100)-2x1:H surface (Fig.16) by comparing with a surface structure model shown in Fig.17.

The bright spots of Fig.15 don't image the silicon atom site, because the distance 0.32 ± 0.01nm between bright spots forming dimer structure of Fig.15 is lager than the distance between silicon atoms of dimer structure model as shown in Fig.17 (a) (Maximum distance between alternating upper Si atoms on asymmetric dimer structure is 0.292nm). This result suggest that the chemical bonding interaction strongly works between the tilted dangling bond out of the silicon dimer and the dangling bond out of silicon tip apex. As a result, the dimer structure with the lager distance than that between Si dimmer is obtained. Beside, as shown in the cross-sectional line profile of Fig.15, the frequency shift rapidly increases around imaged dimmer. It also suggests that the strong chemical interaction occurs between dangling bonds.

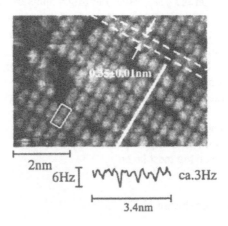

Fig.16 NC-AFM image of Si(100) 2x1-H surface. White dotted lines show the dimer arrangement. The cross-sectional line profile along a white solid line. Averaging frequency shift was set to be <Δv> = -11 Hz (Yokoyama et al., 2000).

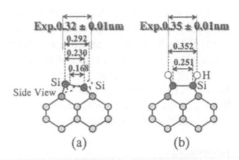

Fig.17 Structure models of (a) a Si(100)2x1 surface and (b) a Si(100)2x1:H monohydride surface. In Si(100)2x1 surface, the alternative buckling configurations are shown by solid and dashed circles. The upper-upper, upper-lower, and lower-lower distance between silicon atoms in two configurations is 0.292nm, 0.230nm, and 0.168nm, respectively (Chadi, 1979a; 1979b). In Si(100)-2x1:H surface, the distance between hydrogen atoms is 0.352nm (Chabal and Raghavachari, 1984).

On the other hand, the bright spots of Fig.16 seem to be located at hydrogen atom sites on Si(100)-2x1:H surface, because the distance between bright spots 0.35 ± 0.01nm forming dimer structure approximately agrees with the distance between the hydrogen, i.e., 0.352nm as shown in Fig.17 (b).

On Si(100)-2×1:H surface, the dangling bond on silicon surface is terminated by a hydrogen atom, and the hydrogen atom on top most layer don't have a chemical reactivity like a silicon atom on Si(100)-2x1 surface. Therefore, the interaction between the hydrogen atom on top most layer and the silicon tip apex is not the chemical bonding interaction with strong direction dependence as in Si(100)2x1 surface. As a result, each bright spot of NC-AFM image corresponds to the individual hydrogen atom site on top most layer.

8. Conclusion

We succeeded in mapping atomic force three-dimensionally by two different methods; one by measuring site-dependent frequency-shift curves on an atomic scale as a function of the tip-sample distance between a clean Si(111)7x7 surface and a clean Si tip, and the other by measuring the tip-sample distance dependence of atomically resolved NC-AFM image on Si(111)$\sqrt{3}$x$\sqrt{3}$-Ag sample surface. Further, by placing a suitable atom on the tip apex, we investigated the possibility to control the interaction force between the tip and sample atoms on an atomic scale. In case of Si(111)7x7 surface, we succeeded to suppress the strong chemical force by using the oxidized Si tip instead of a clean Si tip, while, in case of Si(111)$\sqrt{3}$x$\sqrt{3}$-Ag surface, we succeeded to observe individual Ag atoms by placing Ag atoms on the clean Si tip apex. From these three-dimensional mapping and control of atomic force, we can investigate atomic scale origins of force interaction. Further, we clarified the difference of the NC-AFM imaging mechanisms between Si(100)2x1 reactive surface and Si(100)2x1:H inactive surface. Thus the NC-AFM has a potential to investigate atomic scale origins of force interaction which is very important to clarify and control atomic-scale friction mechanism.

References

Albrecht, T.R., Grütter, P., Horne, D. and Rugar, D. (1991), "Frequency modulation detection using high-Q cantilevers for enhanced force microscope sensitivity", J.Appl.Phys., **69**, 668-673.

Allers, W., Schwarz, A., Schwarz, U.D. and Wiesendanger, R. (1999), "Dynamic scanning force microscopy at low temperatures on a van der Waals surface: graphite (0001)", Appl.Surf.Sci., **140**, 247-252.

Bammerlin, M., Lüthi, R., Meyer, E., Baratoff, A., Lü, J., Guggisberg, M., Gerber, Ch., Howald, L. and Güntherodt, H.-J. (1997), "True Atomic Resolution on the Surface of an Insulator via Ultrahigh Vacuum Dynamic Force Microscopy", Probe Microscopy, **1**, 3-9.

Binnig, G., Quate, C.F. and Gerber, Ch. (1986), "Atomic Force Microscope", Phys.Rev.Lett. **56**, 930-933.

Chabal, Y.J. and Raghavachari, K. (1984), "Surface Infrared Study of Si(100)-(2x1)H", Phys.Rev.Lett., **53**, 282-285.

Chadi, D.J. (1979a), "Atomic and Electronic Structures of Reconstructed Si(100) Surfaces", Phys.Rev.Lett., **43**, 43-47.

Chadi, D.J. (1979b), "Si(100) surfaces: Atomic and electronic structures", J.Vac.Sci.Technol., **16**, 1290-1296.

Eng, L.M., Bammerlin, M., Loppacher, Ch., Guggisberg, M., Bennewitz, R., Lüthi, R., Meyer, E. and Güntherodt, H.-J. (1999), "Surface morphology, chemical contrast, and ferroelectric domains in TGS bulk single crystals differentiated with UHV non-contact force microscopy", Appl.Surf.Sci., **140**, 253-258.

Fukui, K., Onishi, H. and Iwasawa, Y. (1999), "Imaging of atomic-scale structure of oxide surface and adsorbed molecules by noncontact atomic force microscopy", Appl.Surf.Sci., **140**, 259-264.

Giessibl, F.J. (1995), "Atomic Resolution of the Silicon (111)-(7x7) Surface by Atomic Force Microscopy", Science, **267**, 68-71.

Giessibl, F.J. (1997), "Forces and frequency shifts in atomic-resolution dynamic-force microscopy", Phys.Rev.B, **56**, 16010-16015.

Hosoi, H., Sueoka, K., Hayakawa, K. and Mukasa, K. (2000), "Atomic resolved imaging of cleaved NiO(100) surfaces by NC-AFM", Appl.Surf.Sci., **157**, 218-221.

Ishizaka, T., Sugawara, Y., Kumagai, K. and Morita, Seizo. (1990), "Anomalous Force Dependence of AFM Corrugation Height of a Graphite Surface in Air", Jpn.J.Appl.Phys., **29**, L1196-L1198.

Katayama, M., Williams, R.S., Kato, M., Nomura, E. and Aono, M. (1991), "Structure analysis of the Si(111) sqrt 3 x sqrt 3 R30° -Ag surface", Phys.Rev.Lett., **66**, 2762-2765.

Kitamura, S. and Iwatsuki, M. (1995), "Observation of 7x7 Reconstructed Structure on the Silicon (111) Surface using Ultrahigh Vacuum Noncontact Atomic Force Microscopy", Jpn.J.Appl.Phys., **34**, L145-L148.

Kobayashi, K., Yamada, H., Horiuchi, T. and Matsushige, K. (1999), "Investigations of C_{60} molecules deposited on Si(111) by noncontact atomic force microscopy", Appl.Surf.Sci., **140**, 281-286.

Lantz, M.A., Hug, H.J., van Schendel, A., Hoffmann, R., Martin, S., Baratoff, A., Abdurixit, A., Güntherodt, H.-J. and Gerber, Ch. (2000), "Low Temperature Scanning Force Microscopy of the Si(111)7x7 Surface", Phys.Rev.Lett., **84**, 2642-2645.

Loppacher, Ch., Bammerlin, M., Guggisberg, M., Battiston, F., Bennewith, R., Rast, S., Barratoff, A., Meyer, E. and Güntherodt, H.-J. (1999), "Phase variation experiments in non-contact dynamic force microscopy using phase locked loop techniques", Appl.Surf.Sci., **140**, 287-292.

Minobe, T., Uchihashi, T., Tsukamoto, T., Orisaka, S., Sugawara, Y. and Morita, S. (1999), "Distance dependence of noncontact-AFM image constrast on Si(111) $\sqrt{3} x \sqrt{3}$-Ag structure", Appl.Surf.Sci., **140**, 298-303.

Morita, S., Fujisawa, S. and Sugawara, Y. (1996), "Spatially quantized friction with a lattice periodicity", Surface Science Reports, **23**, 1-42.

Morita, S. and Sugawara, Y. (1999), "Guidelines for the achievement of true atomic resolution with noncontact atomic force microscopy", Appl.Sur.Sci., **140**, 406-410.

Morita, S., Abe, M., Tokoyama, K. and Sugawara, Y. (2000a), "Defects and their charge imaging on semiconductor surfaces by noncontact atomic force microscopy and spectroscopy", Journal of Crystal Growth, **210**, 408-415.

Morita, S., Sugawara, Y., Yokoyama, K. and Uchihashi, T. (2000b), "Correlation of frequency shift discontinuity to atomic positions on a Si(111)7x7 surface by noncontact atomic force microscopy", Nanotechnology, **11**, 120-123.

Ohta, M., Konishi, T., Sugawara, Y., Morita, S., Suzuki, M. and Enomoto, Y. (1993), "Observation of Atomic Defects on LiF(100) Surface with Ultrahigh Vacuum Atomic Force Microscope (UHV AFM)", Jpn.J.Appl.Phys., **32**, 2980-2982.

Ohta, M., Sugawara, Y., Osaka, F., Ohkouchi, S., Suzuki, M., Mishima, S., Okada, T. and Morita, S. (1995), "Atomically resolved image of cleaved surfaces of compound semiconductors observed with an ultrahigh vacuum atomic force microscope", J.Vac.Sci.Technol.B, **13**, 1265-1267.

Orisaka, S., Minobe, T., Uchihashi, T., Sugawara, Y. and Morita, S. (1999), "The atomic resolution imaging of metallic Ag(111) surface by noncontact atomic force microscope", Appl.Surf.Sci., **140**, 243-246.

Pérez, R., Payne, M.C., tich, I. And Terakura, K. (1997), "Role of Covalent Tip-Surface Interactions in Noncontact Atomic Force Microscopy on Reactive Surfaces", Phys.Rev.Lett., **78**, 678-681.

Raza, H., Pang, C.L., Haycock, S.A. and Thornton, G. (1999), "Non-contact atomic force microscopy imaging of TiO_2(100) surfaces", Appl.Surf.Sci., **140**, 271-275.

Sasaki, N. and Tsukada, M. (1999), "Theory for the effect of the tip-surface interaction potential on atomic resolution in forced vibration system of noncontact AFM", Appl.Surf.Sci., **140**, 339-343.

Sugawara, Y., Ohta, M., Hontani, K., Morita, S., Osaka, F., Ohkouchi, S., Suzuki, M., Nagaoka, H., Mishima, S. and Okada, T. (1994), "Observation of GaAs(110) Surface by an Ultrahigh-Vacuum Atomic-Force Microscope", Jpn.J.Appl.Phys., **33**, 3739-3742.

Sugawara, Y., Ohta, M., Ueyama, H. and Morita, S. (1995), "Defect Motion on an InP(110) Surface Observed with Noncontact Atomic Force Microscopy", Science, **270**, 1646-1648.

Sugawara, Y., Uchihashi, T., Abe, M. and Morita, S. (1999), "True atomic resolution imaging of surface structure and surface charge on the GaAs(110)", Appl.Surf.Sci., **140**, 371-375.

Uchihashi, T., Sugawara, Y., Tsukamoto, T., Ohta, M., Morita, S. and Suzuki, M. (1997), "Role of a covalent

120

bonding interaction in noncontact-mode atomic-force microscopy on Si(111)7x7", Phys.Rev.B, **56**, 9834-9840.

Uchihashi, T., Sugawara, Y., Tsukamoto, T., Minobe, T., Orisaka, S., Okada, T. and Morita, S. (1999a), "Imaging of chemical reactivity and buckled dimmers on Si(100)2x1 reconstructed surface with noncontact AFM", Appl.Surf.Sci., **140**, 304-308.

Uchihashi, T., Okada, T., Sugawara, Y., Yokoyama, K. and Morita, S. (1999b), "Self-assembled monolayer of adenine base on graphite studied by noncontact atomic force microscopy", Phys.Rev.B, **60**, 8309-8313.

Ueyama, H., Ohta, M., Sugawara, Y. and Morita, S. (1995), "Atomically Resolved InP(110) Surface Observed with Noncontact Ultrahigh Vacuum Atomic Force Microscope", Jpn.J.Appl.Phys., **34**, L1086-L1088.

Ueyama, H., Sugawara, Y. and Morita, S. (1998), "Stable operation mode for dynamic noncontact atomic force microscopy", Appl.Phys.A, **66**, S295-S297.

Watanabe, S., Aono, M. and Tsukada, M. (1991), "Theoretical calculations of the scanning-tunneling-microscopy images of the Si(111) sqrt3xsqrt 3 –Ag surface", Phys.Rev.B, **44**, 8330-8333.

Yokoyama, K., Ochi, T., Sugawara, Y. and Morita, S. (1999), "Atomically Resolved Silver Imaging on the Si(111)-($\sqrt{3}$x$\sqrt{3}$)-Ag Surface Using a Noncontact Atomic Force Microscope", Phys.Rev.Lett., **83**, 5023-5026.

Yokoyama, K., Ochi, T., Yoshimoto, A., Sugawara, Y. and Morita, S. (2000), "Atomic Resolution Imaging on Si(100)2x1 and Si(100)2x1:H Surfaces with Noncontact Atomic Force Microscopy", Jpn.J.Appl.Phys., **39**, L113-L115.

DYNAMIC FRICTION MEASUREMENT WITH THE SCANNING FORCE MICROSCOPE

OTHMAR MARTI AND HANS-ULRICH KROTIL

Experimental Physics, University of Ulm, D-89069 Ulm, Germany
email: *othmar.marti@physik.uni-ulm.de*

Abstract.

Scanning Force Microscopes excert lateral forces on the sample during measurement. The recording of the magnitude of these forces as a function of position gives friction maps. The possible scanning speeds of SFM, however, are far below the velocities of practical devices. The dynamic friction force measurement provides a solution to this problem. The sample is modulated laterally at frequencies up to several kilohertz and with amplitudes in the nanometer range. It is shown that the interaction with the sample is not only determined by friction, but also by the viscoelastic response of the sample. The combination of dynamic friction measurement with the intermittent contact measurement mode, the PulsedForceMode, gives full access to the relevant sample parameters: topography, lateral 'friction' forces, adhesion, sample stiffness and relaxations times.

1. Introduction

The integrating question since the first friction force measurements with atomic (scanning) force microscopy (Mate et al., 1987) is still without any definitive answer: how does friction on an atomic scale correlate to macroscopic friction measurements? Macroscopic measurements average over many contacts between the two rubbing bodies. The detailed interaction mechanisms are only indirectly accessible (Bhushan, 1999a). The handbook from Bhushan (1999b) contains a wealth of articles describing the many models and experimental findings of microtribology. The scanning force microscope is often considered to be a model system for a single asperity contact. The tip of the cantilever is one member of the material

121

B. Bhushan (ed.),
Fundamentals of Tribology and Bridging the Gap between the Macro- and Micro/Nanoscales, 121–135.
© *2001 Kluwer Academic Publishers.*

pairs investigated. Typically one plots of the friction as a function of the (preset) normal force. The optical lever detection system (Marti et al., 1990; Meyer et al., 1990) with its two-dimensional force detection capability is ideally suited for this task. The microscopes are operated in contact mode. The friction signal is calculated by subtracting the lateral force of the backward scan from the lateral force of the forward scan. Unless the microscope has an efficient linearisation accurate to the pixel this method fails. The nonlinearity of the piezo response presents that corresponding pixels on the forward and the backward scan are measured at the same point on the sample. Furthermore the microscope is limited to the very low speed regime ($\leq 10 - 100 \ \mu m$). To overcome the nonlinearity problems it seems to be advantageous to perform a friction experiment locally (Göddenhenrich et al., 1994; Yamanaka et al., 1995; Colchero et al., 1996; Nurdin et al., 1997; Krotil et al., 1999b). The sample is mounted on a piezo-stage and modulated with nanometer-amplitudes perpendicular to the cantilever axis. The lateral forces induce a signal in the lateral force channel of the microscope. The modulation is detected by synchronous detection (Colchero et al., 1996). Since the cantilever is moving only a few nanometers, the nonlinearity is negligible. Furthermore the synchronous detection implicitly subtracts the forward and backward scan lateral force images. With some tips the lateral movement is much smaller than the contact area. Hence it is possible that the experiments in dynamical friction force microscopy can be understood in terms of the dynamical shear testing of materials. The normal force F involved in these experiments is a sum of the externally applied normal force and the adhesion force.

$$F = F_e + F_A \tag{1}$$

The adhesion force F_A is easily measured using the PulsedForceMode (Rosa et al., 1997; Krotil et al., 1999a). In this paper we present a new method which combines dynamical friction mode imaging and PulsedForceMode to achieve a more complete data set on friction.

2. Theoretical Considerations

We investigate the situation, where the sample is modulated laterally with a triangular excitation at the frequency ω. We assume in a first approximation that the tip sticks to the surface until the threshold voltage for static friction, F_{stick}, is reached. The friction force then reduces to F_{slip}, the presumed constant dynamic friction force. The resulting friction signal at

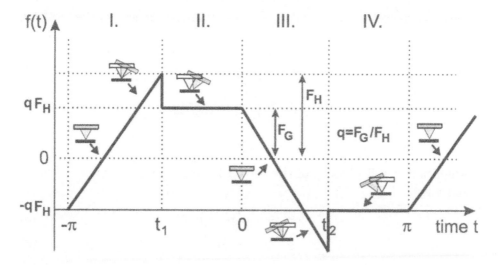

Figure 1. The idealized friction loop shown as a function of time. Two surfaces are moved one past the other. The friction force $f(t)$ increases until the maximum static friction force F_H (stick) is reached. The force level reduces down to F_G (slip) and stays constant until the backward movement reduces the speed to the point where the tip starts to stick again. The same behaviour is also observed on the backward scan.

the detector of the microscope in dependence of the modulation amplitude A_m of the shear piezo is given by the following equations:

$$
\begin{aligned}
f(t) &= \frac{F_H A_m}{2}\left(\cos(t) + 1\right) - qF_H & -\pi \le t \le t_1 \qquad (2)\\
&= qF_H & t_1 < t \le 0\\
&= qF_H - \frac{F_H A_m}{2}\left(1 - \cos(t)\right) & 0 \le t \le t_2\\
&= -qF_H & t_2 < t \le \pi
\end{aligned}
$$

with t_1 and t_2 given by

$$
\begin{aligned}
t_1 &= -\pi + \arccos\left(1 - 2\frac{1+q}{A_m}\right) \qquad (3)\\
t_2 &= \arccos\left(1 - 2\frac{1+q}{A_m}\right).
\end{aligned}
$$

Provided that the modulation amplitude obeys $A_m k_{lat} \le F_{stick}$, the signal is synchronously detected using a lock-in amplifier. This amplifier detects the in-phase component x and the out-of-phase component y of the resulting signal.

$$x = \frac{1}{\pi} \int_{-\pi}^{\pi} f(t) \sin(t) dt \tag{4}$$

$$y = \frac{1}{\pi} \int_{-\pi}^{\pi} f(t) \cos(t) dt$$

Amplitude r and phase ϕ of the detected signal can be calculated from x and y by

$$r = \sqrt{x^2 + y^2} \tag{5}$$

$$\phi = \arctan \frac{x}{y}. \tag{6}$$

The resulting equations look rather complicated:

$$
\begin{aligned}
r &= \sqrt{x^2 + y^2} \tag{7} \\
&= \frac{F_H}{\pi} \left(\left[-4q + \frac{-2 + 4q + 6q^2}{A_m} \right]^2 + \left[\frac{A_m}{2} \arccos \left(1 - \frac{2(1+q)}{A_m} \right) \right. \right. \\
&\quad \left. \left. + \frac{-2 + 6q - A_m}{A_m} \sqrt{-1 - 2q + A_m - q^2 + qA_m} \right]^2 \right)^{\frac{1}{2}}
\end{aligned}
$$

$$
\begin{aligned}
\phi &= \arctan \frac{x}{y} \tag{8} \\
&= \arctan \left(\left[-4 + \frac{-2 + 4q + 6q^2}{A_m} \right] \cdot \left[\frac{A_m}{2} \arccos \left(1 - \frac{2(1+q)}{A_m} \right) \right. \right. \\
&\quad \left. \left. + \frac{-2 + 6q - A_m}{A_m} \sqrt{-1 - 2q + A_m - q^2 + qA_m} \right]^{-1} \right)
\end{aligned}
$$

The signal for a modulation amplitude, small enough to never leave the static friction regime, is proportional to the excitation amplitude. The phase is zero.

$$r = \frac{A_m F_H}{2} \tag{9}$$

$$\phi = 0. \tag{10}$$

At the limit of high excitation amplitudes the signal is again a simple function of the sliding friction force.

$$\lim_{A_m \to \infty} r = \frac{4qF_H}{\pi} = \frac{4}{\pi}F_G \qquad (11)$$

The above theory is valid only for samples with no viscoelastic behaviour. There one can, to a first approximation, interpret the data in terms of static and dynamic friction forces. Viscoelastic surfaces on the other hand are characterized by their moduli and by characteristic time constants. In many experiments the lateral modulation amplitude is small compared to the diameter of the contact zone. Therefore one can view the experiment as a nanoscopic shear compliance testing. The testing of shear compliance involves the three moduli: the bulk modulus B

$$B = \frac{\text{hydrostatic pressure}}{\text{volume strain}} = \frac{\text{hydrostatic pressure}}{\text{volume change per unit volume}} = \frac{pV_0}{\Delta V}, \qquad (12)$$

the Young's modulus or tensile modulus E

$$E = \frac{\text{tensile stress}}{\text{tensile strain}} = \frac{\text{force per unit cross-section of area}}{\text{strain per unit length}} = \frac{F/A}{ln(L/L_0)}, \qquad (13)$$

and the shear modulus or rigidity G

$$G = \frac{\text{shear stress}}{\text{shear strain}} = \frac{\text{shear force per unit area}}{\text{shear per unit distance between shearing surfaces}}$$
$$= \frac{F/A}{S/D}. \qquad (14)$$

These equations deal with purely elastic deformations. However, polymers and many other materials have considerable viscous responses to stimuli. Whereas the elastic response is reversible, the viscous response is plastic.

Molecules are permanently rearranged, their equilibrium is reached only after a relaxation since characteristic relaxation time. For a sinusoidal forced vibration one gets the following equations for the stress and the strain, respectively.

$$\tau = \tau_0 sin\omega t \quad \text{(stress)} \tag{15}$$

$$\gamma = \gamma_0 sin(\omega t - \vartheta) \quad \text{(strain)} \tag{16}$$

We can combine these two quantities into one complex modulus.

$$G' = \frac{\tau_0}{\gamma_0} \cos \vartheta \tag{17}$$

$$G'' = \frac{\tau_0}{\gamma_0} \sin \vartheta \tag{18}$$

The real part of this modulus, G', is the storage modulus (named after the fact that energy is stored reversibly in the deformation). It describes the stiffness of the sample. The imaginary part of the complex modulus, G'', is the loss modulus or viscous modulus. It describes the dissipation of mechanical energy, i.e. the conversion of mechanical energy to heat. In a similar way one can write down a complex shear modulus E. The definition is analogous to the above definition of the complex G. In both cases one can define a loss tangent

$$\frac{G''}{G'} = \tan \delta_G \tag{19}$$

and

$$\frac{E''}{E'} = \tan \delta_E. \tag{20}$$

A dominant relaxation time constant of molecular reorganization under shear conditions shows up as a peak in the imaginary part of the complex shear modulus. This quantity is standard for characterizing polymers and polymer samples.

3. Instrumental Aspects

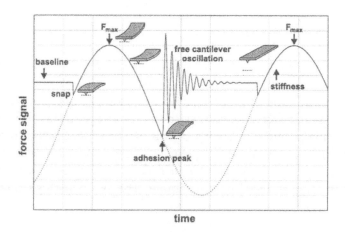

Figure 2. Scheme showing the modulation voltage (dotted line) and the force signal (straight line) over a complete modulation period in PulsedForceMode. The arrows indicate the points where the baseline, adhesion force, stiffness and the maximal applied normal force F_{max} are taken. Time and force are given by arbitrary units.

We use commercial scanning force microscopes (Topometrix, CSEM) in combination with a PulsedForceMode electronics (Witec). In the Pulsed-ForceMode the cantilever position is modulated with a sinusoidal voltage. The amplitude is typically between 10-500 nm. It is large enough that the tip is brought into contact and removed from contact in every cycle. Typically the modulation frequency is between 0.1 to 5 kHz. It is important that it is below the resonance frequency of the cantilever. The nature of the response curve depends on the interaction of the tip with the sample. The stiffness (a not so well defined quantity) of the sample and the viscous behavior are the most important parameters.

If we start at the left side of Figure 2, then the tip is far away from the sample. The force is constant and zero as indicated by the baseline. Shortly before the tip reaches the surface an instability occurs. It is often called "snap in". The size of the force jump is indicative of the interaction potential, but also of double layer forces in electrolytes, for instance. The tip is then pressed against the sample. It follows the indentation curves known from the literature. At the point (time) of maximal indentation the force is measured. The feed back electronics of the scanning force microscope keeps this force constant. Hence the indentation curves are stabilized too. Subsequently the tip is withdrawn from the sample. Adhesion forces permit negative total force.

The break force in the JKR model is given by

$$F_{adhesion} = -3/2\pi W_{1,2}R \qquad (21)$$

where $W_{1,2}$ is Dupré's interfacial energy and R the radius of the tip. Hence the maximum negative force after the indication is determined, within the framework of the JKR theory, by the adhesion of the tip to the sample. The cantilever is then relaxed and starts to oscillate at its resonance frequency. This oscillation permits, in theory, the measurement of mass transfer to the cantilever from the sample. To implement dynamical friction measurements one can measure, while measuring in PulsedForceMode, the lateral force signal modulation induced by the sample. The setup of this experiment is shown in Figure 3.

The electronics shown in Figure 3 is essentially the PulsedForceMode electronics. Lock-in amplifiers for the normal and the lateral force mode image detects the induced modulation signals even though they are not continuous (Krotil et al., 2000b; Krotil et al. 2000a).

Figure 4 shows the setup to measure dynamical friction and adhesion simultaneously. Therefore the AFM is equipped with several piezo actuators which impart oscillations between the probe and the sample such that a vertical oscillation and a relative lateral oscillation are superimposed. These oscillations give the AFM high dynamic components and makes it obvious to name the operating principle **CO**mbined**DY**namic Mode or **CODYMode**®, for short.

The lock-in amplifier generates the modulation signal for the sample piezo. The resulting high frequency signals are detected and phase and amplitude are recorded. It is important to use lock-in amplifiers which have a fast frequency response and a fast calculation of amplitude and phase. The lock-in outputs a signal which starts as soon as the tip gets into contact with the sample. The integration line of the lock-in amplifier determines the rise time of the signals (amplitude and phase). Sample and hold circuits measure the peak values of the induced signals.

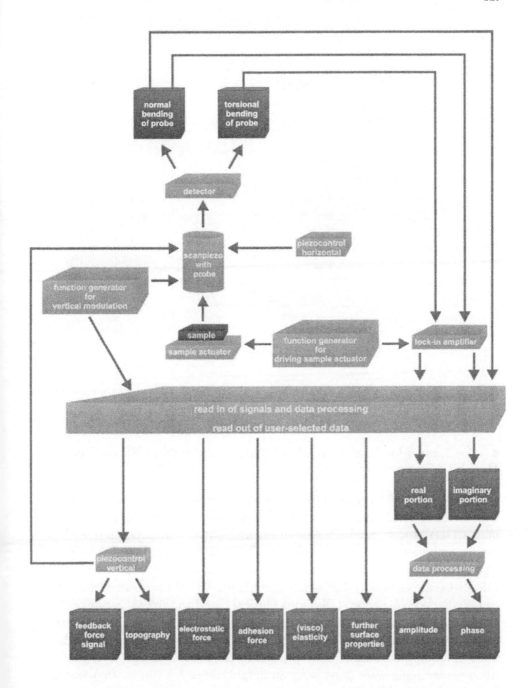

Figure 3. Block Diagram of the CODYMode® scanning force microscope.

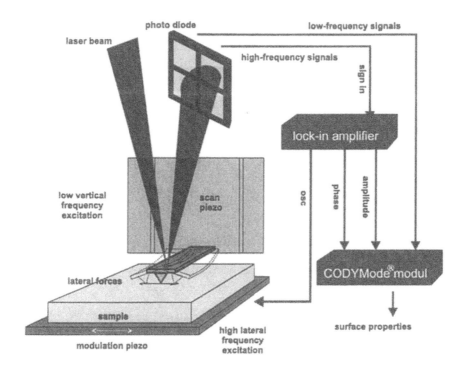

Figure 4. Schematic view of a microscope setup in CODYMode® to measure friction and adhesion simultaneously.

4. An Example: The Analysis of Thin Aminosilane Films Generated by Micocontact Printing

CODYMode® SFM investigations were performed with the atomic scale tribometer (CSEM Instruments, Neuchâtel, Switzerland) with an in-house built CODYMode® module. Amplitude and phase were measured with a lock-in amplifier (SR 844 RF, Stanford Research Systems Inc., CA, USA). The lateral sample excitation was done by a piezoelectric transducer (PIC 155, PI-Ceramics GmbH, Lederhose Germany). All data were acquired with rectangular Si_3N_4 probes (Olympus Optical Co., Ltd. Tokyo, Japan). The spring constant for normal excitation was 0.75 N/m with a fundamental resonance frequency of 85 kHz. The transducer was driven by a synthesized function generator (DS 345, Stanford Research Systems, CA, USA). All measurements were done with an amplitude of 5.5 nm at frequencies from 150 kHz to 700 kHz under ambient conditions (($24 \pm 1)°C$, (42 ± 2) % relative humidity).

Microscopic measurements of frictional forces and shear elasticities were performed on samples prepared by transferring aminosilane ((3-amino--

propyl) triethoxysilane, provided by P. Barth) by microcontact printing (μCP) on silicon wafers. The polydimethylsiloxane (PDMS) stamp (SL-GARD 184 Base silicon elastomer, Dow Corning GmbH, 65201 Wiesbaden, Germany) was controlled inked (Krotil 2000). After plasma oxidation of the PDMS stamp, the aminosilane was vapor deposited for two hours and brought into contact with the substrate immediately. The silicon substrate (pieces of monocrystalline silicon wafer from Wacker GmbH, Germany) were cleaned by heating at $70°C$ for 35 min in Pirhana solution (a mixture 7:3 (v/v) of 98 % H_2SO_4 and 30 % H_2O_2), thoroughly rinsed with deonized water and used immediately. Using this technique a monolayer of aminosilanes was generated, with rings of uncoated silicon exposed. The silicon serves as a hard reference in the analysis of the molecules. The topography of the silicon rings is almost indistinguishable from that of the silane layers. We consider that the coverage of the silicon is mostly one monolayer.

The detection of the adhesion force is independent of the lateral modulation frequency (Krotil, 2000). Figure 5 shows that the adhesion force of the silane covered areas is around (-116±6) nN, that of the silicon (unmodified areas) is (-108±6) nN. The fluctuation of the adhesion force as a function of the excitation frequency is the same for the monolayer and the bare silicon, indicating that other long range forces such as electrostatic forces might be important.

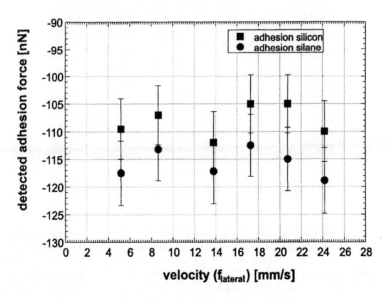

Figure 5. Adhesive forces in CODYMode® SFM. The detection is not affected by the lateral modulation.

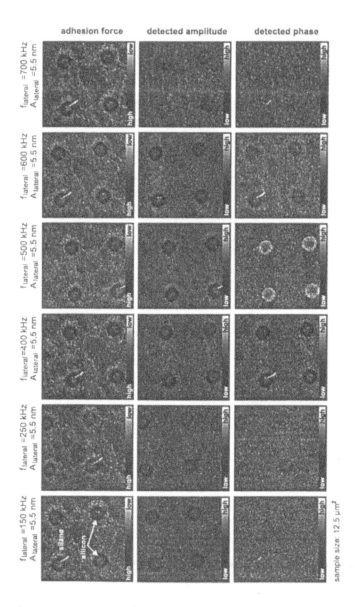

Figure 6. Qualitative overview of high velocity friction investigations of aminosilane, microcontact printed on native silicon. The first column represents the adhesion forces, the second column is the frictional amplitude and the third one the frictional phase. The rows are measurements at different frequencies (from 700 kHz, top row, down to 150 kHz, bottom row). The excitation amplitude was kept constant. All measurements were performed at ambient conditions $((24 \pm 1)^\circ C, (42 \pm 2)$ % relative humidity). A Si_3N_4 tip on a rectangular cantilever was used.

Figure 6 shows a series of measurements at selected frequencies from 150 kHz to 700 kHz. The frictional force (amplitude and phase) depends on the excitation frequency. Figure 7 shows this more clearly.

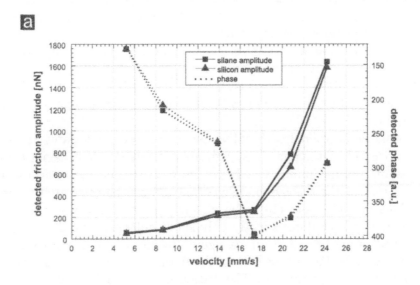

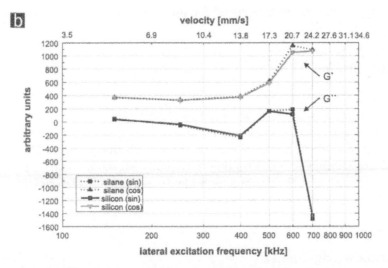

Figure 7. (a) Starting from the detected amplitude and phase spectra as a function of the maximum lateral sliding velocity one can estimate (b) the storage modulus G' and loss modulus G'' of the silane and silicon (water coated).

The amplitude of the frictional signal increases monotonically for the silane coated regions and the silicon (water coated) region. The amplitude on the silane film is slightly larger than that on the silicon. It is expected that the frictional force increases in this velocity range, as a consequence of velocity shear strengthening (Baumberger, 1996). Commonly one assumes that viscous damping is important.

The phase map shows an interesting behaviour. For low velocities (frequencies) the phase shift of the silane is more pronounced than for the silicon. At speeds of 16 mm/s the phase relation increases, leading to an inverted contrast. Finally above 19 mm/s the differences between silanes and silicon vanishes.

Since silicon is covered with native oxide layer there is a water film in the ring areas. Thus, the phase may indicate different elastic behavior between the water and the aminosilane layers. From the amplitude spectra and the phase spectra the storage module G' and G'' can be calculated according to Equations (17) and (18). The qualitative trend is shown in Figure 7. Since the spectra differ slightly only small differences in G' and G'' are observed. Above $f = 400\ kHz$ a significant increase of G'' is observed for both materials. Further increasing the frequency puts more and more energy in the storage module. The dissipation has a minimum at $f = 400\ kHz$, suggesting an optimal speed for minimum friction.

5. Conclusions

Dynamical friction force measurements using CODYMode® allow a simultaneous measurement of topography and adhesion along with friction. The intermittent contact mode of friction measurements prevents the damage of delicate samples. CODYMode® is equally suitable to do mechanical-dynamical testing at the nanometer scale. Improved lateral excitation piezos might increase the useful frequency range to several megahertz.

6. Acknowledgements

The authors kindly acknowledge the help of and discussions with Peter Barth, Volodymyr Senkovsky, Thomas Stifter, Sabine Hild, Martin Pietralla, Bernd Heise, Markus Hackenberg and Gerhard Volswinkler. Some of the works was made with support of the German Science Foundation (SFB 239).

References

Baumberger, T. (1996). *Physics of Sliding Friction* chapter *"Dry Friction Dynamics at Low Friction"*, Kluwer Academic Publishers, 1-26.

Bhushan, B. (1999a). *Handbook of Micro/Nanotribology* chapter *"Introduction - Measurement techniques and applications"*, CRC Series Mechanics and Materials Science, Boca Raton, Florida.

Colchero J., Luna, M., and Baro, A. M. (1996). *"Lock-in technique for measurement friction on a nanometer scale"*, *Appl. Phys. Lett.* **68**, 2896-2898.

Göddenhenrich, T., Müller, S., and Heiden, C. (1994). *"A lateral modulation technique for simultaneous friction and topography measurements with the atomic force microscope"*, *Rev. Sci. Instr.* **65(9)**, 2870-2873.

Krotil, H.-U. (2000). *"CODYMode® scanning force microscopy: The concurrent measurement of adhesion, friction and viscoelasticities"*, *PhD Thesis*, University of Ulm.

Krotil, H.-U., Stifter, Th., and Marti, O. (2000). *"Concurrent measurement of adhesive and elastic surface properties with a new modulation technique for scanning force microscopy"*, *Rev. Sci. Instr.* **7**, 2765-2771.

Krotil, H.-U., Stifter, Th., and Marti, O. (submitted 2000*)*. *"Combined Dynamic adhesion and friction measurement with the scanning force microscope"*, *Applied Physics Letters.*

Krotil, H.-U., Stifter, Th., Waschipky, H., Weishaupt, K., Hild, S., and Marti, O. (1999a). *"PulsedForceMode: A new method for the investigation of surface properties"*, *Surface and Interface Analysis* **27**, 336-340.

Krotil, H.-U., Weilandt, E., Stifter, Th., Marti, O., and Hild, S. (1999b*)*. *"Dynamic friction force measurement with the scanning force microscope"*, *Surface and Interface Analysis* **27**, 341-347.

Marti, O., Colchero, J., and Mlynek, J. (1990). *"Combined scanning force and friction microscopy of mica"*, *Nanotechnology* **1**, 141-144.

Mate, C. M., McCelland, G. M., Erlandson, R., and Chiang, S. (1987). *"Atomic-scale friction of a tungsten tip on a graphite surface"*, *Physical Review Letters* **59**, 1942-1945.

Meyer, G. and Amer, N. M. (1990). *"Simultaneous measurement of lateral and normal forces with an optical-beam-deflection atomic force microscope"*, *Appl. Phys. Lett.* **57**, 2089.

Nurdin, N., Weilandt, E., and Descouts, P. (1997). *"Scanning force microscopy of surface phase separation of polymer blends"*, *Chimia* **51(7)**, 405.

Rosa, A., Weilandt, E., Hild, S., and Marti, O. (1997). *"The simultaneous measurements of elastic, electrostatic and adhesive properties by scanning force microscopy: PulsedForceMode operation"*, *Meas. Sci. Technol.* **8**, 1333-1338.

Yamanaka, K. and Tomita, E. (1995*)*. *"Lateral force modulation atomic force microscope for selective imaging of friction forces"*, *Jpn. J. Appl. Phys.* **34**, 2879-2882.

TOWARDS THE IDEAL NANO-FRICTION EXPERIMENT

J.W.M. FRENKEN,   M. DIENWIEBEL,   J.A. HEIMBERG*,
T. ZIJLSTRA†, E. VAN DER DRIFT†, D.J. SPAANDERMAN‡,
E. DE KUYPER

*Kamerlingh Onnes Laboratory, Leiden University, P.O. Box 9504,
2300 RA Leiden, The Netherlands*
*Naval Research Laboratory, Washington, DC 20375-5342, USA
*†DIMES, Delft University of Technology, 2600 GB Delft, The
Netherlands;*
*‡FOM-Institute for Atomic and Molecular Physics, Kruislaan 407,
1098 SJ Amsterdam, The Netherlands*

1. Introduction

The invention of the Atomic Force Microscope (AFM) has opened the way to
the investigation of the fundamental aspects of friction for extremely small
contacts, with characteristic length scales into the (sub)nanometer regime, both
perpendicular and parallel to the plane of contact (Bhushan et al., 1995). By
putting the AFM in ultrahigh vacuum (UHV) and preparing atomically clean
surfaces and AFM-tips, one can carry out model experiments on atomic-scale
contact formation and friction either in the complete absence of intervening
lubricants or with highly idealized model lubricant monolayers.

For application in nanotribology, AFM's have been constructed that record
and control the force normal to the contact plane – the so-called loading force –
while simultaneously measuring one friction force component (see e.g.
Güntherodt et al., 1995). This is achieved by detecting both the bending and the
torsion of the cantilever, through which the AFM-tip is attached to the rest of
the AFM. These 'traditional' Friction Force Microscopes (FFM's) share several
severe disadvantages. In order to be sensitive to the friction force, the AFM
cantilever has to be made 'oversensitive' to the normal force. As a consequence,
cantilevers used in traditional FFM's suffer from the so-called 'snap-to-contact'.
This is the phenomenon that, at short separations between tip and surface, the
attractive Van der Waals or electrostatic force gradients overcome the spring

137

B. Bhushan (ed.),
Fundamentals of Tribology and Bridging the Gap between the Macro- and Micro/Nanoscales, 137–150.
© 2001 *Kluwer Academic Publishers.*

coefficient of the cantilever, at which point the tip is accelerated into a violent contact with the surface. This not only leads to potential damage of the tip, it also implies that a very interesting range of tip-surface distances cannot be addressed in these microscopes (examples of more suitable cantilevers can be found in Jarvis et al., 2000; Kageshima et al., 1999).

In practice, it is very difficult to measure the spring coefficients of the cantilever accurately and calibrate a traditional FFM, which makes it hard to truly quantify measurements with these instruments (examples of quantitative FFM measurements can be found in Carpick et al., 1996; Schwarz et al., 1997).

Finally, there are almost no measurements in which it is possible to control the shape of the FFM-tip (Cross et al., 1998). This means that of the two surfaces involved in the model nanocontact of an FFM, one usually remains completely uncontrolled on the relevant length scale of the contact diameter.

In this paper, we describe our approach to the solution of these problems and introduce the concepts of a new, quantitative FFM setup that will allow measurements of contact and friction forces over a fully defined and controlled interface. The force sensor and detection system, which are at the heart of this microscope, have been constructed already, and a brief account will be given of the performance of this part of the new instrument (Zijlstra et al., 2000).

2. Philosophy: The theoretician's ideal nano-friction experiment

2.1. THE IDEAL MEASUREMENT

Let us begin by considering what experiment we would like to carry out, if we were not at all hindered by technical limitations. The ideal experiment would be one in which we record all three components of the force between two surfaces in which we know and control 'where all the atoms are'. Some of the fundamental questions that can be addressed with such an experiment are:

1) How does the friction force build up when the distance between the surfaces is decreased? At which distance do we experience the 'onset' of friction?

2) How does the friction force depend on the contact area? In the framework of tribology, the interesting regime is certainly not that of a single-atom contact, but rather that of contact areas ranging from a few atoms to a few thousand atoms.

3) How does the friction force depend on the materials? Of course, the simplest model experiment would be one in which the two surfaces in contact consist of the same material.

4) How does the friction force depend on the relative crystallographic orientation of the two surfaces? Single-atom contacts exhibit pronounced

atomic-scale stick-slip sliding motion. When two rigid lattices are sheared over each other, the friction force should again exhibit high atomic-scale variations when the lattices are oriented such that they fit. However, if the lattices are rotated out of registry, there should be a significant cancellation of the individual contributions to the friction force, leading to a phenomenon that has been predicted under the name 'superlubricity' (see e.g. Sørensen et al., 1996; Hirano et al., 1997). Of course, for larger contacts, this naïve picture should break down, as the two lattices are not perfectly rigid, and a network of misfit dislocations necessarily forms between the two. It is important to find out whether superlubricity exists, how it develops when the contact is made larger than just a few atoms, and how it disappears when the contact area is increased further.

5) How does the friction force depend on the sliding direction with respect to the crystal orientations of the two surfaces?

6) How does the friction force depend on temperature?

7) How does a (model) lubricant change the friction force?

8) How and why does a contact age?

With experimental answers to these questions for ideal, fully controlled contacts, very detailed comparisons can be made with microscopic theories and computer simulations, in search for the energy dissipation mechanisms relevant on different length scales.

2.2. THE IDEAL INSTRUMENT

It need not surprise us that the geometry of our thought experiment turns out to be close to that of large-scale computer simulations of contact formation and friction between two well-defined surfaces (e.g. Sørensen et al., 1996). Based on the above ideas we can directly formulate the technical requirements that our ideal instrument must meet:

1) First of all, we demand complete freedom in the choice of the two materials that are to be brought into contact.

2) Next, we want full control over the contact area. In traditional FFM's, the contact area is determined by the initial radius of the tip and the deformations caused by the forces between the two surfaces (loading + adhesion forces). In our ideal experiment, we want to control the contact area and the loading force independently. This means that we need to go beyond the usual hemispherical tip shape. The tip should end in an atomically flat plane, i.e. a crystal facet, with a controllable radius.

3) Of course, the nanofacet, formed by the end face of the tip, has to be oriented precisely parallel to the crystal surface with which it is to be brought in contact.

4) To complicate matters further, we want control over the precise crystallographic orientations of the two surfaces. This means that we have to specify not only the crystallographic orientation of the surface normal of the tip and the counter surface, but also their azimuthal orientations.

5) The sliding direction has to be adjustable, independently of the azimuthal orientations of tip and counter surface.

6) The force sensing device should be stiff enough to withstand the high force gradients that otherwise lead to the familiar snap-to-contact.

7) At the same time, we require high sensitivity in the measurement of the frictional forces.

8) Needless to add, that we want to measure not only the component of the friction force (anti-)parallel to the sliding direction, but also the component perpendicular to this direction (Morita et al., 1996).

9) Measurements should be possible as a function of temperature.

10) Finally, the instrument should allow us to add controlled overlayers (model lubricants) on each of the two surfaces.

Although these requirements suggest the need for some 'science fiction', we believe that we can meet most of them and make it 'science friction'! The next section briefly describes the design of our FFM.

3. Design

Figure 1 shows the schematic configuration of the FFM that is under development in our laboratory. The instrument will be housed in an ultrahigh vacuum (UHV) chamber, enabling us to produce atomically clean, well-characterized crystal surfaces of various materials, such as metals and semiconductors and to deposit well-defined monolayers of adsorbates that act as model lubricants. There are two features that make our design significantly different from traditional UHV-FFM's.

3.1. FIELD-ION MICROSCOPY

Standard surface-science techniques are frequently used to prepare and characterize the surfaces of solids that are studied in FFM's. These techniques allow one to prepare a broad variety of single crystal surfaces, that are extremely well-ordered, contain a low density of steps, and have impurity concentrations below 1‰ of a monolayer. By contrast, there is almost no information on the other surface in the FFM, namely that of the tip. Our solution to this problem is to view *and shape* the tip by use of field-ion microscopy (Kellog, 1994; Tsong, 1990; Fink, 1986; Fu et al., 1996; Cross et al., 1998).

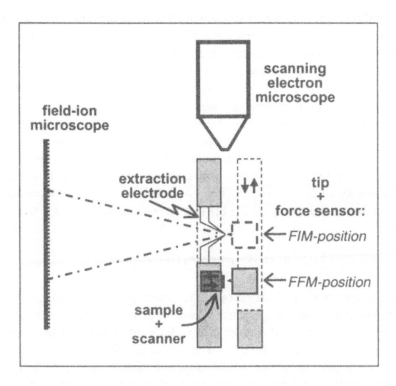

Figure 1. Schematic design of the nano-tribology setup. The unit containing the tip and the force sensor can be positioned either close to the sample, for friction measurements, or close to an extraction electrode, for field-ion microscopy charactarization and shaping of the tip. A scanning electron microscope is focused on the tip-sample contact.

In a field-ion microscope (FIM), a sharp tip, usually of a high-melting-point metal, is imaged by placing it in a high electric field in the presence of a low density of a noble gas, e.g. He. The field strength is highest where the curvature of the tip is at its maximum, which is at or near the apex. The highest fields are sufficient to ionize the noble gas atoms. The ions are accelerated from the ionization position to a phosphor screen, where they build up a strongly enlarged image of the high-curvature regions at the tip apex. In this way, high quality FIM images show the tip apex with atomic resolution, which allows one to directly identify the crystallographic orientation of the tip and to count the number of atoms at the perimeter of the outermost atomic facet, from which one immediately obtains the total number of atoms in that facet. When the tip is brought in contact with the sample surface with a low loading force, these outermost atoms define the contact area.

There are several additional advantages that the FIM provides. The FIM image shows not only which crystallographic plane terminates the tip apex, but also in which precise direction the surface normal is pointing. In our setup, we will be able to rotate the sensor+tip assembly in two independent directions, so that we can align the surface normal of the tip with that of the surface, in order to make these surfaces really parallel. The FIM can further be used to 'shape' the tip and thereby modify the area of the outermost atomic facet. This is achieved by increasing the electrical field to the point where the tip atoms are field-desorbed. In this way, the tip apex is stripped away atom by atom. In addition one can refurnish metal atoms by molecular beam deposition and anneal the tip to form stable, e.g. hexagonal, islands. A beautiful illustration of these methods can be found in (Fu et al., 1996), where tips are produced with perfect hexagonal facets of Ir(111) that have sizes of 7, 19, 37, 61, and 91 Ir atoms.

3.2. FORCE SENSOR AND DETECTION SYSTEM

The force microscopy part of our setup consists of two stages: a fiber head assembly containing a specialized lateral force cantilever with a detection system, and a sample stage that allows for manipulation of the sample with respect to the tip and cantilever. The so-called 'tribolever' is formed by the monocrystalline monolithic silicon structure shown in Figure 2. Four, high aspect ratio legs extend out from a central detection pyramid. A fiber optic interferometer reflects laser light off each pyramid face to track the motion of the pyramid. Combining the four displacements measured by the interferometers, the instrument can follow the three-dimensional motion of the pyramid and thus obtain the three components of the forces on the tip. The scanning tip, which can be etched metal wire, for example tungsten, is threaded through the central hole of the pyramid and extends ~ 50 μm out from the base of the pyramid to interact with the surface.

3.2.1. The Tribolever
The entire tribolever structure is made in a Si wafer by a combination of dry and wet etching and oxidation steps (Zijlstra et al., 2000). The central pyramid of the tribolever is formed via a KOH wet etch of the (100)-oriented Si wafer, which exposes the {111} planes of silicon. A special passivation technique necessary to protect the corners of the convex structure from underetching gives rise to the cross structure within the pyramid (Figure 2), in which an etched metal tip was glued. The pyramid faces are highly reflective, and oriented at well-defined angles, which makes them ideal for interferometry. It is important to note that interferometry does not suffer from the problems associated with reflected beam detection techniques used in AFM: diffraction and dependence

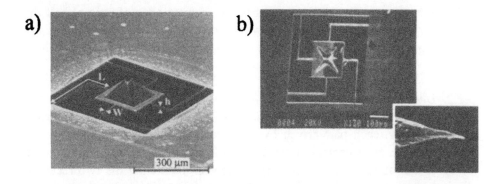

Figure 2. Two SEM micrographs of the tribolever. a) Front side of the sensor pointing away from the sample. The dimensions of the legs are L=350 μm, w=1.4 μm and h=10.6 μm. b) shows the rear side of the tribolever with a W-tip extending approx. 50 μm out from the pyramid base (see inset).

of the measured displacement on the exact placement of the tip on the cantilever and on the beam spot location (2nd order effect). The shape and dimensions of the four legs have been chosen using finite element analysis[*], so that the two lateral spring constants were equal and significantly lower than the torsional spring constant of single board, AFM cantilevers. A typical torsional AFM spring constant is (Rabe et al., 1996)

$$k_x^{AFM} = 71.6 \text{ N/m},$$

whereas the calculated tribolever lateral spring constants were

$$k_x^{tribolever} = k_y^{tribolever} = 1.48 \text{ N/m}.$$

In the vertical direction, typically

$$k_z^{AFM} = 0.2 \text{ N/m},$$

while the calculated value for the tribolever was

$$k_z^{tribolever} = 25.8 \text{ N/m}.$$

Additionally, the calculated coupling between the three orthogonal directions in the tribolever was as low as ~10^{-4}% and the coupling between torsional (out of scanning plane) and normal forces was essentially zero. The tribolever might make the impression of being very sensitive to placement errors of the tip with respect to the precise symmetry center, via the in-plane torques resulting from lateral forces. Nevertheless, the changes in lateral spring constants remain well

---

[*] ANSYS©: finite element analysis software package, distributed by Swanson Analysis Systems, Inc., Houston, TX, USA.

below 5%, even for the maximum possible placement error of a tip of half the hole diameter (20 μm). The maximum rotation angle of such an eccentrically placed tip remains well below 0.05°, even under extreme lateral forces as high as 1 μN.

3.2.2. Detection

The four optical fibers and the pyramid form two pairs of opposing interferometers, each pair driven by a single laser diode. Each fiber pair is sensitive to the displacement in the Z-direction and to one component, either X or Y, of the lateral displacement. By adding or subtracting the two recorded displacements in one such pair, e.g. the X-pair, we obtain the displacement components of the pyramid in the Z-direction and the X-direction respectively.

The distance of the end face of each fiber with respect to the equilibrium position of the pyramid face was adjusted with the use of an inertial piezo motor[*]. The position of the fiber axis can be adjusted in a plane parallel to the pyramid face by the use of flexure hinges in the fiber head, in order to make each fiber aim at the center of its pyramid face (see Figure 3). The silicon chip with the integrated force sensor also contains a kinematic mount (hole-groove-flat combination), which defines the sensor position with respect to three small ruby spheres, glued to the bottom plate of the fiber head. The reproducibility of the sensor position is such that adjustments of the flexure hinges to reposition the fibers are not necessary between tribolevers. A second kinematic mount is used between the detection part (fiber head with tribolever chip) and the sample stage.

3.2.3. Sample stage

The sample is mounted on a scan tube, which rests inside a set of nested inertial motors that allow for four-dimensional motion of the sample with respect to the tip. The scanner is directly coupled to the Z coarse approach motor. This Z motor sits in the center of an X-Y-Φ motor that allows for long-range manipulation of the sample with respect to the tip. The Z and XY motors are similar to motors discussed elsewhere (Hug et al., 1999), but the nested design is new, as is the Φ-motor which allows for rotational adjustment in the plane parallel to the scan. The Φ-motor will be used to rotate the tip and sample lattice planes with respect to each other, in search for alignment and misalignment (e.g. 'superlubricity') effects.

---

[*] Nanomotor®: Klocke Nanotechnik GmbH, Horbacher Str. 128, D-52072 Aachen, Germany.

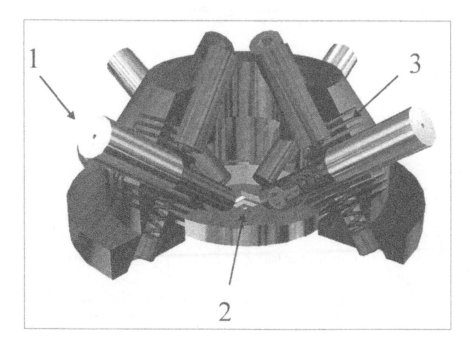

Figure 3. The fiber head assembly. An invar housing contains 1: the Nanomotors®
(positioning of the four glassfibers), 2: the silicon chip with the tribolever, 3: flexure
hinges for coarse adjustment of the Nanomotors® with the glassfibers.

4. Performance

4.1. CALIBRATION

For calibration of the spring constants in the X-, Y- and Z-directions of the
tribolever, the sensor is placed on a special support that can be easily excited
acoustically. By this means we can measure the resonance spectrum of the
cantilever (and the support). We measure the resonance spectrum for a series of
different small masses, placed on the backside of the pyramid. The spring
constants can then be determined by measuring the frequency shifts of the X-,
Y-, and Z-resonances as a function of added mass (Cleveland et al., 1993). A
typical calibration result is shown in Figure 4. The measured spring constants
were

$$k_x^{tribolever} = 1.689 \pm 0.033 \text{ N/m}$$
$$k_y^{tribolever} = 1.685 \pm 0.035 \text{ N/m}$$
$$k_z^{tribolever} = 10.31 \pm 0.123 \text{ N/m}$$

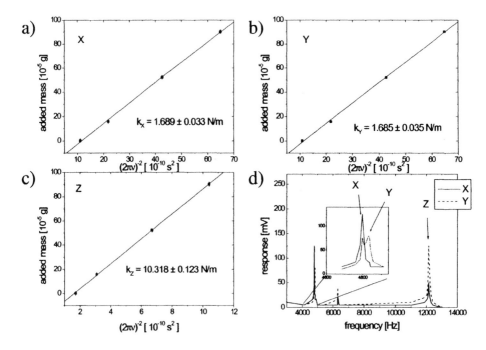

Figure 4. Inverse squared resonance frequencies as function of the added mass for the X- (a), Y- (b) and Z- (c) direction of the tribolever. The slope of each linear fit to the inverse squared frequencies gives the corresponding spring constant. From the inverse-squared frequency at zero added mass we obtain the effective mass. Graph (d) shows a typical resonance spectrum of the tribolever/support system without added masses. The unlabeled peaks do not depend on the added mass and are associated with resonances in the support.

The X- and Y- values[*] are close to the values from the finite element calculations. The Z-value is about a factor 2 lower than calculated. This large deviation between calculated and measured normal spring constants is due to a thin diaphragm that supports pyramid on the silicon chip. This diaphragm is the result of a wet etch step that forms a wide, recessed window to allow room for the detection fibers access to the pyramid.

Due to minor differences in the leg dimensions from sensor to sensor, the spring constants can vary by as much as a factor two. Therefore each sensor is calibrated separately prior to experiments.

---

[*] The slight non-degeneracy in X- and Y-resonance frequency, resulting from minute differences in the precise dimensions of the four legs, allows us to determine both lateral spring coefficients independently.

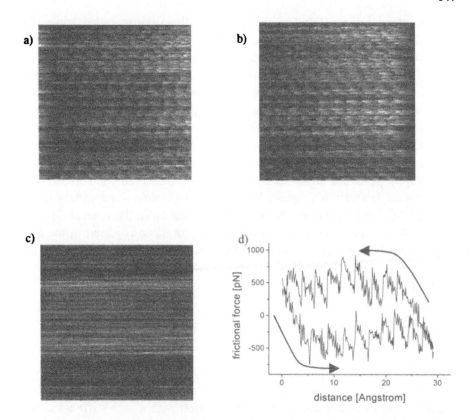

Figure 5. Test measurements for an electrochemically etched W tip on highly oriented pyrolitic graphite. Panel (a) shows a lateral force map in 'forward' scan direction, panel (b) the opposite scan direction. Panel (c) shows the topography image measured simultaneously. Graph (d) shows a line scan through images (a) and (b). Parameters of the images (a)-(d): Scan size 2.9×2.9 nm^2, normal force approx. –35 nN. The gray scale in panels (a) and (b) corresponds to a total force range of 1.08 nN.

4.2. TEST RESULTS

Figure 5 shows results of a measurement of a tungsten tip sliding on a highly oriented pyrolitic graphite (HOPG) sample at a constant (attractive) normal force of approximately –35 nN[*]. Figure 5(a) shows a map of the X-component

---

[*] The spring constant in the Z-direction for the tribolever used during the test measurements was not determined experimentally because the resonance frequency was difficult to excite acoustically in the simple setup used. Instead we estimated the normal spring constant here by calculating it from the measured leg dimensions. The spring constants for the used tribolever are k_x=5.31 N/m, k_Y=5.5 N/m and k_Z≈25 N/m.

of the lateral force parallel to the fast scan direction, recorded while the tip moved forward (from left to right). It shows variations in the frictional force with the unit cell periodicity of graphite. Figure 5(b) shows the friction when the tip was sliding in the opposite direction. A single forward and backward line are plotted in Figure 5(d), showing a frictional force loop with saw-tooth-like force variations, typical for atomic-scale stick/slip motion of the scanning tip (Mate et al., 1987). The total amplitude in the friction force loop is in the order of 1 nN. The present noise level on the lateral force measurement, without any data averaging, is below 50 pN. Finally, Figure 5(c) shows a micrograph of the topography (feedback signal in the Z-direction), measured simultaneously with the forward lateral force image. We cannot see atomic-scale variations in the topography. Note, however, that the force resolution in the normal direction is about 240 pN. The combination of the clear atomic corrugation in the friction signal and the absence of atomic-scale modulation of the Z feedback signal demonstrates the anticipated decoupling of lateral and normal force components in our force detection system.

5. Summary

We have constructed the first stage of a novel special-purpose friction force microscope. First tests in ambient conditions show a high sensitivity, better than 50 pN in both lateral directions, in combination with a high stiffness along the Z-direction. The addition of ultrahigh vacuum and the possibility to view and shape the tip will complete the instrument and allow it to measure contact and friction forces for completely defined geometries. This will make it possible to investigate a collection of fundamental questions that are presently inaccessible to standard FFM's: At which tip-sample separation distance does friction start? What is the dependence of the friction force on the relative orientation of the contacting surfaces? What is the relation between friction force, loading force and contact area? Does the phenomenon of superlubricity really occur? How does friction evolve from the single-atom nanoregime to the continuum regime of contacts consisting of thousands of atoms?

6. Acknowledgments

We are grateful to Dr. H.J. Hug (University of Basel) for his important contributions to the construction of the sample stage and the interferometer. This project is part of the research program of the "Stichting voor Fundamenteel

Onderzoek der Materie (FOM)", which is financially supported by the "Nederlandse Organisatie voor Wetenschappelijk Onderzoek (NWO)".

7. References

Bhushan, B., Israelachvili, J. and Landman, U. (1995), "Nanotribology: Friction, Wear and Lubrication at the Atomic Scale", *Nature* **374**, 607-616.

Carpick, R.W., Agraït, N., Ogletree, D.F. and Salmeron, M. (1996), "Measurement of Interfacial Shear (Friction) with an Ultrahigh Vacuum Atomic Force Microscope", *J. Vac. Sci. Technol. B* **14**, 1289-1295.

Cleveland, J.P., Manne, S., Bocek, D. and Hansma, P.K. "A Nondestructive Method for Determining the Spring Constant of Cantilevers for Scanning Force Microscopy", (1993), *Rev. Sci. Instrum.* **64**, 403-405.

Cross, G., Schirmeisen, A., Stalder, A., Grütter, P., Tschuddy, M. and Dürig, U. (1998), "Adhesion Interaction Between Atomically Defined Tip and Sample", *Phys. Rev. Lett.* **80**, 4685-4688.

Fink, H.-W. (1986), "Mono-Atomic Tips for Scanning Tunneling Microscopy", *IBM Journal of Research and Development* **30**, 460-465.

Fu, T.-Y., Tzeng, Y.-R. and Tsong, T.T. (1996), "Step Edge Diffusion and the Structure of Nanometer-Size Ir Islands on the Ir(111) surface", *Surf. Sci.* **366**, L691-L696.

Güntherodt, H.-J., Ansilmetti, D. and Meyer, E. (eds.), (1995), *Forces in Scanning Probe Microscopy*, NATO ASI Series E, Vol. 286, Kluwer Academic Publ., Dordrecht.

Hirano, M., Shinjo, K., Kaneko, R. and Murata Y. (1997), "Observation of Superlubricity by Scanning Tunneling Microscopy", *Phys. Rev. Lett.* **78**, 1448-1451.

Hug, H.J., Stiefel, B., Van Schendel, P.J.A., Moser, A., Martin, S. and Güntherodt, H.-J. (1999), "A Low Temperature Ultrahigh Vacuum Scanning Force Microscope", *Rev. Sci. Instrum.* **70**, 3625-3640.

Jarvis, S.P., Yamada, H., Kobayashi, K., Toda, A. and Tokumoto, H. (2000), "Normal and Lateral Force Investigation Using Magnetically Activated Force Sensors", *Appl. Surf. Sci.* **157**, 314-319.

Kageshima, M., Ogiso, H., Nakano, S., Lantz, M.A. and Tokumoto, H. (1999), "Atomic Force Microscopy Cantilevers for Sensitive Lateral Force Detection", *Jpn. J. Appl. Phys.* **38**, 3958-3961.

Kellogg, G.L. (1994), "Field Ion Microscope Studies of Single-Atom Surface Diffusion and Cluster Nucleation on Metal Surfaces", *Surf. Sci. Rept.* **21**, 1-88.

Mate, C.M., McClelland, G.M., Erlandsson, R. and Chiang, S. (1987), "Atomic-Scale Friction of a Tungsten Tip on a Graphite Surface", *Phys. Rev. Lett.* **59**, 1942-1945.

Morita, S., Fujisawa, S. and Sugawara Y. (1996), "Spatially Quantized Friction with a Lattice Periodicity", *Surf. Sci. Rept.* **23**, 1-42.

Rabe, U., Janser, K. and Arnold, W. (1996), "Vibrations of Free and Surface-Coupled Atomic Force Microscope Cantilevers: Theory and Experiment", *Rev. Sci. Instrum.* **67**, 3281-3293.

Schwarz, U.D., Zwörner, O., Köster, P. and Wiesendanger, R. (1997), "Friction Force Spectroscopy In the Low-Load Regime With Well-Defined Tips" in *Micro/Nanotribology and Its Applications* (B. Bhushan ed.), NATO ASI Series E, Vol. 330, pp. 233-238, Kluwer Academic Publ., Dordrecht.

Sørensen, M.R., Jacobsen, K.W. and Stoltze, P. (1996), "Simulations of Atomic-Scale Sliding Friction", *Phys. Rev. B* **53**, 2101-2113.

Tsong, T.T. (1990), *Atom Probe Field Ion Microscopy*, Cambridge University Press, Cambridge.

Zijlstra, T., Heimberg, J.A., Van der Drift, E., Glastra van Loon, D., Dienwiebel, M., De Groot, L.E.M. and Frenken J.W.M. (2000), "Fabrication of a Novel Scanning Probe Device for Quantitative Nanotribology", *Sensors and Actuators A: Physical* **84**, 18-24.

INVESTIGATION OF THE MECHANICS OF NANOCONTACTS USING A VIBRATING CANTILEVER TECHNIQUE

U. D. SCHWARZ, H. HÖLSCHER, W. ALLERS, A. SCHWARZ AND
R. WIESENDANGER
*Institute of Applied Physics and Microstructure Research
Center, University of Hamburg, Jungiusstr. 11,
D-20355 Hamburg, Germany*

Abstract. A vibrating cantilever technique is presented, which allows the continuous measurement of the tip-sample interaction force $F_{int}(z)$ in the contact as well as in the non-contact region as a function of the tip-sample distance z. The method relies on the measurement of the frequency difference $\Delta f = f - f_0$ between the eigenfrequency f_0 of the free cantilever and the actual resonance frequency f of the cantilever, which is influenced by the tip-sample interaction potential.

From such frequency shift data, $F_{int}(z)$ can be reconstructed, as we will demonstrate with the example of a silicon tip vibrating near a graphite surface. The resulting $F_{int}(z)$-curves are subsequently used to extract parameters like the adhesion force F_{ad} or the point of contact z_C. A detailed comparison with suitable model interactions additionally opens an elegant way to investigate the mechanics of the nanocontact, which behaves in good approximation as expected from the so-called *Hertz-plus-offset* model.

1. Introduction

For an understanding of the frictional behavior of a macroscopic interface, which consists of a large number of individual nanocontacts, the study of the behavior of an isolated nanocontact is essential. Thus, one of the main research topics in the new field of *nanotribology* concentrates on the investigation of the atomic and nanometer-scale processes, which take place during sliding of such nanocontacts (Bhushan, 1997; Bhushan, 1999). Experimentally, such questions are mostly addressed by means of friction force

151

B. Bhushan (ed.),
Fundamentals of Tribology and Bridging the Gap between the Macro- and Micro/Nanoscales, 151–169.
© 2001 *Kluwer Academic Publishers.*

microscopy (FFM) (Mate et al., 1987; Marti et al., 1990; Meyer and Amer, 1990), where a nanometer-sized tip end moving on an atomically flat surface represents a model system for realistic nanocontacts.

Analyzing friction force microscopical data, however, it turned out that the mechanics of the nanocontact, i.e. the elastic behavior of the contact under load, is one of the decisive issues for the resulting frictional behavior (Hu et al., 1995; Bhushan and Kulkarni, 1996; Carpick et al., 1996; Meyer et al., 1996; Schwarz et al., 1997a, 1997b; Lantz et al., 1997; Enachescu et al., 1998). In particular, it has been found that the frictional force F_f scales as a function of the normal force F_n with $F_f \propto F_n^{2/3}$ in the case of spherical tips with tip radii not exceeding some tens of nanometers sliding on atomically flat surfaces (Schwarz et al., 1997a, 1997b; Enachescu et al., 1998). This tip/sample geometry is usually named *Hertzian contact* geometry (Johnson, 1985).[1] The observed $F_f(F_n)$-dependence can be readily explained with only two assumptions: (1) The assumption that friction is proportional to the effective contact area A (where the *shear stress* $S = F_f/A$ is independent from the mean pressure $p = F_n/A$), and (2) the validity of contact mechanical models, which are derived from continuum elasticity theory, in spite of the atomic-scale dimensions of the contact (Schwarz et al., 1997a, 1997b). For the case of a nanometer-sized Hertzian contact, continuum elasticity theory predicts

$$A = \pi \left(\frac{RF_n}{K} \right)^{2/3}. \tag{1}$$

Here, R is the radius of the tip, and

$$K = \frac{4}{3} \left(\frac{1 - \nu_1^2}{E_1} + \frac{1 - \nu_2^2}{E_2} \right)^{-1} \tag{2}$$

the so-called "effective elastic modulus" (with $E_{1,2}$ and $\nu_{1,2}$ as elastic moduli and Poisson's ratios of sphere and flat, respectively). Additionally, it should be noted that $F_n = F_l + F_{ad}$ consist of two parts, namely the externally

---

[1] Additional requirements for the validity of the $F_f \propto F_n^{2/3}$-dependence are (1) that the materials in contact are not too soft such as, e.g., many polymers or biological materials, and (2) that the surfaces in contact do not exhibit high surface energies. The latter case can often be found for clean crystals investigated in ultrahigh vacuum. Nevertheless, for contacts which do not fit into the parameter range sketched above [which is usually termed as the Derjaguin-Muller-Toporov limit (Derjaguin et al., 1975; Johnson, 1997)], other contact mechanics have to be applied than the "Hertz-plus-offset" model introduced below [see, e.g., Johnson (1997)]. However, also under such circumstances, the line of arguments remains the same as it will be developed in the following; the main advantage of choosing the Hertzian contact geometry in the DMT limit is the mathematically simple form of the resulting equations, which facilitates the subsequent discussion.

applied loading force F_l, which is in a friction force microscope determined by the deflection of the cantilever, and the adhesion force F_{ad}. Therefore, Eq. (1) has sometimes been named "Hertz-plus-offset" model to distinguish it from the original Hertz model, which does not include adhesion (Hertz, 1881; Johnson, 1985), as well as from the more precise Derjaguin-Muller-Toporov (DMT) formalism (Derjaguin et al., 1975).

Inserting Eq. (1) in the above definition of the shear stress leads to

$$F_f = \pi S \left(\frac{R}{K} \right)^{2/3} F_n^{2/3} = \tilde{\mu} R^{2/3} F_n^{2/3}. \tag{3}$$

If the radius of the tip is known [i.e. by a separate electron microscopical inspection of the tip apex (Schwarz et al., 1997c)], we can determine $\tilde{\mu} = \pi S / K^{2/3}$ from an FFM experiment, since such set-ups measure F_f as a function of F_n. The factor $\tilde{\mu}$, which can be regarded as a kind of effective friction coefficient for nanometer-sized Hertzian contacts, obviously depends on S and K simultaneously and thus represents a combination of the intrinsic *frictional* and *elastic* properties of tip and sample. For an extraction of the fundamental frictional behavior of the nanocontact, an independent measurement of the effective elastic modulus K is required, since the actual value of K might for a nanometer-sized tip/sample geometry differ from the theoretical value of K calculated from inserting the corresponding bulk values in Eq. (2). Additionally, a more direct proof for the validity of contact mechanical models than that described in the above paragraph would be desirable.

Since a force microscope is a force sensing instrument, the most obvious method to check the contact mechanical predictions and to determine the effective value of K would be to measure the effective deformation of the tip/sample contact δ in the vertical z-direction as a function of the normal force F_n. The Hertz-plus-offset model predicts in this case (cf. Landau and Lifschitz, 1962)

$$\delta = \frac{F_n^{2/3}}{K^{2/3} R^{1/3}}. \tag{4}$$

A reliable measurement of δ as a function of F_n is, however, only difficult to realize due to several reasons:

1. In a force microscope, the normal force F_n is set by choosing an appropriate deflection of the cantilever. Consequently, for an increment of F_n as small as 1 nN with a typical cantilever possessing a spring constant of 0.1 N/m, the relative position of the sample with respect to the cantilever holder must be diminished by 10 nm. This causes not only an unwanted lateral movement of the tip apex which is not included in the theoretical model, but also a significant additional bending of the cantilever, which disturbs and, in most cases, even dominates the effects

due to the deformation of tip and sample in the optical measurement of the cantilever deflection. As a consequence, much larger *apparent* deformations are measured than they are actually occurring (Mazeran and Loubet, 1997). On the other hand, using stiffer cantilevers makes the signal-to-noise ratio worse (Dürig et al., 1992; Jarvis et al., 1996b).

2. Second, the measurement of the deformation of the tip/sample contact at very low loading forces is complicated by the so-called "jump to contact" (JTC) of the tip upon approach to the sample surface due to attractive surface forces (Burnham et al, 1989, 1991; Jarvis and Tokumoto, 1997). Usually, only a limited part of the "overjumped" distance range is accessible during cantilever retraction.

3. Finally, the non-linearity of the optical sensor for the measurement of the cantilever deflection might complicate exact comparisons with the $\delta \propto F_n^{2/3}$-law introduced above.

A possible method to overcome at least parts of the problem described above is to mount a small magnet at the end of the cantilever (Jarvis et al., 1996a; Mazeran and Loubet, 1997). Then, by adjusting an external magnetic field, a certain normal force F_n can be applied without the need to change the relative distance between the sample surface and the cantilever holder. Using an additional feedback, even the JTC can be prevented (Jarvis et al., 1996b). In this article, however, we will present an alternative approach by means of a vibrating cantilever technique, which allows the continuous determination of the tip-sample interaction force as a function of the vertical position z in contact as well as in the non-contact regime with high accuracy and excellent signal-to-noise ratio.

2. General principle of the vibrating cantilever technique

In the above section, we have seen that a reliable measurement of the deformation of the tip-sample contact as a function of the normal force is problematic not only due to calibration problems; an additional disadvantage is the JTC which usually prevents any measurement at very low loading forces as well as in the non-contact regime close to the sample surface. It has been shown, however, that it is possible to approach the tip at any distance to the surface without JTC by vibrating the cantilever with a large enough amplitude (Giessibl, 1997). Unfortunately, due to the oscillation of the cantilever, we lose the possibility to directly detect the tip-sample interaction force by measuring the cantilever deflection, and an alternative detection principle has to be applied. For the measurements presented here, which have been performed in ultrahigh vacuum to ensure clean conditions at the tip-sample interface, we chose the so-called *frequency modulation* (FM) technique introduced by Albrecht et al. (1991). This detection scheme is to-

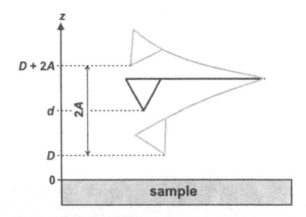

Figure 1. Scheme which illustrates the notations used in this work. The cantilever oscillates with the amplitude A, which is defined as the distance between the upper and the lower turning point of the oscillation divided by 2. During each cycle, the tip approaches the sample to the nearest tip-sample distance D. The distance between the tip and the sample at the point where the cantilever is undeflected is called the support-sample distance d. Please note that under the influence of an attractive tip-sample interaction, the oscillation range $[D, D + 2A]$ moves somewhat closer to the sample surface compared with the undisturbed case, i.e. $D < d - A$ for $F_{int}(z) < 0$ within $[D, D + 2A]$.

day routinely used in dynamic force microcopy (DFM), which is also known as non-contact atomic force microscopy (NC-AFM).[2]

The principle of FM detection and the notations used in this work are shown in Fig. 1. A cantilever (spring constant c) is vibrated with a fixed *resonance amplitude A* at its *resonance frequency f*. This resonance frequency is different from the *eigenfrequency f_0* of the free cantilever due to the influence of the interaction force between the tip and sample and changes while changing the resonance amplitude A or the distance d between the sample and the tip. Since this distance is varied in the experiment by moving the position of the cantilever support relative to the sample surface, d will be called the *support-sample distance* in the following. Additionally, the tip-sample distance at the point of closest approach is denoted as D.

The reason for the shift of the cantilever resonance frequency can be easily understood by looking at the potentials plotted in Fig. 2. If the cantilever is far away from the sample surface, the tip moves in a parabolic potential $V_{spring}(z) = c(z - d)^2/2$ (dotted line), and its oscillation is harmonic. In such a case, the tip motion is sinusoidal and the resonance frequency is given by the eigenfrequency f_0 of the cantilever. If, however, the support-sample distance d is reduced, the potential which determines the tip oscillation is modified and given by an effective potential V_{eff} (solid line)

---

[2]For an overview on the current status of the developments in NC-AFM see, e.g., Morita and Tsukada, 1999, or Bennewitz et al., 2000.

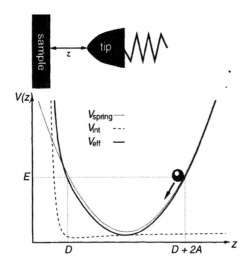

Figure 2. Graph illustrating the origin for the frequency shift observed in dynamic force microscopy (see text). The ball represents the tip, which moves in the effective potential (solid line).

represented by the sum of the parabolic potential and the tip-sample interaction potential V_{int} (dashed line). This effective potential differs from the original parabolic potential and shows an asymmetric shape, contrary to the symmetric shape of the parabolic potential. Consequently, the resulting tip oscillation becomes anharmonic, and the resonance frequency of the oscillation now depends on the oscillation amplitude A. Since the effective potential experienced by the tip changes with the support-sample distance d, the frequency shift Δf depends on both parameters.

3. Determination of tip-sample interactions

3.1. ANALYSIS OF FREQUENCY SHIFT VERSUS AMPLITUDE CURVES

After the above analysis of the general principle of the vibrating cantilever technique, this section 3 will be devoted to the central question addressed in this article, namely how information on the tip-sample interaction potential can be recovered from the measurement of the frequency shift.[3] Since Δf depends only on A, d, $V_{int}(z)$, and $V_{spring}(z)$, one might expect that it should be possible for a given cantilever/sample system (i.e. for constant c and f_0 during the measurement) to reconstruct $V_{int}(z)$ either if Δf is recorded as a function of d for constant A or, vice versa, as a function of A keeping

---

[3]For the following considerations, we will always assume undamped oscillation of the cantilever. In UHV, the dissipated energy is even if tip and sample are in slight contact in most cases small compared with the total energy stored in the system as long as the oscillation amplitudes are significantly larger than the range of the tip-sample interaction. Thus, the error due to dissipation is small; on the other hand, only *conservative* forces can be determined from such frequency shift data (Gotsmann et al., 1999b; Hölscher et al., 2000a).

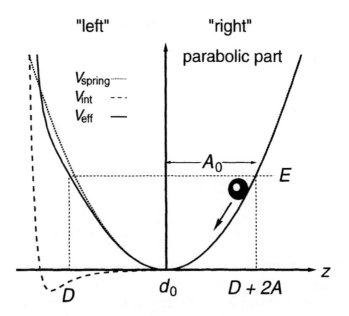

Figure 3. The principle of the mathematical analysis of the $\Delta f(A)$-curves. The ball represents the tip, which moves in the effective potential $V_{\text{eff}}(z)$. Analogous to Fig. 2, the effective potential represents the sum from the unperturbed parabolic potential of the cantilever $V_{\text{spring}}(z)$ and the tip-sample potential $V_{\text{int}}(z)$; the point d_0 is defined as $d_0 := \min\{V_{\text{eff}}(z)\}$. In the experiment, the oscillation amplitude A = (upper turning point - lower turning point)/2 is measured, which is different from the distance A_0 indicated in the figure (cf. the caption of Fig. 1).

d fixed. Unfortunately, the corresponding mathematical procedure – which can be addressed both numerically (Gotsmann et al., 1999a) or analytically (Hölscher et al., 1999; Dürig, 2000) – is not straightforward. The difficulty is that $f = 1/T$ is determined by the integral

$$T(E) = \sqrt{2m} \int\limits_{D}^{D+2A} \frac{\mathrm{d}z}{\sqrt{E - V_{\text{eff}}(z)}}, \tag{5}$$

where T is the oscillation time, $m = c/(2\pi f_0)^2$ the effective mass of the cantilever, and E represents the total energy stored in the system. In order to determine $V_{\text{eff}}(z)$, Eq. (5) has to be inverted. This is generally not possible because $z(V_{\text{eff}})$ is a two-valued function, i.e. every value of V_{eff} has two associated solutions for z (see Fig. 3).

The problem can be solved by using a trick, which has been described by Hölscher et al. (1999): Since typically occurring tip-sample interaction potentials can be neglected for large enough distances from the sample surface in comparison with the unperturbed parabolic potential of the cantilever, the "right" side of the effective potential in Fig. 3 with $z > d_0$, where $d_0 := \min\{V_{\text{eff}}(z)\}$, may be replaced by the parabolic potential $c(z - d_0)^2/2 \approx c(z - d)^2/2 = V_{\text{spring}}(z)$. Then, the integral can be divided into two parts (a "left" and a "right" part):

$$T(E) \approx \sqrt{2m} \int_D^{d_0} \frac{dz}{\sqrt{E - V_{\text{eff}}(z)}} + \sqrt{2m} \int_{d_0}^{D+2A} \frac{dz}{\sqrt{E - c(z - d_0)^2/2}}. \qquad (6)$$

Since the oscillation time of the parabolic potential depends only on c, but not on A, Eq. (6) can be written as

$$T(E) \approx \sqrt{2m} \int_D^{d_0} \frac{dz}{\sqrt{E - V_{\text{eff}}(z)}} + \tfrac{1}{2}T_0, \qquad (7)$$

where T_0 is the oscillation time of the unperturbed cantilever.

Within the interval $[D, d_0]$, V_{eff} is single-valued, and Eq. (7) can now be rewritten according to a procedure described by Landau and Lifschitz (1962):

$$D(V_{\text{eff}}) = d_0 + \frac{1}{\pi\sqrt{2m}} \int_0^{V_{\text{eff}}} \frac{\tfrac{1}{2}T_0 - T(E')}{\sqrt{V_{\text{eff}} - E'}} \, dE'. \qquad (8)$$

At this stage, it has to be considered that frequencies and amplitudes are measured instead of oscillation times and energies in the experiment. Thus, $T(E)$ and E have to be substituted by $\Delta f(A_0)$ and A_0 (see Fig. 3 for the definition of A_0), and we obtain after some transformations ($E' = \frac{1}{2} \cdot cA_0'^2$, $V_{\text{eff}} = \frac{1}{2} \cdot cA_0^2$, $T = \frac{1}{f} = \frac{1}{f_0 + \Delta f}$, $T_0 = \frac{1}{f_0}$)

$$D(A_0) = d_0 - \int_0^{A_0} \frac{f_0 - \Delta f(A_0')}{f_0 + \Delta f(A_0')} \frac{A_0'}{\sqrt{A_0^2 - A_0'^2}} \, dA_0'. \qquad (9)$$

Using this result, we can calculate the desired value of the tip-sample potential $V_{\text{int}}(z)$ at the position $z = D$:

$$V_{\text{int}}(D) = \tfrac{1}{2} \cdot c(A_0(D)^2 - (d_0 - D)^2). \qquad (10)$$

If this method is applied to frequency shift versus amplitude curves, we additionally have to take into account that in a real experiment, the oscillation amplitude A is measured, which is different from the distance A_0 (cf. Fig. 3). Therefore, we implemented a simple search algorithm which varies A_0 until $A_0 = D(A_0) + 2A - d_0$.

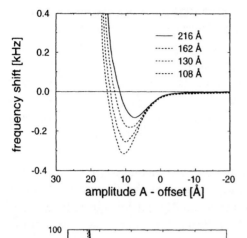

Figure 4. Four different frequency shift versus amplitude curves acquired at different distances, but the same silicon cantilever on a graphite sample at $T = 294$ K. The curves are individually shifted along the z-axis in order to fit all into the same graph; the subtracted offset is 216, 162, 130, and 108 Å, respectively.

a)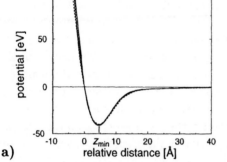

b)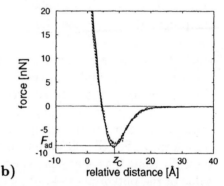

Figure 5. a) The reconstructed tip-sample potential, calculated according to the procedure described in the text from the experimental data shown in Fig. 4. The reproducibility is good in spite of the different shapes of the $\Delta f(A)$-curves displayed in Fig. 4. The zero point of the z-axis has been arbitrarily chosen. b) The corresponding tip-sample forces; the zero point of the z-scale is identical with the one in a). Assuming a contact point of tip and sample at the position where the force has its minimum (see Sec. 4 for details), it follows that the tip-sample force can be recorded continuously and with high resolution from large tip-sample distances down to an interaction range where already intimate tip-sample contact (and thus significant elastic deformation) has occurred.

Fig. 4 shows exemplarily four $\Delta f(A)$-curves recorded on a graphite(0001) surface at different tip-sample distances using a rectangularly-shaped cantilever made from monocrystalline silicon with a spring constant $c = 38$ N/m and an eigenfrequency $f_0 = 177$ kHz. The measurements were performed in ultrahigh vacuum and at room temperature with a home-built atomic force microscope, which has been described in detail by Allers et al. (1998). The graphite sample was cleaved *in situ*; zero bias voltage has been applied between tip and sample during data acquisition. Obviously, larger amplitudes A lead to smaller minimum values of Δf. Such a behavior is expected because the total time that the tip moves under the influence of the sample potential is reduced.

From these curves, the tip-sample potential has been reconstructed according to the mathematical procedure described above (see Fig. 5a). We find an excellent agreement of all curves, regardless their different tip-sample distances.[4] Due to the good signal-to-noise ratio, we can additionally calculate the corresponding tip-sample interaction force (Fig. 5b). From this plot, the adhesion force of the contact $F_{ad} = \min\{F_{int}(z)\}$ can be determined to $F_{ad} \approx 9$ nN. In section 4, we will see that z_C, which is defined by $F_{ad} = F_{int}(z_C)$, can additionally be identified as the *point of contact* between tip and sample surface.

Finally, we note that the tip-sample potential displayed in Fig. 5a reaches a minimum value of $\approx$ -40 eV, which exceeds by far the typical binding energies within a solid body and can consequently not be caused by a single atom at the tip apex. Comparing the difference between the minimum of the potential at z_{min} and the point of contact z_C determined in the force plot of $z_C - z_{min} \approx 4.5$ Å, it is evident that the potential minimum is only reached *after already significant deformation of the tip apex has occurred*. From the reproducibility of the force measurements, we can conclude that this deformation is purely *elastical*, i.e. after certain plastic deformations occurring at the beginning of the measurement series due to the intimate contact between tip and sample during each cycle, a stable tip apex has been formed. For this situation, the Hertz theory predicts a contact area of typically some tens of atoms, which readily explains the observed minimum value of the potential of -40 eV.

3.2. ANALYSIS OF FREQUENCY SHIFT VERSUS DISTANCE CURVES

As argued at the beginning of the previous section, a reconstruction of the tip-sample interaction potential should also be possible by measuring Δf as a function of d while keeping A constant. A possible procedure for this case has indeed already been suggested by Dürig (1999a). Employing variational methods and Fourier expansion of the tip motion, he could derive general expressions which at this first stage relate arbitrary tip-sample interaction forces F_{int} to frequency shifts Δf. For interactions with a range much shorter than the vibration amplitude, i.e. with essentially the same restriction as it was made in Sec. 3.1, Δf can then be expressed as

$$\Delta f = \frac{1}{\sqrt{2\pi}} \frac{f_0}{cA^{3/2}} \int\limits_{D}^{\infty} \frac{F_{int}(z)}{\sqrt{z-D}} dz. \tag{11}$$

[4]It has, however, to be noted that the curves for different tip-sample distances were individually shifted along the z-axis until they showed maximum overlap since no absolute values for d_0 can be determined from the experiment.

The advantage of this integral equation is that it can again be inverted, which leads to the following analytic expression for the interaction potential (Dürig, 1999a)

$$V_{\text{int}}(z) = \sqrt{2} \frac{cA^{3/2}}{f_0} \int\limits_D^{\infty} \frac{\Delta f(z)}{\sqrt{z-D}} dz. \tag{12}$$

We can thus directly calculate the interaction force $F_{\text{int}}(z)$ from frequency shift versus distance data by numerically solving the equation

$$F_{\text{int}}(z) = \sqrt{2} \frac{cA^{3/2}}{f_0} \frac{\partial}{\partial D} \int\limits_D^{\infty} \frac{\Delta f(z)}{\sqrt{z-D}} dz. \tag{13}$$

An application of the above described method to experimental data is presented in Fig. 6. The measurements were carried out with the same instrument as already introduced in Sec. 3.1; in this case, however, sample was cooled down to $T = 80$ K. The rectangularly-shaped cantilever used for this study was of the same type as the one in Sec. 3.1 with $c = 38$ N/m and $f_0 = 171$ kHz. The tip was sputtered *in situ* with Argon prior to the measurements. The graphite sample was cleaved *in situ* at room temperature at a pressure below 10^{-9} mbar and immediately inserted into the cooled microscope. The experimental data was again recorded with a zero bias voltage between tip and sample.

Exemplary $\Delta f(z)$-curves obtained with oscillation amplitudes between 54 Å and 180 Å are displayed in Fig. 6. All curves show the same typical behavior already known from Fig. 4: If the cantilever approaches the sample surface, the frequency shift decreases and reaches a minimum. With a further reduction of the support-sample distance, the frequency shift increases again and shifts to positive values.

The interaction force as determined from these data applying Eq. (13) is plotted in Fig. 6b. The good agreement of all data points, regardless their different vibration amplitudes, illustrates the effectiveness of the method.

4. Quantitative analysis by comparison with model interactions

With the extraction of complete data sets of the tip-sample interaction force as a function of the distance, we have everything that is needed for a detailed investigation of the mechanical properties of the tip/sample system in contact as well as in the non-contact regime. Taking, e.g., the data displayed in Fig. 6b as basis for closer analysis, we first note that the minimum force, which corresponds to the adhesion force of the contact, is at $F_{\text{ad}} = -6.7$ nN. If a perfectly spherical tip with known tip radius would

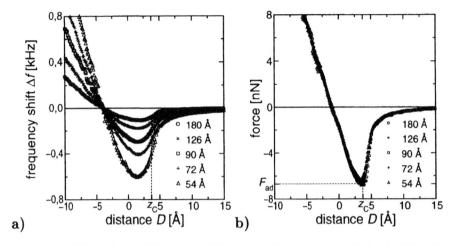

Figure 6. a) Experimental frequency shift versus distance curves obtained with a silicon tip and a graphite(0001) sample and displayed by symbols for a total of five different oscillation amplitudes A. All curves are shifted along the z-axis; the zero point has been chosen in accordance to the force law Eq. (16) defined below. b) The tip-sample force calculated from the data displayed in a) applying Eq. (13). From this representation, the adhesion force $F_{ad} := \min\{F_{int}(z)\} = -6.7$ nN and the point of contact $z_C = 3.7$ Å can be determined (cf. Secs. 3.1 and 4). Comparison with a) reveals that contact occurs in the $\Delta f(z)$-curves significantly *before* the minimum of the frequency has been reached during approach.

have been used (Schwarz 1997a, 1997c), this value would allow the determination of the effective surface energy (Dupré energy)[5] γ of the contact applying of the equation $\gamma = F_{ad}/2R$, which has been derived within the DMT theory (Derjaguin et al., 1975). This has, however, not yet been the case in these preliminary experiments; nevertheless, assuming a typical order of magnitude for R of about 10-30 nm at least leads to reasonable values for γ around 100-300 mJ/m^2.

To obtain more information on the mechanics of the contact, it is necessary to compare the experimental data with an appropriate model interaction force F_{int}. For this purpose, different types of analytic expressions for the interaction law have been proposed by various authors (cf., e.g., Giessibl, 1997; Sokolov et al., 1997; Dürig, 1999b; Guggisberg et al., 2000; Hölscher et al., 2000). An important issue in this context is that the non-contact and the contact regime have to be analyzed separately, as we will see below.

If tip and sample are far away from each other, the macroscopic properties of tip and sample are most prominent, and the tip-sample force is given by long-range van der Waals forces F_{vdW}. The mathematical form of

---

[5]γ is actually defined as $\gamma = \gamma_1 + \gamma_2 - \gamma_{12}$, where $\gamma_{1,2}$ are the surface energies of tip and sample, respectively, and γ_{12} represents the energy of the tip-sample interface.

a corresponding force law depends on the exact geometry of the tip and the sample surface; formulas for a variety of different tip shapes are given by Giessibl (1997) or Guggisberg et al. (2000). Here, we will choose the simplest geometry as a first approximation to the realistic case, namely the geometry of a sphere over an atomically flat surface. Under these circumstances, F_{vdW} can be written as

$$F_{vdW} = \frac{A_H R}{6z^2}, \tag{14}$$

where A_H represents the Hamaker constant. If tip and sample come closer together, the interatomic forces between the foremost atoms of the tip apex and the sample surface atoms become more important. This short range force between the foremost atoms of the tip and nearest sample surface atoms is for non-reactive surfaces such as graphite often considered by the force of a Lennard-Jones potential

$$F_{LJ} = \frac{12E_0}{r_0} \left(\left(\frac{r_0}{z} \right)^{13} - \left(\frac{r_0}{z} \right)^7 \right), \tag{15}$$

where E_0 represents the binding energy and r_0 the equilibrium distance of the Lennard-Jones potential (Giessibl, 1997; Sokolov et al., 1997). Consequently, the effective non-contact force between tip and sample can be described by the sum of these two forces and is given by

$$F_{NC} = F_{vdW} + F_{LJ}. \tag{16}$$

With further approach of the tip to the sample, the two bodies come into contact and the *elasticity* of both bodies has to be taken into account. It can be shown that the elastic contact force F_C between a sample and a regularly curved (but not necessarily spherical) tip can be described by (Landau and Lifschitz, 1975)

$$F_C = g_0(z_0 - z)^{3/2} + F_0 \quad \text{for} \quad z < z_0, \tag{17}$$

where z_0 is the point of contact and g_0 a constant value. F_0 has to be added in order to consider an eventual offset due to non-contact forces.

A fit of both F_{NC} as well as F_C to the experimental data presented in Fig. 6b is shown in Fig. 7. It is obvious that F_{NC} (dashed-dotted line) describes the tip-sample interaction force quite accurately, but only to its minimum at $z = z_C$.[6] The parameters used are $A_H R = 2.4 \times 10^{-27}$ Jm,

---

[6]The zero point of the z-scale is chosen in such a way that maximum agreement between F_{NC} and the experimental data is achieved.

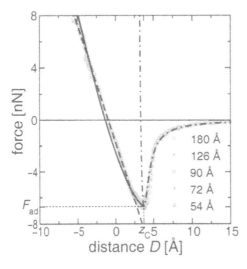

Figure 7. Comparison of the experimenally obtained data presented in Fig. 6 with the model forces described in the text. Dashed-dotted line: Best fit for F_{NC} Eq. (16); solid line: best fit for F_C Eq. (17); dashed line: fit for a linear force law for z-values smaller than $z = 2.2$ Å. It is obvious that F_{NC} shows large deviations from the experimentally observed data points in the contact region due to the influence of elastic deformation. z_C marks the border between the contact- and the non-contact regime.

$E_0 = 3$ eV, and $r_0 = 3.4$ Å. The latter value can be compared with the equilibrium distance between two individual graphite layers of 3.44 Å and thus seems to be of a reasonable order. For z-values smaller than $z = z_C$, however, a good agreement is obtained using the contact force law F_C with $g_0 = 5.8 \cdot 10^5$ N/m$^{3/2}$, $z_0 = z_C$, and $F_0 = F_{ad} = \min\{F_{NC}(z)\}$.

The good agreement between experiment and simulation demonstrates that the introduced interaction models are suitable to describe the tip-sample force. For the present purpose, the elastic part of the interaction force is of special interest. Calculations performed solely with a combination of short- and long-range potentials *without* consideration of elastic forces never showed an acceptable quality of the corresponding fits, but always resulted in large deviations between simulation and experimental data from a certain distance to the surface on (see Fig. 7). This demonstrates that *elastic forces* dominate over all other interactions from the distance z_0 on, which can be identified in agreement with its definition in Eq. (17) as the *point of contact* (Hölscher et al., 2000b). Thus, since $z_C = z_0$, we see that our preliminary interpretation of z_C in this sense (cf. Sec. 3) has been correct.

In order to derive additional information from $F_{int}(z)$, tips featuring a well-defined tip apex have to be used which allow a further specification of the contact force law Eq. (17). In the limit where the Hertz-plus-offset model applies, $g_0 = K\sqrt{R}$ is found by comparison of Eq. (17) with Eq. (4). Application of this formula would lead to $K = 5.8$ GPa with the assumption of a tip radius of 10 nm, as it is specified by the manufacturer[7]. This value

---

[7]Nanosensors, Aidlingen, Germany.

is somewhat lower than the value calculated with Eq. (2): Taking $E_{silicon} =$ 129.2 GPa, $\nu_{silicon} = 0.279$, $E_{graphite} = 36.4$ GPa, and $\nu_{graphite} = 0.012$ from the literature, we find $K = 38.5$ GPa. However, considering the fact that our guess for R of $R = 10$ nm might be very unprecise and that *no* tip with well-defined spherical tip apex has been used for these preliminary experiments, as we have already stated above, this discrepancy can be easily explained.

An other point which might have a certain influence in this context is that the "effective" value of K which applies for a tip/sample contact with nanometer dimensions might differ from the value which accounts if macroscopic properties are described. The structure of the tip end, e.g., could be somewhat disordered, making the contact "softer". Additionally, the geometry of the tip, which features a sharp apex, might lead to apparently softer contacts, since atoms located close to the tip apex could possibly more easily be displaced than atoms situated well within the bulk. Nevertheless, it can be expected that if tips with a well-defined spherical tip end are used as, e.g., the ones introduced by Schwarz et al. (1997c), reliable values for the "effective K's" of the contact can be determined from measurements as described above. It will be the aim of future work to relate such results with the corresponding values of K derived from their bulk properties applying Eq. (2).

Finally, we should note that the proposed technique is also useful to check whether the Hertz-plus-offset model applies or if deviations occur due to the atomic-scale dimensions of the contact. For our data, a linear force law fits equivalently well as the Hertz-plus-offset model Eq. (17) for indentations larger than $\approx$ 1-2 Å (see the dashed line in Fig. 7), which leads to a contact stiffness of approximately 18 N/m. Consequently, it might be interesting to compare future experimental results obtained with well-defined tips with more sophisticated models of contact mechanics as, e.g., the theories proposed by Muller et al. (1980, 1982), Burgess et al. (1987), or Maugis (1992).

5. Conclusions

To summarize, we introduced a vibrating cantilever technique and discussed mathematical procedures to reconstruct the complete tip-sample interaction force $F_{int}(z)$ and the corresponding potential $V_{int}(z)$ in both the contact and the non-contact regime. Experimentally, this is realized by measuring the deviation $\Delta f = f - f_0$ of the actual resonance frequency f of the cantilever in the force field of the surface from the eigenfrequency f_0 of the undisturbed cantilever as a function of the oscillation amplitude A or the tip-sample distance d, respectively. After acquisition, the obtained data can be converted to $F_{\text{int}}(z)$ or $V_{\text{int}}(z)$.

Experiments performed with a silicon tip on a graphite sample in ultrahigh vacuum were presented which show the potential of the method: Among other parameters, the adhesion force F_{ad} and the point of initial contact between tip and sample z_C could be determined with great accuracy. Additionally, by analysis of the interaction force in the contact regime, direct evidence for the validity of the Hertz-plus-offset model has been given. Future experiments carried out with tips featuring well-defined spherical apexes are expected to provide reliable values for surface energies γ and effective elastic moduli K as well as to enable a detailed verification of contact mechanical models.

Acknowledgements

We would like to thank B. Gotsmann for helpful discussions. Financial support from the Deutsche Forschungsgemeinschaft (Grants No. SCHW 641/1-1 and WI 1277/13-1), the Graduiertenkolleg "Physik nanostrukturierter Festkörper", as well as from the BMBF (Grant No. 13N 7694/8) is gratefully acknowledged.

References

Albrecht, T. R., Grütter, P., Horne, D., and Rugar, D. (1991), "Frequency modulation detection using high-Q cantilevers for enhanced force microscope sensitivity", *J. Appl. Phys.* **69**, 668-673.

Allers, W., Schwarz, A., Schwarz, U. D., and Wiesendanger, R. (1998), "A scanning force microscope with atomic resolution in ultrahigh vacuum and at low temperatures", *Rev. Sci. Instrum.* **69**, 221-225.

Bennewitz, R., Gerber, Ch., and Meyer, E., Eds. (2000), "Proceedings of the Second International Workshop on Noncontact Atomic Force Microscopy", *Appl. Surf. Sci.* **157**.

Bhushan, B., and Kulkarni, A. V. (1996), "Effect of normal load on microscale friction measurements", *Thin Solid Films* **278**, 49-56.

Bhushan, B., Ed. (1997), *Micro/Nanotribology and Its Applications*, NATO ASI Series E: Applied Sciences, Vol. 330, Kluwer Academic Publishers, Dordrecht, The Netherlands.

Bhushan, B., Ed. (1999), *Handbook of Micro/Nanotribology*, Second Edition, CRC Press, Boca Raton, Florida.

Burgess, A. K., Hughes, B. D., and White, L. R. (1987), "Adhesive Contact of Elastic Solids", unpublished.

Burnham, N. A., and Colton, R. J. (1989), "Measuring the nanomechanical properties and surface forces of materials using an atomic force microscope", *J. Vac. Sci. Technol. A* **7**, 2906-2913.

Burnham, N. A., Colton, R. J., and Pollock, H. M. (1991), "Interpretation issues in force microscopy", *J. Vac. Sci. Technol. A* **9**, 2548-2556.

Carpick, R. W., Agrait, N., Ogletree, D. F., and Salmeron, M. (1996), "Measurement of interfacial shear (friction) with an ultrahigh vacuum atomic force microscope", *J. Vac. Sci. Technol. B* **14**, 1289-1295.

Derjaguin, B. V., Muller, V. M., Toporov, Y. P. (1975), "Effect of Contact Deformations on the Adhesion of Particles", *J. Colloid Interface Sci.* **53**, 314-326.

Dürig, U., Züger, O., and Stalder, A. (1992), "Interaction force detection in scanning probe microscopy: Methods and applications", *J. Appl. Phys.* **72**, 1778-1798.

Dürig, U. (1999a), "Relations between interaction force and frequency shift in large-amplitude dynamic force microscopy", *Appl. Phys. Lett.* **75**, 433-435.

Dürig, U. (1999b) "Conservative and Dissipative Interactions in Dynamic Force Spectroscopy", *Surf. Interface Anal.* **27**, 467-473.

Dürig, U. (2000), "Extracting interaction forces and complementary observables in dynamic probe microscopy", *Appl. Phys. Lett.* **76**, 1203-1205.

Enachescu, M., van den Oetelaar, R. J. A., Carpick, R. W., Ogletree, D. F., Flipse, C. F. J., and Salmeron, M. (1998), "Atomic Force Microscopy Study of an Ideally Hard Contact: The Diamond(111)/Tungsten Carbide Interface", *Phys. Rev. Lett.* **81**, 1877-1880.

Giessibl, F. J. (1997), "Forces and frequency shifts in atomic-resolution dynamic-force microscopy", *Phys. Rev. B* **56**, 16010-16015.

Gotsmann, B., Anczykowski, B., Seidel, C., and Fuchs, H. (1999a), "Determination of tip-sample interaction forces from measured dynamic force spectroscopy curves", *Appl. Surf. Sci.* **140**, 314-319 (1999).

Gotsmann, B., Seidel, C., Anczykowski, B., and Fuchs, H. (1999b), "Conservative and dissipative tip-sample interaction forces probed with dynamic AFM, *Phys. Rev. B* **60**, 11051-11061.

Guggisberg, M., Bammerlin, M., Loppacher, Ch., Pfeiffer, O., Abdurixit, A., Barwich, V., Bennewitz, R., Baratoff, A., Meyer, E., and Güntherodt, H.-J. (2000), "Separation of interactions by noncontact force microscopy", *Phys. Rev. B* **61**, 11151-11155.

Hertz, H. (1881), "Ueber die Berührung fester elastischer Körper", *J. Reine Angew. Math.* **92**, 156-171.

Hölscher, H., Allers, W., Schwarz, U. D., Schwarz, A., and Wiesendanger, R. (1999), "Determination of Tip-Sample Interaction Potentials by Dynamic Force Spectroscopy", *Phys. Rev. Lett.* **83**, 4780-4783 (1999).

Hölscher, H., Gotsmann, B., Schwarz, A., Allers, W., Schwarz, U. D., Fuchs, H., and Wiesendanger, R. (2000a), manuscript in preparation.

168

Hölscher, H., Schwarz, A., Allers, W., Schwarz, U. D., and Wiesendanger, R. (2000b) "Quantitative Analysis of dynamic-force-spectroscopy data on graphite(0001) in the contact and noncontact regimes", *Phys. Rev. B* **61**, 12678-12681.

Hu, J., Xiao, X.-D., Ogletree, D. F., and Salmeron, M. (1995), "Atomic scale friction and wear of mica", *Surface Science* **327**, 358-370.

Jarvis, S. P., Yamada, H., Ymamoto, S.-I., and Tokumoto, H. (1996a) "A new force controlled atomic force microscope for use in ultrahigh vacuum", *Rev. Sci. Instrum.* **67**, 2281-2285.

Jarvis, S. P., Yamada, H., Yamamoto, S.-I., Tokumoto, H., and Pethica, J. B. (1996b) "Direct measurement of interatomic potentials", *Nature* **384**, 247-249.

Jarvis, S. P., and Tokumoto, H. (1997), "Measurement and Interpretation of Forces in the Atomic Force Microscope", *Probe Microscopy* **1**, 65-79.

Johnson, K. L. (1985), *Contact Mechanics*, Cambridge University Press, Cambridge, United Kingdom.

Johnson, K. L. (1997), "Adhesion and friction between a smooth elastic spherical asperity and a plane surface", *Proc. R. Soc. Lond. A* **453**, 163-179.

Landau, L., and Lifschitz, E. M. (1962), *Lehrbuch der Theoretischen Physik, Band 1: Mechanik*, Akademie-Verlag, Berlin.

Landau, L., and Lifschitz, E. M. (1975), *Lehrbuch der Theoretischen Physik, Band 7: Elastizitätstheorie*, Akademie-Verlag, Berlin, pp. 34-39.

Lantz, M. A., O'Shea, S. J., Welland, M. E., and Johnson, K. L. (1997), "Atomic-force microscope study of contact area and friction on $NbSe_2$", *Phys. Rev. B* **55**, 10776-10785.

Maugis, D. (1992), "Adhesion of Spheres: The JKR-DMT Transition Using a Dugdale Model", *J. Coll. Interface Sci.* **150**, 243-269.

Mazeran, P.-E., and Loubet, J.-L. (1997), "Force modulation with a scanning force microscope: an analysis", *Tribol. Lett.* **3**, 125-132.

Mate, C. M., McClelland, G. M., Erlandsson, R., and Chiang, S. (1987), "Atomic-Scale Friction of a Tungsten Tip on a Graphite Surface", *Phys. Rev. Lett.* **59**, 1942-1945.

Marti, O., Colchero, J., and Mlynek, J. (1990), "Combined scanning force and friction microscopy of mica", *Nanotechnology* **1**, 141-144.

Meyer, G. and Amer, N. M. (1990), "Simultaneous measurement of lateral and normal forces with an optical-beam-deflection atomic force microscope", *Appl. Phys. Lett.* **57**, 2089-2091.

Meyer, E., Lüthi, R., Howald, L., Bammerlin, M., Guggisberg, M., Güntherodt, H.-J., Scandella, L., Gobrecht, J., Schumacher, A., and Prins, R. (1996), "Friction Force Spectroscopy", in *Physics of Sliding Friction* (B. N. J. Persson and E. Tosatti, eds.), pp. 349-367, NATO ASI Series E:

Applied Sciences, Vol. 311, Kluwer Academic Publishers, Dordrecht, The Netherlands.

Morita, S., and Tsukada, M., Eds. (1999), "Proceedings of the First International Workshop on Noncontact Atomic Force Microscopy", *Appl. Surf. Sci.* **140**.

Muller, V. M., Yushchenko, V. S., and Derjaguin, B. V. (1980), "On the Influence of Molecular Forces on the Deformation of an Elastic Sphere and Its Sticking to a Rigide Plane", *J. Coll. Interface Sci.* **77**, 91-101.

Muller, V. M., Yushchenko, V. S., and Derjaguin, B. V. (1980), "General Theoretical Consideration of the Influence of Surface Forces on Contact Deformations and the Reciprocal Adhesion of Elastic Spherical Particles", *J. Coll. Interface Sci.* **92**, 92-101.

Schwarz, U. D., Zwörner, O., Köster, P., and Wiesendanger, R. (1997a), "Quantitative analysis of the frictional properties of solid materials at low loads. I. Carbon compounds.", *Phys. Rev. B* **56**, 6987-6996.

Schwarz, U. D., Zwörner, O., Köster, P., and Wiesendanger, R. (1997b), "Quantitative analysis of the frictional properties of solid materials at low loads. II. Mica and germanium sulfide.", *Phys. Rev. B* **56**, 6997-7000.

Schwarz, U. D., Zwörner, O., Köster, P., and Wiesendanger, R. (1997c), "Preparation of probe tips with well-defined spherical apexes for quantitative scanning force spectroscopy", *J. Vac. Sci. Technol. B* **15**, 1527-1530.

Sokolov, I. Yu., Henderson, G. S., and Wicks, F. J. (1997), "The contrast mechanism for true atomic resolution by AFM in non-contact mode: quasi-non-contact mode?", *Surf. Sci. Lett.* **381**, L558-L562.

A SCANNING PROBE AND QUARTZ CRYSTAL MICROBALANCE STUDY OF C$_{60}$ ON MICA AND SILVER(111) SURFACES

T. COFFEY, M. ABDELMAKSOUD AND J. KRIM
North Carolina State University Physics Department Box 8202 Raleigh, NC 27695

Abstract.
   Experimental investigations of friction, lubrication, and adhesion at nanometer length scales have traditionally been performed by employing scanning probe microscopy (SPM), surface forces apparatus, (SFA), or quartz crystal microbalance (QCM) techniques. While collectively these techniques have yielded much useful information, their results have never been mutually cross-referenced. In order to perform such a study, we have investigated the changes in interfacial friction of toluene on single crystal substrates in both the presence and absence of C$_{60}$ adsorbed layers.

1. Introduction

Three techniques have traditionally been used to study friction, lubrication, and adhesion at nanometer length scales: scanning probe microscopy (SPM), which includes atomic force microscopy (AFM) and scanning tunneling microscopy (STM), the surface forces apparatus (SFA), and the quartz crystal microbalance (QCM). These techniques have given new insights into micro- and nanotribology. Each of these techniques, however, covers a different range of shear stresses, length scales, time scales, and sliding speeds, and their results have never been mutually cross-referenced. Such a cross-referencing would give a deeper understanding of how these very different techniques relate to one another and exactly how each technique measures friction. It would also enable better selection of the appropriate technique to use for studying any given system.
   In order to mutually cross reference the results of various nanotribological probes, we have performed AFM and QCM investigations of the changes in interfacial friction of toluene on single crystal substrates in both the presence and absence of C$_{60}$ adsorbed layers. This system was chosen so as to be able to correlate our results with previous SFA measurements. (Campbell *et al.*, 1996) Campbell and his colleagues reported that the C$_{60}$ adsorbed as monolayers on the mica surfaces immersed in liquid toluene. In the presence of the adsorbed layers, fluid flow between the two mica surfaces exhibited full-slip boundary conditions. For the pure toluene between the

171

B. Bhushan (ed.),
Fundamentals of Tribology and Bridging the Gap between the Macro- and Micro/Nanoscales, 171–176.
© 2001 *Kluwer Academic Publishers.*

mica surfaces, the fluid exhibited a typical no-slip boundary condition. It was also found, however, that the adsorbed C_{60} layers possessed unusually high fluidity, and they were easily pushed out of the way when the surfaces were approached slowly together.

2. AFM Experimental Details and Results

For our contact mode AFM measurements, we used a Molecular Imaging microscope with RHK control electronics and software and standard oxide-sharpened silicon nitride triangular cantilevers and microprobes from Digital Instruments. Our measurements were conducted at room temperature and pressure. For liquid measurements, the AFM cantilever was completely submerged in the liquid, to avoid capillary effects. Our measurements were conducted at low normal loads (smaller than 50 nN).

The high-purity C_{60} (99.5% C_{60}) was purchased from Alfa Aesar in powder form. For the AFM measurements, a solution of toluene and C_{60} was prepared by dissolving 0.1 mg C_{60} for every 1 mL of toluene. The lateral force microscopy (LFM) measurements of mica under toluene and mica under the C_{60}/toluene solution were carried out successively with the same cantilever on the same day. A different cantilever was used to acquire the LFM measurements comparing bare mica with mica under toluene.

Atomic-scale AFM images of toluene on mica and the C_{60}/toluene solution on mica are shown in figure 1. As shown in figure 1, AFM atomic-scale images acquired under pure toluene show mica atoms. For the C_{60} to have an effect on the lateral force, they must form a barrier layer between the AFM tip and the mica. In our measurements, however, the C_{60} did not form the barrier layer, and were instead pushed out of the way by the AFM tip. For AFM images acquired under the C_{60}/toluene solution, the periodicity of the atoms in the image corresponds to the interatomic spacing of mica atoms, which is much smaller than the spacing of C_{60} molecules. As our AFM images do not show trapped C_{60} layers, our results agree with those of Campbell et al. in that we show that for the slow-moving AFM, the C_{60} monolayers on the mica are easily disrupted, or pushed out of the way.

Figure 1. AFM image of mica atoms submerged in toluene (left) and AFM image of mica atoms in the C_{60}/toluene solution (right)

It is not surprising, therefore, that our lateral force measurements (figure 2) show no difference in the lateral force between pure toluene on mica and the C_{60}/toluene solution on mica. We see a large difference in friction between bare mica and mica under toluene. The bare mica has much higher friction than the toluene on the mica, probably due to the capillary forces of adsorbed water layers on the AFM tip.

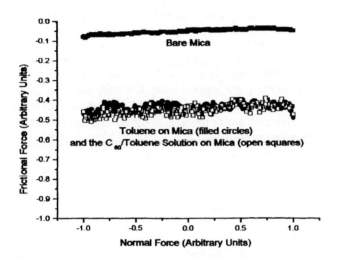

Figure 2. Frictional force vs. normal force for bare mica, toluene on mica, and the C_{60}/toluene solution on mica. Note that the toluene on mica and C_{60}/toluene solution on mica data sets are indistinguishable from one another

3. QCM Experimental Details and Results

As the AFM, STM, and SFA techniques are described extensively elsewhere in these proceedings, we will focus on a description of the QCM technique. The QCM has been used for decades for micro weighing purposes (Lu *et al.*, 1984), and was adapted for friction measurements in 1986-88 by Widom and Krim (Krim *et al.*, 1986; Widom *et al.*, 1986; Watts *et al.*, 1990). A QCM consists of a single crystal of quartz which oscillates in transverse shear motion with a quality factor Q near 10^5. Adsorption onto the microbalance produces shifts in both the frequency f_o and the quality factor Q, which are indicative of the degree to which the adsorbate is able to track the oscillatory motion of the underlying substrate. Characteristic slip times τ, and friction coefficients (i.e. shear stresses per unit velocity) η, are determined via the relations (Krim *et al.*, 1986):

$$\delta(Q^{-1}) = 4\pi\tau\delta f_o \qquad \eta = \frac{\rho_2}{\tau} \qquad (1)$$

where ρ_2 is the mass per unit area of the adsorbate. In terms of separate phonon and electron-hole slip times, τ_{ph} and τ_{eh}, the slip time τ can ideally be written as

$$\frac{1}{\tau} = \frac{1}{\tau_{ph}} + \frac{1}{\tau_{eh}}. \tag{2}$$

Our QCM measurements were conducted in UHV to avoid monolayers of water and other surface contaminants. For the C_{60} experiment, we first deposited 30 nm of Ag(111) onto an 8 MHz crystal at $\tilde{1}0^{-8}$ Torr. We then adsorbed toluene onto the Ag(111) surface and monitored both the frequency shift and change in the quality factor. We then again evacuated the chamber to UHV conditions, and deposited two monolayers of C_{60} onto the Ag(111). We then again adsorbed toluene onto the C_{60}/Ag(111) surface and observed changes in the frequency and quality factor.

Our QCM in UHV results are shown in figure 3. The slip time of toluene on Ag(111) is a factor two longer than the slip time of toluene on C_{60}/Ag(111) (figure 3). This means that the friction for toluene sliding on C_{60} is higher than the friction for toluene sliding on Ag(111). Our measured frequency shift of toluene sliding on C_{60}/Ag(111) was much larger than the frequency shift for toluene sliding on Ag(111).

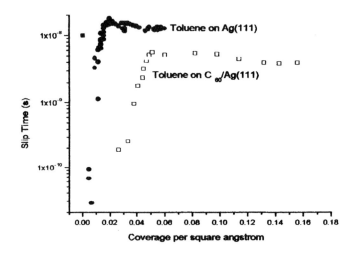

Figure 3. QCM slip times for toluene on Ag(111) and toluene on C_{60}/Ag(111)

4. Discussion

Our AFM and LFM results are well correlated with previous SFA results. (Campbell *et al.*, 1996) Although our LFM results do not show a significant change in the lateral force between toluene on mica and the solution of C_{60}/toluene on mica as is seen in the SFA, both SFA and AFM results show that the monolayers of C_{60} on mica are easily disrupted.

Our QCM in UHV measures slip times which are due to both the electronic and phononic contributions to friction. For the Ag(111) surface, the atomic surface corrugation is small, which means that the phononic contribution to friction will be relatively small. The large number of conduction electrons implies, however, that the electronic contribution to friction might be relatively large. When monolayers of C_{60} molecules are deposited on the Ag(111) surface, the frictional properties change dramatically. Now, the number of conduction electrons is small, which means the electronic contribution to friction will also be small. The surface corrugation, however, is quite large, which causes the phononic contribution to friction to increase. As we see the friction increase when C_{60} is present, we conclude that phononic friction is dominant for this system.

For our QCM in UHV results, we found that toluene sliding on monolayers of C_{60} deposited onto Ag(111) has a higher friction than that of toluene sliding on Ag(111). The distinguishing feature between our results and previous SFA results is the fact that C_{60} is known to chemisorb onto silver and physisorb onto mica (Altman et al., 1993; Sakurai et al., 1995). Therefore the C_{60} molecules can freely slip on a mica surface, but cannot on a silver surface. This is confirmed by the fact that we saw no change in the quality factor between the bare Ag(111) surface and the layers of C_{60} on the Ag(111) surface, which indicates that the C_{60} is not slipping on the Ag(111). Our results also show that the full-slip boundary condition found in previous SFA measurements (Campbell et al., 1996) is not due to the slipping of the toluene on the C_{60} molecules. It is instead due to the slipping of the C_{60} on the mica surface.

5. Future and Related Works

The operation of a QCM in a liquid environment has become routine within the last decade. (Grate et al., 1993) When the electrode operates in contact with a liquid, the shear motion of the surface generates motion in the liquid near the interface. If the surface is sufficiently smooth (as are the electrodes on our QCMs), then the oscillating surface generates plane-parallel laminar flow in the contacting liquid. The response of the oscillator in such conditions depends critically on the viscosity of the fluid adjacent to the electrode, which may well exhibit a structure and viscosity which is from that of the bulk, and also on whether slippage is occurring at the interface. A variety of models have been developed to predict the oscillator response in a liquid environment, incorporating the possibility of liquid structure and interfacial slip. (Lea et al., 1985; Duncan-Hewitt et al., 1992; Yang et al., 1993)

We have made preliminary measurements of a QCM with a gold electrode completely submerged in both pure toluene and the C_{60}/toluene solution. We compared the frequency shifts of the QCM between air/toluene and air/solution of C_{60}/toluene. The frequency shift of the QCM from air to the solution of C_{60}/toluene was a factor two larger than the frequency shift of air/toluene, completely consistent with the QCM in UHV measurements. This suggests that we can correlate bulk, or macroscale, QCM measurements with the monolayer, or nanoscale, measurements acquired in UHV.

We are also planning to study the C_{60}/Ag(111) system using a combination of STM and QCM. (Borovsky, Mason et al., 2000) We are currently using the STM/QCM combination to study tertiary butyl phenyl phosphate, a common commercial additive with known lubrication properties. (Borovsky *et al.*, 2000) The STM allows a single asperity contact to be formed, and allows the buried contact to be imaged in both stationary and sliding conditions. The QCM, whose surface is oscillating in transverse shear motion at speeds near 1 m/s, can be employed to measure the uptake rate and frictional properties of the adsorbed lubricant species, as described above. The STM/QCM combination will allow us to cross-reference the atomic-scale imaging of the AFM and STM techniques with the slippage measurements of the QCM. The realistic sliding speeds will also allow for comparison with macroscopic techniques.

6. Acknowledgements

This work has been supported by NSF grant No. DMR0072030 and AFOSR grant No. F49620-98-1-0201 and a Department of Education GAANN Fellowship.

References

Altman, E.I. and R.J. Colton, (1993) "The Interaction of C_{60} with Noble Metal Surfaces", *Surface Science* **295**, 13.

Borovsky, M. Abdelmaksoud, and J. Krim (2000) "STM-QCM Studies of Vapor Phase Lubricants", to be published.

Borovsky, B., Mason, B.L. and Krim, J. (2000) "Scanning Tunneling Microscope Measurements of the Amplitude of a Quartz Crystal Microbalance", to be published in *Journal of Applied Physics* **88**, **7**

Campbell, S.E., Luengo, G., Srdanov, V.I., Wudl, F., and Israelachvili, J.I. (1996) "Very Low Viscosity at the Solid-Liquid Interface Induced by Adsorbed C_{60} Monolayers", *Nature* **382**, 520–522.

Duncan-Hewitt, W.C. and Thompson, M. (1992) "4-Layer Theory for the Acoustic Shear Wave Sensor in Liquids Incorporating Interfacial Slip and Liquid Structure" *Anal. Chem.* **64**, 94.

Grate, J.W., Martin, S.J., and White, R.M. (1993) "Acoustic Wave Microsensors 2" *Anal. Chem.* **65**, 987.

Krim, J. and Widom, A. (1986) "Damping of a Crystal Oscillator by an Adsorbed Monolayer and its Relation to Interfacial Viscosity" *Physical Review B* **38**, 12184.

Lea, M.J. and Fozooni, P. (1985) "The Transverse Impedance of an Inhomogeneous Viscous Fluid" *Ultrasonics* **41**, 133.

Lu, C. and Czanderna, eds. (1984) *Applications of Piezoelectric Quartz Crystal Microbalances* Elsevier, Amsterdam.

T. Sakurai *et al.*, (1995) "Adsorption of Fullerenes on Cu(111) and Ag(111) Substrates", *Applied Surface Science* **87/88**, 405.

Watts, E.T., Krim, J., and Widom, A., (1990) "Experimental Observations of Interfacial Slippage at the Boundary of Molecularly Thin Films with Gold Substrates" *Physical Review B* **41**, 3466.

Widom, A. and Krim, J., (1986) "Q Factors of Quartz Oscillators as a Probe of Submonolayer Film Dynamics" *Physical Review B* **34**, R3.

Yang, M., Thompson, M., and Duncan-Hewitt, W.C. (1993) "Surface Morphology and the Response of the Thickness-Shear Mode Acoustic Wave Sensor in Liquids" *Langmuir* **9**, 802.

INTERACTIONS, FRICTION AND LUBRICATION BETWEEN POLYMER-BEARING SURFACES

JACOB KLEIN

Weizmann Institute of Science, Rehovot 76100, Israel

Abstract: Polymers that are adsorbed on a surface, or grafted to it (so-called polymer brushes), generally adopt extended configurations, For this reason forces between polymer-bearing surfaces are long-ranged, and are largely determined by steric interactions between molecules on the opposing layers. When two polymer-bearing surfaces are compressed and made to slide past each other, the frictional forces between them are determined by several different factors. The reduction of friction by polymer brushes occurs because such brushes can support a large load while maintaining a very fluid interface due to limited mutual interpenetration. At the highest pressures, slip reverts from the mid-plane to the polymer-substrate interface. Functionalized polymers with specifically attractive groups can lead to frictional dissipation on sliding which is modulated by the breaking and reforming of bonds. Detailed understanding of polymer-modulated friction may be obtained by taking due account of the polymer dynamics, surface structure and topology, and their specific surface attachment.

1. Introduction

Surfaces interactions in liquid media control effects ranging from adhesion to wetting, and to lubrication (Bowden and Tabor, 1964). These forces are frequently modified by attaching polymer chains to the surfaces, and the resulting long-ranged steric interactions are often also used to stabilize (or occasionally to destabilize) colloidal dispersions (Vincent, 1974; Goodwin, 1982; Napper, 1983; de Gennes, 1987; Klein, 1988; Patel and Tirrell, 1989; Fleer et al., 1993). The range, magnitude and sign of these forces are determined by the size and flexibility of the polymer chains, and especially by the nature of the solvent-polymer and surface-polymer interactions. Surfaces to which flexible polymer chains are attached are commonly encountered in a wide range of synthetic materials and processes, and are found in almost all natural and living systems. Indeed, the pressures relevant to interactions between solvated polymer-bearing surfaces cover the

177

B. Bhushan (ed.),
Fundamentals of Tribology and Bridging the Gap between the Macro- and Micro/Nanoscales, 177–198.
© 2001 *Kluwer Academic Publishers.*

range commonly found in biological systems: Sliding of biopolyelectrolyte-coated cells past blood-vessels or connective tissue (Zamir et al., 2000), or lubrication at the interface of articular cartilage in mammalian joints (McCutchen, 1959), are biotribological phenomena that come to mind. The mechanisms underlying the frictional response when two solvated polymer-bearing surfaces slide past each other are very different to those underlying friction between sliding (bulk) polymer surfaces (Briscoe, 1981). They also differ conceptually from the shear and rheological properties of thin polymer melt films (Van Alsten and Granick, 1990; Luengo et al., 1997). These are frequently analysed in terms of bulk viscoelastic models that may not be appropriate for a situation where all the chains are attached to one surface or another, so that there is no 'bulk', but rather only surface effects. In what follows we treat polymeric surface layers, and their frictional interactions, primarily from the point of view of the structure and dynamics of the polymers attached to the surfaces. The basic reason why polymers adsorb readily to surfaces, and why they adhere tenaciously and form extended layers, is a result of their size and flexibility, as discussed below.

1.1. CONFIGURATIONAL ENTROPY AND EXCLUDED VOLUME EFFECTS IN FLEXIBLE POLYMERS

A neutral, flexible, freely-jointed polymer molecule with N monomers, each of size a, can adopt a huge number of configurations in solution, all of which have essentially the same internal energy (N is typically $10^3 - 10^4$). This results in a large configurational entropy contribution to the free energy of such a molecule. Any reduction in the number of possible configurations, for example by excluding the volume available to a monomer, either by an impermeable wall or by the presence of another monomer, results in a loss of configurational entropy and a corresponding increase in the free energy (Flory, 1953; de Gennes, 1979). This is the excluded volume or steric effect. Interactions between monomers along the chain are a combination of van der Waals attraction and repulsions resulting from excluded volume, but it is possible to adjust the interactions of monomers and solvents to eliminate the net mutual interaction. In this case (known as ideal or θ(theta)-solvent conditions) the polymer adopts an ideal or random walk configuration, of mean square coil size $R^2_0 = Na^2$, as illustrated in fig. 1.

In a so-called good-solvent, the net van der Waals attractions between monomers are weaker and do not compensate for excluded volume effects. This leads to a net repulsion between monomers, and polymer dimensions are on average greater, the mean square size of a coil being the Flory dimension, $R^2_F = N^{3/5}a^2$ (in poor solvents, the van der Waals attraction between monomers exceeds their excluded volume repulsion, which results

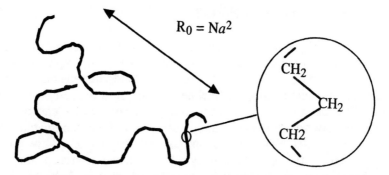

Figure 1: Schematically showing a randomly-coiled polyethylene molecule, $(CH_2)_N$

in phase separation or chain collapse). Confining a flexible molecule to a region of size D smaller than its equilibrium size R_0 (or R_F in the case of a good solvent) – for example a gap of thickness D between two impermeable surfaces – increases its free energy because of the resulting configurational entropy loss. For an ideal coil the increase is $\Delta F_{conf} \cong k_BT(R_0/D)^2$, where k_B is Boltzmann's constant and T the absolute temperature (de Gennes, 1979).

1.2.   ADSORBED AND END-TETHERED POLYMERS

When a flexible polymer *adsorbs* on a substrate, there will be two opposing effects: on the one hand the monomers want to adhere to the substrate (with net adhesion energy $-\delta k_BT$ per monomer-surface contact), which tends to make the polymer flatten on the surface. On the other hand, flattening (i.e. smaller D) is resisted because of the increase in ΔF_{conf}. The situation is illustrated in fig. 2a., where a polymer is adsorbed to equilibrium thickness D.

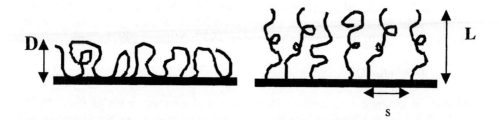

Figure 2:   a) Adsorbed polymer.   b) A polymer brush

A simple minimization of the total energy of attachment ΔF of a single polymer chain to the surface (the sum of its sticking + confinement energies) suggests the following relationships at equilibrium (de Gennes, 1979):

$$D = (a/\delta) \text{ and } \Delta F = - N\delta^2 k_B T. \tag{1}$$

In general the net sticking energy for common polymers adsorbing from a good solvent onto solid surfaces is of order $0.01 - 0.1$ $k_B T$ per monomer, i.e. $|\delta|$ is in the range $0.01 - 0.1$. Since monomer dimensions are typically around 5Å, this tells us that the thickness D of an adsorbed layer can be of order 100Å or more – a rather 'fluffy' layer which ensures that the interaction between polymer-coated surfaces will be dominated by steric rather than van der Waals attractions as they approach. At the same time, since $N = 10^3 - 10^4$, eq. (1) tells us that the total attachment energy ΔF of long chains may greatly exceed the thermal energy $k_B T$ of the chain, ensuring the polymer is strongly and essentially irreversibly adsorbed. Although the above results are valid for single isolated chains, the general features – thick adsorbed layers and strong adsorption - apply also for the more usual case of overlapping many-chain adsorption (de Gennes, 1981; Ingersent et al., 1986; Ingersent et al., 1990). In practice, the effective thickness D of adsorbed layers of overlapping chains is determined by the size of the largest loops or tails, and varies as $D \approx R_0$ or R_F, depending on the solvency condition.

For the case where the *solvent* molecules, rather than the monomers, are preferentially attracted to the surface, the polymer will not adsorb; it may however be made to adhere to the surface by one end, creating a polymer 'brush' as illustrated in fig. 2b. The end-attachment may be a 'sticky' end-group (such as a zwitterion); a copolymer moiety which does adsorb, and so anchors the non-adsorbing 'tail'; or a covalent bond attaching the end-monomer of the otherwise non-adsorbing chain to reactive groups on the surface. In a good solvent, the excluded volume repulsion between monomers causes the chains to stretch out from the surface so as to avoid each other; at the same time the stretching is resisted by the elastic entropy cost (whose origin, the reduction of available chain conformations, is identical to that of ΔF_{conf}). In equilibrium, the brush thickness L in a good solvent depends on the degree of polymerization N and on the interanchor spacing s (fig. 2b) as follows (Alexander, 1977; Milner, 1991):

$$L \cong Na(a/s)^{2/3} \tag{2}$$

That is, in contrast to adsorbed layers whose thickness varies as $R_0 = N^{1/2}a$ or $R_F = N^{3/5}a$, the brush thickness varies linearly with N, so that brushes are generally more extended normal to the surface than are the corresponding adsorbed layers. Normal interactions between polymer brushes in good solvents are strongly repulsive, as the excluded volume interactions lead to a high osmotic repulsion when the brush layers are compressed against each other.

1.3. STERIC FORCES BETWEEN POLYMER-COATED SURFACES

Steric forces have been extensively studied, driven by the practical need to stabilize colloidal systems, and over the past two decades or so the equilibrium force-distance laws between polymer-bearing surfaces have been measured directly in a wide range of conditions (Klein, 1988; Patel a n d Tirrell, 1989; Luckham, 1991; Fleer et al., 1993). Such forces are reasonably well understood (de Gennes, 1987; Pincus, 1989; Luckham, 1991; Klein, 1992) and are summarised schematically in figures 3a and 3b for the case of low and high adsorbance of the chains. Figure 3 emphasises the differences between adsorbed polymer layers - where each monomer has a propensity to adhere to the surface - and grafted layers, or polymer brushes, where the monomers do not adsorb and the chains are attached to the surface by one end only.

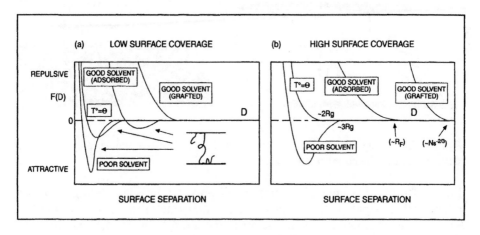

Figure 3: Equilibrium normal forces between polymer-bearing surfaces at (a) low adsorbance and at (b) high adsorbance of chains on the surfaces, showing the importance of bridging for adsorbed (but not for grafted) chains at low surface coverage (from (Klein, 1996)).

2. Frictional forces between sliding, polymer-bearing surfaces

As seen above, interactions between surfaces with adsorbed or grafted chains can be very much longer-ranged than van der Waals attractions between the underlying substrates, or than double-layer electrostatic repulsions between charged surfaces in aqueous salt solutions. This affects not only the normal interactions, but also their frictional behaviour (Klein, 1996). Different factors that may contribute to frictional forces between sliding surfaces are illustrated in fig. 4.

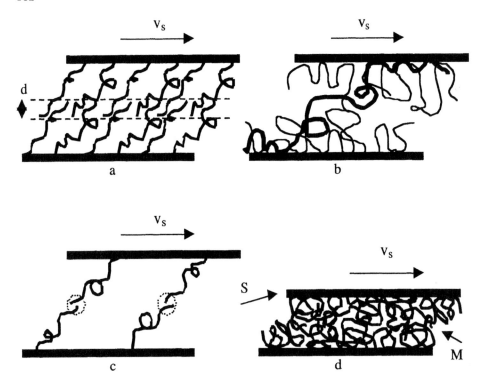

Figure 4: Illustrating different possible modes of friction between sliding polymer-bearing surfaces. For description see text below.

Fig. 4(a) shows two mutually-compressed polymer brushes sliding past each other, with dissipation taking place in an interpenetration zone of width d. This case is considered in more detail in section 4.1. In fig. 4(b) is illustrated the case of a polymer (highlighted in bold) bridging the gap between two adsorbing surfaces; such a bridge resists being stretched on initial shear, and its two adsorbing moieties may then drag along the surfaces, resulting in a frictional resistance to sliding. Fig. 4(c) shows how specific interactions between chains (indicated in broken circles) can also lead to bridge formation, and to frictional dissipation on sliding when bridges break and reform. This situation may arise when living cells - with highly specific ligand-receptor interactions at the cell membrane – slide past biological surfaces (Zamir et al., 2000); we consider it further in section 5. Under high loads and high compressions, where the chains may be strongly entangled, as indicated in fig. 4(d) very large frictional forces can result on shear, and the plane of sliding may then shift from the mid-plane M between the two polymer layers to the polymer/substrate interface S. This situation is discussed in section 4.2.

The modes of friction illustrated schematically in fig. 4 are examined in the following sections in the light of recent experiments. It is important to note, however, the many insights provided also by computer simulation work, but whose description is beyond the scope of this review (Thompson et al., 1995; Grest, 1996; Neelov et al., 1998).

3. Measurement of ultra-weak shear forces between surfaces

Frictional forces between sliding polymer-bearing surfaces can, in certain regimes, be orders of magnitude smaller than the corresponding forces between conventional solids or between solid surfaces across small molecule lubricants (Klein et al., 1994; Schorr et al., 1999). To measure such weak shear forces we developed a new design of the mica surface force balance (SFB), as recently described (Klein et al., 1991; Klein and Kumacheva, 1998). The essential features of the SFB are illustrated in fig. 5.

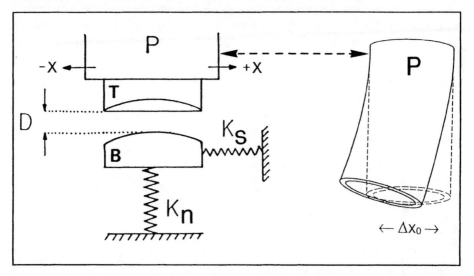

Figure 5. Schematic illustration of the main features of the SFB used to measure ultraweak shear forces between sliding surfaces (Klein and Kumacheva, 1998). The (backsilvered) mica sheets are glued on crossed cylindrical lenses, with the top one (T) mounted on a sectored piezocrystal P which can move laterally as well as up and down (illustrated on RHS). The bending of two orthogonal springs (K_n and K_S) measure normal and shear forces between the surfaces as they approach/separate (changes in D) or slide past each other (in the x direction). The bending can be monitored to ±2Å and ±3Å using optical interferometry and changes in capacitance respectively. The shear-spring stiffness K_S can be varied by up to 2 orders of magnitude during an experiment.

In terms of the shear stresses measurable, this instrument has a sensitivity and resolution that is several orders of magnitude better than scanning probe techniques (such as friction force microscopy or AFM in the lateral mode (Klein and Kumacheva, 1998)). These capabilities enable the very weak frictional forces and subtle relaxation effects arising from the presence of surface-attached polymers to be investigated.

4. Friction mediated by polymer brushes

4.1. THE LOW FRICTION REGIME

The friction coefficient for solid surfaces sliding across a simple liquid can be very large due to the fact that, under a load, the liquid between contacting asperities squeezes out to the last few monolayers, which may then behave in a solid-like fashion (Israelachvili et al., 1988; Granick, 1991; Klein and Kumacheva, 1995; Kumacheva and Klein, 1998). The presence of solvated polymer brushes on the surfaces (in a good solvent) can reduce this friction dramatically (Klein et al., 1994; Schorr et al., 1999). This is demonstrated in fig. 6 which shows the sliding friction force F_{shear} between bare mica

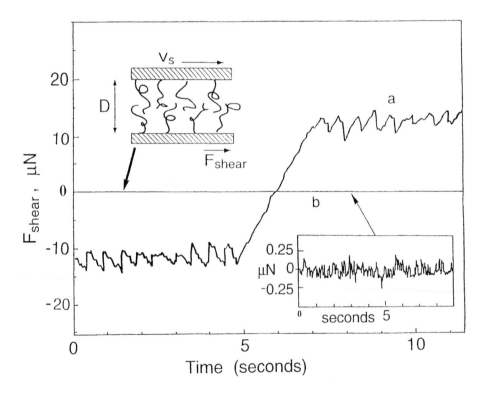

Figure 6: Traces of frictional force F_{shear} between sliding surfaces. Trace a: bare mica across toluene, showing the characteristic stick-slip response; the effective sliding friction coefficient is μ = ca. 0.7. Trace b: following attachment of a polystyrene brush. The friction coefficient is now $\mu < 0.005$ (taken from (Klein et al., 1994)).

surfaces sliding under a given load across the simple solvent toluene (fig. 6, trace a), and following attachment of a polymer brush to the mica surfaces (trace b). The presence of the brush results in reduction in the friction coefficient by at least a hundred-fold. Fig. 7 shows the variation of the sliding friction force and of the normal load with surface separation D, for polystyrene brushes of 3 different molecular weights.

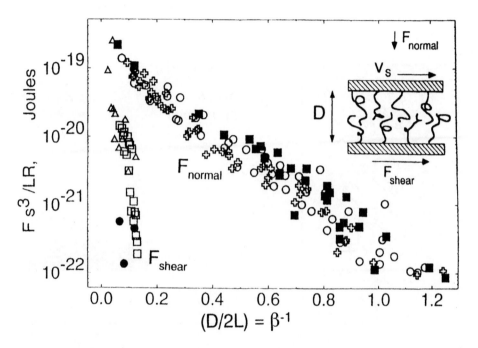

Figure 7: Variation of shear forces F_{shear} and normal loads F_{normal} with normalised surface separation (D/2L) for 3 different polystyrene brushes (different symbols), where D is the surface separation and 2L the separation at onset of interaction. For compressions (D/2L) > 0.15, F_{shear} is below the resolution of the SFB.

At lower loads and compressions, the frictional forces F_{shear}, shown in fig. 7 as the top surface slides past the lower one, are below the resolution of our force balance over the range of compressions (D/2L) > ca. 0.15 (this is also seen in trace b in fig. 6). At the point where the friction begins to rise, at a compression (D/2L) ≤ ca. 0.15, the effective friction coefficient is down to 0.001 or less; the normal pressure over the effective area of contact is then roughly 1 MPa or 10 atmospheres. This is a remarkable lubricating effect whose origins are considered below.

In fig. 8 we draw schematically, but more or less to scale for the PS brush in fig. 6, the relative configurations as the opposing brush layers first come into overlap, and as they appear when shear forces first become measurable at a compression ratio $\beta \approx 7$ (the compression ratio $\beta \equiv (2L/D)$ where L is the unperturbed thickness of each brush, as in fig. 2b)..

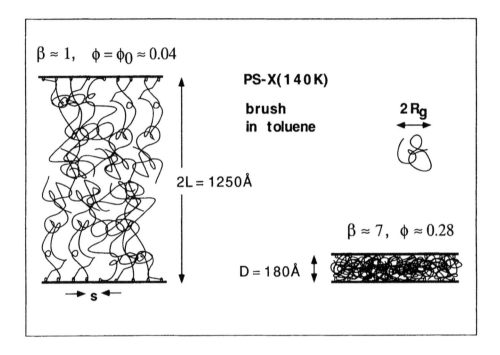

Figure 8: Showing schematically the relative configuration of two polymer brushes (based on PS-X(140k) brush layers) when repulsion is first measured (left) and when shear forces first become measurable (right). The unperturbed chain (radii of gyration $R_g = R_0$) is shown for comparison.

We note particularly the seven-fold increase in monomer volume fraction in the intersurface gap, from ca. 0.04 in the unperturbed brushes to ca. 0.3 when shear forces are first detected on sliding.

The reason for the strong lubrication effect afforded by the polymer brushes is because – over a substantial regime of normal pressures - they are able to support a large normal load while maintaining a very fluid interfacial region between them when sliding. This results in very little frictional drag opposing the sliding, hence the low friction coefficients. This effect is better seen by considering the extent of interpenetration of two polymer brush layers, as illustrated in fig. 9. Due to the strong excluded volume repulsion that an outside chain encounters when penetrating into an existing brush, the interpenetration of two brush layers is strongly resisted. It can be shown

(Witten et al., 1990; Wijmans et al., 1994; Klein, 1996) that the extent of interpenetration d (fig. 9a) depends on the extent of compression very weakly, as

$$d \propto D^{-1/3} \qquad\qquad (D < 2L) \qquad\qquad (3)$$

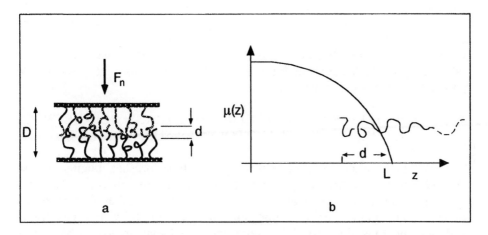

Figure 9: (a) Two compressed polymer brushes have a mutual interpenetration zone of extent d, where the frictional dissipation takes place on sliding. (b) Showing schematically the parabolic potential field $\mu(z)$ associated with a polymer brush a distance z from the anchoring surface, resulting from excluded volume interactions, and the extent to which a chain penetrating into the brush layer feels this field.

Thus, a fivefold-compression of the brushes, while requiring a large normal load, will only increase the width of the interpenetration zone by some 70%. A simple model, which takes into account the drag of the polymer moieties from the opposing brush layers within the interpenetration zone, provides an expression for the shear stress σ_s between two brushes shearing past each other. Over a compression range where the polymer concentration remains in the semi-dilute regime, one finds (Klein, 1996):

$$\sigma_s = (6\pi\eta_{eff}v_s\beta^{7/4})/s \qquad\qquad (4)$$

Here η_{eff} is the effective viscosity responsible for the drag on the polymer moieties sliding through the interpenetrated region when the surfaces are sliding at velocity v_s, at a compression ratio β, and a mean interanchor spacing s (fig. 2b). Over a substantial range of compressions, i.e. loads, d remains small, the interpentrated polymer moieties remain below their entanglement length, and the magnitude of η_{eff} is not far above that of the pure solvent itself. Hence the low shear stress required for sliding, and the very low friction coefficients. At sufficiently high loads and thus compressions, the two opposing brushes will interpenetrate substantially and eq. (4) no longer holds. Thus beyond a sufficiently high compression, $\beta*$ say, the entire gap will be mutually interpenetrated by the chains. If we make the reasonable assumption that on initial overlap $d \approx s$, then $\beta = \beta*$ when $d = D$, yielding at once from eq. (3) that $\beta* \cong (2L/s)^{3/4}$. For the system illustrated in fig. 6, a polystyrene brush in toluene, $2L = 1250\text{Å}$, $s = 85\text{Å}$, giving $\beta* \approx 7 - 8$. That is, for compressions of this brush system by a factor of order 7-fold or greater from overlap, the gap will be fully interpenetrated by the two opposing brushes. It is of interest that this is the compression ratio beyond which the shear force begins to grow measurably (within the parameters of our experiments).

4.2.   THE HIGH FRICTION REGIME: THE TRANSITION BETWEEN
        POLYMER-POLYMER AND POLYMER-SURFACE SLIDING

The friction between surfaces bearing polystyrene brushes begins to rise sharply beyond a certain compression ratio, typically for values $\beta \equiv (2L/D) >$ ca. 7 or 8, as seen in figs. 8 and 10. This is due to a number of factors. At these strong compressions the opposing brushes are fully interpenetrated and heavily entangled, so that within the interfacial region, which now occupies the entire gap D, sliding is resisted by strong viscous drag. At the same time, for the particular case of polytyrene, the concentration of monomers in the gap attains a substantial fraction of the glassy concentration, which is around $70 - 80\%$ for this polymer at room temperature. At these concentrations the monomeric friction, which determines the viscous drag on the chains as they are sheared past each other, begins to diverge. To obtain further insight into the mechanism by which the surfaces slide past each other in the high friction regime, we examine the nature of the friction response at progressivelly higher compressions. This is shown in fig. 10, where the back-and-forth motion of the driven upper brush-bearing surface is shown together with the corresponding shear force transmitted to the lower surface, as D decreases progressively.

The traces in fig. 10a, at surface separations $D = 160 - 150\text{Å}$, reveal the onset of shear forces just measurable above the noise, and these progressively increase as D decreases, as shown by the F_{shear} traces in figs. 10b and 10c. Closer examination shows that the drive (upper Δx_0 traces in each of figs. 10a-c) and the response (lower F_{shear} traces) are typical of a dissipative viscous process, in being close to 90^0 out of phase with each other, and considerable sliding between the surfaces takes place (the amplitude of motion of the lower surface is a small fraction of that of the upper surface). For the traces of fig. 10d, however, where D has decreased to 120Å or less, the behaviour is very different: examination of the shear response in these traces reveals that in the part of the lower trace marked α, the surfaces stick together and move in tandem, and then – the part of the trace marked Σ - they slide essentially freely past each other. This is a stick-

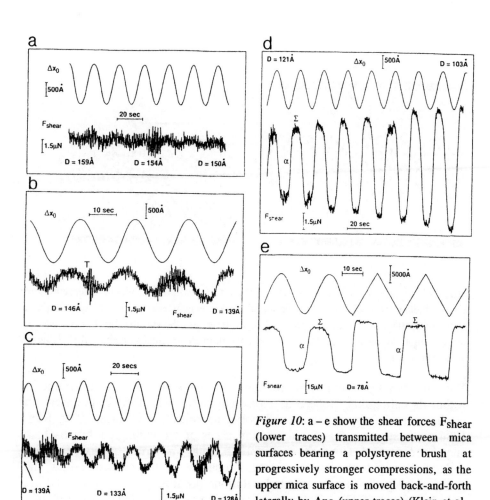

Figure 10: a – e show the shear forces F_{shear} (lower traces) transmitted between mica surfaces bearing a polystyrene brush at progressively stronger compressions, as the upper mica surface is moved back-and-forth laterally by Δx_0 (upper traces) (Klein et al., 1998).

slip response resembling that between two solid surfaces (Bowden and Tabor, 1964). This type of response is even more marked in fig. 10e, where D has been reduced by further compression to 78Å, and the frictional behaviour is very clearly one of sticking followed by sliding.

An understanding of the frictional process is provided by considering the viscous drag on the interacting chains from the two mutually interpenetrated brushes. The key point here is that the chains in each brush are tethered by one end to a surface, while their dangling tails are highly entangled with the other chains within the interpenetration zone, whose extent d may cover the entire intersurface gap. This is illustrated in fig. 11, where for clarity only one chain is shown.

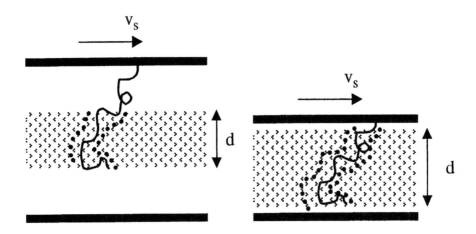

Figure 11: Illustrating the entanglement of the brushes within the mutual interpenetration zone of width d (shaded), which may be partial (left) or may extend across the gap D (right) at high compressions. For clarity only a single brush is shown; the brush can relax its configuration within this zone by retracting back through its entanglement 'tube', indicated by broken lines.

The only way by which such an entangled chain may relax is by retraction within its entangling 'tube', as indicated in fig. 11. This is precisely the mechanism whereby star-branched polymers relax (de Gennes, 1979; Klein, 1986); this suggests that the relaxation times of the entangled brush-chains may be estimated from the appropriate data (in terms of concentration, molecular weight etc.) on the viscous properties of the corresponding star-branched polymer in concentrated solutions (Graessley et al., 1967; Utracki and Roovers, 1973). Relaxation times may be related to diffusion coefficients and through these, via the fluctuation-dissipation theorem, to the effective friction on such an entangled chain (Witten et al., 1990; Joanny, 1992). One may therefore, starting with the bulk viscosity data on concentrated solutions of the corresponding star-polymers, estimate the drag on the mutually interpenetrated brushes. This has been done, using a model based on the above considerations, with due allowance for the brush density and height, the sliding velocity v_s, and for the geometry of the interacting crossed-cylindrical surfaces on which the brushes are mounted. The predicted (calculated) variation of the shear force with surface separation is shown in fig. 12, together with the corresponding experimental data (from traces such as in fig. 10). The particular system considered is a PS-X(140k) brush on mica in toluene, where PS denotes polystyrene and X is a zwitterionic end-group which sticks to the mica. The point to note in fig. 12 is that the variation of the calculated frictional forces with decreasing surface separation follows quite closely that of the experimentally measured ones,

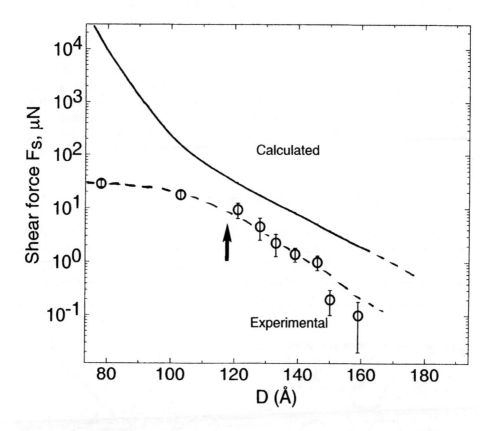

Figure 12: The calculated and measured frictional force required to slide two surfaces bearing a given polystyrene brush at known sliding velocities and different surface separations D. The arrow marks the point at which a significant and growing deviation appears between the calculated and measured results.

down to separations $D \approx 120\text{Å}$. The quantitative agreement, within an order of magnitude or so in this range, is reasonable in view of the assumptions in the model. However, for $D < 120\text{Å}$ there is increasing deviation; in particular the leveling off of the shear force at smaller D values and correspondingly higher loads. This suggests that the mechanism of friction is changing from that assumed in the model, where interfacial slip is assumed to take place at the mid-plane between the surfaces. The slip plane reverts rather to the *polymer/mica* interface, as indicated earlier in fig. 2d. That is, as the increasing load and compression force the mutually entangled brushes to higher concentration and so to increasing frictional drag, the shear stress required to pull them past each other exceeds that required to shear the polymer/solid interface. The slip plane then shifts from the brush-brush interface to the polymer-solid one. Further indications that this is happening are provided by a detailed examination of the onset of the stick-slip behaviour, and by an estimate of the shear stress at the point (arrow in fig. 12) where the surface slip begins: this is of the order of, though somewhat smaller than the stress required to overcome the zwitterion-mica adhesion. It is of interest to note that the effective friction coefficient in the wall-slip regime (D $\approx$ 80Å) is still quite low, $\mu \approx 0.06$. This 'cross-over' from polymer-polymer to polymer-wall slip provides a cutoff on the progressive increase of the frictional forces as the surfaces are compressed, and may have interesting practical implications.

5. Friction mediated by telechelic polymers

The large reduction in the friction coefficient afforded by polymer brushes indicates that two polymer 'loop' layers might provide even better lubrication. This is because the interpenetration of such loops is unfavourable not only for osmotic reasons arising from excluded volume

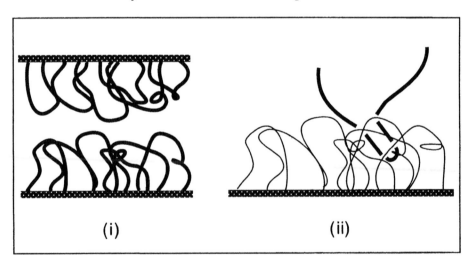

Figure 13: Two interacting polymer loop layers (i) resist interpenetration as the penetration of a loop into the entangled layer of the opposing brush (ii) requires a 'doubling up' of the loop into a hair-pin configuration as shown. This highly improbable configuration ensures very limited loop penetration (from (Klein, 2000)).

interactions, but also for topological reasons (de Gennes, 1979), as illustrated in fig. 13.

Telechelic chains that have surface-adhering groups at *both* their ends, such as polystyrene terminated with a zwitterion at both ends (X-PS-X), are likely candidates to form such loop layers. The zwitterions would be expected to anchor the two ends to the substrate, just as the single-zwitterion-terminated polystyrene, PS-X, forms effective brush layers by anchoring one of its ends as described in the previous section. In practice, however, the behaviour is very different to that expected (Eiser et al., 1999), and is summarised in fig. 14.

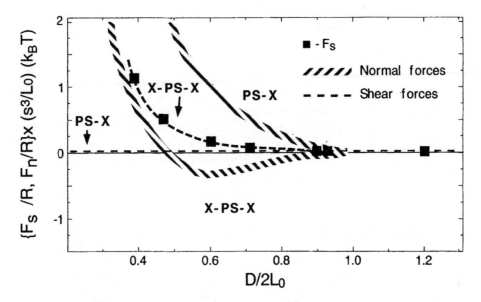

Figure 14: Comparison of the normal interactions (shaded curves) and shear forces (broken curves and data points) between regular brushes (PS-X) and telechelic layers (X-PS-X), based on data from fig. 7 for the regular brushes and from (Eiser et al., 1999) for the telechelics. The surface separation scale is normalised with respect to the distance $2L_0$ for onset of forces (taken from (Eiser et al., 1999)).

In contrast to regular brushes, the telechelic layers, far from repelling strongly as they overlap as figure 13 would suggest, exhibit a marked attractive regime on approach, as indicated in fig. 14 at $(D/2L_0) = 0.6$. In addition, marked frictional forces are observed as soon as attraction develops between the layers, as indicated by the data points in fig. 14. The form of the frictional forces in response to the applied lateral motion is shown in fig. 15. The lower trace in fig. 15 shows that following a rapid increase on initial sliding, the shear force relaxes to a plateau value – which represents the sliding friction for this configuration. A closer examination of both the

194

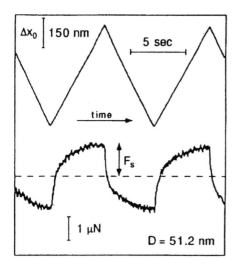

Figure 15: Shear forces between telechelic layers (lower trace) in response to a back-and-forth sliding between the surfaces (upper trace). The surface separation D = 51.2 nm corresponds to the attraction minimum shown in fig. 14. (from (Eiser et al., 1999))

normal and shear force profiles reveals the origin of this behaviour. The configuration of the telechelics is in *not* loop like as suggested in fig. 13: Rather, the mutual attractions between the zwitterion end-groups themselves energetically favours the formation of double or even multiple loop structures, with zwitterion dimers or multimers extending out to solution (Klein, 2000). This is shown in fig. 16.

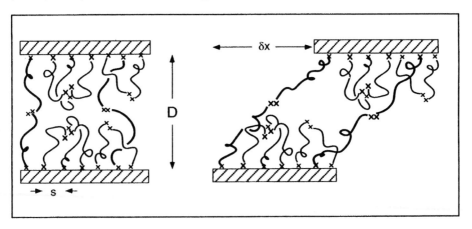

Figure 16: Telechelic polymers form multimer structures rather than simple loops, and when these interact double-bridges form as indicated on the left. On sliding, the bridges initially stretch – hence the rapid rise in the shear force indicated in fig. 15 – followed by breaking and reforming of the bonds which leads to the steady state sliding friction (from (Eiser et al., 1999)).

When two such layers approach, there is a net energy gain (of entropic origin) in forming interlayer 'double-bridges' across the gap, as shown, in preference to intralayer double-loops. This is only of order 0.5 k_BT per bridge formed, or even lower (despite the fact that the absolute magnitude of the zwitterion-zwitterion interaction is itself of order $10k_BT$). This results in the attractive interaction between the telechelic layers indicated in fig. 14 as they come into overlap. When the layers are sheared, the zwitterion-zwitterion bonds break and then reform further down, dissipating energy as the bonds 'snap' irreversibly. The plateau in the shear force represents the steady state value of this dissipation. This mechanism resembles that illustrated schematically in fig. 4c, and may be a simple model for the frictional processes occuring during sliding of biopolymer-coated cells past connective tissue and vessel walls.

6.   Conclusions

Friction forces between sliding, solvated polymer-bearing surfaces cannot be simply described by the extension of ideas from bulk polymer viscoelasticity, but must be considered in terms of the molecular structure and dynamic properties of the surface-attached chains. Several different mechanisms of friction are possible, depending on the structure of the chains on the surfaces, the nature of the solvent, on the monomer concentration within the compressed layers, on the glass transition temperature (T_g) of the chains, and whether they are functionalised or not. Molecular models which take these features into account can provide a detailed description of the friction, and simulation studies can provide further insights.

The role of excluded volume interactions, deriving from the large configurational entropy component of the polymer energies, is crucial. Large osmotic repulsions between compressed polymer brushes in good solvents, which at the same time resist mutual interpenetration, provide remarkable lubrication by supporting a large load while maintaining a very fluid sheared interfacial region. It is very likely that such steric effects play an important role in biolubrication, as in mammalian joints, where pressures – up to a few MPa - are comparable to those within solvated polymers in the semi-dilute and concentrated regime.

The use of high T_g polymers as lubricants, together with differential surface attachment, provides interesting possibilities. This is because at the highest pressures, when most of the solvent is excluded, the slip plane may revert from the polymer-polymer interface to the polymer-substrate interface, a form of self-regulating friction. However, to understand the molecular disentanglement mechanisms during shear, the use of low T_g polymers is necessary; this is because using the high T_g materials, subtle relaxation effects are obscured by the onset of glassiness at high pressures.

196

7. References

Alexander, S. (1977). "Adsorption of chain molecules with a polar head: a scaling description", *J.Phys (Paris)* **38**, 983-987.

Bowden, F. P. and D. Tabor (1964). *The Friction and Lubrication of Solids*. Oxford, Clarendon Press.

Briscoe, B. J. (1981). "Friction and wear of solid polymers and the adhesion model of friction", *Philosophical Mag.* **A43**, 511-527.

de Gennes, P. G. (1979). *Scaling Concepts in Polymer Physics*. N.Y., Cornell Univ. Press, Ithaca.

de Gennes, P. G. (1981). "Polymer solutions near an interface: 1. Adsorption and depletion layers", *Macromolecules* **14**, 1637-1644.

de Gennes, P. G. (1987). "Polymers at an interface: a simplified view", *Adv. Colloid Interface Sci.* **27**, 189-209.

Eiser, E., J. Klein, T. A. Witten and L. J. Fetters (1999). "Shear of telechelic brushes", *Phys. Rev. Lett.* **82**, 5076-5079.

Fleer, G. J., M. A. Cohen-Stuart, J. M. H. M. Scheutjens, T. Cosgrove and B. Vincent (1993). *Polymers at Interfaces*. London, Chapman and Hall.

Flory, P. J. (1953). *Principles of Polymer Chemistry*. Ithaca, Cornell University Press.

Goodwin, J., Ed. (1982). in *Colloidal Dispersions*. Royal Society of Chemistry Special Publications. London, Royal Society of Chemistry.

Graessley, W. W., R. L. Hazleton and L. R. Lindeman (1967). "The shear-rate dependence of viscosity in concentrated solutions of polystyrene", *Trans. Soc. Rheology* **11**, 267 - 285.

Granick, S. (1991). "Motions and relaxations of confined liquids", *Science* **253**, 1374-1379.

Grest, G. S. (1996). "Interfacial sliding of polymer brushes: A molecular dynamics simulation", *Phys. Rev. Lett.* **76**, 4979-4982.

Ingersent, K., J. Klein and P. Pincus (1986). "Interactions between surfaces with adsorbed polymers: poor solvent. 2. Calculations and comparison with experiment", *Macromolecules* **19**, 1374-1381.

Ingersent, K., J. Klein and P. Pincus (1990). "Forces between surfaces with adsorbed polymers .3. Theta-solvent - calculations and comparison with experiment", *Macromolecules* **23**, 548-560.

Israelachvili, J., P. M. McGuiggan and A. M. Homola (1988). "Dynamic properties of molecularly thin liquid-films", *Science* **240**, 189-191.

Joanny, J.-F. (1992). "Lubrication by molten polymer brushes", *Langmuir* **8**, 989 - 995.

Klein, J. (1986). "Dynamics of entangled linear, branched and cyclic polymers", *Macromolecules* **19**, 105-118.

Klein, J. (1988). in *Molecular Conformation and Dynamics of Macromolecules in Condensed Systems*. Amsterdam, Elsevier. p. 333-352.

Klein, J. (1992). "Long-ranged surface forces - the structure and dynamics of polymers at interfaces", *Pure and Applied Chem.* **64**, 1577-1584.

Klein, J. (1996). "Shear, friction, and lubrication forces between polymer-bearing surfaces", *Ann. Rev. Mater. Sci.* **26**, 581-612.

Klein, J. (2000). "Entropic interactions: neutral and end-functionalized chains in confined geometries", *J. of Physics: Condensed Matter* **8A,**, A19-27.

Klein, J. and E. Kumacheva (1995). "Confinement-induced phase-transitions in simple liquids", *Science* **269**, 816-819.

Klein, J. and E. Kumacheva (1998). "Simple liquids confined to molecularly thin layers. I. Confinement-induced liquid-to-solid phase transitions", *J. Chem. Phys.* **108**, 6996-7009.

Klein, J., E. Kumacheva, D. Mahalu, D. Perahia and L. J. Fetters (1994). "Reduction of frictional forces between solid surfaces bearing polymer brushes", *Nature* **370**, 634-636.

Klein, J., E. Kumacheva, D. Perahia and L. J. Fetters (1998). "Shear forces between sliding surfaces coated with polymer brushes: The high friction regime", *Acta Polymerica* **49**, 617-625.

Klein, J., D. Perahia and S. Warburg (1991). "Forces between polymer-bearing surfaces undergoing shear", *Nature* **352**, 143-145.

Kumacheva, E. and J. Klein (1998). "Simple liquids confined to molecularly thin layers. II. Shear and frictional behavior of solidified films", *J. Chem. Phys.* **108**, 7010-7022.

Luckham, P. (1991). "Measurement of the interaction between adsorbed polymer layers - the steric effect", *Adv. Colloid. Interface Sci.* **34**, 191-215.

Luengo, G., F.-J. Schmitt, R. R. Hill and J. Israelachvili (1997). "Thin film rheology and tribology of confined polymer melts: contrasts with bulk properties", *Macromolecules* **30**, 2482-2494.

McCutchen, C. W. (1959). "Mechanism of animal joints", *Nature* **184**, 1284-5.

Milner, S. T. (1991). "Polymer Brushes", *Science* **251**, 905-914.

Napper, D. H. (1983). *Steric Stabilization of Colloidal Dispersions*. London, Academic.

Neelov, I. M., O. V. Borisov and K. Binder (1998). "Shear deformation of two interpenetrating polymer brushes: Stochastic dynamics simulation", *J. Chem. Phys.* **108**, 6973-6988.

Patel, S. and M. Tirrell (1989). "Steric forces between polymer-bearing surfaces", *Ann. Rev. Physical Chem.* **40**, 597-624.

Pincus, P. (1989). in *Lectures on Thermodynamics and Statistical Mechanics*. Singapore, World Scientific. p. 74 - 88.

Schorr, P., S. M. Kilbey and M. Tirrell (1999). "Frictional behavior of self-assembled polymer brushes", *Polymer Preprints (Abstr. Amer. Chem. Soc)* **218: 278-POLY , Part 2 AUG**(Aug. 22 1999), U477.

Thompson, P. A., M. O. Robbins and G. S. Grest (1995). "Structure and shear response in nanometer-thick films", *Isr. J. Chemistry* **35**, 93 - 106.

Utracki, L. A. and J. E. L. Roovers (1973). "Viscosity and normal stresses of linear and star-branched polystyrene solutions. I. Application of corresponding states principle to zero-shear viscosities", *Macromolecules* **6**, 366 - 372.

Van Alsten, J. and S. Granick (1990). "Shear rheology in a confined geometry - polysiloxane melts", *Macromolecules* **23**, 4856-4862.

Vincent, B. (1974). "The effect of adsorbed polymers on dispersion stability", *Adv. Colloid Interface Sci.* **4**, 193-277.

Wijmans, C. M., E. B. Zhulina and G. J. Fleer (1994). "Effect of free polymer on the structure of a polymer brush and interaction between 2 polymer brushes", *Macromolecules* **27**, 3238-3248.

Witten, T., L. Leibler and P. Pincus (1990). "Stress-relaxation in the lamellar copolymer mesophase", *Macromolecules* **23**, 824 - 829.

Zamir, E., M. Katz, Y. Posen, N. Erez, K. M. Yamada, B. Z. Katz, S. Lin, D. C. Lin, A. Bershadsky, Z. Kam and B. Geiger (2000). "Dynamics and segregation of cell-matrix adhesions in cultured fibroblasts", *Nature Cell Biology* **2**(4), 191-196.

EFFECT OF ELECTROSTATIC INTERACTIONS ON FRICTIONAL FORCES IN ELECTROLYTES

L.I. DAIKHIN and M. URBAKH
School of Chemistry, Tel Aviv University, Ramat Aviv, Tel Aviv 69978, Israel

Abstract. We propose a theoretical description of frictional phenomena in nanoscale layers of electrolyte solutions embedded between two plates, one of which is externally driven. It is shown that a presence of nonuniform charge distributions on the plates leads to a space-dependent frictional force, which enters into the equation of motion for the top driven plate. The equation displays a rich spectrum of dynamical behaviors: periodic stick-slip, erratic and intermittent motions, characterized by force fluctuations, and sliding above the critical velocity. Boundary lines separating different regimes of motion in a dynamical phase diagram are determined. The dependencies of the frictional force and regimes of motion on an electrolyte concentration, surface charge distribution and a thickness of the liquid layer are predicted.

1. Introduction

Much attention has been recently developed within the field of nanotribology to the understanding of the nature of friction at a microscopic scale (Singer and Pollack, eds, 1992; Bhushan et al., 1995; Bhushan, ed, 1997; Persson, 1998). Sheared liquids confined between two atomically smooth solid surfaces provide a good example of a system where a broad range of phenomena and new behaviors have been experimentally observed (Yoshizawa et al., 1993a,b; Hu et al. 1991; Demirel et al, 1996; Klein et al., 1995, 1998; Kumacheva et al., 1998; Georges et al., 1996; Crassous et al., 1997). These include dry friction-like behavior observed for atomically thin liquid layers at low driving velocity, transition to a liquid-like sliding with the increase of layer thickness and/or driving velocity, and shear thinning. These and other observations have motivated theoretical efforts, both numerical (Thompson et al., 1990, 1995; Gao et al., 1995, 1996, 1997; Landman et al., 1996; Persson, 1993, 1995; Persson at al., 1996; Rozman et al., 1996, 1997; Braun et al., 1997; Bordarier et al., 1998; Roder et al., 1998) and analytical (Carlson et al., 1996; Urbakh et al., 1995 a,b,c; Elmer, 1997; Weiss et al., 1997; Strunz et al., 1998; Braiman et al., 1996; Baumberger et al., 1995, 1998; Sokoloff, 1990, 1995; Daikhin et al., 1999), but many aspects of friction are still not well understood.

To get new insights that will help establish the basics of nanotribology it is necessary to perform measurements under well-defined conditions and to have a possibility to change interactions in a controlled way. An electrochemical environment

B. Bhushan (ed.),
Fundamentals of Tribology and Bridging the Gap between the Macro- and Micro/Nanoscales, 199–214.
© 2001 *Kluwer Academic Publishers.*

can provide such conditions for nanotribological studies. Electrode surfaces immersed in electrolyte have well defined properties. Dissolution of surface groups leads to charging of these surfaces, resulting in electrostatic interactions between surfaces. All electrostatic interactions are known and well described (Israelachvili, 1991). There are many ways to change interactions in electrochemical systems without changing any other properties of the measurement, for instance by varying the electrolyte concentration and composition. Moreover for conducting surfaces the surface potential can be changed during measurements, which allows distinguishing between different contributions to frictional forces. Friction measurements performed in electrolytic environment (Dhinojwala et al., 1997; Weiland et al., 1997; Wilhelm et al., 1997) have already demonstrated interesting dependencies of frictional dynamics on the electrolyte concentration and on liquid film thickness.

In this paper we propose a theoretical description of frictional phenomena in a thin layer of electrolyte solution confined between two plates. The dependencies of the frictional force and regimes of motion on the electrolyte concentration, surface charge distribution and the thickness of the liquid layer are studied. The proposed model leads to the observed experimental behavior and to predictions that are amenable to experimental tests.

2.   Electrostatic interactions

We consider two plates separated by a thin layer of an electrolyte solution. The top plate of mass M is pulled by a linear spring with a force constant K connected to a stage, which moves with a velocity V. Dissociation of surface groups leads to charging of the plate surfaces (Israelachvili, 1991), resulting in electrostatic interactions between them. The charge on the solid surfaces is obviously not uniformly distributed over the surfaces. The discreteness of the surface charges is a natural source of this nonuniformity.

Motion of the nonuniformly charged top plate gives rise to a reorganization of the ionic distribution in the electrolyte solution, which results in a resistance force acting on the top plate. The relaxation time of the ionic system, τ_D, could be estimated as $\tau_D^{-1} = \kappa^2 D$ (Delahey, 1966), where κ^{-1} is the Debye length and D is the diffusion coefficient of the ions in the solution. For a 1-1 binary electrolyte solution $\kappa^{-1} = (\varepsilon_{el} k_B T / 8\pi n e^2)^{1/2}$, where n is the electrolyte concentration, ε_{el} the dielectric constant of the solvent, e the charge of electron, T the temperature and k_B the Boltzmann constant. In 0.1 - 0.001 M aqueous solutions τ_D^{-1} is typically 10^7-10^9 s^{-1}. The characteristic time related to the motion of the nonuniform surface charge could be estimated as $\tau_m^{-1} = V_{max} / l$, where l is the average distance between charges on the plate surfaces and V_{max} is the maximal velocity of the top plate. The charges on real surfaces are typically 1-10 nm apart from each other on average (Israelachvili, 1991) and V_{max} does not exceed $10^2 v$ (see below), where the velocity of the stage, v, is typically 10^{-2}-1 μm/s. As a result τ_m^{-1} falls in the range 10^2 to

10^5 s^{-1}. Our estimations demonstrate that the relaxation of the ionic atmosphere is much faster than the motion of surface charges (Deborah number is much smaller than 1). In this case the ionic system is in equilibrium at all times and the lateral, frictional, force acting on the top plate, $\mathbf{\Pi}$, is determined by the variation of the free energy of the ionic system Φ,

$$\mathbf{\Pi} = -\frac{\partial \Phi}{\partial \mathbf{X}} \tag{1}$$

where X is the lateral displacement of the top plate with respect to the bottom one and the axis x is chosen to coincide with the direction of motion.

If the ions in the solution are treated as a dilute, ideal gas, the free energy of the electrolyte plasma can be written as (Safran, 1994)

$$\Phi = kT\int d^3\mathbf{r}\{n_+(\mathbf{r})[\ln n_+(\mathbf{r})/n - 1] + n_-(\mathbf{r})[\ln n_-(\mathbf{r})/n - 1] + 2n\} + \frac{\varepsilon_{el}}{8\pi}\int d^3\mathbf{r}\nabla\varphi(\mathbf{r})\nabla\varphi(\mathbf{r}) \tag{2}$$

Here $\varphi(\mathbf{r})$ is the electrostatic potential in the electrolyte and $n_+(\mathbf{r})$ and $n_-(\mathbf{r})$ are the concentrations of positive and negative ions which are related to the potential by the equations

$$n_\pm(\mathbf{r}) = n\exp(\mp e\varphi(\mathbf{r})/k_BT) \tag{3}$$

In order to calculate the force, Π, one needs the distribution of the electrostatic potential, $\varphi(\mathbf{r})$, in the electrolyte. The latter is described by the solution of the Poisson-Boltzmann equation (Israelachvili, 1991; Safran, 1994). As a first step we restrict our consideration by its linearized version, valid for low potentials $\varphi < k_BT/e$:

$$(\nabla^2 - \kappa^2)\varphi(\mathbf{r}) = 0 \tag{4}$$

The solution of Eq.(4) must satisfy the boundary conditions relating the normal component of electrostatic displacement to the surface charge densities at the plates. We describe the surface charge densities at the bottom and the top plates by the functions $\sigma_0(R)$ and $\sigma_d(R+X)$, respectively. The planes $z=0$ and $z=d$ are chosen to coincide with the plate surfaces and $R=(x,y)$ denotes a tangential coordinate. Then, the boundary conditions can be written as

$$\varepsilon_{el}\frac{\partial\varphi(z = 0+, \mathbf{R})}{\partial z} - \varepsilon_{sub}\frac{\partial\varphi(z = 0-, \mathbf{R})}{\partial z} = 4\pi\sigma_0(\mathbf{R}),$$

$$\varepsilon_{sub}\frac{\partial\varphi(z = d+, \mathbf{R})}{\partial z} - \varepsilon_{el}\frac{\partial\varphi(z = d-, \mathbf{R})}{\partial z} = 4\pi\sigma_d(\mathbf{R}+\mathbf{X}), \tag{5}$$

where ε_{sub} is the dielectric constant of the plates. A similar model, with uniform surface charge densities has been used to study a normal pressure between two charged surfaces in an electrolyte solution (Israelachvili, 1991; Safran, 1994).

Here we focus on the effect of lateral nonuniformity of the surface charge density, which plays an essential role in frictional phenomena. We assume that surface charge distributions $\sigma_0(R)$ and $\sigma_d(R+X)$ are frozen and do not depend of the relative displacement of the plates. The influence of fluctuations of surface charges on the interactions between plates has been considered in (Pincus et al., 1998).
The distribution of the electrostatic potential in the solution has been calculated in (Daikhin et al., 1999)

$$\varphi(\mathbf{R},z) = \int \frac{d^2\mathbf{K}}{(2\pi)^2} \frac{4\pi\exp(i\mathbf{KR})}{D(K)} \{\sigma_0(\mathbf{K})[\varepsilon_{el}q_K\cosh(q_K(z-d)) - \varepsilon_{sub}K\sinh(q_K(z-d))]$$

$$+\sigma_d(\mathbf{K})\exp(iK_xX)[\varepsilon_{el}q_K\cosh(q_Kz) + \varepsilon_{sub}K\sinh(q_Kz)]\}$$

$$(6)$$

where
$$D(K) = (\varepsilon_{el}^2q_K^2 + \varepsilon_{sub}^2K^2)\sinh(q_Kd) + 2\varepsilon_{el}\varepsilon_{sub}q_K K\cosh(q_Kd) \qquad (7)$$
In the range of low potentials one may expand ion concentrations, $n_\pm(\mathbf{r})$, in $e\varphi/k_BT$. Then considering in Eq. (2) terms up to the second order in $e\varphi/k_BT$ we can calculate using Eqs.(6), (2) and (1) the final equation for the lateral force acting on the moving top plate

$$\Pi = 4\pi i \int \frac{d^2\mathbf{K}}{(2\pi)^2} \frac{\varepsilon_{el}q_K K_x\sigma_d(\mathbf{K})\sigma_0(-\mathbf{K})\exp(iK_xX)}{D(K)}. \qquad (8)$$

Equation (8) correlates the lateral force with the charge distributions on the plates. As an example, we assume periodically varying charge density distributions along the plate surfaces

$$\sigma_0(\mathbf{R}) = \sigma_d(\mathbf{R}) = \bar{\sigma} + \Delta\sigma\sin\frac{2\pi}{l}x \qquad (9)$$

This leads to the following expression for the space-dependent lateral force

$$\Pi = b\sin(\frac{2\pi}{l}X), \qquad (10)$$

where

$$b = \frac{4\pi^2(\Delta\sigma)^2 S\varepsilon_{el}\sqrt{\kappa^2 + (2\pi/l)^2}}{lD(K = 2\pi/l)}, \qquad (11)$$

S is the area of the plate surfaces. Equation (11) can be simplified taking into consideration that $\varepsilon_{el} \gg \varepsilon_{sub}$

$$b = \frac{4\pi^2 S(\Delta\sigma)^2}{\varepsilon_{el}l\sqrt{\kappa^2 + (2\pi/l)^2}\,\sinh(d\sqrt{\kappa^2 + (2\pi/l)^2})} \qquad (12)$$

Equations (10), (11) and (12) present the dependence of the frictional force on the distance between plates d, the electrolyte concentration n and periodicity of the surface charge distributions l. The amplitude of the frictional force decreases exponentially with the distance between plates. In the range of high electrolyte concentrations, $\kappa \geq 2\pi / l$, we predict a sharp decrease in b as n increases. For lower concentrations, $\kappa < 2\pi / l$, the amplitude, b, depends only slightly on n.

3. Dynamics

The motion of the top driven plate, which is the basic observable in SFA experiments, is determined by the interplay between the electrostatic lateral force, a viscous friction and the external spring force. Taking into consideration the separation of time scales, which correspond to the top plate motion and to the ionic subsystem relaxation, the dynamical equation for the plate can be written in the form

$$M\ddot{X} + \Gamma\dot{X} + K(X - Vt) - b\sin(2\pi X / l) = 0 \qquad (13)$$

The dissipative force, $-\Gamma\dot{X}$, in Eq. (13) describes the viscous friction at the top plate-solution interface, $\Gamma = \eta_{eff} S / d$, where η_{eff} is the effective viscosity in the confined liquid layer. The effective viscosity of the thin layer may differ essentially from the bulk viscosity of the solution (Hu et al., 1991; Demirel et al., 1996).

The important outcome of our electrostatic consideration is that the effective frictional force, $F_{fr} = -\Gamma\dot{X} + b\sin(2\pi X / l)$, in the equation of motion for the macroscopic mechanical degrees of freedom Eq.(13) has to be space-dependent as obtained in Eq.(10). The space dependence of the friction force reflects properties of the microscopic interactions at the surfaces, namely the nonuniformity of surface charge distribution at the plate surfaces taken here as periodic. The typical lateral length scale of the electrostatic interaction, l, reappears in the macroscopic friction force. The independence of the lateral force, Π, of the velocity is a consequence of the fact that the relaxation time of the embedded system, τ_D, is faster than the characteristic time related to the motion of the top plate, τ_m. Similar equations with space-dependent frictional forces emerge also in other systems, for instance, in the case of dry friction (Baumberger et al., 1998; Helman et al., 1994).

It is convenient to introduce dimensionless space and time coordinates $y=2\pi X/l$ and $\tau=t\omega$, where $\omega = (2\pi / l)^{1/2}\sqrt{b / M}$ is the frequency of the small oscillations of the top plate in the minima of the periodic potential, $\dfrac{l}{2\pi}b\cos(\dfrac{2\pi}{l}X)$. Equation (13) can be rewritten then in a dimensionless form as

$$\ddot{y} + \gamma\dot{y} - \sin(y) + \alpha(y - v\tau) = 0 \qquad (14)$$

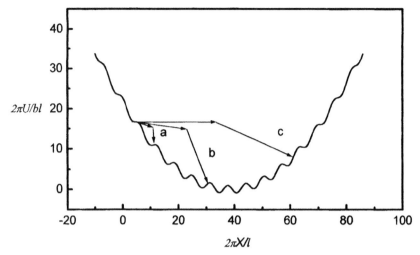

Figure 1. Dimensionless total potential $2\pi U(X,t)/bl$ versus the dimensionless plate coordinate $2\pi X/l$ drawn for $\alpha = 0.03$ and $v\tau = 40$. Arrows indicate a slip motion (a jump) of the plate for three different dynamical regimes (a)-(c) discussed in the text.

The dynamical behavior of the model is determined by the following dimensionless parameters: $\gamma = \Gamma/(M\omega)$ is a dimensionless dissipation constant, $\alpha = (\Omega/\omega)^2$ is the square of the ratio of the frequency of the free oscillations of the top plate $\Omega = \sqrt{K/M}$ to ω, and $v = 2\pi V/(\omega l)$ is the dimensionless stage velocity. The model leads to a number of different regimes of the motion of the top driven plate, which is the experimental observable.

3.1. THE LOW VELOCITY REGIME

The main objective of the SFA experiments is to deduce information on microscopic properties of the system from the observed dynamics of the top plate. For this purpose one needs to understand the dependence of the dynamics on the mechanical (external) parameters and the parameters of the embedded system (internal). First we investigate the motion of the plate for very small velocities of the stage v<<1. In this case, the motion can involve two steps: slow motion (creep) in a local minimum of the total potential $U(X,t)$ (see Figure 1.)

$$U(X,t) = \frac{l}{2\pi} b \cos(\frac{2\pi}{l} X) + \frac{K}{2}(X - Vt)^2 \qquad (15)$$

and a fast slip (sliding) that begins when an instability occurs, i.e. d^2U/dX^2 changes sign. The latter is possible for α<1 only. At the point of instability the spring force reaches a maximum value corresponding to the static friction force, F_s. The static friction equals to the maximum value of the lateral force acting on the top plate, the amplitude b in Eq.(10),

$$F_s = b \qquad (16)$$

During a sliding the spring force, $F=K(X-Vt)$, decreases until it reaches a value, F_k, where the sliding ceases and the top plate is trapped again at a potential minima. Thus a periodic stick-slip motion of the top plate is observed for $v \ll 1$ and $\alpha < 1$. This type of motion has been recently observed in the experiments performed in electrolyte solutions (Wilhelm et al., 1997). For $\alpha > 1$ no instabilities occur, $d^2U / dX^2 \neq 0$, and at all times the plate follows adiabatically the motion of the stage being in a minima of the total potential. Here we concentrate on the dynamics of the system in the most interesting case of $\alpha < 1$, when the electrostatic interaction between plates is stronger than the external spring force.

The dynamics of the stick-slip motion could be analyzed taking into account that the stage is effectively at rest during the fast slip of the plate, $Vt=L_0=const$. The time pattern of the stick-slip motion is determined by the relationship between parameters α and γ. Three regimes can be distinguished (Baumberger et al., 1998): (a) $\gamma^2 / 4 \gg 1$, the system is over damped; (b) $\alpha < \gamma^2 / 4 < 1$, the system is underdamped with respect to the periodic potential and overdamped with respect to the driving spring; and (c) $\gamma^2 / 4 \ll \alpha \ll 1$ the system is under damped.

(a) $\gamma^2 / 4 \gg 1$. In this regime the "slip" motion of the top plate corresponds to the jump between nearest neighbor minima of the potential $U(X,t=L_0/V)$ and the "slip" distance, Δ, is about l, the period of the lateral force Π.

Important information on the nature of stick-slip motion and on a transition to

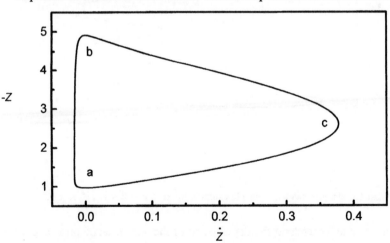

Figure 2. Phase space representation of the plate motion in the over damped regime for $\gamma=4$; $\alpha=0.2$, $v=0.02$.

sliding could be obtained from a phase portrait ($z = y - v\tau + \gamma w / \alpha$ vs $\dot{z} = \dot{y} - v$) that characterizes the oscillatory motion of the system. The phase space representation is closely related to the dependence of the spring force, $\alpha(v\tau-y)$ on the top plate velocity, $\dot{y}$ which has been studied experimentally (Nasuno et al., 1997). Figure 2 shows a representative phase portrait of the over damped regime, $\alpha=0.2$, $\gamma=4$ and $v=0.02$. The interval (a, b) corresponds to a very slow motion (creep) of the top plate located in the minima of the potential $U(X,t)$. The stable point a is the node. The top plate starts to slide when the node transforms into the saddle point b where the instability occurs, approaches the maximal velocity at the point c and comes to rest at a. *(b)* $\alpha < \gamma^2 / 4 < 1$. In this regime the system manifests a qualitatively different dynamical behavior (see Figure 3). The following important features of the top plate motion should be noted: (i) The slip distance increases with the decrease of α and could be much larger than the period l . The number of maxima of $\dot{z}$ as a function of z, seen from the phase portrait, corresponds to a number of periods covered by the plate during the slip. (ii) After sliding, the plate oscillates while approaching the stable equilibrium position that corresponds to the focus. The equilibrium position itself moves in the direction of the next jump, and the nature of the singular point ($\partial^2 U / \partial X^2 = 0$) changes transforming from the focus into the node and then into the saddle point.

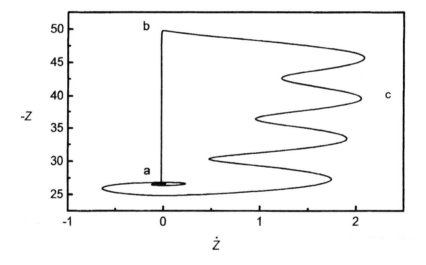

Figure 3. Phase space representation of the plate motion for $\alpha < \gamma^2 / 4 < 1$: $\gamma=0.5$; $\alpha=0.02$, $v=0.02$.

In order to understand the dependence of the slip distance on α and γ we have derived an approximate analytical expression for the top plate velocity, $\dot{X}$, using the energy balance equation

$$\frac{M\dot{X}^2}{2} - \frac{l}{2\pi}b\cos(\frac{2\pi}{l}X) + \frac{K}{2}(X - L_0)^2 = U(X_0, L_0) - \Gamma\int_{X_0}^{X}\dot{X}dX. \tag{17}$$

Equation (17) has been solved using the expression for the dissipated energy, which was found by the solution of Eq.(13) in the zeroth order in b. This approximate solution of Eq.(17) is in a good agreement with the results of numerical calculations of $\dot{X}(X)$ according Eq.(13). The slip distance, Δ, could be determined as a distance between the point $X=X_0$, where the slip starts, and the next turning point, where $\dot{X}=0$. As a result for $\alpha\ll1$ we obtain

$$\Delta \approx \frac{l}{2\pi\alpha}(1 - \sqrt{2}\gamma(1 - \frac{g(\gamma)}{\gamma})), \tag{18}$$

where

$$g(\gamma) = \sqrt{\pi} - \frac{3}{2\gamma} + \frac{\exp(-\sqrt{\pi}\gamma)}{2\gamma}(4 - \exp(-\sqrt{\pi}\gamma)). \tag{19}$$

Equation (18) shows that the slip distance increases proportionally to b/Kl. Figure 4 shows the dependencies of Δ on α for $\gamma=0.5$.

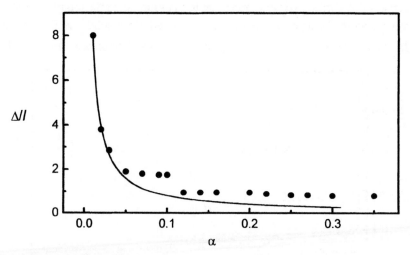

Figure 4. Slip distance, Δ, versus dimensionless spring constant, α. Closed circles show the results of numerical calculations, the solid line corresponds to calculations according Eq.(17). Δ is presented in units of the period of surface charge distribution, l. Dimensionless dissipation constant $\gamma=0.5$, and $\nu=0.02$.

Our calculations show that time patterns of the stick-slip motion depend on (I) the parameters of the embedded system (thickness of the liquid layer, d, concentration of the electrolyte solution, n, lateral length scale, l) and (II) the mechanical parameters (mass of the top plate, M, and spring constant, K). All these parameters are included into the dimensionless quantities α and γ. In the case of the electrolyte solution confined

208

between two plates the parameter α increases exponentially with the distance d between plates and/or electrolyte concentration, n, (see Eqs.10 and 11). Thus the increase of d and/or n, for given K, M and V, should lead to a decrease in the sliding distance and to a transition from stick-slip motion to smooth sliding. These conclusions are in agreement with the preliminary experimental results (Wilhelm et al., 1997). *(c)* $\gamma^2 / 4 \ll \alpha \ll 1$. In this regime the periodic motion of the system includes a stick period followed by slowly attenuated oscillations (Baumberger et al., 1998). The top plate starts to slip from the saddle point, overshoots the lowest well of the total potential $U(X,L_0)$ and bounces a few times across the modulated parabola (see Figure 1) before it slowly comes to rest into one of the pinning wells.

3.2.   TRANSITION FROM STICK-SLIP TO SLIDING

As the stage velocity increases the stick-slip motion of the top plate becomes more erratic and intermittent and then changes to periodically modulated sliding state. Figure 5 shows the dynamical phase diagram (in the α-v plane) that presents regions of parameters, which corresponds to different regimes of motion of the top plate. Stick-slip motion and smooth sliding occur respectively to the left of the solid line, $v_c^{(1)}(\alpha)$, and to the right of the dashed line, $v_c^{(2)}(\alpha)$. The system exhibits an intermittent motion in the

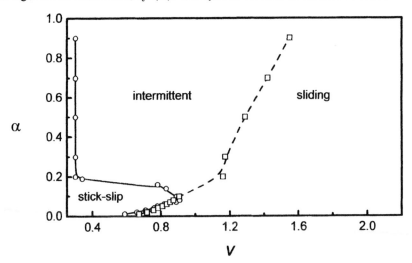

Figure 5. Dynamical phase diagram for the plate motion. The solid line, $v_c^{(1)}(\alpha)$, indicates the boundary between the stick-slip and intermittent motions, the dashed line, $v_c^{(2)}(\alpha)$, is the lower velocity boundary of the smooth sliding. Open circles and squares show the results of numerical calculations of $v_c^{(1)}(\alpha)$ and $v_c^{(2)}(\alpha)$, respectively. Dimensionless dissipation constant γ=0.5.

range of parameters between these two curves. The lines $v_c^{(1)}(\alpha)$ and $v_c^{(2)}(\alpha)$ describe the α-dependencies of the critical velocities corresponding to transitions between different states of motion. They have been found by the analysis of numerical solutions of Eq.(13). The following characteristic features of the phase diagram should be mentioned:

(i) For $0.2 < \alpha < 1$ the critical velocity, $v_c^{(1)}(\alpha)$, separating stick-slip from the intermittent motion depends only slightly on α. Here the stick-slip motion corresponds to jumps between nearest neighbor cells of the total potential U (see Figure 1). The analysis of the phase portrait shows that the stick-slip state holds as long as the relaxation of the top plate to the local minimum of the potential $U(X,t)$ after a slip event is faster than the motion of this minimum. As a result the critical velocity $v_c^{(1)}(\alpha)$ increases with the increase of γ. It should be noted that in this range of parameters we observed a smooth transition from the stick-slip to intermittent motion and therefore the boundary line, $v_c^{(1)}(\alpha)$, is not well defined here. Since for $0.2 < \alpha < 1$ the slip distance lies in a nanometric range $\Delta \approx l$, it should be hard to distinguish experimentally between this type of stick-slip motion and the periodically modulated sliding. (ii) With the decrease of α we observed a steep rise in the critical velocity $v_c^{(1)}(K)$ that parallels the increase of the slip distance. The critical velocity as a function of α has a sharp maximum for $\alpha = \alpha_{max} \approx 0.1$. For smaller values of α, $\alpha < \alpha_{max}$, two boundary lines, $v_c^{(1)}(\alpha)$ and $v_c^{(2)}(\alpha)$, approach each other. Numerical calculations show that for all values of parameters the overdamped regime, $\alpha < \gamma^2 / 4$, lies within the interval $\alpha < \alpha_{max}$ where $v_c^{(1)}(\alpha)$ decreases with decrease of α. For $\alpha < \alpha_{max}$ the critical velocity $v_c^{(1)}(\alpha)$ decreases as γ increases. As it has been already mentioned in (Baumberger et al., 1998) for $\alpha < \alpha_{max}$ the motion bifurcates discontinuously from a periodic stick-slip to an intermittent one.

The line $v_c^{(2)}(\alpha)$, separating the periodically modulated sliding from the intermittent motion could be found analytically by a linear stability analysis of the solutions of Eq.(14) (Weiss et al., 1997; Strunz et al., 1998). For this purpose let us rewrite Eq.(14) in the form

$$\ddot{z} + \gamma\dot{z} + \alpha z + \sin(z)\cos(v\tau - v\gamma / \alpha) + \cos(z)\sin(v\tau - v\gamma / \alpha) = 0, \tag{20}$$

where $z = y - v\tau + \gamma v / \alpha$. Equation (20) describes a damped harmonic oscillator that is driven parametrically by the external force $f = -\sin(z)\cos(v\tau) - \cos(z)\sin(v\tau)$. For high driving velocities $v \gg 1$ the solution of Eqs.(14) has the form of the periodically modulated sliding state (Helman et al., 1994)

$$y = v\tau + \gamma v / \alpha + \sin(v\tau) / v^2. \tag{21}$$

Besides the sliding state of Eq.(21) modulated with the frequency v we observed also sub-harmonic oscillations with frequencies v/n (n=2, 3, ...) that arise due to parametric resonances (Weiss et al., 1997; Strunz et al., 1998). Parametric resonance is an instability phenomenon. In (Daikhin et al., 1999) we find velocity intervals which correspond to these instabilities. For $\gamma \approx 1$ the system exhibits the first-order (n=2) paramentric resonance only. The critical velocity $v_c^{(2)}(\alpha)$ is defined as the largest driving velocity at which the first-order parametric resonance is able to destabilize the sliding state (21). This condition yields the equation (Daikhin et al., 1999)

$$v_c^{(2)}(\alpha) = 2\left\{\alpha - \gamma^2/2 + \frac{1}{2}\sqrt{\gamma^4 - 4\gamma^2\alpha + 1}\right\}^{1/2},\qquad(22)$$

which approximates well the numerical solution $v_c^{(2)}(\alpha)$ presented in Figure 5. When the driving velocity decreases and becomes lower than $v_c^{(2)}(\alpha)$ the period of the top plate oscillations is doubled and their amplitude increases sharply. For a low dissipation constants $\gamma \ll 1$ we also observe an instability corresponding to the n-th order parametric resonances, with $n>2$. These resonances could exist for velocities $v > v_c^{(2)}(\alpha)$. The period of top plate oscillations equals to $2\pi n/v$ in the vicinity of the resonances.

As the driving velocity varies from $v_c^{(1)}(\alpha)$ to $v_c^{(2)}(\alpha)$ the motion of the top plate bifurcates from the periodic stick-slip to modulated sliding. Above $v_c^{(1)}(\alpha)$ the stick-slip motion becomes erratic and intermittent. For a wide range of system parameters we find that the motion is chaotic. Figure 6 shows various examples of phase portraits and time series of the spring force as one passes from the stick-slip motion to sliding. The amplitude and the period of force oscillations decrease drastically as the driving velocity increases. The results presented have been calculated for the case of $\alpha \ll 1$ when the slip distance is much larger than the period of the plate potential. The system exhibits a rich spectrum of behaviors within the interval $(v_c^{(1)}(\alpha)), v_c^{(2)}(\alpha))$ even though it is very narrow for chosen parameters. This makes it clear that a single critical velocity (a single boundary line) cannot characterize the transition from the periodic stick-slip to smooth sliding. The dynamical phase diagram should include two boundary lines $v_c^{(1)}(\alpha)$ and $v_c^{(2)}(\alpha)$ in order to account for the region of the intermittent motion, which is essential for the understanding of the frictional dynamics.

4.   Conclusions

We have proposed a theoretical description of frictional phenomena in nanoscale layers of electrolyte solutions. It has been shown that the presence of nonuniform charge distributions on the plates gives rise to a space-dependent frictional force. This force depends strongly on the distance between plates d, electrolyte concentration n, and the

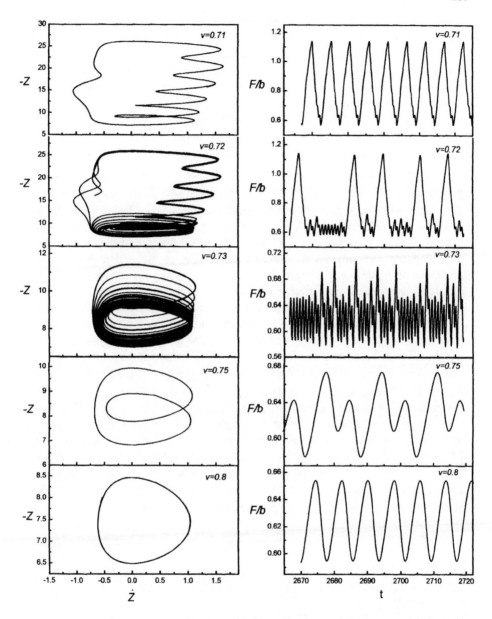

Figure 6. Phase portraits and the time series of the spring force for different stage velocities. Stage velocities are denoted on the graphs; γ=0.5, α=0.03.

lateral length scale of the surface charge distributions l. A separation of time-scales for the plate motion and the relaxation of the ionic subsystem allowed us to derive the equation of motion (13) that includes only one, macroscopic degree of freedom, the displacement of the top driven plate. The microscopic properties of the system enter into this equation through the frictional force derived above. The equation leads to a spectrum of behaviors in the motion of the plate: periodic stick-slip, erratic and intermittent motions, characterized by force fluctuations, and sliding above the critical velocity $v_c^{(2)}(\alpha)$. For a given driving velocity the dynamical properties of the system are determined by two dimensionless parameters: $\alpha = Kl / (2\pi b)$, the ratio of the spring force for the stretching l to the amplitude of the lateral electrostatic interaction between plates, and $\gamma = \Gamma / (2\pi b M / l)^{1/2}$, the dissipation constant. It is usually considered that the stick-slip behavior characterizes the solid-like state of the film situated between two plates of the surface force apparatus. And the smooth sliding relates to the liquid-like state of the film. However, our calculations show that the type of the motion of the top plate depends not only on the internal parameters of the system but also on the properties of the device. For example, changing the stiffness of the spring only, we can translate our system from region of stick-slip to the region of sliding. The behavior of the system under consideration essentially depends on the device, which we use to observe it. Measurements in an electrolytic environment make possible to change these parameters in a controlled way varying the electrolyte composition and concentration and the distance between plates (see Eqs.(11), (12)). For conducting surfaces the surface charge densities σ_0 and σ_d could be easily changed that will strongly influence the parameter α. We have shown that the increase of d and/or n should result in decrease of the amplitude of stick-slip oscillations, the slip distance and in transition from the stick-slip to intermittent motion and to smooth sliding. These conclusions are in agreement with the preliminary results of experimental measurements in aqueous solutions (Wilhelm et al., 1997). Our calculations also demonstrated that a variation of the electrolyte concentration and the distance between plates strongly affects the dynamical phase diagram (leads to a shift of the boundary lines $v_c^{(1)}(\alpha)$ and $v_c^{(2)}(\alpha)$), which could be determined experimentally (Baumberger et al., 1994).

5. References

Baumberger T., Caroli C., Perrin B. and Ronsin O. (1995), "Nonlinear Analysis of the Stick-Slip Bifurcation in the Creep-Controlled Regime of Dry Friction", *Phys. Rev. E* **51**, 4005-4010.

Baumberger T. and Caroli C. (1998), "A Phenomenology of Boundary Lubrication: the Lumped Junction Model", *Eur.Phys.J. B* **4**, 13-23.

Berman A.D., Ducker W.A. and Israelachvili J. (1996), "Origin and Characterization of Different Stick-Slip Friction Mechanisms", *Langmuir* **12**, 4559-4563.

Bhushan B., Israelachvili J. and Landman U. (1995), "Nanotribology-Friction, Wear and Lubrication at the Atomic Scale", *Nature* **374**, 607-616.

Bhushan B. (ed) (1997), *Micro/Nanotribology and Its Application*, Vol. **330** of *NATO Advanced Study Institute, Series E: Applied Sciences*, Kluver Academic, Dordrecht.

Bordarier P., Schoen M. and Fuchs A.H. (1998),"Stick-Slip Phase Transitions in Confined Solidlike Films from an Equilibrium Perspective", *Phys. Rev.E* **57**, 1621-1635.

Braiman Y., Family F. and Hentschel (1996), "Nonlinear Friction in the Periodic Stick-Slip Motion of Coupled Oscillators", *Phys. Rev. B* **55**, 5491-5504.

Braun O., Dauxois T. and Peyrard M. (1997), "Friction in a Thin Commensurate Contact", *Phys. Rev. B*. **56**, 4987-4995.

Carlson J.M. and Batista A.A. (1996), "Constitutive Relation for the Friction between Lubricated Surfaces", *Phys. Rev. E* **53**, 4153-4165.

Crassous J., Charlaix E. and Loubet J.L. (1997), "Nanoscale Investigation of Wetting Dynamics with a Surface Force Apparatus", *Phys. Rev. Lett.* **78**, 2425-2428.

Daikhin L.I. and Urbakh M. (1999), "Frictional Forces in an Electrolytic Environment", *Phys. Rev. E* **59**, 1921-1931.

Delahey P. (1966), *Double Layer and Electrode Kinetics*, Wiley, New York.

Demirel A.L.and Granick S. (1996), "Friction Fluctuation and Friction Memory in Stick-Slip Motion", *Phys. Rev. Lett.* **77**, 4330-4333.

Dhinojwala A. and Granick S. (1997), "Relaxation Time of Confined Aqueous Films under Shear", *J. Am. Chem. Soc.* **119**, 241-242.

Elmer F.J. (1997), "Nonlinear Dynamics of Dry Friction", *J. Phys. A* **30**, 6057-6063.

Gao J.P., Luedtke W.D. and Landman U. (1995), "Nano-Elastohydrodynamics – Structure, Dynamics and Flow in Nonuniform Lubricated Junctions", *Science* **270**, 605-608.

Georges J.M., Tonck A. and Loubet J.L. (1996), "Rheology and Friction of Compressed Polymer Layers Adsorbed on Solid Surfaces", *J. Phys. II* **6**, 57-76.

Helman J.S., Baltensperger W. and Holyst J.A. (1994), "Simple Model of Dry Friction", *Phys. Rev. B* **49**, 3831-3838.

Hu H.W., Carson G.A. and Granick S. (1991), "Relaxation Time of Confined Liquids under Shear", *Phys. Rev. Lett.* **66**, 2758-2761.

Israelachvili J. (1991), *Intermolecular and Surface Forces*, 2nd ed., Academic, London.

Klein J. and Kumacheva E. (1995), "Interfacial Shear of Polymeric Surface Phases", *Science* **269**, 816-819.

Klein J. and Kumacheva E. (1998), "Simple Liquids Confined to molecularly thin layers. I. Confinement induced liquid to solid phase transitions", *J. Chem. Phys.* **108**, 6996-7009.

Kumacheva E. and Klein J. (1998), "Simple Liquids Confined to Molecularly Thin Layers. II. Shear and Frictional Behavior of Solidified Films", *J. Chem. Phys.* **108**, 7010-7022.

Landman U., Luedtke W.D. and Gao J.P. (1996), "Atomic-Scale Issues in Tribology: Interfacial Junctions and Nano-Elastohydrodynamics", *Langmuir* **12**, 4514-4528.

Nasuno S., Kudrolli A. and Gollub J. (1997), "Friction in Granular Layers. Hysteresis and Precursors", *Phys. Rev. Lett.* **79**. 949-052.

Persson B.N.J. (1998), *Sliding Friction. Physical Principles and Applications*, Springer-Verlag, Berlin.

Pincus P.A. and Safran S.A. (1998), "Charge Fluctuations and Membrane Attractions", *Europhys. Lett.* **42**, 103-108.

Rozman M.G., Urbakh M. and Klafter J. (1996), "Origin of Stick-Slip Motion in a Driven Two-Wave Potential", *Phys. Rev. E* **54**, 6485-6494.

Rozman M.G., Urbakh M. and Klafter J. (1997), "Stick-Slip Dynamics as a Probe of Frictional Forces", *Europhys. Lett.* **39**, 183-188.

Roder J., Hammerberg J.E., Holian B.L. and Bishop A.R. (1998), "Multichain Frenkel-Kontorova Model for Interfacial Slip", *Nature* **374**, 607-616.

Safran S.A. (1994), *Statistical Thermodynamics of Surfaces, Interfaces and Membranes*, Addison-Wesley, reading, MA.

Singer I.L. and Pollock H.M. (eds.) (1992), *Fundamentals of Friction*, Vol. **220** of NATO Advanced Study Institute, Series E: Applied Sciences, Kluwer Academic, Dordrecht.

Sokoloff J.B. (1990), "Theory of Energy-Dissipation in Sliding Crystal Surfaces", *Phys. Rev. B.* **42**, 760-765.

Sokoloff J.B. (1995), "Microscopic Mechanisms for Kinetic Friction Nearly Frictionless Sliding for Small Solids", *Phys. Rev. B* **52**, 7205-7214.

Strunz T. and Elmer F.J. (1998), "Driven Frenkel-Kontorova Model. I. Uniform Sliding and Dynamical Domains of Different Particle Densities", *Phys. Rev. E* **58**, 1601-1611.

Tompson P.A. and Robbins M.O. (1990), "Origin of Stick-Slip Motion in Boundary Lubrication", *Science* **250**, 792-794.

Tompson P.A., Robbins M.O. and Grest G.S. (1995), "Structure and Shear Response in Nanometer-Thick Films", *Israel J. Chem.* **35**, 93-106.

Urbakh M., Daikhin L. and Klafter J. (1995a), "Dynamics of Confined Liquids under Shear", *Phys. Rev. E* **51**, 2137-2141.

Urbakh M., Daikhin L. and Klafter J. (1995b), "Velocity Profiles and the Brinkman Equation in Nanoconfined Liquids", *Europhys. Lett.* **32**, 125-130.

Urbakh M., Daikhin L. and Klafter J. (1995c), "Sheared Liquids in the Nanoscale Range", *J. Chem. Phys.* **103**, 10707-10713.

Weiss M. and Elmer F.J. (1997), "Dry Friction in the Frenkel-Kontorova Model: Dynamical Properties", *Z. Phys. B* **104**, 55-69.

Weiland M., Zink B., Shifter T. and Marti O. (1997), "Nanotribology in Electrolytic Environment, in Micro/Nanotribology and Its Applications", in Vol **330** of *NATO Advanced Study Institute, Series E: Applied Sciences*, (B.Bhushan, ed), Kluver Academic, Dordrecht.

Wilhelm M. and Klein J. (1997), *(private communication)*.

Yoshizawa H., McGuiggan P. and Israelachvili J. (1993a), "Identification of a Second Dynamic State During Stick-Slip Motion", *Science* **259**, 1305-1308.

Yoshizawa H., Chen Y.L. and Israelachvili J. (1993b), "Fundamental Mechanisms of Interfacial Friction. 1. Relation between Adhesion and Friction", *J. Phys. Chem.* **97**, 4128-4140.

ADSORPTION OF THIN LIQUID FILMS ON SOLID SURFACES AND ITS RELEVANCE FOR TRIBOLOGY

J. COLCHERO, A. GIL, P.J. DE PABLO, M. LUNA, J. GÓMEZ
AND A.M. BARÓ
Departamento de Física de la Materia Condensada
Universidad Autónoma de Madrid, E-28049 Madrid

Abstract. In many applications the sliding surfaces are exposed to ambient conditions and thus to humid air. It is well known that in air a liquid film adsorbs on surfaces, which will severely influence its tribological properties. In the present paper, a brief introduction to the physics of surfaces in equilibrium with vapor will be presented and experiments related to water adsorption on surfaces using Scanning Force Microscopy will be described. Tip-sample interaction has been measured and capillary condensation of water between tip and sample has been observed. By careful adjustment of tip-sample interaction a surface can be imaged extremely gently and the effect of water adsorption on solid surfaces can be visualized.

1. Introduction

Tribology is, without doubt, applied Surface Science. A better understanding of friction and wear is an issue of fundamental importance: a huge amount of energy would be saved if devices could work with less energy dissipation and many resources could be used more efficiently if machines lasted longer due to reduced wear. In the past few years a new field, Nanotribology, has emerged with vigor (Bhushan et al., 1995). The aim of Nanotribology is to describe and explain Tribology on a nanometer scale. It seems that the task of understanding friction is easier on this scale, since this length scale is nearer to the atomic scale where many "fundamental" processes occur which are known from material and surface science. The hope of many scientists in this field is that macroscopic friction could be connected directly to some "fundamental" tribological processes on a nanometer scale, possibly through some appropriate averaging mechanism. Progress in Nanotribology has been achieved recently theoretically as well as experimentally. Experimental advances in Nanotribology have been

B. Bhushan (ed.),
Fundamentals of Tribology and Bridging the Gap between the Macro- and Micro/Nanoscales, 215–234.
© 2001 *Kluwer Academic Publishers.*

driven by essentially two techniques: the Surface Force Apparatus on the one hand (Israelachvili, 1992), which allows detailed studies of the interaction of surfaces separated by distances in the micron to (sub-) nanometer range, and by Scanning Probe Microscopy (SPM) on the other. Application of Scanning Force Microscopy (SFM) to Nanotribology was pioneered the group of McClelland (Mate et al., *1987*). With the development of SPM-techniques, even atomic and nanometer scale tribological studies are possible.

Scanning Probe Microscopy, Tribology and Surface Science in general can be investigated in essentially two extreme situations: under Ultra High Vacuum (UHV) conditions or in ambient air. The first case is experimentally more demanding, but much easier to interpret and to compare with theoretical results due to the well-controlled surface conditions. Working in ambient air is normally much easier experimentally, but the data is much more difficult to understand due to the complexity of surfaces exposed to ambient air (Charvolin et al., 1990). The present work is dedicated to try to understand from a point of view of SPM the precise nature of surfaces in ambient conditions, that is, exposed to some gas, which in general will be water vapor. We believe that this is a very interesting topic in itself, but also a fundamental topic for tribology, since many tribological processes occur in air.

The motivation for this work derives partly also from an intellectual uneasiness after quite some years in the field of SFM. In fact, after many images of different substances and systems in ambient air one ends up with the feeling that the surface of any body is just the boundary of an ideal solid, as one would indeed expect for an ideal surface in UHV. However, it is well known that different substances, and specially water, adsorb on solid surfaces if the vapor pressure is high enough. Therefore, as a Scanning Probe Microscopist one might wonder:

Where is this liquid layer, and why does one not see this layer in most SPM applications in air?

The question posed above may also be reversed, that is:

Can we detect, observe and investigate the effects of water adsorption with SPM techniques?

As we will try demonstrate in the present paper the answer is a hopeful "Yes, but very carefully". Moreover, we believe that progress towards a better understanding of the processes involved in SPM in ambient conditions will come from investigating wetting (De Gennes, 1985) of surfaces on a nanometer scale (Herminghaus et al., 1998). On the other hand, we also believe that our knowledge about wetting on a nanometer scale will benefit from the results obtained with SPM-techniques, thus the benefits will be mutual.

2. Statistical Mechanics and some basic aspects of Wetting

Under ambient conditions, surfaces are exposed to vapor of different substances and in particular to water vapor. In this case the total system may be composed of three different phases: the solid surface, the vapor, and an adsorbed liquid film. The last two phases are open with respect to each other, that is, they can exchange energy and particles among themselves. The necessary framework to describe the corresponding physics is Thermodynamics (Kittel, 1969). It can be shown that if the two phases are in equilibrium, the Gibbs Free Energy $G \equiv E - TS + pV$ of the total system is a minimum. Here, E is the internal energy, T the temperature, S the entropy, p the pressure and V the Volume. The Gibbs Free Energy is related by $G(N, T, p) = N \cdot \mu(T, p)$, where N is the number of particles in the total system, to the Chemical Potential μ, which is another fundamental thermodynamical quantity. The Chemical Potential can be interpreted as the mean free energy per particle. Finally the entropy of a system

$$\sigma = -\sum_{i=1}^{g} p_i \ln(p_i) \tag{1}$$

is another fundamental quantity from which many termodynamical properties can be derived. It describes the degree of disorder. A system is said to be in equilibrium if its entropy, and thus its degree of disorder, is maximum. If two systems are brought together in a way that they can exchange particles and energy, one can show that in equilibrium the following quantities

$$\frac{1}{kT} \equiv \frac{\partial \sigma}{\partial E} \quad \text{and} \quad \mu \equiv -kT \frac{\partial \sigma}{\partial N} \tag{2}$$

defined respectively as, kT: Thermal Energy and μ: Chemical Potential, are equal in both systems. Two systems with different Thermal Energy and Chemical Potential will exchange energy and particles, until these two quantities are equal and the total system is thus in equilibrium. Thermal Energy flows from the system with higher Thermal Energy to that with lower Thermal Energy. The same is true for the Chemical Potential.

2.1. IDEAL GASES, GASES IN AN EXTERNAL POTENTIAL AND REAL GASES

For an ideal gas without interactions the Chemical Potential of its constituent particles can be shown to be (Kittel, 1969)

$$\mu(n, T) = kT \ln(n/n_{sat}) \ ,$$

where n is the particle density and n_{sat} the saturation density. If the molecules are in an external potential $V(x)$ this relation has to be modified to take into account the extra energy supplied by this potential. The correct relation is then

$$\mu = kT \ln (n(x)/n_{sat}) + V(x) \ , \qquad (3)$$

therefore the particle density will vary locally: since μ has to be constant for all molecules within the whole system, this implies that a local variation of the potential energy $V(x)$ has to be compensated by a corresponding variation $n(x)$ of the particle density. It is straightforward to rewrite eq. 3 as

$$n(x) = n_0 \ e^{-V(x)/kT} \ ,$$

where n_0 is some reference density where the potential vanishes. This relation describes, for example, the variation of atmospheric pressure with altitude in the gravitational potential $V(x) = mgh$. For molecules in the proximity of an attractive surface[1] the density of the particles increases with decreasing distance from the surface. The adsorption of liquid films on surfaces can be understood in the following terms: let $n_0 < n_{sat}$ be the pressure far away from the surface. At the particle density n_0 a free liquid phase is not stable. However, molecules near the surface will feel an attraction by the surface and therefore near the surface the vapor density increases. If the attraction is strong enough, the local pressure may exceed n_{sat} and a liquid film will condense on the surface. The exact height will depend, as will be shown below, on the density n_0 and on the interaction $V(z)$.

Equation 3 can also be used to describe "normal" condensation of a free liquid. In fact, molecules in a real gas are subject to mutual interactions,

$$\mu_{gas}(n) = kT \ln (n/n_{sat}) + \langle V(d(n)) \rangle \ ,$$

where $\langle V(d(n)) \rangle$ is the mean potential which the molecules feel due to their interaction. As the density of the gas is increased, the mean separation between the molecules decreases and correspondingly the mean potential energy which the particles feel varies. At a certain density the particles may choose between two different configurations which have the same Chemical Potential: the gas phase with little (negative) potential energy and low density (thus very negative $kT \ln(n(x))$-term) or the liquid phase with a high (negative) potential energy and a high density (and thus a less negative $kT \ln(n(x))$-term).

2.2. LIQUIDS ON SOLID SURFACES

One of the most evident questions regarding liquids on solid surfaces is: how does a liquid arrange itself on a solid surface? For a non-volatile liquid, that is, for a liquid whose volume does not change during experimental time, this issue was resolved almost two centuries ago by Young (1804) and Laplace

---

[1]With attractive we mean: $F(z) = -V'(z) < 0$.

(1805). The liquid spreads on the surface if $\Delta\gamma \equiv \gamma_S - \gamma_{SL} > \gamma_L$, where γ_S, γ_{SL} and γ_L are the surface energies of the solid-vacuum interface, of the solid-liquid interface and of the liquid-vacuum interface respectively. Otherwise the liquid forms a drop of contact angle $\cos(\vartheta) = \Delta\gamma/\gamma_L$.

What happens if the liquid in question is volatile, as for example water? Then the volume of the liquid is not a conserved quantity since the molecules may evaporate from the liquid, or adsorb on it. The fundamental quantity in this case is the Chemical Potential of the particles in the gas and liquid phases, which is constant in equilibrium. In the rest of the present work we will concentrate on the case of volatile liquids which adsorb on surfaces and are in equilibrium with its vapor. As will be shown below, two main effects determine the structure and equilibrium properties of adsorbed liquid films on a nanometer scale: the so called "Laplace Pressure" and the "Disjoining Pressure". For the sake of clarity, these concepts will first be introduced separately.

2.2.1. Laplace Pressure.

Consider a free spherical drop of radius r. If γ_L is the surface energy of the liquid, the total energy of the drop is $E_{surf} = 4\pi r^2 \gamma_L$. To minimize this energy, the drop will tend to shrink. The energy variation associated with a change of volume is $dE_{surf} = 2 \cdot 4\pi r \gamma_L \, dr = (2\gamma_L/r)(4\pi r^2 \, dr) = (2\gamma_L/r)dV$. Since dE/dV is a pressure, we obtain the relation

$$P_L = -2\gamma_L/r \tag{4}$$

for the so-called Laplace Pressure of the drop. This pressure is related to the tendency of the drop to reduce its surface area, and the minus sign takes into account precisely that the pressure in this case is oriented inwards (that is, contrary to the radius). If the surface of the liquid is not spherical, the geometrical entity which describes the variation of area is the mean curvature $m(r_a, r_b) \equiv \kappa^{-1} = r_a^{-1} + r_b^{-1}$, where r_a and r_b are the radii of curvature of the principal axis of the surface and κ is the so-called Kelvin Radius. For a general surface the Laplace Pressure is therefore $P_L = \gamma_L/\kappa$. The Laplace Pressure modifies the Chemical Potential of a curved liquid volume. This can be seen as follows: the surface energy of a liquid may be interpreted as a potential energy. If the surface is flat, a displacement of the surface does not "cost" any energy. This is however not true for a curved interface. In this case the displacement of the surface will "cost" some energy which has to be supplied by the internal energy of the total system. From the definition of the Chemical Potential, eq. 2, one obtains

$$\mu_{liq}(\kappa) \equiv -kT \frac{\partial \sigma}{\partial N} = -kT \frac{\partial \sigma}{\partial E} \frac{\partial E}{\partial V} \frac{\partial V}{\partial N} = +\frac{1}{n_l} \frac{\gamma_L}{\kappa} , \tag{5}$$

where the relations 2, 4 and $\partial V/\partial N = 1/n_l$, with n_l number density of the liquid, have been used. In equilibrium with its gas phase a liquid surface has to satisfy

$$\mu_{liq}(\kappa) = \frac{1}{n_l}\frac{\gamma_L}{\kappa} = kT \ln(n/n_{sat}) = \mu_{gas} .$$

Therefore, we find that in equilibrium a liquid surface must have a mean curvature

$$\kappa = \frac{\gamma_L}{n_l\, kT}\frac{1}{\ln(n/n_{sat})} = \frac{0.05}{\ln(n/n_{sat})} nm . \tag{6}$$

The numerical value gives the Kelvin radius for water at room temperature. For typical dry $(n/n_{sat} = 0.1)$ and humid $(n/n_{sat} = 0.95)$ environments one obtains -0.05 nm and -1nm respectively. The corresponding functional dependence is plotted in fig. 1(a). From eq. 6 it follows that drops and convex surfaces in general need oversaturated vapor to be stable, a flat surface is stable in saturated vapor and that concave surfaces are stable in undersaturated vapor. This latter case is the basis of capillary condensation. In this context it is worthwhile to visualize the space between tip and sample as a narrow gap where condensation of a liquid from its vapor may occur. It is straightforward to show by elementary geometry that, to a very good approximation, capillary condensation between two flat surfaces will occur if their distance d is small enough:

$$d < \kappa\left(\cos(\vartheta_1) + \cos(\vartheta_2)\right) , \tag{7}$$

where ϑ_1 and ϑ_2 are the contact angles of the liquid on the two surfaces.

2.2.2. Disjoining Pressure.

Long-range attractive forces play an important role in many fields of science (Israelachvili, 1992). These interactions are generally called "Dispersion Forces" and are ultimately of quantum mechanical nature. For two semi-infinite solids separated by distance D and a dispersion force of Van der Waals type the interaction energy per surface area is

$$w(D) = A_{AMB}/D^2 , \tag{8}$$

where A_{AMB} is the Hamaker constant for the two materials A and B through the medium M. The interaction depends critically not only on the materials of the two solids, but also on the medium M in-between. Equation 8 can be applied directly to the interaction of liquids on flat solid surfaces. For this it is convenient to interpret the liquid gas interface as the boundary of a semi-infinite body, namely the gas, and the liquid in-between as the medium through which the gas interacts with the solid. The interaction between the two interfaces is then $w(h) = A_{GMS}/h^2$, which induces a pressure, the Disjoining Pressure,

$$\Pi(h) = -\frac{\partial w}{\partial h} = \frac{A_{GLS}}{h^3} = \frac{\partial}{\partial h}\left(\frac{Energy}{Area}\right) = \frac{Force}{Area} , \tag{9}$$

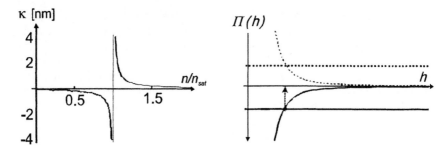

Figure 1.

where A_{GLS} is the Hamaker Constant for the gas-liquid-solid system. This pressure tries to displace the liquid interface. If $A_{GLS} > 0$ the interaction is said to be repulsive, the Disjoining Pressure tries to push the gas-liquid interface away from the solid surface and the surface is said to wet. Correspondingly, for $A_{GLS} < 0$ the Disjoining Pressure tries to push the gas-liquid interface towards the solid surface and the surface is said not to wet.

If the surface wets, one might wonder what prevents the film from thickening to macroscopic dimensions? Growing of the liquid film "costs" some energy, which has to be supplied by the internal energy of the total system. In analogy to the reasoning leading to eq. 5, this energy will induce a variation of the Chemical Potential of the liquid

$$\mu\left(h\right) \equiv -kT\frac{\partial\sigma}{\partial N} = -kT\frac{\partial\sigma}{\partial E}\frac{\partial E}{\partial V}\frac{\partial V}{\partial N} = -\frac{\Pi\left(h\right)}{n_l} \ .$$

Since the Chemical Potential has to be constant throughout the total system we obtain the relation

$$\mu_{gas}\left(n\right) = kT\ln\left(n/n_{sat}\right) = -\Pi\left(h\right)/n_l$$

between the Disjoining Pressure and the relative number density of the vapor. This is an implicit equation for the equilibrium thickness of the adsorbed liquid film as a function of the relative vapor pressure. Its graphic solution is shown in fig. 1(b): the equilibrium thickness is found from the intersection of the constant line $y(h) = kT\ln\left(n/n_{sat}\right)$, and the curve corresponding to the Disjoining Pressure. Only for the wetting case (repulsive potentials, $A > 0$) a stable point is found.

2.2.3. Adsorption of liquids on curved solid surfaces.

In the two preceding sections, the effect of Disjoining Pressure and Laplace Pressure has been discussed separately. The Disjoining Pressure was analyzed assuming a flat substrate, and the Laplace Pressure for a case of

a free liquid surface. Solid surfaces are however generally rough. Therefore the question arises how liquids adsorb on such surfaces. This issue is
specially important on a nanometer scale, since both the Disjoining Pressure as well as the Laplace Pressure are effects that are more pronounced
on a small scale. To a very good approximation, Disjoining Pressure and
Laplace Pressure are independent effects, thus one can assume that their
effects simply add, therefore

$$\mu_{liquid} = \frac{\gamma_L/\kappa\left(x,y\right) - \Pi\left(h\left(x,y\right)\right)}{n_l} = kT\ln\left(n/n_{sat}\right) = \mu_{gas} , \qquad (10)$$

where $1/\kappa\left(x,y\right)$ is the local mean curvature of the liquid film of height
$h(x,y)$. If one considers only one lateral dimension it can be shown that the
mean curvature is related to the second derivative of a profile by $h''\left(x\right) =
1/\kappa\left(h\left(x\right)\right)$, so that eq. 10 may be rewritten as

$$\frac{\gamma_L \cdot h''\left(x,n\right) - \Pi\left(h\left(x,n\right)\right)}{n_l} = kT\ln\left(n/n_{sat}\right) ,$$

where the dependence of the film height on the of the number density n
has been written explicitly: $h = h\left(x,n\right)$. For a given number density, this
is a non-linear and implicit equation for the equilibrium height $h\left(x,n\right)$.
In physical terms, a graphical solution can be constructed as in fig. 1(b).
However, now the additional term arising from the Laplace Pressure has
to be taken into account. In fact, for a positive curvature the line $y(h) =
kT\ln(n/n_{sat}) - \gamma_L\, h''$ lies lower than the corresponding line for a flat surface
and the height of the film is smaller. For positive curvature the surface may
be thought of as being locally dryer. The same argument shows that for
negative curvature the film is thicker and can be thought of as being locally
wetter. The combination of Laplace Pressure and Disjoining Pressure will
therefore even out the corrugations of the underlying substrate, which is
what is expected from an adsorbed film, due to its tendency to minimize
its surface energy.

3. Experimental Results

3.1. TIP-SAMPLE INTERACTION

Tip-sample interaction is fundamental in SPM since it determines the exact
behavior of any SPM during imaging acquisition. In addition, tip-sample
interaction in a SPM-setup can be considered a model system for many
other fields: in tribology as a single asperity contact or in the field of intermolecular forces as two bodies of nanometer dimensions whose interaction
is to be studied. Figure 2 shows a model system for a SFM-setup. The
cantilever as well as the interaction potential are visualized as springs. In
the case of the interaction potential the spring is highly nonlinear. Three
different distances are relevant in the system: the tip-sample distance z, the

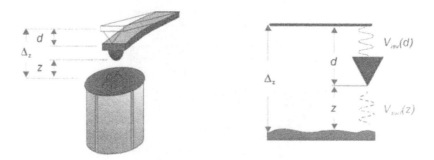

Figure 2. Simple model for an SFM-setup. The elastic potential of the spring and the interaction potential are both represented by springs. Note that, as discussed in the main text, only two of the three distances which occur in the model are independent.

deflection d and the separation Δ between the base of the cantilever and the sample[2]. The separation Δ is controlled experimentally, the deflection d is measured and the tip-sample distance z is generally needed to compare data with theoretical predictions. Only two distances are independent variables, since $\Delta = d + z$. A convenient choice for describing the behavior of the system is z and Δ. Then the fundamental equations of the tip-sample system are:

$$V_{tot}(z, \Delta) \equiv V_{surf}(z) + \frac{c}{2}(\Delta - z)^2 , \tag{11}$$

$$\text{Equilibrium} \quad F(z, \Delta) \equiv -\frac{\partial V_{tot}}{\partial z} = -\frac{\partial V_{surf}}{\partial z} + c(\Delta - z) = 0 ,$$

$$\text{Stability} \quad c(z, \Delta) \equiv \frac{\partial^2 V_{tot}}{\partial z^2} = \frac{\partial^2 V_{surf}}{\partial z^2} + c > 0 ,$$

where c is the force constant of the cantilever, and $V_{surf}(z)$ the surface potential. The equilibrium condition is an implicit equation for the tip sample distance: for a given separation Δ, its solution is $z(\Delta)$. For attractive potentials the stability condition is not satisfied if the cantilever is too soft, that is, if: $V''_{surf}(z) < -c$. Then the solution $z(\Delta)$ is multi-valued which is the origin of hysteresis and energy dissipation. In a SFM-experiment this implies that at a certain distance, the cantilever will not be able to "hold" the tip within the attractive potential. The tip will then "fall" into this

---

[2]In principle, also other distances might be needed to describe the system: tip and sample may deform and each deformation would have to be parametrised by an appropiate distance. In what follows, we will neglect these defromations, which is correct if the force constant of the cantilever is considerably "softer" than the effective force constant of tip and sample.

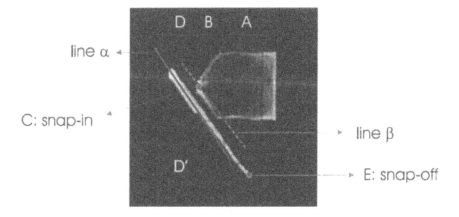

Figure 3.

potential and onto the sample until it is stabilized by the strong repulsive forces due to mechanical contact with the solid surface. A detailed discussion on modelling a SFM-setup can be found for example in ref. (Colchero et al., 1996).

From an experimental point of view, tip-sample interaction can be measured in essentially two different ways: statically and dynamically. In a static measurement, the deflection of the cantilever is recorded as the tip approaches the sample surface (Weisenhorn et al., 1989). In a dynamic measurement, the cantilever is usually oscillated near the resonance frequency and the variation of the resonant properties of the tip-sample system is studied as the separation is varied (Dürig et. al, 1992). In most aspects, the dynamical method is superior to the static one, since the interaction is deduced from a frequency measurement, which is generally more precise. However, if an instability as described above occurs a dynamic measurement has the important disadvantage that the tip-sample distance cannot be determined. If a force vs. distance curve is acquired statically, the distance which the tip has "jumped" during the instability can be calculated from the deflection of the cantilever before and after the instability. The best approach is to combine static and dynamic approaches to study tip sample interaction. In the following sections we will describe two experimental methods which combine static and dynamical measurement of the interaction between tip and sample

3.1.1. *Force vs. Distance curves with an oscillating cantilever.*

The experiments described in this section have been performed by continuous acquisition of force vs. distance curves while the tip is oscillated near its resonance frequency with different amplitudes (van der Werft et al., 1994;

De Pablo et al., 2000). The sample is fixed to a piezo tube that allows motion along the X, Y and Z directions. The cantilever is glued to a small piezo plate to excite mechanical oscillations. To analyze the precise dependence of the cantilever oscillation with tip-sample distance, Jumping Mode is used (De Pablo et al., *1998*). In brief, Jumping Mode is a succession of force vs. distance curves (F vs. Z)[3] with a feedback period between them at a given normal force set-point. The data were taken using an oscilloscope in the X-Y mode. The sample motion was input through the horizontal channel and the signal corresponding to the tip motion was input through the vertical channel. The oscilloscope image is then recorded using a digital video camera.

To help the forthcoming discussion several labels discussed in the figure caption have been included in fig. 3 to mark the relevant regions. The lines α and β are drawn as guidelines and are fundamental for the correct interpretation of the experimental data. Line α (solid line) goes through points where the tip is in contact with the surface. This line defines the surface position neglecting deformations of tip and sample[4]. For any data point on the force vs. distance curve, the corresponding tip-sample distance can be read off as the horizontal distance to this line. Line β (dashed line) is defined by the lower turning point of the cantilever oscillation. The fundamental feature that is observed in many of our experiments is that, as the vibration amplitude of the cantilever is reduced between points B and C, the tip does not touch the surface. In fact, between points B and C the force vs. distance curve, and more precisely line β, does not reach line α, which defines the position of the surface. Another important feature in our data is that, if the oscillation amplitude a_{osc} is low enough, the maximum restoring force F_{osc} of the cantilever during oscillation is lower than the adhesion force F_{ad} which is measured during the receding part of the force vs. distance curves, i.e.: $F_{osc} = c \cdot a_{osc} < F_{ad}$. This observation is not consistent with the formation of a contact between tip and sample, since the tip would then stick to the surface and the oscillation would stop. Therefore we again conclude that tip-sample contact does not occur in this case.

3.1.2. *"Imaging" tip-sample interaction by simultaneous measurement of its dynamic and static properties.*

From what has been discussed so far, it seems evident that tip-sample contact can be avoided when the tip is oscillated. This is a somewhat surprising result, interesting in itself and very important for imaging applications (see below). The names "Tapping Mode" or "Intermitted Contact Mode" do not seem to be the very appropriate terms in view of this result. In what follows we will use the naming "Dynamic Mode" SFM instead. For a precise understanding of tip-sample interaction the method described in

---

[3]With force we mean in the present context cantilever deflection.

[4]For the points on line α, tip and sample are in contact, therefore they move together. That is: tip movement = sample movement.

the previous section has the disadvantage that a relatively large oscillation amplitude has to be used. This leads to a very high non-linearity in tip-sample interaction, which is extremely difficult to interpret. To reduce this non-linearity, the oscillation amplitude has to be reduced until it is smaller than a typical length scale related to the variation of curvature of the interaction potential. In this section we describe a technique to measure the tip-sample interaction with small amplitudes but still with high precision. Essentially, the force exerted on the cantilever is acquired simultaneously with a spectrum of the cantilever. To understand this technique in detail, we first recall that for most applications, a SFM-setup can be approximated by a harmonic oscillator. If the tip feels no interaction, the resonance frequency of the cantilever is $\omega_{00} = c_{lever}/m_{eff}$, where c_{lever} is the force constant of the cantilever and $m_{eff} = m_{tip} + 0.24\, m_{lever}$ an effective mass taking into account that the mass is distributed along the whole cantilever and not only at its free end. With an interaction potential $V(z)$, the resonance frequency of the free cantilever is modified according to (Dürig et. al, 1992)

$$\omega_0\left(z\right) = \sqrt{c_{lever} + V''(z)/m_{eff}} = \omega_{00}\sqrt{1 + V''(z)/c_{lever}} \;, \qquad (12)$$

where $V''(z)$ the force gradient of the interaction potential. Essentially, the force gradient and the force constant of the cantilever add to an effective force constant of the system. For attractive potentials, that is, for potentials which induce forces that pull the tip towards the surface, this force gradient is negative and therefore the resonance frequency shifts towards lower values. An interesting feature of eq. 12 is that for $V''(z) = -c$, the resonance frequency is zero , and that for $V''(z) < -c$, it is imaginary, and is therefore physically not defined. This is related to the mechanical instability for $V''(z) = -c$. Finally, we note that the oscillation amplitude of the system driven by a harmonic driving force $F(t) = F_0 e^{iwt}$ is described by a Lorenz function

$$a\left(\omega\right) = \frac{a_0}{1 - \left(\omega/\omega_0\right)^2 + i\omega/\left(Q\omega_0\right)} \qquad (13)$$

where Q is the quality factor of the oscillator and $a_0 = F_0/c_{lever}$ is the amplitude at zero frequency, that is, the deflection of the cantilever due to a D.C. force of magnitude F_0. In our experiments the resonance frequency of the tip sample system is determined by acquiring a spectrum of the cantilever. The resonance frequency of the tip-sample system can then be deduced from a fit to the resonance curve. The experiments were performed with a SFM by slightly varying the typical setup used for imaging applications. More precisely, the ramp which is normally used for the fast scan direction (typically "x"-axis) was fed into a voltage controlled oscillator, which then provided a frequency sweep of fixed voltage amplitude. Its sinusoidal voltage was applied to a small piezo on which the cantilever was mounted to excite it mechanically. Typical oscillation amplitudes of the cantilever were 0.2 to 1 nm at resonance. The slow scan signal which is used typically for varying the "y"-position of the tip in imaging applications was used in our setup to vary the separation Δ between the sample and the base of the cantilever. For each experiment, two "images" were taken simultaneously, one corresponding to the output of the lock-in amplifier, which is proportional to the oscillation amplitude of the cantilever (fig. 4a), and the other one corresponding to the normal force (fig. 4b). The first image reflects how the resonance frequency varies with the tip sample distance. Every line of this image represents a spectrum of the cantilever at a fixed tip sample distance. The image corresponding to the normal force on the other hand is essentially a force vs. distance curve: apart from a small variation due to the oscillation, the normal force signal should be constant along each horizontal line. A vertical cut of the normal force image is simply a force vs. distance curve. The two raw images are processed as follows: In the case of the normal force image, each line is averaged to give a single force value. The whole image therefore gives a typical force vs. distance curve of high precision due to the averaging process. Each line of the image corresponding to the output of the lock-in amplifier is fitted to a Lorenz curve (eq. 13) to give the three parameters a_0, Q and ω_o. This image therefore results in three curves: a_0 vs. distance, Q vs. distance and ω_0 vs. distance. Only the latter will be considered here.

Mica, Gold and HOPG have been used as sample surfaces, but only data corresponding to HOPG will be discussed here and is shown in fig. 4. In the image corresponding to the oscillation amplitude (4a) we find that the resonance peak of the system shifts towards lower frequencies (left) as the tip approaches the surface. This is expected for an attractive potential. Approximately in the center of the images the oscillation suddenly vanishes. The corresponding line in the normal force image (4b) shows a discontinuous drop of the force indicating jump to contact. From the raw data shown, a resonance vs. distance curve and a force vs. distance curve are calculated (fig 4c,d). The bigger points in fig. 4c correspond to the measured resonance frequency at the corresponding tip-sample distance. The solid curve represents a least square fit to these points assuming a Van der Waals interaction potential $V(z) = -AR/6z$. The z-scale in this graph is adjusted so that $z = 0$ corresponds to the pole of the interaction potential, that is, to the position of the surface as predicted by the fit. To guide the eye, this position is marked by a vertical solid line. The significance of the dotted vertical line is discussed below. The smaller points scattered around the dotted horizontal line are proportional to the error between the data and the fit. Since this error is without tendency, we conclude that the fit and therefore the assumption of pure Van der Waals interaction is essentially correct. The black points in figure 1 d correspond to the force vs. distance curve obtained from the normal force image.

Before interpreting the experimental data, we would like to describe the type of data which is expected for an ideal surface without liquid layer. As discussed above, if the attractive potential is strong enough, the resonance

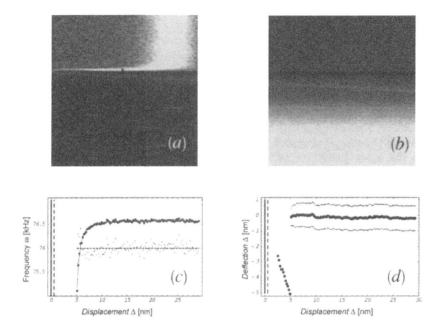

Figure 4. Top: raw "images" of the tip sample interaction. The left image corresponds to the oscillation amplitude of the cantilever, the right one to its (mean) static deflection. Bottom: Frequency vs. distance and force vs. distance curves calculated from the raw data as described in the main text.

frequency of the system becomes zero at some tip-sample distance. At this distance, the tip jumps onto the surface. For pure Van der Waals interaction this happens at a critical distance $z_{jump} = (AR/3c)^{1/3}$. In this case, the resonance frequency is

$$\omega_0(z) = \omega_{00}\sqrt{1 - z_{jump}^3/z^3} . \qquad (14)$$

The critical distance z_{jump}, which is calculated from a fit to the resonance frequency $\omega_0(z)$, is marked with a dotted vertical line in both graphs. An important point in the present context is that the jump to contact instability in a force vs. distance curve should be at the same tip-sample position as the zero frequency point in the resonance vs. distance curve (dotted line). Moreover, the (complex) pole at $z = 0$ of eq. 14, which is marked by the black vertical line, should agree with the position of the surface as deduced from the force vs. distance curve. This position is deduced by tracing a horizontal line from the last point before the instability to the increasing linear part of the force vs. distance curve corresponding to the tip-sample contact. However, in our experimental results we find a behavior which is completely different from the one just described: jump to contact occurs at a frequency shift of only $\Delta\omega \simeq 1\%$, and at a distance of about 3.5 nm, instead of the distance $z_{jump} = 0.7$ nm predicted by the fit.

This experiment is explained as follows. As discussed in detail in sections 2.2.1 and 2.2.2 in ambient conditions the effect of relative humidity is twofold: on the one hand, a liquid film may condense and on the other hand, a liquid meniscus can form around the tip-sample contact. We believe that the $z = 0$ position calculated from the frequency data corresponds to the liquid-vapor interface, while the $z = 0$ position deduced from force vs. distance curve corresponds to the position of the solid surface, since the tip is pulled through the liquid film. The resonance data therefore "sees" the liquid-vapor interface and the force vs. distance curve "sees" the solid-liquid interface. Finally, when the tip is at about twice the Kelvin radius (see eq. 7) we believe that a liquid meniscus forms spontaneously between tip and sample which then pulls the tip onto the sample because the force constant of the cantilever used in our experiment is not strong enough to hold the tip within the liquid neck.

3.2. IMAGING APPLICATIONS

SFM studies of weakly adsorbed structures on surfaces, and in particular of liquid films, is difficult, mainly due to the fact that in ambient conditions most SFM modes work in the contact regime, that is, tip and sample are in mechanical contact. Therefore, structures that are weakly adhered to the surface are usually destroyed or moved away. Recently, approaches based on SFM techniques have been developed which allow reproducible and stable imaging in the non-contact regime. Scanning Force Polarization Microscopy (Hu et al., 1995), which is based on the electrostatic interaction between tip and sample, is one of them; and Dynamic Mode SFM (DM-SFM) is another one, if the parameters which control the interaction are set carefully (Luna et al., 1998; Pompe et al., 1998). As discussed above, the oscillation amplitude may decrease, even though no mechanical contact occurs. Therefore non-contact images of samples can be obtained if a lock-in technique is used to measure the oscillation amplitude of the cantilever and a feedback loop is enabled to keep an appropriate set-point (a_{set}) for the oscillation. An important point in our experiments is that we can assure that tip-sample contact is avoided during image acquisition by applying a small oscillation amplitude a_{free}. In fact, if the restoring force of the cantilever at its lower turning point is smaller than the adhesion force F_{ad}, that is, if $c_{lever} \cdot a_{free} < F_{ad}$ then the tip will stick to the surface and the oscillation will stop as soon as the tip touches the surface. When feedback is enabled for imaging applications, this results in a retraction of the piezo until the cantilever detaches from the surface. Then oscillation starts again and the tip is engaged to find the adjusted set-point. If the chosen set-point is inappropriate for non-contact imaging this process is cyclic and results in completely noisy and unstable imaging, signaling that contact is occurring

 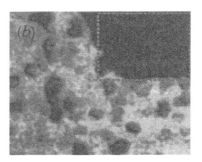

Figure 5. Images showing the effect of water adsorption on mica at high relative humidity. The left image (a) corresponds to the topography, the right one (b) to the phase. Image size is $1.3 \times 1.1 \mu m^2$. The total gray scale of the topographic image corresponds to about 1.2nm. The phase image is in arbitrary units. Previously to the image shown a contact mode image was acquired in the region marked with the rectangle.

and that more appropriate operating conditions have to be adjusted.

To demonstrate the potential of dynamic SFM to study liquid structures on surfaces we have imaged adsorption of water on three different surfaces: mica, gold and graphite (Gil et al., 2000). In what follows we will discuss the effect of water adsorbtion on the first two surfaces. Experimental parameters were chosen as follows: cantilevers with a force constant of 0.75N/m, a resonance frequency of about 75 kHz, oscillation amplitudes between 10 and 30 nm, and set-points a_{set}/a_{free} in the range of 0.8-0.99.

3.2.1. *Mica.*

Figure 5 shows an image of water layers adsorbed on a mica surface. Mica was freshly cleaved and purified water was pulverized on the mica surface in order to eliminate electric charges. Subsequently the surface was dried with nitrogen. After this process, the surface was clean as observed with SFM, in contact as well as in non-contact dynamic SFM. Then the relative humidity in the chamber was risen up to 90% and all the measurements were done in these conditions. In a consecutive series of images, a first pair of images (not shown) was taken in DM-SFM on a region were no tip-sample contact had been established previously. The next image was taken on the same spot, but in contact mode (not shown). Only the flat mica surface is observed in this contact image. Finally a pair of images was acquired again in DM-SFM (fig. 5). The homogeneous region in the upper right of this image is due to the modification of the surface induced by the contact scan. The layered structures are observed only in dynamic SFM and are thus composed of some soft material, which is weakly adsorbed to the surface. These layered structures are severely modified by the scanning motion of

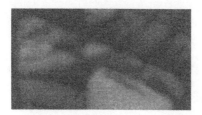

Figure 6. Topographic images of the Au (111) showing the evolution structures related to the adsorption of water. The height of these structures is about 0.2nm. The images were taken in DM-SFM at a relative humidity of about 40%. Image size: 600x400nm², z-scale: 1nm for the left image, and 1.5nm for the right one.

the tip during contact, which induces a spreading of these structures. If the relative humidity is low, these layered structures are not observed. Therefore we conclude that these structures are composed of water, and are not due to any other undefined contamination layer.

The topographic height of the layered structures is about 0.7 nm for the first layer, and another 0.7 nm for the second one. We believe that the water molecules adsorb on mica in an ordered structure and that the topographic height of 0.7 nm measured for the two observed layers correspond to one bi-layer of ice each. This is in agreement with other experimental (Miranda et al., 1998) and theoretical results (Odelius et al., 1998). Another interesting feature observed in our experiments is the strong difference in phase signal which is measured on the two different water layers (fig. 5b). We strongly believe that this phase contrast is due to electrostatic interaction induced by a variation of surface charge density. Surprisingly, this local variation of tip-sample interaction is stronger between the two water layers, than between the mica substrate and the first water layer.

3.2.2. *Gold.*

Water adsorption on flat Au (111). Figure 6 shows a series of images acquired on the gold (111) surface in DM-SFM. The experiment was started at an initial relative humidity of about 35%. Au (111) terraces are observed, and the water layer adsorbed on the gold is not continuous. Structures of 0.2 nm height grow in preferential directions, perpendicular to the edges of the terraces and thus forming 120°among them. After some time, the relative humidity is risen and for values over 65%, these structures are not observed any more and the water layer becomes homogenous on the gold surface. The height of 0.2 nm agrees with the diameter of a single water molecule.

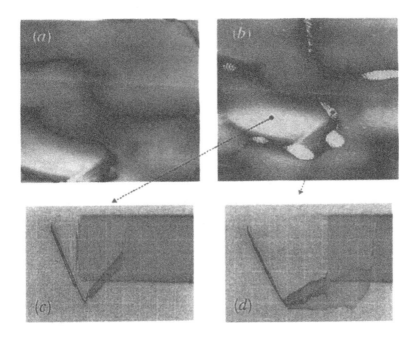

Figure 7. Top: Images of gold grains taken at different relative humidity; the left image (a) at about 40% and the right image (b) at more than 95%. For both images the total scan size is $2\mu m^2$ and the total gray scale corresponds to about 35 nm. Bottom: force vs. distances curves measured in Jumping Mode. Total piezo displacement for both curves is 100nm. These curves were obtained at high relative humidity with a digital video camera during image acquisition. Left curve (c) on the terraces, right curve (d) at a grain boundary.

Effects of water adsorption at grain boundaries. Figure 7 shows the topography of the gold surface measured in DM-SFM at different relative humidity: in (a) at 40% and a temperature of 20°C and in (b) at 85% and 7°C. In the second image drop-like structures appear which we associate to the adsorption of water. In fact, when the relative humidity is decreased, and the temperature is increased, these structures disappear. At a first glance, these structures might be interpreted as nanometer sized drops. However we believe that this interpretation is not correct. In fact, as can be observed in fig. 7(a) and (b), feedback on these drop-like structures is unstable and shows an oscillatory behavior. We relate this behavior to the formation of water necks between tip and sample. To confirm this interpretation we have studied force vs. distance curves at different positions of the sample with an oscillating cantilever for low and high relative humidity. A curve obtained for low relative humidity is shown in fig. 3. For high relative humidity however, we found essentially two typical kind of curves: one on the gold terraces 7(c), and the other at the grain boundaries 7(d). On the terraces we find a region, similar to region B in fig. 3, where oscillation amplitude decreases due to tip and sample interaction. However, at high relative humidity this region shows sudden drops in the oscillation amplitude. We attribute this to the formation of small "pre"-necks of water: at these points a small liquid neck forms between tip and sample which is too weak to induce snap-in, but strong enough to take some energy out of the oscillation, therefore its amplitude is reduced.

From a collection of force vs. distance curves taken during image acquisition while scanning over a grain boundary, we have tried to deduce the position of the Au surface, the height of the snap-in point, as well as the length of the water neck at the snap-off. In this context, an important issue is where exactly the water air-interface is located. The surface position of the liquid surface has to be located somewhere between the position of the solid Au surface and the position where the jump to contact occurs. We recall that the jump-in distance corresponds to the sum of the height of the neck formed during tip approach and the thickness of the flat water layer on the tip and the sample. Up to now we have found no method to measure these two quantities individually. Therefore, we have not yet been able to resolve another important question: what is the exact shape of the liquid-gas interface at the grain boundary: convex (as drops) or concave (as meniscus). From what was discussed in section 2.2.3 one would expect a concave shape, since the local relative humidity around the tip is lower than 100% and thus a drop shape should not be thermodynamically stable. However, more work, both experimental and theoretical, is needed to solve this issue on a nanometer scale.

4. Epilog

In conclusion, we believe that adsorption of water on solid surfaces is interesting not only due to its evident applications in many other disciplines, and particularly in Tribology, but that it is a fascinating topic in itself. Moreover we think that the application of SPM-techniques to study wetting on a nanometer scale has only started and that there is still a long way ahead both experimentally and theoretically until, hopefully, surfaces in ambient conditions are as well understood as in UHV-environment.

The authors thank S. Herminghaus and J.J. Saenz for very interesting discussions. We acknowledge support from Ministerio de Educación y Cultura through a CYCIT Project No. PB95-0169 and a contract to J. Colchero. A. Gil acknowledges support from the FGUAM through a scholarship from the project Nanodigital.

234

References

Bhushan, B., Israelachvili, J. and Landman, U. (1995), "Nanotribology: friction, wear and lubrication at the atomic scale", Nature **374**, 607-616.

Charvolin, J., Joanny, J.F. and Zinn-Justin, J. (1990), "Liquids at Interfaces", NATO ASI Les Houches, Session XLVIII, Elsevier Science Publishers.

Colchero, J., Baró, A.M. and Marti, O. (1996), "Energy dissipation in Scanning Force Microscopy-Friction on an atomic scale", Tribology Letters **2**, 327-343.

Colchero, J., Storch, A., Luna, M., Gómez-Herrero, J. and Baró, A.M. (1998), "Observation of liquid neck formation with Scanning Force Microscopy techniques", Langmuir **14**, 2230-2234.

De Gennes, P.G. (1985), "Wetting: statics and dynamics", Reviews of Modern Physics **57**, 827-863.

De Pablo, P.J , Colchero, J., Luna, M., Gómez-Herrero, J. and Baró, A.M. (1998), "Jumping Mode Scanning Force Microscopy", Appl. Phys. Lett. **73**, 3300-3302.

De Pablo, P.J., Colchero, J. , Luna, M., Gómez-Herrero, J. and Baró, A.M. (2000), "Tip-Sample Interaction in Scanning Force Microscopy", Physical Review B **61**, 14179-14183.

Dürig, U., Züger, O. and Stalder, A. (1992), "Interaction Force Detection in Scanning Probe Microscopy: Methods and Applications", J. Appl. Phys., **72**, 1778-1798.

Gil, A., Colchero, J., Luna, M., Gómez-Herrero, J. and Baró, A.M. (2000), "Adsorption of Water on Solid Surfaces by Scanning Force Microscopy", Langmuir **16**, 5086-5092.

Herminghaus, S., Jacobs, K., Mecke, K., Bischof, J., Fery, A., Ibn-Elhaj, M. and Schlagowski, S. (1998), "Spinodal Dewetting of Liquid Crystal and Liquid Metal Films", Science **282**, 916-918.

Hu, J., Xiao, X-D., Ogletree, D.F., Salmeron, M. (1995), "Imaging the condensation and evaporation of molecular thin films of water with nanometer resolution", Science **168**, 267-269.

Israelachvili, J.N. (1992), "Intermolecular and Surface Forces", Academic Press, San Diego.

Kittel, C. (1969), "Thermal Physics", John Wiley & Sous, Inc. New York.

Luna, M., Colchero, J. and Baró, A.M. (1998), "Intermittent Contact Scanning Force Microscopy: the role of liquid necks", Appl. Phys. Lett. **73**, 3461-3463.

Mate, C.M., McClelland, G.M., Erlandsson, R. and Chiang, S. (1987), "Atomic-Scale Friction of a Tungsten tip on a Graphite surface", Physical Review Letters, **59**, 1942-1945.

Miranda, P.B., Xu, L., Shen, Y. R. and Salmeron, M. (1998), "Icelike Water Monolayer Adsorbed on mica at Room Temperature", Phys. Rev. Lett. **81**, 26, 5876-5879.

Odelius, M., Bernasconi, M. and Parrinello, M. (1998), "Two Dimensional Ice Adsorbed on Mica Surface", Phys. Rev. Lett. **78**, 14, 2855-2858.

Pompe, T, Fery, A, and Herminghaus, S. (1998), "Imaging Liquid Structures on inhomogeneous surfaces by Scanning Force Microscopy", Langmuir **14**, 2585-2589.

Van der Werft, K.O., Putman, C. A. J., De Grooth, B.G., and Greve, J. (1994), "Adhesion force imaging in air and liquid by adhesion mode atomic force microscopy", Appl. Phys. Lett. **65**, 9, 1195-1197.

Weisenhorn, A.L, Hansma, P.K and Albrecht, T. R. (1989), "Forces in atomic force microscopy in air and water", Appl. Phys. Lett. **54**, 26, 2651-2653.

THEORY AND SIMULATIONS OF FRICTION BETWEEN FLAT SURFACES LUBRICATED BY SUBMONOLAYERS

MARTIN H. MÜSER
Institut für Physik, Universität Mainz, 55099 Mainz, Germany

ABSTRACT. Recent simulations suggest that wearless friction between two solid surfaces can only be obtained if the two surfaces are commensurate or if they are lubricated by a film. Some simple theoretical arguments are given why the presence of a submonolayer film between two solids leads to friction. Possible implications of the symmetry of the confining walls on the tribological properties of the system are then investigated in the presence of a thin film by means of molecular dynamics simulation. Erratic stick-slip motion of the incommensurate system and oscillating friction forces for the commensurate system in the sliding regime are observed.

KEYWORDS. Friction, thin films, Frenkel-Kontorova-Tomlinson model

1. Introduction

The origin of static and kinetic friction between incommensurate solids is often explained in terms of the Frenkel-Kontorova-Tomlinson (FKT) model (Weiss, 1996, Robbins and Müser, 2001): At least one of the two solids in contact has to be compliant enough to accommodate the (surface) density modulation of the other solid via elastic deformations leading to many mechanical stable states. Such multistable elasticity ultimately leads to friction. Recently, computer simulations have challenged the applicability of this theory to the explanation of dry, wearless friction: No atomistic simulations have been reported in which wearless friction was seen - unless the surfaces were commensurate. Commensurability, however, is rare in experiments albeit a frequently imposed artifact in computer simulations. This artifact automatically invokes geometric pinning between the surfaces (Müser et al., 2000). Strong adhesion between two solids, plastic deformation (Müser, 2000), and abrasive wear (Sørensen et al., 1996) also lead to friction in computer simula-

235

B. Bhushan (ed.),
Fundamentals of Tribology and Bridging the Gap between the Macro- and Micro/Nanoscales, 235–240.
© 2001 Kluwer Academic Publishers.

tions. These are strongly irreversible mechanisms, which are far from being wearless. The only simulations of wearless (non-geometric) friction between two solids include airborn atoms and/or molecules between the two surfaces (He et al., 1999, Müser et al., 2000). Shedding further light on this friction mechanism is important, because it might play the predominant role in many tribological experiments such as in surface force apparatus (SFA) and atomic force microscope (AFM) experiments: In both techniques tribological properties often depend crucially on the atmospheric conditions (Homola et al., 1990, Putman et al., 1995) and no plastic deformation, material mixing, generation of debris, etc. are believed to occur.

Here, we intend to discuss general issues of the recently suggested mechanism leading to finite shear forces due to airborn atoms (He et al., 1999), instead of focusing on one particular system. This is done by discussing a simple, generic model, in which the system has to pay an exponentially growing energy penalty with increasing overlap of the surfaces and by computer simulations using crystalline, elastic walls and Lennard Jones interactions. The numerical studies focus particularly on the implication of the symmetry of the confining walls on the frictional properties of the system. The main qualitative difference between commensurate and incommensurate walls in the presence of a thin film is the following: There is a well-defined free-energy profile for the upper wall in the commensurate case. The profile is periodic with the lattice constant - or double periodic along one coordinate if the top wall makes a zig-zag motion. For an incommensurate system, the free energy only depends on the spacing between the two solids but not on their relative lateral displacement.

2. Theory

The most simple model for the interaction between two solids is to consider their surfaces as rigid and impenetrable. In order to be more realistic, one may consider the surfaces as penetrable instead of elastically deformable as assumed in FKT models. The new model should of course capture the effect of a quickly increasing potential energy with growing overlap of the two surfaces. The most efficient way to do this is to assume an exponential increase of the potential energy V with the local overlap (Müser et al., 2000a)

$$V = \int dx\, dy\, \exp\left[-\frac{1}{\xi}(z_t - z_b)\right],$$ (1)

where z_t and z_b define where the upper and lower walls start, respectively. z_t and z_b are functions of the lateral coordinates (x, y). Here, the vector normal to the surfaces is supposed to point in z-direction and ξ is a free parameter. It can easily be shown that this model not only predicts Amontons' law for the maximum lateral force, which corresponds to the static friction F_s, but it predicts a linear relationship between load L and any arbitrary lateral force F.

The main effect of "between-sorbed" atoms is to decrease the effective distance at their lateral positions by their diameter. From detailed atomistic simulations it can be inferred that the direct wall-wall interaction becomes negligible as soon as the surfaces are sufficiently covered - even for submonolayer lubrication. This limit is easy to study analytically, because the direct fluid-fluid interaction can be neglected as well. It is then possible to show that two flat walls have a well-defined friction coefficient μ in the limit of large contact areas A. In contrast to this result, bare disordered surfaces show $\mu \propto 1/\sqrt{A}$ (Müser et al., 2000a). We want to note in passing that the exact surface coverage has relatively little effects on the friction coefficient in our simulations, in which wall-wall and fluid-fluid interactions are included as well. Exceptions are coverages close to a complete monolayer, which often invokes crystallization of the film.

The reason why the two surfaces are pinned when they are covered by some small airborn atoms comes from the fact that the airborn atoms typically occupy positions in the interface where they satisfy the interaction with both walls simultaneously. Hence, the Fourier transform of the density (including the phase factors!) is correlated with both walls. Having such correlated components in the Fourier transform of the surface density has been shown to be a necessary condition for the existence of static friction between two bare walls (Müser et al., 2000a). While it is easy to treat static friction within this model, it is far from being trivial to apply the model to sliding friction.

From an FKT point of view, one may say that the airborn atoms correspond to a film with infinite compliance. Solids with infinite compliance always pin. While this explanation is certainly valid, one has to keep in mind that there is a well defined topology of the atoms in the FKT model, while no such constraints are imposed on airborn atoms. This difference makes it difficult to apply the FKT model to non-dry friction. There have been some mixed theoretical numerical attempts to describe the frictional properties between two commensurate walls that are lubricated by one single atom (Rozman et al., 1998) or one finite Frenkel-Kontorova chain (Rozman et al., 1998a). Such an analysis is certainly an important first step, but it does not address a couple of interesting questions, such as, whether the order of the transition from stick-slip to sliding depends on the commensurability of the walls. This and other issues will be investigated in the simulations.

3. Simulations

Simulations are performed for a system consisting of two walls and a "between-sorbed" quarter layer of point atoms. Interactions within one wall are modeled such that each atom is coupled to its ideal lattice site via a harmonic spring. This is equivalent to the treatment of crystals in the Tomlinson model. All other interactions, wall-wall, wall-fluid, and fluid-fluid, are Lennard Jones (LJ) interactions. In all simulations presented here, the LJ parameters for energy ϵ_0, for the length scale σ_0, and the mass m_0 of individual atoms is set to unity. The bottom wall's center of mass is kept fixed and the top wall is pulled at

238

a constant velocity v via an elastic spring with stiffness k. For more details on the model we refer to (Müser et al., 2000).

Our first study focuses on the stick-slip transition. It is well known that both real experimental systems as well as FKT models exhibit a transition from stick-slip motion to smooth sliding. This transition can be invoked either by increasing the spring constant k or the velocity v with which the upper solid is pulled over the substrate. For $v > v_c$ and for $k > k_c$ sliding is smooth. Of course, v_c depends on k and k_c depends on v. In simulations, it can be more appropriate to go through the stick-slip transition by varying k instead of varying v. In this study, v was kept constant at $v = 0.01$ in LJ units, which translates to about $1\,m/s$ in real world units. This is a relatively large velocity as compared to some experimentally used velocities. However, the use of small inertias m_{wall} (about 10^{-13} of the mass of an AFM tip!) associated with the upper wall increases the intrinsic frequencies of the system by many orders of magnitude (typically $\Omega_{intrinsic} \propto \sqrt{m}$). This effect bridges most of the gap between simulation and experiment.

In Fig. 1, the average pulling/friction force F is shown as a function of the stiffness k of the pulling spring. For our model system, there does not seem to be a discontinuity in the lateral force for either symmetry. However, the increase in F from large to small k values is much more dramatic for the commensurate system than for the incommensurate system. Here, system sizes of $N \approx 900$ atoms in one wall layer were used for the commensurate system, which has only small finite-size effects; the incommensurate system was made four times larger.

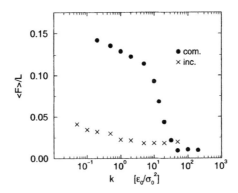

Figure 1: Medium lateral force $\langle F \rangle$ over normal load L for commensurate (circles) and incommensurate (crosses) walls in the presense of a thin film as a function of the spring stiffness k (per atom in one wall layer) with which the top wall is pulled.

For the interpretation of Fig. 1, it is instructive to investigate the time dependence of the external driving force F as a function of the symmetry and the sliding state, see Fig. 2. In the stick-slip regime, commensurate walls show both long-scale ($k = 1$) and atomic-scale ($k = 10$) slips. The atomic-scale slips reflect the periodicity of the substrate. The incommensurate system shows similar behavior, however, the stick-slips ($k = 0.5$) are erratic. As k exceeds k_c, the force between commensurate walls becomes more and more sinusoidal, despite the presence of the film. Negative and positive contributions cancel out each other leading to very small net forces, similar to the behavior of dry, commensurate surfaces. This behavior is often seen in computer simulations but not in experiment. In contrast, the incommensurate system shows a positive friction force in the sliding regime, resembling typical experimental situations.

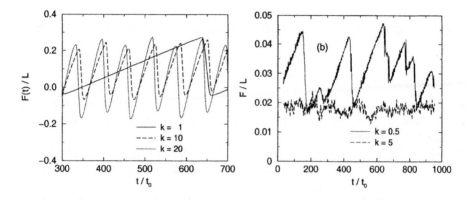

Figure 2: Lateral force F in units of the load L as a function of time for various stiffnesses k of the pulling spring. a) Commensurate walls and b) incommensurate walls. In both cases, the pulling velocities is $v = 0.01$ in LJ units.

What can we learn from these curves for AFM experiments, where the tip is amorphous and the substrate crystalline? At low loads, only a few atoms from the tip are in contact with the substrate. Hence, the mismatch between the surfaces does not play an important role and the system can be treated as if it was commensurate. Therefore, atomic resolution can be expected to be observed at low loads. For large normal loads, the area of contact is large as well and the geometric mismatch between tip and substrate makes the system more incommensurate. This results in reduced resolution of the surfaces.

4. Conclusions

It has been shown that the presence of a thin film automatically leads to static friction between two solids even if their surfaces are incommensurate. However, there is a qualitative difference between sliding friction of commensurate walls and incommensurate walls in the presence of a thin film: In the commensurate case, the lateral force oscillates with (half of) the period of the lattice constant in the sliding regime, while for incommensurate systems (like in experiment) no such oscillations can be observed. It was shown that the transition from smooth sliding to stick-slip for commensurate walls is accompanied by an anomalous large increase in the average lateral force. A theoretical analysis in terms of an FKT model for two (in)commensurate surfaces with one infinitely compliant elastic solid in between might shed further light on the friction mechanism of submonolayer lubrication.

Acknowledgment

This work has been supported throug the Israeli-German D.I.P.-Project No 352-101.

5. References

He G., Müser M. H. and Robbins M. O. (1999), "Adsorbed Layers and the Origin of Static Friction", Science **284**, 1650-1652.

Homola A. M., Israelachvili, J. N., McGuiggan, P. M. and M. L. Gee (1990), "From "Interfacial" Friction of Undamaged Molecularly Smooth Surfaces to "Normal" Friction With Wear", Wear 136, 65-83.

Müser, M. H. (2000), "Dry Friction Between Flat Surfaces: Wearless Multistable Elasticity vs. Material Transfer and Plastic Deformation", submitted to Tribology Letters.

Müser, M. H. and Robbins, M. O. (2000), "Conditions for static friction between flat, crystalline surfaces", Phys. Rev. B 61, 2335-2342.

Müser, M. H., Wenning, L. and Robbins, M. O. (2000a), "Simple Microscopic Theory of Amontons' Laws For Static Friction", cond-mat/0004494 and submitted to Phys. Rev. Lett.

Putman, C. A. J., Igarashi, M. and Kaneko, R (1995), "Single-Asperity Friction in Friction Force Microscopy: The Composite Tip Model" Appl. Phys. Lett. 66, 3221-3223

Robbins, M. O. and Müser, M. H. (2001), "Computer Simulations of Friction, Lubrication and Wear", in *Modern Tribology Handbook* (in press) CRC Press, New York.

Rozman, M. G., Urbakh, M., and Klafter, J. (1998), "Controlling Chaotic Frictional Forces", Phys. Rev. E 57, 7340-7343.

Rozman, M. G., Urbakh, M., Klafter, J. and Elmer, F.-J. (1998a), "Atomic Scale Friction and Different Phases of Motion of Embedded Molecular Systems", J. Phys. Chem. B 102, 7924-7930.

Sørensen, M. R., Jacobsen K. W. and Stoltze P. (1996), "Simulations of atomic-scale sliding friction", Phys Rev. B 53, 2101-2113.

Weiss, M. and Elmer, F.-J. (1996), "Dry friction in the Frenkel-Kontorova-Tomlinson model: Static properties", Phys. Rev. B 53, 7539-7549.

FRICTION MECHANISMS AND MODELING ON THE MACROSCALE

P. J. Blau
Oak Ridge National Laboratory
Oak Ridge, Tennessee USA

As simple as possible, but not simpler.

A. Einstein

Nature has neither kernel nor shell.

J. W. von Goethe

The desire to understand friction on a fundamental level arises because friction exists so widely in common experience and because controlling friction can improve the quality and safety of our lives, enable the advance of mechanical technology, and create jobs in industries like transportation, manufacturing, oil drilling, and power generation. The earliest philosophical explorations of friction originated from macroscopic observations, but as physical sciences and engineering matured, explanations began to center on events happening on much smaller size scales. Now, machines themselves are approaching the sizes of the conceptual building blocks of frictional behavior. This chapter reviews explanations of 'macroscopic' frictional phenomena that are based on mm- to μm-scaled objects and events. Finer-scale phenomena are described elsewhere in this book.

Coulomb's simple definition of the friction coefficient as the ratio of the tangential force to the normal force has led some to envision the existence of a continuum of frictional phenomena, but in fact none may exist. So it seems bold of us to assert that a single reductionism paradigm will explain all manner of frictional phenomena from about 0.3 nm (the approach between individual atoms in a solid) to several km (as in the sliding of glaciers over the ground).

A modified viewpoint, called *stratified reductionism*, may be better suited for friction problems because it enables us to define a size range that applies to the current problem of interest without having to reconcile phenomena that occur on vastly different scales. To aid in this approach, we can then introduce an additional level of interaction between the macroscale and the mechanism scale called the *process level*. Considering processes of interaction will help to tie nano-/and micro-scale phenomena to surface features that might ordinarily be described in qualitative terms or by using drawings instead of mathematical expressions.

This chapter describes how tribosystems can be defined at various levels and how the level of interaction can change with time, requiring a new definition of the friction problem.

B. Bhushan (ed.),
Fundamentals of Tribology and Bridging the Gap between the Macro- and Micro/Nanoscales, 241–260.
© 2001 *Kluwer Academic Publishers.*

1.0   Frictional mechanisms and processes

The term *mechanism* will be used here to describe physical or chemical interactions between molecules or groups of molecules. The term *process* will be used to indicate the operation of two or more mechanisms so as to produce observable interfacial features. Dividing frictional phenomena into mechanisms and processes enables us to begin to construct an ordered, hierarchical view of additive or synergistic frictional contributions.

1.1 THE INFLUENCE OF SURFACE ANALYSIS INSTRUMENTS

Elucidation of basic friction processes and mechanisms has been facilitated by progress in instrumentation and imaging during the last 50-70 years. New techniques of importance to tribology include roughness measurement, topographic mapping, scanning electron microscopy, surface chemical analysis, and point-probe microscopy, including lateral force measurements. Since friction models are influenced by perceptions of the nature of surfaces, it is interesting to speculate how different the state of friction science would be today had past investigators like da Vinci, Newton, Coulomb, Amontons, Tomlinson, Holm, and Deryagin had access to modern surface imaging and analysis instruments. Would they have viewed the problem in different terms and come up with different models for the normal force-friction force relationship? Yet, Ludema (1998) has asserted that even today some friction theorists develop elaborate models without ever having looked at a real contact surface through a microscope.

　　The challenge for theorists is to determine how simply frictional phenomena can be modeled without leaving out any essential elements, like those responsible for the features can be observed on contact surfaces. In the following sections, friction will be considered at both the mechanistic and process levels to illustrate how different kinds of friction problems require different approaches.

2.0   Frictional energy transformation and dissipation

To truly understand friction, it is necessary to understand the manner of its energy transformation and dissipation. Lord Rumford (a.k.a. Sir Benjamin Thompson) is well known for his eighteenth century (1784-98) research on the nature of frictional heating [see Dowson (1978)]. However, all the energy developed during frictional contact does not transform into heat. Other energy sinks include plastic deformation, fracture, and the stimulation of noise and other vibrations.

　　The energy associated with frictional processes has been considered by past and present investigators [e.g., Czichos (1978)]. Approaches to understanding the energy flows in friction have utilized thermodynamics approaches [e.g., Rymuza (1996)]. The central questions in these energy-based approaches to friction conceptualization and modeling are:

　　1) How do heat and work enter, accumulate in, and exit a given tribosystem?
　　2) How is the energy used (partitioned) within a given tribosystem?
　　3) Do energy balances reach a steady-state or not?

It is experimentally difficult to determine what fraction of frictional energy goes into heat versus the other possibilities. Nevertheless, this kind of work has been conducted for more than fifteen years by researchers like J. Sadowski and Z. Rymuza in Poland. Likewise, researchers in machining science have analyzed energy partition to better understand and control grinding procedures [e.g., Gao et al. (2000)].

Time-dependent changes in the modes of transformation and dissipation of frictional energy are suggested both by the long-period trends and the fine-structure of fluctuations in the friction force. These fluctuations reflect an ongoing competition between frictional processes. The role of time is especially important in cases like break-in, periodic instabilities (e.g., stick slip), or friction-vibration interactions. The dynamics of frictional phenomena can place limitations on the ability of equilibrium thermodynamic approaches to predict tribological events. Classical thermodynamic relationships indicate the tendency for certain events to occur, but time of interaction (kinetics) either permits or denies these events from reaching completion.

2.1 THE RELATIONSHIP BETWEEN FRICTION AND WEAR

The relationship between friction and wear is a consequence of the partition of the available frictional energy into heat, vibrations, deformation, and fracture (creation of new surfaces). Different material pairs that may have similar sliding friction coefficients under the same applied conditions can nevertheless exhibit greatly different wear rates because the frictional energy is partitioned differently between creating wear particles and generating heat.

3.0   Complexity and its dependence on applied conditions

A simple thought experiment illustrates how the changes in tribosystem variables, like normal force and speed, can increase the complexity of friction analysis [Blau (1996)]. Assume that a block of mass M rests on the planar surface of a material of lesser hardness. If a horizontal force H is applied, the block will either remain in place, tip over, or slip. If we force the block to slide for some distance, stop and examine the contact surfaces of block and plane under a low-power magnifying glass, we may see no difference in appearance. Perhaps, friction force F in that case was largely governed by the shear strength of adsorbed water or other boundary lubricating layers, maybe with some minor contribution from the surface roughness. The sources of friction are not obvious in this case.

If M is increased and a higher H is required to initiate and maintain slip, then additional processes, like overcoming adhesive bonds or deforming asperities on the microscale may operate. Magnified views of the surfaces of the block and/or plane may now show evidence for abrasion or plowing. The contribution of adsorbed layers, deformation, transfer of counterface materials, and even fracture, may also be present. Now, no single process adequately accounts for the behavior. In fact, several frictional processes can operate in different proportions and at different locations on the nominal area of contact. If M is once more increased, additional factors, such as the tendency for the leading edge of the block to dig into the surface (as in machining) is added to F. Therefore, for each additional process P_i, np_i additional mechanisms must be added to the analysis.

244

The same mechanism(s) may contribute to two or more processes. For example, the atomic bond breaking necessary for the advance of a moving crack in a delamination-type wear process, may also occur in a process involving the fracture of sharp asperities. Thus, for each tribosystem, there exists a set of all operable friction processes Sp {P1, P2, P3...} and each process contains within it a set of mechanisms: Sp_1 {m$_1$, m$_2$, m$_3$, ...}. The same mechanism can be a member of more than one process set.

In the foregoing example, an increase in the normal force **P**, induced an increase in process complexity. That argument can be extended to the effects of sliding speed and temperature.

Increasing the magnitude of a given applied variable does not necessarily imply an increase in process complexity. For example, increasing the sliding velocity will raise the frictional heating and the energy that must be dissipated per unit time. At lower speeds, the shear and fracture of interacting asperities may consume energy, but at some point, melting may begin, thereby reducing the dominant processes to viscous shear and the regeneration of the molten layer.

The complexity of several interactive and simultaneous frictional contributions can be treated by introducing the concept of a *friction stack*.

4.0   The Friction Stack and the Layering of Friction Processes.

A friction stack (FS) is a conceptual cross-section of a sliding interface composed of various layers that offer distinct contributions to the sliding resistance (Figure 1). It is a simplified, 2-dimensional slice of the 4-dimensional phenomenon [Blau (1993)].

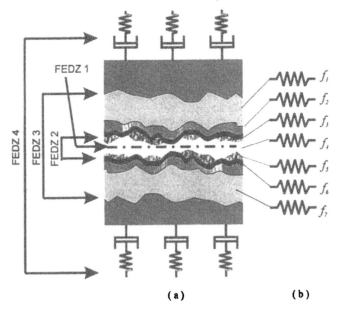

(a)   (b)

Figure 1. (a) Friction stack showing the zones of energy dissipation. (b) An electrical analog to the friction stack showing sliding resistances in parallel.

Most macroscopic friction models tend to focus on only the weakest interface in the stack, but suppose that several interfaces are nearly alike in shear strength (as in the shearing a stack of playing cards), or that seizure occurs momentarily in one slipping interface, thereby shifting relative motion to another level. Therefore, relative motion can occur at several levels within the same stack, either alternately or simultaneously.

A first approach to modeling the behavior of a FS is to apply an electrical analog as in Fig. 1(b) :

$$\frac{1}{F} = \sum_{i=1}^{n}\left(\frac{1}{f_i}\right) \tag{1}$$

where there are n possible means to accommodate relative motion in the stack. The instantaneous friction force F reflects all levels in the FS that operate at that time. If the resistance of most layers is relatively large, most of the terms in Eqn. [2] become small, leaving only 1 or 2 or 3 dominant terms in the final expression.

All operable or potentially-operable layers in a stack can be called its *critical layers*. If the contact area A between all critical layers in the stack is equal, then we can write Eqn. [1] in terms of layer shear strengths τ :

$$\frac{1}{F} = \frac{1}{A}\sum_{i=1}^{n}\left(\frac{1}{\tau_i}\right) \tag{2}$$

As n becomes large, and if τ's are about equal, then the friction force becomes very small. This is the case for a thick liquid film in which shear can take place between many hundreds or thousands of layers of equal shear strength.

If the only critical layer is the interfacial medium (Friction Energy Dissipation Zone I in Figure 1), then models for the behavior of the stack should be based on lubrication theory. If the deformation of thin films or boundary layers contributes, then models should consider FEDZ II. If the deformation of the contiguous solid(s) under the boundary layers adds to the friction force, then models should incorporate processes in FEDZ III. Finally, if the rigidity and vibration of the structure is a consideration, then FEDZ IV processes must be incorporated. In general, the scale of processes and mechanisms increases from FEDZ I to IV, and the behavior of discrete microscale events tends to be combined into 'bulk' or stochastic values of material properties.

The concept of *velocity accommodation*, described by Berthier, et al (1986), is a useful concept to apply to a FS. The operable processes in a stack will depend on where the minimization of free energy dictates that the relative motion (velocity) can be best accommodated. As Berthier et al. described, at each location, there are multiple ways in which the relative motion can be accommodated. Alternatively, one could say that at each location, there will exist at least one dominant, thermodynamically- and kinetically-favored frictional process consisting of a combination of two or more mechanisms. For example, if shear were occurring in a polycrystalline metallic layer adjacent to a frictional interface, then the set of operable mechanisms could include a combination of elastic deformation, plastic deformation, work hardening (crystallographic point- and line-defect generation), phonon generation (vibrations), and the conversion of excess mechanical energy into thermal energy. The relative influence of these mechanisms in the friction processes depends on the set of conditions imposed on the tribosystem. The natural selection of processes that operate in the tribosystem is dependent upon the external conditions that induce relative motion and the properties of materials that reside at the potential locations for velocity accommodation to occur.

5.0   Modeling Mechanisms and Processes

Progress will be made toward process-level modeling by identifying all the possible sources of frictional resistance. Given a compilation of the possible mechanisms and processes, subsets can be defined to enable the modeling of specific cases. One can use a list like that in Table 1 to identify the possible building blocks for friction models.

TABLE 1. Descriptions of several common friction processes.

Process	Description
Homogeneous shear of a continuous layer	shear within a homogeneous layer of material (solid or liquid); thin lubricating films
Homogeneous shear of a stack of continuous layers	simultaneous shear of homogeneous, parallel layers of material within the friction stack
Heterogeneous shear of continuous or discontinuous layers	shear within a heterogeneous layer of composite material or particles; deformation is complicated by discontinuities in layer's mechanical properties or structural features
Transfer	adhesion (adherence) between asperities, rupture of the adhesive junctions permits mass to be transferred to the opposing surface
Plowing	sliding resistance associated with the displacement of softer material, including debris layers and/or bulk surface material
Highly-deformed layer formation	creation of a near-surface layer with characteristics and structure distinct from the underlying material
Tribochemical film formation	reaction of tribo-activate surfaces with the surrounding chemical species so as to form films having friction-modifying properties
Friction-vibration interactions	the stimulation of vibrations by fluctuations in friction, and the influence of those vibrations 'reflected' back to the interface by the surrounding structure
Layer-interface lock-up and release	a process in which sliding interfaces can seize and suddenly release, forcing velocity accommodation to move from one location to another within the friction stack
Microlubrication regimes	microscale, localized transitions among hydrodynamic, mixed film, or boundary lubricating conditions
Chatter mark development	a stick-slip related process that results in uniformly-spaced fine-scale features similar to 'chatter marks' in machining processes

The same friction coefficient can result from different combinations of processes. Therefore, as shown in Table 2, materials with greatly different mechanical properties can, under the right circumstances, exhibit similar friction coefficients [Blau (1998)]. All the material combinations in Table 2 were tested in low-speed, pin-on-disk

247

tribometers. The materials range from high hardness ceramics to visco-elastic polymers and relatively-ductile metals.

TABLE 2. Materials couples with similar reported friction coefficients.
(all materials were tested at low speeds with pin-on-disk tribometers)

Material Pair	Type of Pair	μ_k	Reference
titanium on titanium	metal on metal	0.40	Budinski (1991)
α-alumina on α-alumina	ceramic on ceramic	0.38 - 0.42	Jahanmir and Dong (1992)
rubber on volcanic breccia	elastomer on ceramic	0.40	Yarnell (1970)
carbon steel on Nylon	metal on polymer	0.38	Steijn (1967)
carbon steel on polyvinyl chloride	metal on polymer	0.38	Steijn (1967)
TiN-coated 440C stainless steel on Al	hard coating on ductile metal	0.40	Bhushan and Gupta (1991)

Different process combinations can produce the same friction coefficient. For example, ceramic materials can exhibit similar sliding friction coefficients at low and high temperatures (**See Fig. 2**). At room temperature, the friction of alumina on alumina is lowered by adsorption of water films onto the frictional surfaces, but at high temperature it is affected by the formation of a layer of oxide particles. Moisture desorption raises the friction at intermediate temperatures.

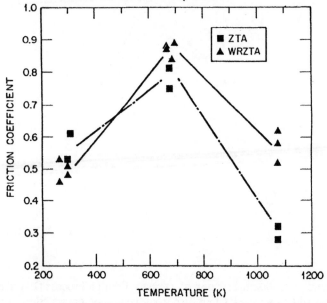

Figure 2. Effect of temperature on μ_k of zirconia-toughened alumina (ZTA) and whisker-reinforced zirconia-toughened alumina (WRZTA). [Yust (1994)]

248

Since friction processes can combine in different ways and lead to transitions in the behavior of tribosystems, a means to graphically depict the changes in friction processes was developed. This construction is called a Friction Process Diagram.

6.0 The Friction Process Diagram

The construction of Friction Process Diagrams (FPDs) was described previously [Blau (1994, 1996)] and will be only summarized here.

A FPD is the graphical representation of a linear rule of mixtures for two or more friction processes operating in a given tribosystem. It is plotted in units of friction coefficient. Each process, consisting of a set of mechanisms, is represented by one apex or axis on the diagram. The value assigned to each apex is the friction coefficient for the tribosystem if its friction were completely controlled by that process. The axes are scaled linearly and a set of lines of equal μ_k ('isotribes') are constructed. The current operating point can be represented on a FPD as a circle or ellipse whose borders represent the variability of μ_k at that time.   As the system ages, the operating point can move within the FPD (Figure 3).   If the temperature changes, the apex values may change, necessitating the addition of a temperature axis to the FPD   (Figure 4).

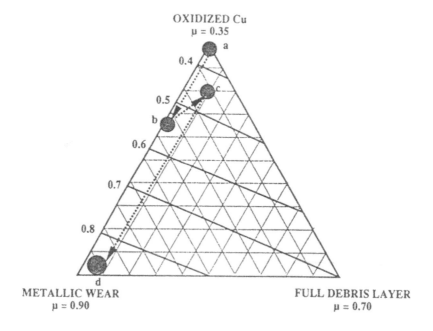

Figure 3. Isothermal section of a ternary Friction Process Diagram for brass sliding on steel. Clockwise from the top, processes correspond to oxide film shear, third-body layer shear and metallic contact.

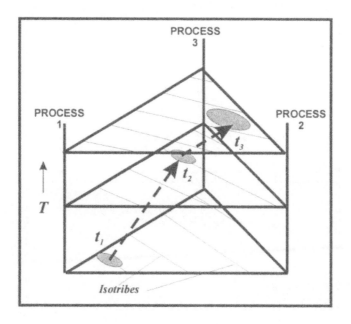

Figure 4. Ternary Friction Process Diagram that includes temperature. The time dependent trajectory of the operating point is shown.

7.0 Friction modeling

Four challenges confront those striving to develop predictive models for friction:

(1) to define the tribosystem's boundaries, key processes, and scales of interaction
(2) to comprehend the time-dependence of simultaneously operating processes
(3) to model the mechanisms that contribute to each contributing process
(4) to establish rules of scaling that link processes in four dimensions.

Process scale and time are related. Figure 5 shows schematically how the scale of operating friction processes, can change as a function of time.

Friction modeling depends on defining the tribosystem in such a way as to enable the establishment of relationships between the known variables and the observable attributes of that tribosystem. If two theorists disagree on what has been observed and begin with different definitions for the problem, it is unlikely that they will arrive at the same model, and even if they agree on a common definition of the tribosystem, different sets of assumptions may still be chosen during the derivation, and the resultant models will also differ.

250

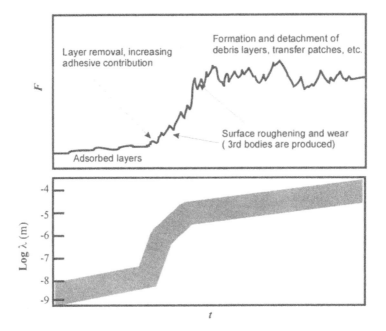

Figure 5. The scale of dominant friction processes can increase with time. Initially, thin adsorbed films may dominate, but as they are wiped away, adhesive transfer occurs, and the deformation of the material, extending tens of micrometers into the surface, dominates.

In friction modeling, the boundary conditions of the problem can change. For example, sliding friction can result in wear that alters the surface topography and produces third-body layers that densify in the interface. What began as an analysis of the solids covered by films becomes a problem of third-body creation and agglomeration.

7.1 TYPES OF FRICTION MODELS

There are four types of friction models: (1) descriptive, non-mathematical models that show pictures or conceptual diagrams of frictional interactions, (2) empirical models which are based on fitting data, (3) first-principles models based on surface physics, chemistry, and mechanics, and (4) hybrid models that combine (2) and (3).

Model building starts with conceptualizing the tribosystem and identifying which of many possible variables are to be included. Then rules and relationships, either assumed or learned from experiments, are included. At last a final model emerges. The model must be tested by comparing calculated results with experimental findings.

There have been numerous reviews of friction models in the literature, and their details will not be repeated here. The reader is referred instead to the Appendix to this chapter. For the remainder of this discussion we will consider certain existing models in light of the friction stack concept.

8.0 Building a Friction Stack

As discussed earlier, the behavior of a friction stack depends on where the relative velocity is accommodated and on the relative contributions to frictional work. This can be illustrated by considering simple models that relate to different layers in the FS.

8.1 FRICTION CONTROL BY LIQUID SHEAR

If the tribosurfaces are separated and all the relative motion and energy dissipation takes place in an interfacial fluid, at first order, the friction force F can be expressed in terms of the shear of a fluid between planar surfaces. Eschbach (1966), for example, gives the following expression:

$$F = \left(\frac{\eta v}{h} \right) A \qquad [3]$$

where, η = the lubricant viscosity, v = relative velocity of the surfaces, h = film thickness, and A = load-bearing area. A question arises here as to whether the variables contained in the parentheses are independent. In fact, the velocity, viscosity and film thickness are interrelated (as indicated in the Reynolds equation) and it has been recognized that properties of fluids in confined spaces can change significantly when the thickness of the lubricant film becomes small [Granick (1992)].

8.2 LIQUIDS CONTAINING PARTICLES

When solid particles are present in sufficient concentration in interfacial fluids, and when the particle size approaches the film thickness, then a different form of model should be used. Recent work, like that of Mizuhara, et al. (2000), has attempted to account for these effects. Mizuhara et al. assumed that the rise in friction must be proportional to the number (or concentration) of particles in the fluid as well as their size relative to the lubricant film thickness. Table 3 lists selected experimental results that show how the concentration of particles and the velocity of relative motion affect the friction force. Experiments were performed using alumina particles in oil films between 0.45% C steel cylinders and a rotating 0.45% C steel disk. The friction force was lowest at low concentrations and high speeds, and highest at high concentrations and low speeds.

TABLE 3. Effects of alumina particle concentration on friction force at different sliding speeds [twin cylinders-on-disk, alumina particles in synthetic oil flowing at 1 ml/s, 0.9 N normal force].

Particle Concentration (mg/l)	F (N) v = 0.3 m/s	F (N) v = 0.6 m/s	F (N) v = 0.9 m/s
3	0.016	0.009	0.003
6	0.030	0.016	0.006
12	0.044	0.025	0.013

8.3 CONTROL BY PARTICLES AND PARTICLE COMPACTS

When there is a layer of particles in the interface, instead of a liquid, then the modeling situation changes again.   For a non-cohesive granular power, the shear force within the powder F_S can be expressed in a Coulombic form [Brown and Richards (1966)] :

$$F_S = \mu P \tag{4}$$

Bagnold (1966) developed following expression for the average friction force F produced by a particle mass experiencing a velocity gradient (dv/dz):

$$F = \frac{mD^2}{2\lambda}\left(\frac{dv}{dz}\right)^2 F = \frac{mD^2}{2\lambda}\left(\frac{dv}{dz}\right)^2 \tag{5}$$

where m = mass of a particle, D = average particle diameter, and λ = mean distance over which excess energy is dissipated (typically several particle diameters - a function of the packing density). Powder shear models like the above have been employed in powder metal processing, pharmaceuticals, and soil dynamics. Heshmat (1991) has modeled the flow and friction in powder lubricants in terms of quasi-hydrodynamic lubrication.

Other approaches to modeling powder friction have explicitly considered both the internal friction and the friction of the powder mass against the walls. Dawes (1952) constructed a horizontal shear testing box with internal vertical vanes to confine the powder shear to a thin layer between the tips of the vanes. After measuring the shear force of limestone powder under different normal loads, he replaced the upper part of the box with a flat plate faced with sandpaper and repeated the measurements (Fig. 6). These two linear plots suggest an Amontons-like relationship between tangential force and normal force. Given the same contact area, the ratio of the slopes indicates the ratio of wall shear strength to shear strength within the powder (0.643 / 0.823 = 0.781).

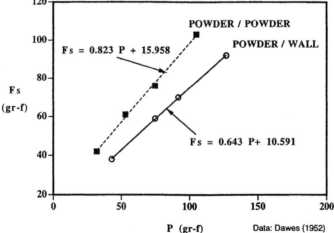

Figure 6. Proportionality of forces for the shear of limestone powder within a compact and at the walls of the compact [adapted from Dawes (1952)].

By using Eqn. [1], the effective friction force for a stack with two possible shear locations operating in parallel is:

$$\frac{1}{F_{eff}} = \frac{1}{F_{powder}} + \frac{1}{F_{wall}}$$ [6]

or,

$$F_{eff} = \frac{F_{powder} F_{wall}}{F_{powder} + F_{wall}} = \frac{\tau_{powder} \tau_{wall} A^2}{\left(\tau_{powder} + \tau_{wall}\right) A}$$ [6a]

Assuming that the shear strength of the wall is 0.781 times that of the powder (from the Dawes results) and that the area of the FS is the same for each layer, then for a FS containing limestone in the interface,

$$F_{eff} = \frac{0.781 \cdot \tau_{powder} A}{1.781} = 0.439 \cdot \tau_{powder} A$$ [6b]

Therefore, when both shearing at the wall and internal shearing of the powder are operating, the friction of the stack can be reduced by over 50%. If, however, the plane of shear shifts back and forth between the wall and the interior of the powder, then the instantaneous friction force varies with the fraction of yielding at each location.

Brown and Richards (1966) expressed wall friction shear stress S_w as a sum:

$$S_w = A + \mu_w P$$ [7]

where A = adhesion of the powder to the wall (units of pressure), μ_w = coefficient of wall friction. In a more rigorous treatment, Adams (1992) derived an expression for the wall friction force F of an assemblage of uni-sized particles between parallel planes acting over area A :

$$F = \left[\frac{\pi \tau_o}{\left(4E^* s / 3r_p\right)^{2/3}} W^{2/3} + \alpha W \right] A$$ [8]

τ_o = shear strength of the compact, s = a particle packing parameter, α = pressure coefficient of shear stress, W = normal load per unit area, r_p = particle radius, and E^* is the composite modulus, a function of the elastic moduli and Poissons ratios of the particles and the wall material as follows:

$$E^* = \left(\frac{1 - v_1^2}{E_1} + \frac{1 - v_2^2}{E_2} \right)$$ [9]

Eqn. [8] correctly predicts the dilatancy behavior commonly observed in loose particulate bodies. That is, as the volume of the particle assemblage expands, the particle spacing increases, and the friction decreases. Complications arise for mixtures of variously-sized particles of different compositions and mechanical properties.

8.4 SOFT HOMOGENEOUS FILMS

A analogous form of Eqn [4] for the shear of liquid layers can be written. Here, all the material properties are conveniently lumped into a single parameter, the interfacial shear stress τ that acts over an area A :

$$F = \tau A \qquad [10]$$

The effect of contract pressure, p , on the shear strength of the layer can be accommodated by including a pressure coefficient α, as described by:

$$F = \left(\tau_o + \alpha p \right) A \qquad [11]$$

This form has been used for modeling the response of solid lubricants and polymeric surface films. [See the review by Briscoe and Stolarski (1993).]

The shear of thin layers can exhibit certain characteristics resembling the shape of Stribeck (1902) relationship for liquids. In that case, friction force first decreases then increases as a function of the product of viscosity and velocity, and inversely with the pressure. In a classic study of In films, Bowden and Tabor (1950) plotted film thickness versus μ and found the same kind of initial frictional decrease to a minimum, followed by a gradual increase. At layers thinner than about 0.1 μm, the film failed to lubricate effectively. A thicker soft film will spread out more, forcing a wider track width, and therefore a higher τA product.

8.5 SOLID-ON-SOLID FRICTION

If liquids, liquids containing particles, dry particle layers, and thin homogeneous films cannot exclusively account for the resistance to sliding, the effects of solid sliding on solid must be considered. Adhesion models for friction imply that the friction force represents the resistance to the shear of a solid-solid adhesive junction with a contact area A that is subjected to a certain contact pressure. The shearing force F, acting over A is expressed as the shear stress τ .

$$\tau = \frac{F}{A} \qquad [12]$$

The value of τ that must be equaled to initiate deformation is the shear strength. Likewise, the contact pressure p, can be expressed as the normal force P per unit area.

$$p = \frac{P}{A} \qquad [13]$$

Thus, the friction coefficient can be expressed as

$$\mu = \frac{F}{P} = \frac{\tau A}{pA} = \frac{\tau}{p} \qquad [14]$$

where p can also be considered to equal the hardness of the softest contact junction.

The contact pressure can be modified to reflect both the contact geometry and the surface energy of adhesion Γ.. Rabinowicz (1965), derived an expression for μ for the sliding of a circular junction of radius r, with an average surface roughness angle of θ,

$$\mu \cong \frac{\tau}{p} \cdot \frac{1}{1 - 2\Gamma \cot(\theta/rp)} \qquad [15]$$

This relationship predicts high friction coefficients for smooth surfaces (low θ).

One of the more widely-accepted approaches to modeling the sliding friction of solids combines adhesive contributions with plowing [Bowden and Tabor (1950), pp. 90-98].

$$F_{total} = F_{adh} + F_{pl} \qquad [16]$$

The adhesion term depicts the interfacial bonding forces that tend to prevent shear and the plowing term accounts for displacing surface material. When the sliding contact is relatively flat (asperities are blunt), one might expect a dominant contribution from the first term, but when the asperities are sharp and hard, as for scratching, a significant contribution from deformation is expected.

Zwicker and Jirgal (1964) compared the form of the adhesion and plowing terms in three models for friction (see Table 4). Differences in evaluating the true area of contact, a long-standing debate in friction theory, and differences in the assumptions regarding elastic and plastic behavior led to differences in the forms of the adhesion and plowing terms.

TABLE 4. Comparison of adhesion and plowing terms in three friction models [after Zwicker and Jirgal (1964)], P = normal force, τ = shear strength, p = yield pressure, α = a constant (2/3 α 1), other constants K, A, B, C, D.

Theory	Adhesion term	Plowing term
Bowden and Tabor (1950)	$\tau\,(P/p)$	$K\,(P^3/p)^{1/2}$
Archard (1953)	$\tau\,(P^\alpha A)$	$B_p\,P^{(3\alpha/2)}$
Rubenstein (1956)	$C\tau\,(P/p)^\alpha$	$D\,(P/p)^{(3\alpha/2)}$

Elsewhere in this book, the author describes considerations in measuring and defining the friction coefficient when a sharp indenter is involved. The adhesive and plowing components of friction in classical models provide a means to interpret forces developed during scratching. Yet there are still questions as to when friction experiments

256

become micromachining experiments, and when the reported 'friction coefficient' is more a characteristic of the contact geometry than the combination of materials.

Since the 1950's and 1960's, refinements have been made to the adhesion and plowing term model. Models of the 1940's and 50's tended to focus on bulk properties, but work over the last 40 years has explored the effects of crystallographic orientation [C. H. Riesz and H. S. Weber (1964), R. P. Steijn (1964), D. F. Buckley (1981), Weick and Bhushan (2000)], phases in composites [Roberts (1985)], and other microstructural features. Mechanically-mixed layers can greatly affect friction [e.g., Figure 7, Blau (1985)]. Therefore, physically consistent friction models must be able to accommodate the microstructural changes associated with tribological transitions.

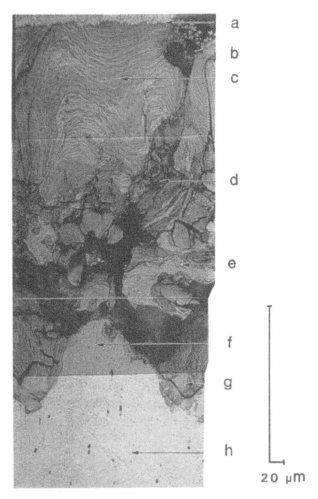

Figure 7. Polished taper-section of the microstructure of a mechanically-mixed layer developed on a Cu block that was slid against a steel ring. Note the rich complexity and the intricate lamellar structure of the metal-oxide mixture. [The labelled features are described in more detail by Blau (1985).]

The effects of asperity interaction time is particularly relevant to modeling friction of elastomers and other polymers, whose response is affected by visco-elasticity and hysteresis. [See the review in Moore (1975), cited in the Appendix.] Microstructure-based models must also consider the effects of strain rate on shear strength.

8.6 THE MACROMECHANICAL SYSTEM

FEDZ 4 considers the macro-mechanical response of the machine in which the friction couple resides. Examples include modeling of friction-vibration interactions in automotive and aircraft brakes, and stick-slip phenomena. The interface is seen as a 'stimulus' and the response of the system to that stimulus is modeled analytically or numerically. Parameters such as natural frequency, stiffness, and damping capacity (loss tangents) are major variables in these models. Such models are characterized by the computation of non-dimensional friction spaces or phase plane diagrams [Brockley and Ko (1970].

Macromechanical modellers tend not to concern themselves with materials microstructures, but rather focus on generalized inputs and outputs to the system. It should be noted, however, that stimulus-response approaches have also been applied on the nano-scale, assuming that the atom-atom responses are modelable as a assembly of springs and/or visco-elastic elements in an idealized lattice or sub-element of a lattice [e.g., the review by McClelland and Glosli (1992)].

9. Summary

Modeling friction in its many forms requires a multi-dimensional perspective and a recognition that the scale of contributing factors can change with time. The differentiation of frictional contributions into processes and mechanisms permits one to link models on a fundamental and phenomenological macroscale. The "friction stack" is one way to assimilate the roles of materials and structures. Understanding friction in the presence of wear is an important aspect of many practical tribosystems.

Recognizing that the friction coefficient is not necessarily a single-valued characteristic of a tribosystem enables one to understand transitions and momentary changes in operative processes. Friction process diagrams can be used to depict changes in frictional contributions with time.

Friction models for a variety of processes, built up from basic mechanisms, can be developed to suit the characteristics of individual tribosystems or classes of problems.

Frictional modeling requires one to first identify the important variables in the tribosystem of interest, and then to select submodels for the relevant processes or mechanisms to enable the attributes of friction force, like the time-dependency, variability, and average magnitude, to be fully described. Energy dissipation can serve as a guide to identifying relevant submodels.

A future challenge in friction modeling in practical machinery is to be able to account for conditions of operation that change over the lifetime of the contact. Examples include, start-up and shut down, intermittant periods of high-speed, and periods of dormancy when corrosion can occur on contact surfaces. Not only the average value of the friction coefficient, but its changes with time must be modeled.

258

Acknowledgements

The author wishes to gratefully acknowledge the support of the conference organizers and the U.S. Department of Energy, Office of Transportation Technologies, in enabling his participation in this Advanced Studies Institute. The help of author's wife, E. R. Blau, in proof-reading the camera-ready copy and suggesting changes to clarify its wording, is also greatly appreciated.

References

Adams, M. (1992) " Friction of Granular Non-Metals," in Fundamentals of Friction: Macroscopic and Microscopic Processes, ed. I. L. Singer and H. M. Pollock, NATO ASI Series, Vol. 220, Kluwer Academic Publishers, Dordrecht, 183-207.

Archard, J. F. (1953) *J. of Applied Phys.*, **24**, 981-988.

Bagnold, R. A. (1966) "The shearing and dilation of dry sand and the singing mechanism," *Proc. of the Royal Society*, **A 295**, 219-232.

Berthier, Y., Godet, M., and Brendle, M. (1986) "Velocity Accommodation in Friction," Soc. of Tribologists and Lubr. Engr., Preprint 88-TC-3A-2.

Bhushan, B. and Gupta, B. K. (1991) Handbook of Tribology, Table 14.16a.

Blau, P. J. (1985) "Measurements and Interpretations of Sliding Wear Damage in Metals," *J. of Tribology*, **107**, 483.

Blau, P. J. (1989), Friction and Wear Transitions of Materials, Noyes Pub., Park Ridge, NJ.

Blau, P. J. (1993) "Friction - A Multidimensional Phenomenon," at the Symposium on Tribology and Surface Engineering, Massachusetts Institute of Technology, Cambridge, MA, June 29-30.

Blau, P. J. (1994) "Friction Process Diagrams for Analyzing Interfacially-Complex Sliding Contacts," *Tribology Transactions*, vol. 37(4), pp. 751-756.

Blau, P. J. (1996) Friction Science and Technology, Marcel Dekker, New York.

Blau, P. J. (1998) "Four Great Challenges Confronting our Understanding and Modeling of Sliding Friction," in Tribology for Energy Conservation, ed. D. Dowson, Elsevier, UK.

Blau, P. J. (2001) "Experimental Aspects of Friction Research on the Macroscale" in this volume.

Bowden, F. P. and Tabor, D. (1950) Friction and Lubrication of Solids, Clarendon Press, Oxford.

Briscoe, B. J. and Stolarski, T. A. (1993) "Friction," in Characterization of Tribomaterials, ed. W. A. Glaeser, Butterworth-Heinemann, Boston, pp. 31-64.

Brockley, C. A. and Ko, P. L. (1970) "Quasi-harmonic friction-induced vibration," *ASME Transactions*, October (1970), pp. 543-549.

Brown, R. L. and Richards, J. C. (1966) *Principles of Powder Mechanics*, Pergamon Press, Oxford.

Buckley, D. F. (1981) Surface Effects in Adhesion, Friction, Wear, and Lubrication, Elsevier, New York.

Budinski, K. G. (1991) Proc. International Conference on Wear of Materials, ASME, New York, 289.

Czichos, H. (1978) Tribology - A systems approach, Elsevier Pub., Amsterdam

Dawes, J. G. (1952) "Dispersion of dust deposits by blasts of air, Part I," *Safety in Mines Research Establishment, Report 36*, Ministry of Fuel and Power, Sheffield, UK.

Dowson, D. (1978) History of Tribology, Longman, London

Gao, C., Wu, Y., Varhese, V., and Malkin, S. (2000) "Temperatures and Energy Partition for Grinding with CBN Wheels," Abrasives Magazine, June/July, 28-33.

Granick, S.(1992) "Molecular Tribology of Films," in Fundamentals of Friction: Macroscopic and Microscopic Processes, ed. I. L. Singer and H. M. Pollock, Kluwer Pub., Dordrecht, 387-401.

Eschbach, O. W. (ed.) (1966) Handbook of Engineering Fundamentals, 2nd edition, John Wiley and Sons, New York, 4-53.

Heshmat, H. (1991) "The rheology and hydrodynamics of dry powder lubrication, *Trib. Trans.*, **34** (3), 433-39.

Jahanmir, S. and Dong, X. (1992) "Wear mechanisms of aluminum oxide ceramics, in Friction and Wear of Ceramics, ed. S. Jahanmir, Marcel Dekker, New York,15-49.

Ludema, K. C. (1997), University of Michigan, private communication.

McClelland, G. M and Glosli, J. N. (1992) "Friction at the Atomic Scale," in Fundamentals of Friction: Macroscopic and Microscopic Processes, ed. I. L. Singer and H. M. Pollock, Kluwer Pub., Dordrecht, 405-425.

Mizuhara, K., Tomimoto, M. and T. Yamamoto, T. (2000), "Effect of Particles on Lubricated Friction," *Trib. Trans*, **43** (1) 51-56.

Riesz, C. H. and Weber, H. S. (1964) "Friction and Wear of Sapphire," *Mechanisms of Solid Friction* ed. P. J. Bryant, M. Lavik, and G. Salomon, Elsevier, Amsterdam, 67-81.

Rubenstein, C. (1956) Proc. Royal Soc. (London), **B69**, 921.

Roberts, J. C. (1985) Surface morphology studies in polymer-graphite epoxy sliding," *ASLE Transactions*, vol. **28** (4), 503-510.

Rymuza, Z. (1996) "Energy concept of the coefficient of friction," *Wear*, **199**, 187-196.

Steijn, R. P. (1964) "Friction and Wear of Single Crystals," in Mechanisms of Solid Friction, P. J. Bryant, M. Lavik, and G. Salomon (ed.), Elsevier, Amsterdam, 48-66.

Steijn, R. P. (1967) *Metall. Engr. Quarterly*, **7**, 9.

Stribeck, R. (1902), "Die Wesentlichen Eigenschaften der Gleit- und Rollenlager, Z. Verein. Deut. Ing., **46** (38) 1341-1348.

Weick, B. L. and Bhushan, B. (2000) "Grain Boundary and Crystallographic Orientation Effects on Friction," *Trib. Trans*, **43** (1), 33-38.

Yarnell, A. G. (1970) "A new theory of hysteretic sliding friction," *Wear*, **17**, 229-244.

Yust, C. S. (1994) "Tribological Behavior of Whisker-Reinforced Ceramic Composite Materials," in Friction and Wear of Ceramics, ed. S. Jahanmir, Marcel Dekker, New York, 199-223.

Zwicker, E. and Jirgal Jr, G. H. (1964) "The Effects of X-ray Irradiation on Self-Friction of Potassium Chloride," in Mechanisms of Solid Friction , P. J. Bryant, M. Lavik, and G. Salomon, Elsevier, Amsterdam, 39-47.

Appendix

Selected Publications Related to Friction Modeling

Blau, P. J. (1996) *Friction Science and Technology*, Marcel Dekker Inc., New York (1996).

Bowden, F. P. and Tabor, D.(1986)*The Friction and Lubrication of Solids*, 2nd ed., Clarendon Press, Oxford.

Bryant, P. J., Lavik, M. and Salomon, G. editors (1964) *Mechanisms of Solid Friction*,), Elsevier Publishing Company, Amsterdam.

Coulomb, C. A. (1809)*Theorie des machines simples eu ayont au frottment de laures et a la rainder das cardages*, Paris.

Glaeser, W. A. ed. (1993) "Friction," B. J. Briscoe and T. A. Stolarski, in *Characterization of Tribomaterials*, Butterworth-Heineman, Boston.

Hutchings, I. M.(1992) *Tribology - Friction and Wear of Engineering Materials*, CRC Press, Boca Raton, Florida.

Kragelsky, I.V., Dobychin, M. N. and Kombalov, V. S. (1982) *Friction and Wear Calculation Methods*, Pergamon Press, Oxford; English translation of *Osnovy Raschetov na Trenie i Iznos* (1977).

Ludema, K. C. (1996) *Friction, Wear, and Lubrication*, K. C. Ludema, CRC Press, Boca Raton, Florida.

Moore, D. F. (1975)*Principles and Applications of Tribology*, Pergamon Press, Oxford.

Persson, B. N. J. (1998) *Sliding Friction*, B. N. J. Persson, Springer-Verlag, Berlin.

Rabinowicz, E. (1965)*Friction and Wear of Materials*, John Wiley and Sons, New York.

Reynolds, O. (1876) "On rolling friction,"*Philosophical Transactions of the Royal Society*, **166**.

Singer, I. L. and Pollock, H. M. eds.(1992) *Fundamentals of Friction: Macroscopic and Microscopic Processes* NATO ASI Series Vol. 220, Klewer Academic Publishers, Dordrecht

Suh, N. P. and Saka, N. (1980) *Fundamentals of Tribology*, MIT Press, Cambridge, Massachusetts.

Tomlinson, G. A. (1929) "The molecular theory of friction," *Philosophical Magazine*, **11** (7).

EXPERIMENTAL ASPECTS OF FRICTION RESEARCH ON THE MACROSCALE

P. J. Blau
Oak Ridge National Laboratory
Oak Ridge, Tennessee USA

The earliest investigations of sliding friction were prompted by engineering problems like the launching of ships and the braking of wagon wheels. Experiments were first performed on the objects themselves, and later, by building replicas that could be more conveniently managed. Concepts for early friction measuring instruments are credited to Leonardo da Vinci who documented his designs for various sleds, pulleys, journals and inclined planes during the period of 1452-1519 [Dowson (1979)]. Recent work has benefited from the development of sensors capable of measuring and recording friction forces with high precision and accuracy. Yet, even in the face of such advances, friction coefficients are still usually reported to only one or two significant figures. Additional measures of frictional behavior are now readily available. These will be discussed later.

Experiments involving contact dimensions greater than about 10 μm will be considered 'macroscale' in the present context. At that scale, friction measurements on tribocontacts involve the combined contributions of many thousands or millions of atoms that are distributed across the real area of contact. The reasons for acquiring friction data influence the choice of the test method and affect how the data are reduced for analysis. Therefore, this chapter reviews devices and techniques used to measure friction on the macroscale for both research and applications.

1. Introduction

Students are commonly introduced to the measurement of friction forces and friction coefficients during a general science or physics course. Their first brief exposure might consist of the definition of the static and kinetic friction coefficients and the solution of a few exercises, perhaps making use of a table of 'typical' friction coefficients. Rarely do discussions of friction extend beyond two or three pages in elementary physics textbooks. Therefore, students get the impression that there is nothing very mysterious or subtle about friction: solid friction is simply due to electrostatic forces between atoms on the mating surfaces, and it is straightforward to measure friction coefficients. While it is good to introduce young people to friction early in their schooling, that initial introduction may give them a lasting, simplistic impression of the subject.

Table 1 compares friction coefficients for the same materials combinations that were printed in several textbooks and handbooks. Sometimes the friction coefficient is given as a single value having one or two significant figures, and less commonly (but more correctly) as a range of values.

B. Bhushan (ed.),
Fundamentals of Tribology and Bridging the Gap between the Macro- and Micro/Nanoscales, 261–278.
© 2001 *Kluwer Academic Publishers*.

TABLE 1. Static and Kinetic Friction Coefficients Listed in Introductory
Physics and Mechanics Textbooks and Handbooks.

Material Combination	μ_s	μ_k	Table Reference
Wood on wood	0.25 - 0.5	0.19 -	1
Wood on wood (dry)	0.25 - 0.5	0.38*	2
Wood on wood	0.30 - 0.70	-	3
Oak on oak (para. to grain)	0.62	-	4
Oak on oak (perp. to grain)	0.54	0.48	4
Wood on wood	0.6	0.32	5
Wood on wood	0.6	0.5	6
Wood on wood	0.4	0.2	7
Steel on steel (unlubricated)	0.58	-	2
Steel on steel	0.78	0.42	4
Steel on steel (dry)	0.15	0.12	5
Steel on steel	0.7	0.6	7
Ice on ice	0.05 - 0.15	0.02	2
Ice on ice	0.05	0.04	4
Ice on ice	0.1	0.03	7
Metal on ice	0.03 - 0.05	-	3
Steel on ice	0.03	0.014	5
Steel on wet ice	0.03	0.01	6
Rubber on concrete	0.6 - 0.9	045 - 0.68*	1
Rubber on dry concrete	1.0	0.8	6
Rubber on dry concrete	1.0	0.8	7
Rubber on wet concrete	0.7	0.5	6
Rubber on wet concrete	0.7	0.5	7

* Text authors suggest reducing μ_s by 25% to get μ_k.

Table References
1 Beer, F. P. and Johnston Jr., E. R. (1962) *Vector Mechanics of Engineers*,
McGraw Hill, New York, 275.
2. Anon. (1967-8) *Handbook of Chemistry and Physics*, 48th ed., CRC Press,
Cleveland, F-14, 15.
3. Hibbeler, R. C. (1983) *Engineering Mechanics*, 3rd ed., Macmillan Pub., 285.
4. Frauschi, S. C., Olenick, R. P., Apostol, T. M. and Goodstein, D. L. (1986) *The
Mechanical Universe*, Cambridge Univ. Press, 181.
5. Anon. (1993) *Automotive Handbook*, 3rd ed., Bosch, Dusseldorf, 49.
6. Zafiratas, C. D. (1985) *Physics*, 2nd ed., J. Wiley, New York, 166.
7. Giancoli, D. C. (1991) Physics, 3rd ed., Prentice Hall, Englewood Cliffs, 84.

The methods used to obtain 'typical' friction coefficients, like those shown in
Table 1, are usually omitted from the textbooks and handbooks, and the materials are
only vaguely described as 'metal' or 'wood.' In only one case [Frautschi et al., Table
reference 4], did the authors imply a possible effect of material structure on the friction
coefficient by indicating that oak differed in friction depending on its grain orientation
with respect to sliding. Clearly, there is a need for better explanations of the
significance and use of friction coefficients, even on an introductory level.

When mechanical or materials engineers are faced with a friction problem, their early training prompts them to seek a table of friction coefficients, expecting that the number they seek can simply be looked up. If they happen to be ambitious enough to consult more than one table, they will probably find several different values, and that will inevitably prompt the question: "Which friction coefficient value is the *correct* one for these materials?" In this question there lies an implicit, but incorrect, assumption that the friction coefficient between any two materials is a single-valued, basic property of that materials pair.

It is now widely recognized by the tribology community that the friction coefficient is a function of more than materials properties alone. Hence, there is a need for different kinds of friction measuring instruments, whose choice is dictated by the problem at hand. This chapter will provide an overview of various methods for measuring friction on the macroscale and indicate the reasons for selecting different methods.

1.1 REASONS TO CONDUCT FRICTION EXPERIMENTS

Laboratory-scale investigations of friction are conducted for several reasons. Information and friction data that are obtained for one purpose may not necessarily be suitable for another:

(1) *Material Selection*: To select from among a group of candidate materials, lubricants, or coatings, in order to meet the needs of a specific, friction-critical engineering application. This includes developing friction data to go into computer models.

(2) *Materials Characterization and Development*: To compile non-system-specific (so-called "generic") friction data during the development of new materials, lubricants, or coatings to optimize their tribological characteristics. Data may or may not be relevant for use in selecting materials for specific applications.

(3) *Basic Research*: to study the fundamental nature of friction of solids or lubricated solids by carefully controlling various influences like normal force, velocity, surface roughness, temperature, humidity level, lubricant composition, etc.

No single test can simulate all types of frictional situations, and the method of testing must be selected to address the needs of the investigator. It is therefore risky to use friction coefficients listed in general tabulations for application-specific or for the computer-aided modeling of machine behavior.

1.2 DEFINITION OF TRIBOMETER

The 1989 edition of the Oxford English Dictionary defined a tribometer as "An instrument for estimating sliding friction." It is interesting that the verb *estimating*, rather than *determining* or *measuring* was used, and that the term *friction* was used in the non-quantitative sense (c.f. *friction coefficient* or *friction force*). The earliest reference provided by that dictionary is to the 1774 writings of Goldsmith who used the word *tribometre* to mean a "measurer of friction." In 1877, Knight defined it in rather specific terms as "an apparatus resembling a sled, used in estimating the friction of rubbing surfaces." In the present case, we shall consider a *tribometer* as being any apparatus that is designed for the purpose of estimating or directly measuring friction forces or their effects.

Measuring solid or lubricated friction might initially appear to be a simple matter of measuring forces in certain directions. In the simplest case, just the force opposing relative motion is needed. If a friction coefficient, as defined by Coulomb, is to be determined, then the magnitude of a second force, orthogonal to the first, is also needed. However, there are many cases in which this challenge is complicated by the contact geometry or the environment in which the measurements must be made. For example, the construction of a high-temperature, or controlled atmosphere tribometer may require a more sophisticated design to protect the force sensor from thermal damage or to permit the transducer signal to be conducted outside of the chamber. Sometimes the friction forces in a system do not act colinearly, but rather act over an arc of contact. Examples include journal bearings, brake discs, face seals, metal working presses, and textiles wrapped over rollers. In such cases, friction can be related to the torque rather than measured directly.

Dowson (1979) has provided a rich historical treatment of the measurement of friction and the present discussion will not probe that subject in detail. However, it is interesting to note that many of the basic principles recognized and used by early investigators have survived to influence the designs of modern friction measuring apparatus and industry-wide friction measurement standards.

1.3 THE SIZE RANGE OF TRIBOMETRY EXPERIMENTS

The scale of terrestrial frictional phenomena spans at least twelve orders of magnitude, ranging from molecular-dimensioned contacts to the scale of earthquake fault structures and glaciers creeping slowly over the underlying soil and rock. In practice, however, experimental tribometry spans a smaller size range of contact dimensions than the frictional phenomena occurring in nature (only about seven orders of magnitude). Friction measurements at the lower size ranges of tribometry are becoming more commonplace as precision instrumentation has become available, and MEMS (micro-electro-mechanical systems) have developed into an important technology area. Nevertheless, friction tests used to simulate machine behavior are more likely to be able to achieve comparable contact dimensions than tests aimed at understanding larger-scale natural phenomena, like geological fault-drag structures and subducting tectonic plates.

1.3.1 *Terminology Issues*
Unfortunately, there is some ambiguity in tribology terminology, as regards the meaning of 'nanoscale', 'microscale', 'mesoscale', and 'macroscale.' In the present context, 'macroscale' will arbitrarily be associated with nominal contact areas that exceed about 10 μm in breadth. Imposing this arbitrary limit does not imply that the finer-scale phenomena are not of interest. Rather it means that their net contributions are treated as the combined effect of many discrete events. Even so, it is possible to detect fine-scale asperity interactions as fluctuations in the detailed friction records of what ordinarily might be called 'macroscopic' experiments.

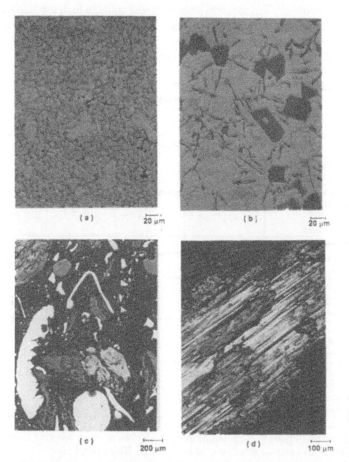

Figure 1. The size and morphology of microstructural constituents relative to the tribo-contact size can affect the variability of friction forces. (a) die-cast Mg-9 wt.% Al- 1 wt% Zn alloy showing a cellular dendritic structure, (b) cast Al-18 wt% Si alloy showing angular Si particles and Mg-Si phases (medium gray), (c) commercial automotive brake pad material showing metallic, ceramic, and polymeric constituents of many sizes and shapes, (c) ductile features on the wear track of an oxidized Cu-70wt% Zn alloy showing exposure of the bright metal below the oxide coating.

Recognizing the frictional size-effect, a tribometer with a typical contact diameter of 5-20 µm was developed to study friction on the microstructural scale [Blau (1990)]. Contact dimensions lie between the point contact probes of nanotribology and typical nominal contact dimensions of macroscale tests. Table 2 shows that the average friction coefficient tends to rise as the SiC content increases, and that the spacing of peaks on the friction trace tends to decrease when the spacing of the whiskers decreases. These results highlight some additional considerations in small-scale friction testing; namely, that adsorbed moisture can govern the friction process and equalize differences in friction between different phases, and secondly, that polishing relief, which tends to cause harder particles to stand-up above the softer ones, can increase their relative percentage contact and influence results. Here, SiC is harder than alumina and only 10 vol.% SiC whiskers could still protrude enough to dominate the friction in the contact.

TABLE 2. Comparison of the Hertzian contact areas with the areas of microstructural constituents tested with the ORNL friction microprobe.*

Quantity	Alumina*	Alumina + 5 vol% SiC whiskers	Alumina + 10 vol% SiC whiskers	SiC
Ave. SiC whisker spacing in the composite (μm)	not applicable	12.8	8.7	not applicable
Ratio of Hertzian contact diameter to whisker spacing	not applicable	1.47	1.39	not applicable
Ratio of ave. whisker area to Hertzian contact area	not applicable	0.077	0.082	not applicable
Ave. friction coefficient along the 400 μm sliding path	0.17	0.18	0.23	0.24
Average peak-to-peak spacing on the friction force trace (μm sliding distance)	not applicable	35.8	12.9	not applicable

* 1.0 mm diameter 440C stainless steel AFBMA Grade 5 bearing ball sliding at 10 μm/s at room temperature in air, 98 mN load. Average Hertzian contact diameter on alumina = 12.8 μm. Ref.: Blau (1990)

When the sliding contact area is large relative to the microstructural constituents, their individual effects tend to be averaged. For example, when testing automotive disc brake materials, the frictional contact typically extends over a nominal area of more than 50 cm^2. If the average diameter of a brake additive particle is estimated to be 100 μm, then approximately 12,700 particles will be within the nominal contact area at any given time. Furthermore, the degradation of the pad material tends to produce a transfer film on the brake disc, and that film may eventually govern frictional behavior. Therefore, the brake frictional torque is insensitive to the behavior of individual constituents bur rather tends to average all the contributions within the contact.

Attempts have been made to model friction with a linear rule of mixtures that requires measuring friction coefficients for each constituent phase. For example, Blau (1982) used a mixtures rule for Si and Al to predict the running-in friction changes for Al-Si alloys sliding against 52100 bearing steel. The fraction of each phase exposed at the surface changed between initial contact (100 area % Si, as-etched) and the run-in condition (about 12 area % Si coverage). Results like these support the development of Friction Process Diagrams, described by the author elsewhere in this volume.

Contact area considerations also affect the influence of frictional heating. As friction force and velocity increase, the contact temperature tends to rise. The ratio of the contact area to the total sliding track length affects the time available to dissipate heat before the next pass of the slider. Therefore, contact area considerations enter into macroscale testing in terms of their influence on the heat flow near the interface.

Contact area considerations are also associated with specimen cleaning. It has been shown in pin-on-disk tests that the friction force is more sensitive to finger prints on the disk specimen than on the pin tip. As might be expected, the pin material is quickly removed during rubbing, while the disk, with its larger track length, retains the lubricating effects of perspiration much longer.

2.0 Design and Selection of Tribometers

2.1 MACROSCALE TRIBOMETER DESIGNS

Laboratory methods for friction measurement can be divided into six categories, as shown in Table 3. Gravitation-based devices are used for static friction coefficient measurements. Sled-type devices, having varying levels of sophistication, are used for both static and kinetic measurements. The linear force measurement category includes common pin-on-disk and flat-block-on-ring tests where the friction force acts tangentially to the contact even though the wear track on one of the specimens may be curved. Torque measurements are used when the friction forces are distributed over an arc of contact. Figure 2 and Figure 3 show examples of testing geometries from the first four categories. Specialized tests, like those to measure friction in pivot bearings (Figure 4) and bolts are used in mechanical engineering design.

TABLE 3. Basic types of friction measuring devices.

Category	Examples
1) Gravitation-based devices	tilting-plane apparatus, wrapped cylinders with hanging weights
2) Direct linear force measurement	'sled'-type tests, spring scales, tension or compression load cells
3) Torque measurement	torque transducers, outrigger devices, face seals, disc brakes
4) Tension-wrap devices	differential tension measurements using load cells
5) Oscillation decrement devices	pendulum-type devices
6) Indirect indications	motor current monitors, vibration sensors

2.2 TRIBOMETER DESIGNS FOR ENGINEERING

Historically, friction testing devices have been designed to meet specific engineering needs. The geometry of such devices is chosen to mimic the contact geometry, contact pressures, motions, and environment of the machine to be simulated. They encompass a wide range, such as the James machine for measuring friction of footwear-on-flooring (e.g., ASTM Standard Test Method F-695), microscale tribometers on silicon wafers [e.g. Dugger (2000)], tire traction machines (e.g., ASTM Standard Test Method E 670), and inertial dynamometers used to evaluate vehicle brakes. A list of standard friction testing devices has been provided earlier by Budinski (1992).

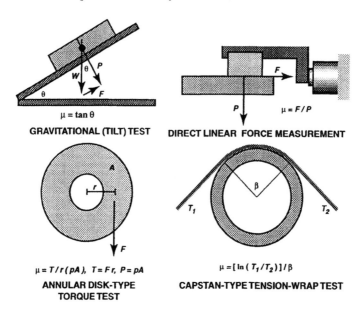

Figure 2. Four types of tribometers and the relationships commonly-used to compute friction coefficients.

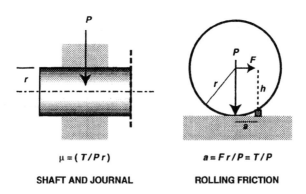

Figure 3. Friction measurement in journal bearings (left), and rolling elements (right) in which the analysis is based on continually rolling over an obstruction that is a distance "a" from the contact centerline.

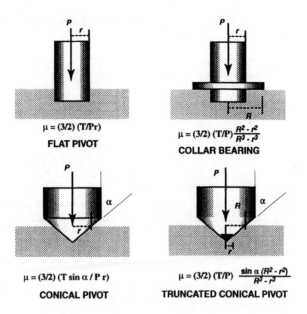

Figure 4. Friction coefficient computation in various pivot bearings.

There are far too many different tribometers designed for engineering applications to be described in this chapter, but the main issue for experimentalists who work in applied friction science is: "How much does the laboratory tribometer have to resemble the engineering application in order for the data to be relevant and useful?" There is no simple answer to that question, but it is sometimes surprising how well basic bench-scale friction data can sometimes correlate with larger, macroscale test results.

For example, consider one case in which the author compared friction coefficients obtained with a pin-on-disk tribometer to those from a full-scale, inertial brake dynamometer [Jang et al. (1997), Blau (1996)]. Table 4 summarizes the characteristics of these two quite different test devices. Table 5 compares the friction coefficients obtained on these machines for the same material pairs. Brake pad material "J" was the normal commercial composition. Pad material "C" had none of the usual abrasive additive, but twice the normal amount of lubricant, and pad material "F" had no lubricant additives, but twice the normal amount of abrasive. While the characteristics of the testing machines were clearly quite different, the pin-on-disk test was nevertheless capable of indicating relative trends in frictional behavior produced by modifying the compositions of the pad materials. This illustrates the value of conducting laboratory tests to uncover trends in frictional behavior, even though the friction coefficients are not identical for these two machines. Pin-on-disk results do not always correlate as well as they did in the foregoing example and a good correlation should not be assumed without further comparative testing.

TABLE 4. Comparison of two tribometers used to measure the friction of brake pad materials against actual cast iron rotor material.

Characteristic	Machine "A"	Machine "B"
specimens	full-sized brake discs and pads	6.35 mm diameter pins on 25 mm diameter disks
position of materials	full-scale pads mounted on calipers and clamped on full-sized rotor discs	cast iron pins machined from rotors, on disks cut from brake pads
contact area (nominal)	5307 mm^2 (pad area)	1.2-1.7 mm^2 (pin wear scar)
contact pressure	0.315 - 0.914 MPa	1.73 - 2.48 MPa
equivalent vehicle speed	96.6 to 0 km/hr (full-stops)	4.2 - 8.9 km/hr (constant drag)
environment	room-temperature air	room temperature air
friction computation	based on measured torque and air pressure to the brake actuator	load cell for friction force, dead-weight load for normal force

TABLE 5. Comparison of friction coefficients from two tribometers, and the ratio of friction coefficients of given pad materials to material 'J'.

Pad Material	Machine "A" μ (μ/μ mat'l. J)	Machine "B" μ (μ/μ mat'l. J)
"J" (normal pad composition))	0.37 (1.00)	0.308 (1.00)
"C" (2x lubricant, no abrasive)	0.34 (0.92)	0.253s (0.82)
"F" (2x abrasive, no lubricant)	0.39 (1.05)	0.312 (1.01)

2.3 INFLUENCE OF MATERIAL CHARACTERISTICS

Sometimes the properties of the materials in the friction pair influence the design of the friction test device. For example, measuring the friction of ice requires attention to temperature control and measuring the friction of fibers and unsupported plastic films requires attention to material elasticity (stretching) when applying the normal force.

In another article in this book, the author describes how frictional energy is dissipated within a series of zones that extend to both sides of the interface. Depending on how the friction forces are transmitted from one zone to another, the requirements for the tribometer configuration will be affected. For example, in a well-lubricated friction test, the frictional energy is mainly dissipated in fluid shear, and within reason, the stiffness of the typical apparatus fixtures should exhibit no undue influence on the results. In friction testing of dry, clean surfaces, higher friction forces could be transmitted into the supporting fixtures to a level whether their mechanical response (feedback) becomes important. The stiffness and damping capacity of the fixtures could induce stick-slip, for example. Consequently, tribometer design should be considered on the basis of the degree to which transmitted friction forces are 'felt' by the structure.

2.3.1 *Influence of Surface Cleanliness and Time Prior to Testing*

Since frictional response can be influenced by the presence of adsorbed films, lubricating layers, or contaminants, special attention must be paid to preparing surfaces of the test pieces and this is especially true for static friction determinations. Some materials easily adsorb moisture or hydrocarbons from the surrounding environment, react with oxygen in the air, or remain relatively unaffected. The collected data in Table 6 show how the static coefficient of friction can change by as much as a factor of six depending on method of cleaning and testing.

TABLE 6. Effects of cleaning method on static friction measurements of self-mated metals. [Blau (1996), collected data, p. 167]

Metal	μ_S (50% RH, flat-on-flat)	μ_S (pure metals/outgassed, flat-on-flat)
Ni	0.50	3.0
Au	0.50	2.8
Al	0.55	1.8
Cu	0.60	1.5
Ag	0.50	1.5
Fe	0.55	1.2

The time between surface preparation and testing can be important, particularly when measuring the static friction coefficient. Early experiments by Akhmatov (1939) on cleaved NaCl surfaces illustrate the effects of time before testing on static friction coefficients (Table 7).

TABLE 7. Effects of time prior to testing on the static friction between cleaved salt crystals. [Akhmatov (1939)]

Time of Exposure to Air (min)	μ_S
1	2.5
2	1.9
7	0.8
14	0.5
24	0.4
37	0.4

In long-term friction experiments or those conducted at high contact pressures, the influence of surface preparation may not extend beyond a brief running-in period, if at all. If, on the other hand, sensitive, low-load, low contact pressures are involved, careful surface preparation and cleaning become critical. There are some instances of friction testing in which surfaces must be tested in the 'as-manufactured' condition. Examples include the friction of textile fabrics and paper products.

2.3.2 *Friction Testing of Materials in Powder Form*

Specialized tribometers have been designed to measure the friction of powder layers or the friction between confined powder particles. Applications include copying machines, pharmaceutical manufacturing, powder metallurgy processing, and bearing surfaces with entrapped third bodies. Powder friction testers tend to use either direct linear force measurements in what are known as "shear cells" [Brown and Richards (1966)], or measurements of torque in annular cup-like devices.

The friction forces developed along a die wall during the compression of metal powders can be calculated using a cylindrical piston device. German (1994) gave the following expression for the friction coefficient:

$$\mu = \frac{F}{p(\pi z D \cdot dH)} \qquad [1]$$

where F = wall friction force (axial), p = pressure applied to the top of the compact, D = cylinder diameter, z = a proportionality constant that depends on powder density, and dH = increment of distance along the cylindrical wall.

Another measure of friction in inter-particle contacts uses the principle of the 'angle of repose.' Brown and Richards (1966) describe several devices used to measure the angle of repose, and hence indirectly the interparticle friction of powders. By pouring powder through a funnel onto a round, flat platform and allowing the excess to drain over a sharp surrounding edge, a conical heap will be formed. The angle of the slope relative to the horizontal plane is the angle of repose. Sample values for this test method are given in Table 8. There are several variations of the method, such as slowly rotating the platform (dynamic angle of repose), or using the inside of a cylinder with the particles draining through a hole in the base.

TABLE 8. Angles of repose for conical piles of powdered material.
[Ref. Brown and Richards (1966), p. 29]

Powder material	Shape Description	Angle of repose (degrees)
Beads	smooth, spherical	17.5 - 20
Sand	rough surfaces, but nearly spherical	32.5
Sand	angular grains	35.5-27
Rice	grains	35
Coal	rough particles of similar size	35.5 - 38.5
Coal	mixture of coarse and fine particles	54

3.0   Contact Geometry Effects of Friction Test Data

The contact geometry used in a friction test will significantly affect the following:

1) Depth of penetration by the slider under a given normal force
2) Entrainment of lubricants and third bodies into the contact
3) Heat transfer characteristics
4) Stress distribution in both bodies

From the viewpoint of contact geometry, it is interesting to consider the definition of the friction coefficient in the case of scratch testing. A scratch test-like geometry is commonly involved when reporting 'friction coefficients' in micro-scale experiments. At the opposite extreme, scratch-like geometries can occur at large scales - an example being the kilometer-long 'ice scours' produced when huge protrusions on the bottoms of glacial ice 'scratch' the bottoms of ancient lake beds.

Figure 5 represents cross-sectional views of a diamond (upper specimen) moving along a surface of a ductile metal 'M' under normal load **P**. The width of contact is, say, ~10-100 μm and the depth might be at a similar scale.

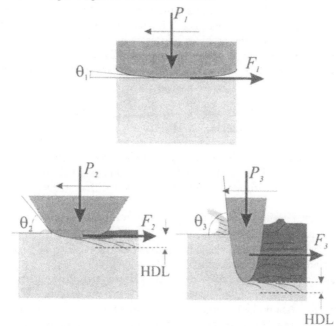

Figure 5. The (F/P) ratio for a diamond scratching a ductile surface is influenced greatly by tip sharpness and normal force, necessitating a reconsideration of what is meant by the 'friction coefficient' in scratching.

For each case shown in Figure 5, one can calculate a value for (**F/P**). The contact radius in the upper center illustration is relatively large, and force F_1 is controlled primarily by the shear strength of adsorbed layers and oxides ('shear films'). At a higher load, or using a sharper point (lower left illustration), adsorbed films or oxides are penetrated and the contribution of shearing the the highly-deformed layer (HDL) of metal is also felt. There is a relatively deep penetration for a sharp point (lower right) even at modest values of **P**. If the stylus point does not break, the work needed to displace material should represent a major component of F_3, not the friction between the sides of the stylus material and the material on the side walls. Clearly, the sharper the point and the deeper its penetration, the higher the magnitude of **F**, and higher the corresponding (**F/P**) ratio - until the point breaks. This raises the following questions:

1) Is the value of μ_k obtained in scratching with a diamond stylus really reportable as the friction coefficient for diamond against M?

2) If the answer to 1) is 'no', then what is the significance of (**F/P**) in scratching, and what should that quantity be called?

3) At what value of θ, or depth of penetration, does the definition for μ_k of diamond on M take on a different meaning?

Irrespective of scale, there are questions about the meaning of 'friction coefficients' reported in scratch tests. One way to avoid this issue is to define a new quantity, such as the 'stylus drag coefficient.' Even so, there is a need to precisely define the critical geometric conditions (i.e., tip 'sharpness') beyond which the term 'friction coefficient' ceases to apply, and when the problem bifurcates into one of sliding friction along the stylus surfaces coupled with plowing. Bowden and Tabor (1950, p. 91) considered the sliding of spheres, cylinders, and blades, but they always referred to the tangential force as the 'friction force.'

4.0   Friction Data Collection Methods

The purpose for measuring friction dictates both the type of data that are collected and the extent to which the data are subjected to further analysis and interpretation. Friction data from macroscale experiments can be reported with different levels of detail:

1) Average ('nominal') value of the μ_s or μ_k for the experiment

2) Average, standard deviation, and/or range of μ_k at steady state conditions

3) Statistical distribution of the friction force values over some interval of contact (including Fourier transforms)

4) Characteristic friction coefficients for a tribosystem that undergoes transitional behavior (e.g., initial μ_k, steady-state μ_k, and pre-seizure μ_k).

5) Friction data in the conetxt of other types of data (e.g., specimen displacement due to wear or transfer, acoustic emission, audible noise, wear debris characteristics)

Fundamental studies may require detailed schemes for acquiring and analyzing friction force data. By contrast, some studies of coating durability only require knowing when the kinetic friction coefficient exceeds an arbitrarily-chosen 'failure' value.

In tribology research, recording detailed fluctuations of the friction force is desirable because these indicate the stability of the interfacial processes. To observe these fluctuations, the data acquisition rate and sensitivity of the force sensors must be adequate. Calibration of the instruments and signal drift compensation are important. The stiffness of the sensor itself can affect results. Piezo-electric force sensors may be more responsive for high-frequency measurements, but their finite time constants may affect their ability to accurately measure very gradual changes in friction force.

Blau (1989) has identified three different attributes of time-dependent friction force (or friction coefficient) behavior.. There are (1) general trends, (2) the duration of certain events, and (3) short term amplitude of fluctuations. (See Figure 6.)

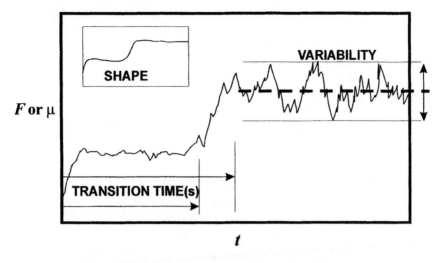

Figure 6. Three attributes of friction versus time behavior.

1) *General trends*. General trends are documented by periodic data sampling, or by observing the general shape of the friction force versus time plot.

2) *Duration of features in general trends*. Studying the time needed to reach certain prominent features in the friction force versus time record can help to understand running-in or transitional behavior.

3) *Amplitude of short-term frictional fluctuations*. The characteristics of fluctuations can be a rich source of detailed information about the stability and competition of frictional processes in the interface.

In certain tribosystems, friction/time attributes are very repeatable. In that case, the investigator can conduct interrupted tests to study the surface conditions just prior to, during, or after transitions. Learning to interpret these 'frictional signatures' can enable a better understanding of the causes for stable and unstable frictional behavior.

Not all tribosystems exhibit repeatable friction/time attributes. In some tribosystems, the general trends might be repeatable, but the duration and amplitude of friction force fluctuations may not. For example, Figure 7 shows that the time required for the friction coefficient to reach a steady state is affected by the size of abrasive particles in two-body sliding (Type 440 stainless steel on SiC abrasive papers). The higher the 'grit number,' the smaller the particles. The shape of the curve is similar for all three grit sizes, but the time to attain steady-state frictional conditions is shorter for finer grits. The filling of the spaces between the grits (called 'loading'), and the time needed to expose the SiC grit points above their adhesive binder material can affect the time needed to reach steady-state. In this case, friction tests with abrasion can display similar curve attributes to sliding tests, but for different underlying reasons.

276

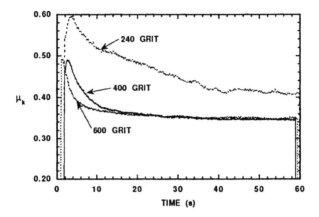

Figure 7. Changes in friction observed during the abrasive running-in of 440C stainless steel balls against dry SiC paper in a pin-on-disk configuration. (2 N load, 0.2 m/s, in air).

When collecting detailed friction data, the effects of tribometer design must be considered. The more sensitive the recording system or the higher the recording rate, the greater the potential to detect artifacts from machine construction. It is prudent in such cases to be aware of the natural frequencies of the testing equipment. Tribometer design effects can be particularly noticeable in reciprocating tests where the direction of the friction force reverses periodically of if the moving portion has substantial mass. Accounting must be made for the elasticity of the fixtures and the inertia of the moving parts during direction changes. It is also advisable to calibrate the friction force sensors in both directions when conducting reciprocating friction tests.

5.0   Macroscale Friction Testing - Future Work

This chapter has considered a number of issues associated with the selection and use of macro-scale friction tests. As long as new mechanical designs emerge and the demands placed on tribocomponents and materials continue to change, there will be a need for such tests. Several questions deserving further study are:

1)   How can we improve the quantitative correlation between laboratory friction data and machinery performance data?
2)   How can we better define the friction coefficient and related terms to avoid ambiguities in special cases like scratching with styli, or when measuring forces at molecular-level size scales?

3) How can we make enlightened use of the attributes of friction-time data to study the subtleties of interfacial behavior at various scales?
4) Can we develop more effective means to differentiate whether the variability in friction data is due to the materials, the lubricant, the tribosystem design, the operator's testing technique, or some combination of these things?
5) Can we develop standard requirements for characterizing our tribometers that should always be included with published friction data?

A close correspondence between macroscale and micro- or nano-scale results can exist only if similar friction processes operate in those tribosystems. Studies of the features on contact surfaces can help to indicate the extent to which the sliding processes are similar from one scale to another. Coupling macro-scale friction data with micro- and nano-scale surface observations and analysis will continue to help better define the size-ranges of phenomena that control different kinds of friction behavior.

6.0 Summary

Macroscale friction tests are conducted for different reasons, and the set of mechanical, geometrical, thermal, chemical, and materials variables must be selected appropriately in each case. Understanding the characteristics of the tribometer, itself, is an essential element in interpreting the test results.

Tables of 'generic' friction coefficients are often incomplete because they ignore the characteristics of running-in and other friction transitions.

Friction data that are useful for engineering can be obtained on relatively simple laboratory instruments provided that the key variables in the contact environment are identified and properly simulated.

Friction coefficients from laboratory-scale tests may not agree exactly with those measured in operating machinery; however, laboratory tests can reveal important trends in material and lubricant behavior.

Learning to interpret friction/time signatures can enable a better understanding of friction processes and mechanisms. (See the author's second paper in this volume.)

As the size scale of machines becomes smaller and smaller, the machines themselves approach the size of fundamental frictional phenomena. In that case, engineering improvements and basic studies can, in principle, be done with the same type of apparatus. This interesting convergence of science and engineering is not always true, however. Consequently, there will continue to be a need to conduct friction experiments over a wide range of size scales.

Acknowledgments

The author gratefully acknowledges the support of the conference organizers and the U.S. Department of Energy, Office of Transportation Technologies, for enabling his participation in this Advanced Studies Institute. The help of author's wife, E. R. Blau, in proof-reading the early drafts of this paper and suggesting changes to clarify the wording is greatly appreciated.

278

References

Akhmatov, A. S. (1939) "Some items in the investigation of the external friction of solids," as cited by Kragheski, I. V. (1965) *Friction and Wear*, Butterworths, London, 159.

Blau, P. J. (1982) "Test of a Rule of Mixtures for Dry Sliding of 52100 Steel on an Al-Si-Cu Alloy, *Wear*, **81**, 187-188.

Blau, P. J. (1990) "Friction Microprobe Studies of Composite Surfaces," in *Tribology of Composite Materials*, ed. P. Rohatgi, P. J. Blau and C. S. Yust, ASM International, Materials Park, Ohio, 59-68.

Blau, P. J. (1989) *Friction and Wear Transitions of Materials*, Noyes Publications, Park Ridge, NJ.

Blau, P. J. (1996) Oak Ridge National Laboratory, Technical Report M-5824, Oak Ridge, Tennessee.

Bowden, F. P. and Tabor, D. (1950) *Friction and Lubrication of Solids*, Clarendon Press, Oxford, 91.

Brown, R. L. and Richards, J. C. (1966) *Principles of Powder Mechanics*, Pergamon Press, Oxford.

Budinski, K. G. (1992) "Laboratory Testing Methods for Solid Friction," in ASM Handbook, Volume 18 *Friction, Lubrication, and Wear Technology*, ASM International, Materials Park, Ohio, 45-58.

Dowson, D. (1979) *History of Tribology*, Longman, London.

Dugger, M., Poulter, D. A. and Ohlhausen, J. A. (2000) "Development of a Surface Micromachined Device to Quantify Friction on Polysilicon Sidewall Surfaces" presented at the Society of Tribologists and Lubrication Engineers, Annual Meeting, Nashville, TN, May 10.

German, R. M. (1994) *Powder Metallurgy Science*, 2nd ed., Metal Powder Industries Federation, Princeton, New Jersey, 213-216.

Jang, H. and Kim, S. J. (2000) "The effcts of antimony trisulfide and zirconium silicate in the automotive brake friction material on friction characteristics," *Wear*, **239**, 229-236.

Sergeant, P. M. (1985) "Use of the Indentation Size Effect on Microhardness in Materials Characterization," in *Microindentation Hardness Testing for Research and Applications*, ed. P. J. Blau and B. R. Lawn, ASTM Special Tech. Pub. 889, 160-174.

Steijn, R. P. (1964) "Friction and Wear of Single Crystals," in *Mechanisms of Solid Friction*, ed. P. J. Bryant, M. Lavik, and G. Salomon, Elsevier Pub. Co., Amsterdam, 48-66.

THE ANISOTROPIC FRICTION CHARACTERISTICS OF CRYSTALLINE MATERIALS: A REVIEW

BRIAN L. WEICK
Department of Mechanical Engineering
University of the Pacific
Stockton, California 95211 U.S.A.

BHARAT BHUSHAN
Computer Microtribology and Contamination Laboratory
Department of Mechanical Engineering
The Ohio State University
Columbus, Ohio 43210-1107 U.S.A.

Abstract. A thorough understanding of friction mechanisms is needed for the successful development of future microelectromechanical systems (MEMS). Standard macro as well as advanced micro/nano friction experiments are reviewed, which address anisotropic friction characteristics of crystalline materials such as polycrystalline silicon (polysilicon) used for MEMS devices. Theoretical studies that predict these characteristics using the fundamentals of mechanics, thermodynamics, and associated molecular dynamics simulations are also reviewed. Results are discussed in light of our current understanding of fundamental friction mechanisms developed from both a macrotribological standpoint as well as a micro/nanotribological standpoint. The important roles of crystal orientation and grain boundaries in determining the magnitude of friction are considered.

1. Introduction

1.1. NANOTRIBOLOGICAL STUDIES OF MATERIALS USED FOR MEMS

Many advanced microelectromechanical systems (MEMS) are comprised of surfaces that slide against one another. Tribological issues associated with the development and operation of these devices are therefore critical, and have been discussed at a recent workshop (Bhushan, 1998). Nanotribological characteristics of polycrystalline silicon (polysilicon) and SiC materials used for MEMS applications have been presented by Sundararajan and Bhushan (1998). They also provide an overview of MEMS devices in the introduction to their paper. These devices include silicon-based acceleration sensors for the auto industry, micromachined pressure sensors for the medical, aerospace, and automotive industries, and devices for information storage systems. As an example, miniature wobble motors and actuators for tip-based recording have been developed, and these MEMS devices pose challenges to tribologists. As described by Bhushan (1999a) there is intermittent

279

B. Bhushan (ed.),
Fundamentals of Tribology and Bridging the Gap between the Macro- and Micro/Nanoscales, 279–297.
© 2001 *Kluwer Academic Publishers.*

contact at the rotor-stator interface of these tiny motors, and physical contact at the rotor-hub interface. Friction limits the repeatability of operation of these motors, and static friction (or stiction) prevents operation of the motor altogether. In addition, there is a need to develop bearing/bushing materials that are both compatible with MEMS fabrication processes and provide superior tribological performance. Note that friction characteristics must also be understood in valves used for flow control in MEMS. Mating valve surfaces should be smooth to seal, but a certain roughness is required to maintain low adhesion/friction. Other microdevices also suffer from tribological concerns, and it is likely that such issues will continue to cause problems in future MEM machines. Bhushan and Gupta (in press) have summarized a number of studies to measure static friction in MEMS systems such as micromotors. Efforts have also been made to determine the tribological characteristics of bulk silicon and polysilicon materials (Bhushan and Venkatesan, 1993a, b; Venkatesan and Bhushan, 1993, 1994; Gupta et al., 1993, 1994; Bhushan and Koinkar, 1994; Bhushan and Li, 1997; Sundararajan and Bhushan, 1998; Weick and Bhushan, 2000).

1.2. ADHESION AND FRICTION ISSUES ASSOCIATED WITH MEMS

Developers of MEMS devices also recognize the need to understand tribological issues associated with their micromachines. Adhesion and friction issues are of particular concern. Bhushan (1999a) and Tas et al. (1997) address such issues for MEMS applications. Tas et al. (1997) describe how large adhesion forces can develop when the extremely smooth polysilicon structures are brought together. Adhesion energies between a surface micromachined beam and substrate were measured to be on the order of 0.1 J/m^2, which is significant when compared to a surface energy of 20 mJ/m^2 for a low-adhesive material such as PTFE (Tas et al., 1997). This means that large friction forces must be used to initiate sliding. Furthermore, to minimize friction in MEMS, Tas et al. (1997) state that the real area of contact must be minimized and suggest that increasing the surface roughness will reduce adhesion forces in micromotors. They also recognize that MEMS structures have a high surface-to-volume ratio, and as discussed by Sundararajan and Bhushan (1998) friction forces are proportional to an increase in area a thousand times more than they are proportional to an increase in volume.

Maboudian (1998) also addresses adhesion and friction issues associated with MEMS. She points out that the large surface area to volume ratio and related adhesion problem is also problematic during fabrication. This adhesion problem is referred to as release-related stiction, and prevents the surface microstructure from being released from the substrate. Maboudian (1998) relates this problem to capillary, electrostatic and van der Waals forces, and suggests that chemical modification of polysilicon surfaces could be done in addition to or in lieu of roughening. Gardos (1998) has also discussed the properties of Si-based materials for MEMS applications, and states that high friction is caused by the low cohesive energy density and surface chemistry of Si crystallites such as undoped Si (100), Si (111), and polycrystalline silicon.

As pointed-out by Burnham and Kulik (1999), the growing field of nanotribology provides an opportunity to understand tribological processes at a more fundamental level. More traditional continuum mechanics-type models have been used to explain macroscopic processes, but their ability to explain the micro/nanoscopic processes is still in question.

1.3. INTERATOMIC FRICTION AND MATERIAL STRUCTURE

Early work by Tomlinson (1929) suggested the possibility that single-crystal, defect-free, atomically flat solids could move with respect to one another without wear. Frictional forces would still be present, and would be related to interactions between atoms at the interface between the contacting materials. Bluhm et al. (1995) recognized this, and stated that the surface potential (or energy) of a crystal is determined by the atoms at the surface. The type of atoms and arrangement of these atoms contribute to the magnitude of the friction force. Therefore, friction should be anisotropic and depend on the crystallographic directions in which the two crystals are moving relative to one another. This is particularly true at the nanoscopic level encountered in MEMS devices.

Czichos (1978) also pointed out that the well-known models for friction involving both deformation and adhesive processes are indeed useful for making a qualitative assessment of the causes of friction, but cannot be used to make a quantitative calculation unless simplifying assumptions are made. In addition, for polycrystalline materials such as metals, Czichos (1978) and Bhushan (1999b) note that surface orientation, lattice spacing, and grain boundaries all contribute to the nature of the interface in sliding friction. Buckley (1982) also recognized that crystallographic orientation and grain boundaries can have an effect on friction. One of the objectives of this paper is to review such macroscale friction studies and discuss theoretical models developed to explain these friction characteristics.

The invention of the atomic force microscope (AFM) in the 1980s enabled Mate et al. (1987) to show how the atomic structure of a graphite surface influences the frictional dynamics of a tungsten tip sliding on the surface. The AFM and present-day friction force microscopes (FFM) enable researchers to carry-out friction studies on the micro/nanoscale, and understand the friction characteristics of MEMS as well as any anisotropy in friction that could be associated with MEMS devices. Such anisotropic friction characteristics could limit the operation of MEMS devices due to an inability to slide or rotate parts efficiently in one direction versus another. Therefore, another objective of this paper is to review micro/nanoscale friction studies and discuss theoretical models developed to explain anisotropic friction characteristics that are relevant to the development of MEMS devices.

2. Macroscale Friction Studies – Experimental Studies of the Effects of Crystal Orientation and Grain Boundaries on Friction

2.1. PIN-ON-DISK EXPERIMENTS TO EXAMINE EFFECTS OF CRYSTAL ORIENTATION AND GRAIN BOUNDARIES ON FRICTION

Classic friction experiments involve the use of pin-on-disk testers in which a spherically-tipped pin slides against a rotating or oscillating disk. Many researchers use sapphire (Al_2O_3) or diamond as a "standard" pin material, however other materials are often used depending on the focus of the study. Some of the fundamental research performed by Buckley (1982) to examine the effects of crystal orientation and grain boundaries on friction will be discussed next. He utilized sapphire pins sliding against a well-characterized tungsten surface, and copper materials were used in an additional study. More recent studies by Weick and Bhushan (2000) will also be presented. They examined the effects of crystal

282

orientation and grain boundaries on the friction behavior of both tungsten and polysilicon. The polysilicon material was chosen for its applicability to MEMS devices.

2.1.1. *Tungsten and Copper Experiments*

Buckley (1982) has addressed the effects of grain boundaries and crystallographic orientation on friction. His fundamental studies utilized a 10 mm diameter sapphire ball rubbing against a pure tungsten disk under air and vacuum conditions. In this fundamental work, Buckley (1982) chose to use a polycrystalline tungsten (polytungsten) sample with a relatively large grain size. The influence of crossing grain boundaries could then be examined in addition to the crystal orientation effects. As shown in Figs. 1a and 1b,

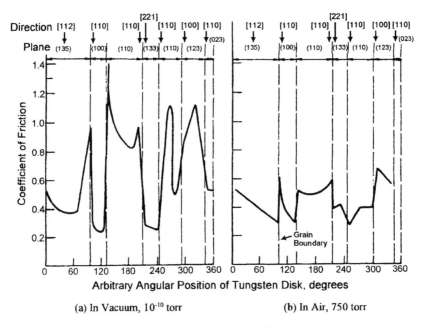

(a) In Vacuum, 10^{-10} torr   (b) In Air, 750 torr

Figure 1. Coefficient of friction of sapphire ($10\overline{1}0$) plane sliding in [0001] direction of polycrystalline tungsten. Load, 500 grams; sliding velocity, 0.013 centimeters per second (Buckley, 1982).

Buckley (1982) discovered that changes in the coefficient of friction as high as 1.0 occurred in vacuum when the sapphire ball moved from one grain to another. In air, this change in the coefficient of friction decreased to about 0.2.

Other research by Buckley showed that changes in friction occur when a polycrystalline copper slider is rubbed against a copper bicrystal. His results shown in Fig. 2 indicated that this change is due to the presence of the grain boundary. Such boundaries are known to be high energy sites that have a pronounced influence on friction. To a lesser extent, Buckley's results in Figs. 1 and 2 seemed to indicate that there is an influence of crystallographic orientation on friction. In other words, not only did the coefficient of friction change because the sapphire slider passed over a grain boundary, but the friction increased or decreased depending on the crystallographic orientation of the grain. Such measurements lead one to expect that grain boundaries influence the friction behavior of metals, and there

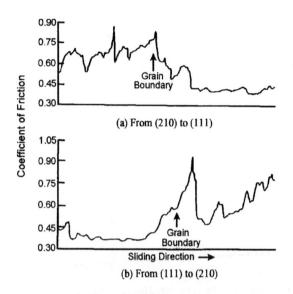

(a) From (210) to (111)

(b) From (111) to (210)

Figure 2. Recorder tracings of friction force for copper slider sliding across grain boundary on copper bicrystal. Load, 100 grams; sliding speed, 1.4 millimeters per minute (Buckley, 1982).

is an effect of crystallographic orientation on friction.

To explain his results, Buckley discussed how different crystallographic directions will have different atomic densities, and there will be a corresponding difference in friction. He noted that friction coefficients measured when a slider travels in directions with high atomic densities will be lower than friction coefficients measured in directions with low atomic densities. As shown in Fig. 1, this was found to be true for the (100) plane in tungsten. When a sapphire slider is rubbed against the (100) plane, he found that friction was a minimum when the slider was moved in the [100] direction and a maximum in the [110] direction. (Buckley noted that the [100] direction has a higher atomic density than the [110] direction.) Furthermore, as a slider moves out of a grain, across a boundary, and into another grain, the orientation of the crystallographic slip planes will change, and this could also be associated with changes in friction. As a result, the relative maxima and minima in friction may be located within the grain itself or along a grain boundary. Whether or not peaks in friction occur at the grain boundaries depends on both the increased work of adhesion at the grain boundaries due to the higher surface energies, and whether or not this increased work of adhesion exceeds the friction associated with the particular crystallographic direction in which the slider is traveling on either side of the grain boundary.

2.1.2. Polytungsten and Polysilicon Experiments

Polytungsten experiments. In a recent study, a polycrystalline tungsten (polytungsten) sample was used by Weick and Bhushan (2000) to examine the relationship between friction, grain boundaries, and crystal orientation. A diamond slider with a 20 μm radius was used as the pin in a single pass pin-on-disk test apparatus. A detailed optical micrograph of the etched polytungsten sample surface and associated friction measurements are shown in Figs. 3a and 3b. Weick and Bhushan (2000) describe how the sample was uniformly polished prior to the friction experiment, and then etched to bring-out the grain structure after the friction experiments. The micrograph was obtained after etching the sample, and friction measurements were made along the scan line drawn on the micrograph. Grain boundary locations are shown as dotted lines drawn between Figs. 3a and 3b.

Figures 3a and 3b show that changes in friction indeed occur when grain boundaries are crossed. Although these changes appear to be mainly peaks in friction at the grain

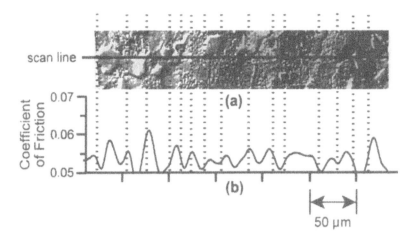

scan line

Figure 3. Detailed view of the friction data for a 20 μm radius diamond tip sliding on polytungsten showing (a) the optical micrograph of tthe etched polytungsten surface, and (b) friction data acquired along the scan line. The polytungsten surface was etched after the friction experiments, and grain boundaries are shown by the dotted lines between the figures. (Weick and Bhushan, 2000).

boundaries, there are some boundaries where decreases in friction appear to occur, and others where the friction appears to be highest within the grain itself. Since the grain boundaries are known to be high energy sites in polycrystalline metals, one would expect an increase in the work of adhesion to be associated with passing over grain boundaries. This would in turn cause an increase or "peak" in friction. In addition, peaks in friction could occur within the grains themselves due to differences in the crystal orientation of the grain. Buckley's (1982) results shown in Fig. 1 appear to be consistent with those presented by Weick and Bhushan (2000) in Fig. 3, however the magnitudes of friction are different due to the different sliders used (sapphire versus diamond).

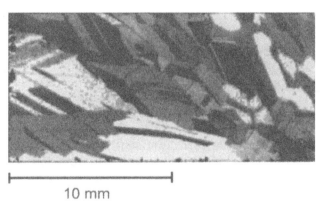

10 mm

Figure 4. Polysilicon wafer used for friction studies. Large grains on the bottom left-hand side of the wafer were used for friction measurements. (Weick and Bhushan, 2000).

Polysilicon experiments. A photograph of the entire polysilicon wafer used for friction experiments is shown in Fig. 4. Note that this is the unpolished back side of a 300 μm thick wafer, and grains on the order of 3 to 5 mm are clearly visible. The front side of the wafer used for friction studies is unetched and highly polished. Grain boundaries are not visible on the front side of the polysilicon wafer.

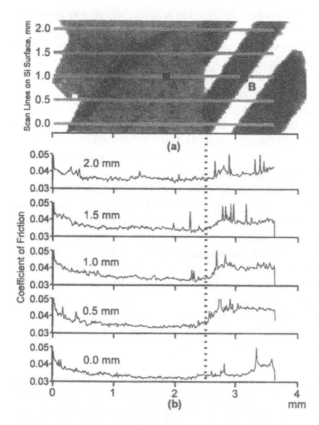

(a)

(b)

Figure 5. Coef. of friction measurements for a 20 μm radius diamond tip sliding on polysilicon (Weick and Bhushan, 2000).
  (a) Polysilicon surface showing scan lines where friction measurements were made. Locations where AFM/FFM data were acquired are labeled as A and B.
  (b) Friction data acquired along the 0.0, 0.5, 1.0, 1.5, and 2.0 mm scan lines. The location of the major grain boundary is shown as a dotted line.

However, the grain boundary locations on the front side were determined to be consistent with those on the back side using an optical microscope with a DIC prism. In essence, the back side of the wafer served as a "map" for determining where to perform friction measurements on the front side of the wafer.

Since large grains are present in the lower left portion of the wafer, friction measurements were made in this region along the scan paths shown in Fig. 5a. White regions, gray regions, and black regions each represent parts of the wafer with differing crystallographic orientations. A total of five friction measurements were made in 0.5 mm intervals over a 3.5 mm sliding distance. All the friction scans pass through the three crystallographic regions, and changes in friction are clearly distinguishable. These friction measurements are shown in Fig. 5b, and the magnitude of the coefficient of friction always increases after approximately 2.5 mm have been traversed. This is the boundary between the two large grains shown as gray and black regions in Fig. 5a. Note that the location of this boundary is shown as a dotted line in Fig. 5b. Therefore, as observed for polytungsten, there is clearly a change in the magnitude of friction as a grain boundary is traversed.

In addition to the relationship between grain boundaries and friction observed for both polytungsten and polysilicon, there also appears to be an effect of crystallographic orientation on friction for polysilicon. For this material, the magnitude of the coefficient of friction is different depending on the particular grain being traversed. For instance, as shown in Fig. 5b the coefficient of friction is between 0.04 and 0.05 at the start of each scan, and then decreases to between 0.03 and 0.04. Although part of this initial higher friction measurement could be due to the difference between static and kinetic friction, the rapid change in friction can be attributed to a difference between measurements on one grain versus another. This influence of crystallographic orientation on friction was even more prevalent when the diamond slider passed over the boundary at 2.5 mm. This caused a significant change in friction from a range of 0.03 - 0.04 to a range of 0.04 - 0.05. Furthermore, the friction measured after passing over the grain boundary is always higher, and only changes as white regions with a different crystallographic orientation are traversed. As a result, the effect of crystallographic orientation on friction is clearly demonstrated for a polysilicon material in addition to the effect of grain boundaries on friction.

Bhushan and Li (1997) published friction data on bulk polysilicon. They also used a 20 μm radius diamond tip for some of their studies and found that the coefficient of friction was 0.06. This is slightly higher than the 0.03 to 0.05 friction coefficient measurements shown in Fig. 5b for diamond rubbing against polysilicon. Results for polysilicon friction experiments appear to be quite different if the countersurface is not diamond. As an example, Bhushan and Li (1997) reported that the coefficient of friction for a 3 mm diameter sapphire ball rubbing against bulk polysilicon is almost an order of magnitude higher with values ranging from 0.30 to 0.46. These sapphire-on-polysilicon experiments were performed under the same test conditions that they used for the diamond-on-polysilicon experiments. In addition, Gabriel et al. (1990) have estimated that dynamic friction coefficients for polysilicon-on-silicon range from 0.25 to 0.35 in MEMS devices, and Tai and Muller (1990) have performed studies to show that friction coefficients for polysilicon surfaces adjacent to silicon nitride surfaces in MEMS range from 0.21 to 0.38.

2.2. EFFECTS OF LATTICE MISFIT ANGLE ON FRICTION

Hirano et al. (1991) studied the effects of crystal orientation on the frictional behavior of two mica surfaces. Their fundamental work provides insight into interatomic friction mechanisms that could occur in MEMS devices. In lieu of using a pin-on-disk type of test system, they used smooth mica sheets in a unique test apparatus. One sheet was attached to a cylindrically curved substrate with an 8 mm radius and 5 mm length, and the other was attached to a substrate with a 10 mm diameter. Lattice orientations were determined prior to friction testing, and this knowledge enabled Hirano et al. (1991) to use their test apparatus to study the effects of lattice misfit angle on friction. The lattice misfit angle, θ, can be defined as the angle between crystal lattices of two opposing surfaces when they are in contact with one another. The schematic diagram at the top of Fig. 6 shows the crystal lattice of one surface sliding against the crystal lattice of another surface, and θ is the angle between the two lattices. As shown in Fig. 6, when the misfit angle is approximately 0° or 60° the two specimens are said to be in commensurate contact, which means the crystal lattice orientation of the surfaces is the same. If the misfit angle is 30° the mica materials are said to be in incommensurate contact. This means that the crystal lattice of one surface is rotated 30° with respect to the other surface. Figure 6 clearly shows that commensurate contact at a misfit angle of 0° or 60° leads to an increase in friction; whereas incommensurate contact at 30° leads to a reduction in friction between the surfaces. Note that the tests were performed at elevated temperature (130 °C) and ambient (20 °C) conditions. The water vapor pressure, p/p_o, was approximately 9 x 10^{-5} at the upper temperature, and 1.0 at the lower temperature. Therefore, the effect of misfit angle is more pronounced at dry, elevated temperature conditions than at the relatively dirty and wet ambient conditions.

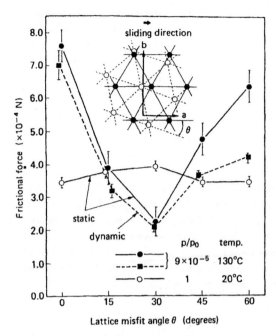

Figure 6. The change in the measured static and dynamic frictional forces as a function of the lattice misfit angle θ between two contacting mica lattices. The misfit angle is approximately 0° when the two specimens are brought into commensurate contact without rotation of the lower specimen (Hirano et al., 1991).

Martin et al. (1994) demonstrated that MoS_2 surfaces also show frictional anisotropy during intercrystalline slip, and this behavior could be attributed to the degree of commensurability between the two surfaces. Sheehan and Lieber (1996) utilized the concept of commensurate vs. incommensurate surfaces to help explain the friction characteristics of MoO_3 sliding against MoS_2. Their work was carried out with an atomic force microscope (AFM), and utilized the tip of the AFM to push the MoO_3 nanocrystals on the single crystal MoS_2 surface. Sheehan and Lieber (1996) showed that the observed sliding directions of the MoO_3 crystals coincided with the crystal lattice directions of the MoS_2 substrate, and the extent of the friction anisotropy could be related to the degree of commensurability of the two surfaces. The qualitative model they developed to explain their experimental results showed that the MoO_3 surface atoms can slide in one direction, in channels defined by the sulfur atom rows of the MoS_2 surface (Sheehan and Lieber, 1996). Motion along other lattice directions required the interfacial atoms to move over one another leading to a larger friction force. Sheehan and Lieber (1996) used these anisotropic friction characteristics to fabricate nanostructures with interlocking pieces. Latches cut from one MoO_3 nanocrystal could be slid into notches cut in another nanocrystal. Due to the preferential sliding directions of the nanocrystals on the MoS_2 substrate, the locked nanocrystals could not be moved until the latch was pulled out of the notch.

2.3. ADDITIONAL RESEARCH TO SHOW RELATIONSHIPS BETWEEN FRICTION AND CRYSTAL ORIENTATION

Layered materials such as MoS_2 and graphite have well-documented preferential sliding directions that can be related to frictional anisotropy (Martin et al., 1994). In the case of graphite, covalently bonded planes are held together by weak van der Waals attractions. The planes can therefore slide over one another leading to lower friction characteristics in that preferential direction.

Frictional anisotropy of diamond has also been studied. Schmitt et al. (1999) studied the effects of crystal orientation on the friction characteristics of diamond-coated materials in a pin-on-disk apparatus. The pin substrate was diamond, and the disk substrate was steel. Diamond coatings with two different growth directions were used: [111] and [100]. In their research, complex tribological behavior was observed for the two different diamond coatings. They observed that oxide and metal particles that transfer to and possibly adhere to the countersurface lead to differences in friction between the two different crystal orientations. Graphitization of the diamond material and the presence of condensed phases also lead to such differences in friction.

The industry-driven necessity to understand the diamond polishing process has also led to a better understanding of the role of crystal orientation on frictional behavior. Grillo et al. (2000) performed friction measurements on the (100) plane in the [100] and [110] directions using a specialized polishing apparatus equipped with friction measurement capabilities. They found that the experiments in the [100] direction yielded a dependence of the friction coefficient on load. For the 2 to 15 N load range, the friction coefficient varied from approximately 0.1 to 0.3. In contrast, the friction coefficient measured in the [110] direction remained constant with applied load, and stayed at 0.1. Based on these results, Grillo et al. (2000) state that more power is dissipated at higher loads when polishing is performed in the [100] direction. This is explained by considering the high anisotropy in the elastic constants of diamond. Bond deformation (particularly bending) is much greater in the [100] direction.

3. Micro/Nanoscale Friction Studies – Experimental Studies of the Effects of Crystal Orientation on Friction

3.1. AFM/FFM MEASUREMENTS

3.1.1. *Silicon and Polysilicon Experiments*
The development of the atomic force microscope (AFM) and friction force microscope (FFM) has enabled tribologists to perform experiments on the micro/nanoscale (Bhushan et al, 1995). Bhushan (1999a, b) provides a review of the development of these devices. Although there has been some discussion as to whether or not the AFM/FFM tip is a single or multiple asperity contact, the device is clearly capable of friction measurements to discern differences in crystal orientation. Mate (1987) showed this by demonstrating a correspondence between frictional forces and the atomic periodicity of a graphitic surface.

Sundararajan and Bhushan (1998) used an AFM/FFM to measure friction coefficients for Si (100) single crystals and both doped and undoped polysilicon materials. Their results showed that friction coefficients tend to be lower for the polysilicon materials when compared to the single crystal materials. Weick and Bhushan (2000) also used an AFM/FFM to make micro/nanoscale friction measurements using the polysilicon sample shown in Fig. 4. The measurements were made in two 25 x 25 μm scan regions, and these regions are labeled as A and B in Fig. 5a. For region A, the coefficient of friction was determined to be 0.0314 with a standard deviation of 0.0056. For region B, the coefficient of friction was determined to be 0.0366 with a standard deviation of 0.0061. Although the standard deviations are rather high, the magnitude of the coefficients measured using the AFM/FFM is consistent with the macroscale measurements shown in Fig. 5b. Furthermore, despite the rather large standard deviations, the AFM/FFM results also show that the coefficient of friction measured in region B appears to be higher than the friction measured in region A. This is also consistent with the macroscale friction measurements made with the pin-on-disk test apparatus. Since regions A and B are in different grains with different crystal orientations, it is reasonable to state that the difference in friction measured in these regions could be attributed to the different crystal structures present.

3.1.2. *Triglycine Sulfate Experiments*

Bluhm et al. (1995) and Czajka et al. (1997) performed experiments with triglycine sulfate (TGS) surfaces and measured anisotropic friction characteristics with an AFM/FFM on the (010) surface of this material. They noted that frictional differences occur between domains of different polarity as well as inside the domains. Furthermore, they stated that the asymmetric arrangement of molecules at the surface causes an asymmetric surface potential, which causes differences in friction that are measured by the FFM tip (Bluhm et al., 1995). In their discussion, Bluhm et al. (1995) also explained their findings using an interatomic stick-slip mechanism that has been summarized by Colchero et al. (1999). In this mechanism, the tip of the AFM/FFM will not only "stick" to points along the scan line, but "look" for favorable sticking points off the scan line (Colchero et al., 1999). This leads to a two-dimensional stick-slip effect, and Bluhm et al. (1995) used this concept to explain the anisotropic friction measurements for TGS.

3.1.3. *Determination of Chemical Composition*

Due to the fact that energy dissipation occurs at the interface between an AFM/FFM tip-sample surface in a localized region, it is reasonable to state that this device can be used to determine the composition of surfaces used in semiconductors and MEMS. Sundararajan and Bhushan (1998) detected differences in friction between doped and undoped polysilicon. Dopants in semiconductors affect the crystal structure in addition to changing the conductivity of the material. Therefore, one would expect a small change in chemical composition to also influence the friction measured for MEMS. Results presented by Garcia et al. (1997) further demonstrate this relationship between friction force and small changes in chemical composition. Their results show that frictional forces can change when an $In_xGa_{1-x}As$ sample with differing amounts of indium is traversed. As the indium composition (x) is changed from 0.1 to 0.6, friction decreases.

3.2 OTHER MEASUREMENT METHODS

Micro/nanoscale friction force measurements have typically been obtained using AFM/FFM devices in the past ten years. However, other types of equipment have been used to study friction characteristics at this level including the surface force apparatus (SFA). Tabor and Winterton (1969) and Israelachvili and Tabor (1972) developed such apparatuses for measuring van der Waals forces between two molecularly smooth mica surfaces as a function of separation in air or vacuum. SFAs have also proved their usefulness to study static and dynamic properties of molecularly thin liquid films sandwiched between two molecularly smooth surfaces. They could provide insights into crystal orientation effects on friction, but samples would have to be prepared to meet the SFA geometry.

Ohmae (1997) used field ion microscopy (FIM) to study adhesion and friction, and was able to observe relationships between friction and crystal orientation with this device. In the

FIM technique used by Ohmae (1997), the initial structure of a tungsten surface is produced as a photograph showing crystallographic planes. After a single pass sliding contact with a Ni surface, a field evaporation technique is used to reveal successive layers of the material. Ohmae (1997) describes the slip and dislocation phenomena that occur due to friction in great detail. He also notes that dislocations are generated in the tungsten material without affecting the grain boundaries.

4. Theoretical Friction Models – Prediction and Explanation of the Effects of Crystal Orientation and Grain Boundaries on Friction

4.1. CLASSIC MECHANICALLY-BASED MACROSCOPIC FRICTION MODELS

Reviews of classic friction models are prevalent in the available tribology textbooks such as those by Czichos (1978) and Bhushan (1999b). Early models were developed by DaVinci (1452 - 1519), Amontons (1706), and Coulomb (1785) among others. They defined the friction coefficient as the ratio of the friction force to the normal load, and it was stated by Amontons that this coefficient is independent of apparent contact area, load, and sliding velocity. Based on theories of interlocking asperities, friction force was also attributed to the force needed to pull a load W up an inclined plane of slope θ, which led to the mathematical statement that $F = W \tan\theta$. Although these models still find their utility in explaining macroscopic friction phenomena, it is difficult to apply these models to the current task of explaining the influence of crystal orientation on friction. Perhaps the reduction in friction observed during incommensurate contact between contacting surface lattice structures could be explained by a reduction of interlocking asperities, but this approach is rather limiting.

A mid-twentieth century friction model based on macroscopic contact was developed by Bowden and Tabor (1950). Their two-component model stated that frictional forces are developed by asperity interaction at real areas of contact. Adhesion and plowing of asperities at real areas of contact lead to the development of frictional forces:

$$F = F_{adhesion} + F_{plowing} = A_r \tau + A'_r H \qquad (1)$$

where,
F = total friction force equal to adhesive and plowing components
A_r, A'_r = real areas of contact for adhesion and plowing junctions, respectively
τ = bulk shear strength of the weaker material
H = pressure required to cause plastic flow (hardness)

Although Eq. 1 is a highly simplified friction model, it is important to use as a starting point for understanding the effects of grain boundaries and crystal orientation on friction. As a simplifying assumption, let us assume that the friction forces developed are due to adhesive forces only. (This is an appropriate assumption for friction mechanisms proposed in MEMS.) With this assumption, the development of a friction force is due to contacting of asperities at real areas of contact followed by shear of the weaker material. Note that interfacial shear can also occur at the junction between the two asperities without causing bulk shear of either material. In this case, only interfacial bonds between atoms on the opposing surfaces are broken, and as recognized by Tomlinson (1929) friction of this type can be attributed to energy transfer between rows of atoms on the interacting surfaces. This type of interfacial friction will be discussed in Section 4.2.

For the moment, let us assume that the adhesive friction involves shearing of the weaker material as originally proposed by

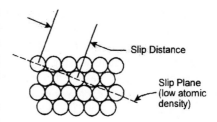

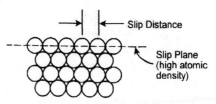

Figure 7. Slip is more difficult along a low atomic density slip plane than along a high atomic density slip plane.

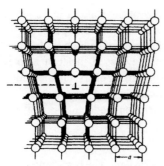

(a) Schematic diagram of an edge dislocation

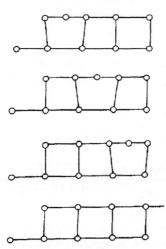

(b) Schematic diagram showing the propagation of a dislocation.

Figure 8. Friction forces developed between materials could be related to the presence of defects in crystal structures such as (a) edge dislocations. Subsurface shear could be attributed to (b) propagation of these dislocations. (Adapted from Shackelford, 1996.)

Bowden and Tabor (1950). This could indeed be the case for the macroscopic experiments already discussed. Relationships between friction forces and crystal orientation have been observed, and these relationships could be due to differences in subsurface shear strength for one crystal orientation versus another. In the absence of defects in the lattice structure, as shown in Fig. 7 this difference in shear strength could be attributed to the presence and orientation of slip planes as well as the atomic density along these slip planes. Slip is more difficult along a low atomic density slip plane versus a high atomic density slip plane. Therefore, higher shear strengths would be present, and higher frictional forces would develop if adhesive friction is attributed to subsurface shear along a low atomic density slip plane.

If defects such as dislocations are present in the crystal structure, subsurface deformation is even easier, and friction forces will be reduced if adhesive friction is assumed to be the only dominant mechanism. As an example, Fig. 8a shows what an edge dislocation looks like in a crystal structure. Figure 8b illustrates the movement of this edge dislocation in a material due to subsurface shear caused by a slider rubbing against the surface. The presence of the edge defect enables slip (and therefore shear) within the bulk of the weaker countersurface material. Without the presence of this defect, slip (and shear) would be more difficult leading to an increase in friction. In a polycrystalline material, picture one grain as having dislocations present to enable subsurface shear and the development of relatively low friction forces, versus another grain with fewer dislocations present and the development of relatively high friction forces.

4.2. FRICTION MODELS – MICRO/NANOSCOPIC

Colchero et al. (1999) provide an excellent review of the state of our understanding of friction on an atomic scale, and it is beyond the scope of the present paper to review all the friction models developed and proposed to explain micro/nanoscopic friction. Instead, emphasis will be placed on models and theories developed that can help researchers understand and predict friction behavior related to the effects of crystal orientation and grain boundaries.

Berman and Israelachvili (1997) presented a friction model that separates the friction force into two components:

$$F = c_1 A + c_2 W \tag{2}$$

where c_1 and c_2 are constants for the specific sliding system. The first component governed by c_1 and the real area of contact, A, is the "adhesion controlled" component, and the second term governed by c_2 and the normal force W is the "load controlled" component. The first term in Eq. 2 is similar to the first term in Eq. 1, however the constant, c_1, does not necessarily equate to shear strength of asperities. c_1 is simply a proportionality constant that depends on the interfacial properties.

Berman and Israelachvili (1997) developed their friction model further, and took a thermodynamic approach to determining how friction at individual contact points, F_i, depends on the specific volume, v, of the surface entities, n_i:

$$F_i = \frac{n_i k_B T}{v} \left(\frac{\partial v}{\partial x} \right) \tag{3}$$

where k_B is Boltzman's constant, T is temperature, and x is the lateral position during sliding. The total friction is the summation of these individual contacts: $F = \sum F_i$. Equation 3 can be interpreted to mean that the friction force at an individual asperity depends on the number of entities (or atoms) at the surface as well as the change in specific volume at the surface with respect to a change in lateral position during sliding. In a polycrystalline material, grains with different crystal orientations will have different specific volumes of the surface atoms depending on the crystal orientation. As an example, a (111) plane in a crystalline material will likely have a different atomic density than a (100) plane. Furthermore, the ability of the specific volume to change with respect to sliding distance ($\partial v/\partial x$) is also influenced by the specific crystal structure and orientation. The change in number of surface atoms, specific volume, and ($\partial v/\partial x$) at the surface also changes at a grain boundary. Impurity atoms are typically present at these boundaries with a completely different specific volume.

It is important to note that Eq. 3 does not depend on the area of contact or asperity height distribution, but rather the interaction of specific entities (or atoms) on the surface. As stated by Berman and Israelachvili (1997), Eq. 3 also does not consider adhesion of the surfaces, and is a more general mechanism for explaining friction between surfaces. The interlocking of surface atoms has been developed into a numerical model by Hirano and Shinjo (1990). Experimental results by Hirano et al. (1991) have already been discussed and presented in Fig. 6. These experimental results supported the theory Hirano and Shinjo (1990) developed to predict interatomic friction between commensurate vs. incommensurate crystalline surfaces. As shown in Fig. 9, their theoretical results were determined for the following interacting planes of α-iron: case (a) the (001) plane sliding against another (001) plane, and case (b) the (110) plane sliding against the (001) plane. Their results predicted

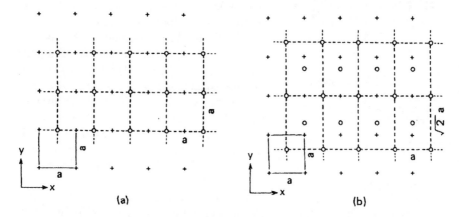

Figure 9. Atomic arrangements at the contact interfaces. The upper body with atoms ($\circ$) is slid over a stationary lower body with atoms (+) in the x direction (adapted from Hirano and Shinjo, 1991).
(a) Simulation of (001) plane sliding against another (001) plane of α-iron.
(b) Simulation of (110) plane sliding against the (001) plane of α-iron.

a higher friction force of 7.6 GPa for case (a), and 3.7 GPa for case (b). These forces were normalized with respect to an elastic contact area. The sliding direction utilized for case (a) represents the commensurate case, whereas the sliding direction for case (b) represents the incommensurate case. Therefore, the model developed by Hirano and Shinjo (1990) supports the findings from their experimental study. In addition, they also state that based on their interatomic locking model, adhesive bonding found from phenomenological studies of contacting surfaces does not appear to be related to interatomic friction.

Friction force microscopes (FFMs) have enabled tribologists to probe surfaces with a single asperity contact and measure friction on the nanoscale. Colchero et al. (1999) have stated that the mechanism responsible for friction on the nanoscale (atomic scale) relates to how the FFM tip overcomes the potential well between adjacent atoms of the surface. Lattice parameters relating to the planar density of atoms on different crystalline planes should therefore correspond with the friction forces measured by an FFM. Atomic scale friction measurements made by Mate et al. (1987) led the researchers to explain their results in terms of stick-slip phenomena between the tip and atoms on the surface. As summarized by Colchero et al. (1999) two-dimensional stick-slip results have been studied in detail by Fujisawa et al. (1993), and related work has been reviewed by Morita et al. (1996). Shimizu et al. (1998) recently performed a molecular dynamics simulation to model this atomic scale

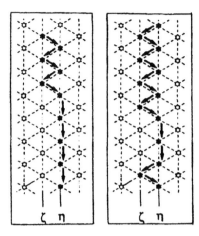

Figure 10. Examples of possible two-dimensional stick-slip motions of an FFM tip on a sample surface (adapted from Morita et al., 1996).

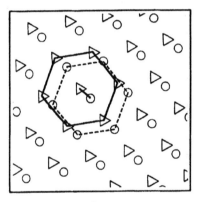

▷ Topography   ○ Friction

Figure 11. Schematic diagram that shows the relative positions of topography maxima and friction maxima for an atomically resolved image of graphite (HOPG). The diagram is based on data obtained with a Si_3N_4 cantilever and a microfabricated tip (Ruan and Bhushan, 1994).

stick-slip phenomenon. In their summary, Colchero et al. (1999) point-out that the surface of a crystal can be described by a two-dimensional potential with a symmetrical arrangement of minima and maxima. To move the tip from an energetically favorable minimum potential condition, a shearing force is required. As the shearing force builds-up, the tip will "jump" to the next energetically favorable minimum. As shown in Fig. 10 this causes the stick-slip phenomenon, which is likely to be a function of atomic spacing and the crystal direction chosen for the FFM scan.

It should be pointed-out that Ruan and Bhushan (1994) acquired topographic and lateral force images of a graphite (HPOG) surface. From their topographic and friction data they determined that the two signals are shifted by one third of a unit cell (see Fig. 11). This indicates that the friction is not necessarily related to the stick-slip phenomenon. Instead, it indicates that the lateral force obtained with the FFM can be decomposed into two components. One component is "conservative", and the other is "nonconservative". The nonconservative component is due to the energy dissipation and can be attributed to the friction force. Any stick-slip that occurs is therefore smaller than the interatomic distance. In their model used to describe the normal and lateral forces, Ruan and Bhushan (1994) include a term that accounts for the atomic corrugation (or topography) of the surface. This implies that the arrangement and density of surface atoms are related to friction, and there is a relationship between crystal orientation and friction.

To clarify the role of crystal orientation, scan path, and corrugation on friction, consider the thermodynamic-based model developed by Rajasekaran et al. (1997). In their model, Rajasekaran et al. (1997) considered a single atom (or tip) passing over an hexagonally closed-packed surface. A schematic diagram of this geometry is shown in Fig. 12. Pathway 1 offers the least friction because the tip slides in a "groove" formed by rows of atoms. Along pathway 4, the tip is required to move directly over substrate atoms, and the frictional force is an order of magnitude greater. Pathways 2 and 3 yield friction results intermediate to those determined along pathways 1 and 4. Note that their thermodynamic model does not include the complexities associated with actual FFM experiments. Instead, it describes the intrinsic anisotropic nature of friction forces between a scanning tip comprised of a single atom and a crystalline surface. In comparison, the two-dimensional stick-slip explanation summarized by Morita et al. (1996) along with the explanation by Ruan and Bhushan (1994) provide insight into the applied problems associated with using an FFM to measure friction characteristics of a crystalline surface. Regardless of whether or not a theoretical or experimental approach is taken, friction on the micro/nanoscale is influenced by the arrangement and orientation of surface atoms.

5. Summary and Conclusions

Based on macroscopic and micro/nanoscopic friction measurements performed with crystalline materials, it is clear that crystallographic orientation and grain boundaries affect friction. The type of atoms and arrangement of these atoms contribute to the magnitude of the friction force. Therefore, friction should be anisotropic and depend on the crystallographic directions in which the two crystals are moving relative to one another. This is particularly true at the nanoscopic level encountered in MEMS, and designers of these devices recognize the problems associated with adhesion and friction during their fabrication and operation.

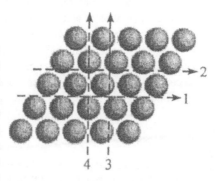

Figure 12. The atomic structure of a close-packed substrate, showing several pathways of the single atom tip corresponding to constant load scans (Rajasekaran et al., 1997).

From macroscopic results, it is evident that adhesion should be minimized to reduce friction. If adhesive junctions form followed by shearing of a weaker material, that weaker material should have a crystal orientation such that the atomic density is high in the shearing direction. Dopants could also be added to produce dislocations that allow for shear to occur more easily, which would result in a reduction in friction. The effects of grain boundaries should also be considered since they are high energy sites. The work of adhesion will increase if a slider passes over the grain boundary leading to an increase in friction.

Shear characteristics of surfaces are also important from a micro/nanoscopic standpoint. However, shear at this level involves breakage of interatomic bonds formed between the surfaces. The sum of these interatomic bonds could be considered to be a real area of contact, and adhesion is directly related to the strength of these interatomic bonds. In other words, the real contact area is still important at the nanoscale level, but must be directly related to interactions between atoms on the opposing surfaces. This is evident in the use of FFMs to measure friction. Stick-slip can occur between the tip and sample surface, which could be directionally-dependent based on the path followed by the probe. The slip can be two-dimensional in nature as the tip jumps between energy minima on the surface. These energy minima are determined by the atomic layout and related energy potentials at the surface, and therefore the crystal orientation plays a direct role in this type of friction. Bonds continuously form and shear between the tip and the surface without causing substantial wear of either surface. In addition, it must be recognized that lateral force measurements obtained with FFM tips can also depict the atomic nature of the surface due to the separation of the lateral force into two components: one that is conservative and not related to friction, and the other nonconservative and directly related to friction.

Both macro and micro/nanoscopic measurements of friction show the importance of having incommensurate versus commensurate surfaces in contact. Note that this situation occurs when two surfaces with specific planar arrangements of atoms interact. Incommensurate surfaces allow atomic planes of a particular crystallographic orientation to slide more easily in low energy "grooves" determined by the orientation of atomic planes on a countersurface. Even thermodynamic models utilizing single atoms sliding on polyatomic surfaces show that the friction will be a minimum if the atoms travel along grooves between rows of atoms on the polyatomic surface.

Sliding surfaces in MEMS devices may both be polyatomic as in the case of wobble motor bearings. Another MEMS device might utilize a tip that takes on the characteristics of a single atom sliding against a polyatomic surface. This latter case could be realized in advanced information storage systems as read/write head sizes continue to decrease. Crystal orientation and grain boundaries can determine the frictional characteristics in both cases.

6. References

Berman, A. and Israelachvili, J. (1997), "Control and Minimization of Friction via Surface Modification," in *Micro/Nanotribology and its Applications* (B. Bhushan, ed.), pp. 317-329, NATO-ASI E330, Kluwer Acad. Publ., Dordrecht, The Netherlands.

Bhushan, B. (1997), "Friction, Scratching/Wear, Indentation and Lubrication on Micro- to Nanoscales," in *Micro/Nanotribology and its Applications* (B. Bhushan, ed.), pp. 169-191, NATO-ASI E330, Kluwer Acad. Publ., Dordrecht, The Netherlands.

Bhushan, B., ed. (1998), *Tribology Issues and Opportunities in MEMS*, Kluwer Acad. Publ., Dordrecht, The Netherlands.

Bhushan, B. (1999a), *Handbook of Micro/Nanotribology, 2nd ed*, CRC Press, Boca Raton, FL.

Bhushan, B. (1999b), *Principles and Applications of Tribology*, Wiley, New York.

Bhushan, B. and Gupta, B. K. (in press), "Macro- and Micromechanical and Tribological Properties," in *Hard Coatings for Wear Reduction, Corrosion/Erosion Protection, and Biomaterials* (R. F. Bunshah, ed.), Noyes Publications, Park Ridge, NJ.

Bhushan, B., Israelachvili, J. N. and Landman, U. (1995), "Nanotribology: Friction, Wear and Lubrication at the Atomic Scale," *Nature* **374**, 607-616.

Bhushan, B. and Koinkar, V. N. (1994), "Tribological Studies of Silicon for Magnetic Recording Applications," *J. Appl. Phys.* **75**, 5741-5746.

Bhushan, B. and Li, X. (1997), "Micromechanical and Tribological Characterization of Doped Single-crystal Silicon and Polysilicon Films for Microelectromechanical Systems," *J. Mater. Res.* **12**, 54-63.

Bhushan, B. and Venkatesan, S. (1993a), "Mechanical and Tribological Properties of Silicon for Micromechanical Applications: a Review," *Adv. Info. Storage Syst.* **5**, 211-239.

Bhushan, B. and Venkatesan, S. (1993b), "Friction and Wear Studies of Silicon in Sliding Contact with Thin-film Magnetic Rigid Disks," *J. Mater. Res.* **8**, 1611-1628.

Bluhm, H., Schwarz, U. D., Meyer, K. P. and Wiesendanger, R. (1995), "Anisotropy of Sliding Friction on the Triglycine Sulfate (010) Surface," *Appl. Phys.* **A61**, 525-533.

Bowden, F. P. and Tabor, D. (1950), *The Friction and Lubrication of Solids – Part 1*, Clarendon, Oxford, U.K.

Buckley, D. H. (1982), "Surface Films and Metallurgy Related to Lubrication and Wear," *Progress in Surface Science* (S. G. Davison, ed.), pp. 1-154, Permagon, NY.

Burnham, N. A. and Kulik, A. J. (1999), "Surface Forces and Adhesion," in *Handbook of Micro/Nanotribology, 2nd ed.* (B. Bhushan, ed.), pp. 247-271, CRC Press, Boca Raton, Florida

Colchero, J., Meyer, E. and Marti, O. (1999), "Friction on an Atomic Scale," in *Handbook of Micro/Nanotribology, 2nd ed.* (B. Bhushan, ed.), pp. 273-333, CRC Press, Boca Raton, Florida.

Czajka, R., Mróz, B., Szuba, S. and Mielcarek, S. (1997), "Investigation of Sliding Friction on the Ferroic Crystals Surface," in *Micro/Nanotribology and its Applications* (B. Bhushan, ed.), pp. 269-273, NATO-ASI E330, Kluwer Acad. Publ., Dordrecht, The Netherlands.

Czichos, H. (1978), *Tribology Series, 1, Tribology a Systems Approach to the Science and Technology of*

Friction, Lubrication, and Wear, Elsevier, NY.

Fujisawa, S., Sugaware, Y., Ito, S., Mishima, S., Okada, T. and Morita, S. (1993), "The Two-dimensional Stick-Slip Phenomenon with Atomic Resolution," *Nanotechnology*, **4**, 138-142.

Garcia, R., Tamayo, J., Gonzalez, L. and Gonzalez, Y. (1997), "Compositional Characterization of III-V Semiconductor Heterostructures by Friction Force Microscopy," in *Micro/Nanotribology and its Applications* (B. Bhushan, ed.), pp. 275-282, NATO-ASI E330, Kluwer Acad. Publ., Dordrecht, The Netherlands.

Gardos, M. N. (1998), "Advantages and Limitations of Silicon as a Bearing Material for Mems Applications," in *Tribology Issues and Opportunities in MEMS* (B. Bhushan, ed.), pp. 341-365, Kluwer Acad. Publ., Dordrecht, The Netherlands.

Grillo, S. E., Field, J. E. and van Bouwelen, F. M. (2000), "Diamond Polishing: the Dependency of Friction and Wear on Load and Crystal Orientation," *J Phys. D: Appl. Phys.* **33**, 985-990.

Gupta, B. K., Bhushan, B. and Chevallier, J. (1994), "Modification of Tribological Properties of Silicon by Boron Ion Implanation," *Tribol. Trans.* **37**, 601-607.

Gupta, B. K., Chevallier, J. and Bhushan, B. (1993), "Tribology of Ion Bombarded Silicon for Micromechanical Applications," *ASME J. Tribol.* **115**, 392-399.

Hirano, M. and Shinjo, K. (1990), "Atomistic Locking and Friction," *Phys. Rev.* **B41**, 11837-11851.

Hirano, M., Kazumasa, S., Kaneko, R. and Murata Y. (1991), "Anisotropy of Frictional Forces in Muscovite Mica," *Phys. Rev. Lett.* **67**, 2642-2645.

Israelachvili, J. N. and Tabor, D. (1972), "The Measurement of Van Der Waals Dispersion Forces in the Range of 1.5 to 130 nm," *Proc. R. Soc. Lond.* **A331**, 19-38.

Maboudian, R. (1998), "Adhesion and Friction Issues Associated with Reliable Operation of MEMS," *MRS Bulletin* **23**, 47-51.

Martin, J. M., Pascal, H., Donnet, C., Mogne Th. L., Loubet, J. L. and Epicier, Th. (1994), "Superlubricity of MoS$_2$: Crystal Orientation Mechanisms," *Surf. and Coat. Tech.* **68/69**, 427-432.

Mate, C. M., McClelland, G. M., Erlandsson, R. and Chiang, S. (1987), "Atomic-Scale Friction of a Tungsten Tip on a Graphite Surface," *Phys. Rev. Lett.* **59**, 1942-1945.

Morita, S., Fujisawa, S. and Yasuhiro, S. (1996), "Spatially Quantized Friction with a Lattice Periodicity," *Surf. Sci. Rep.* **23**, 1-41.

Ohmae, N. (1997), "Adhesion and Friction Using Field Ion Microscopy," in *Micro/Nanotribology and its Applications* (B. Bhushan, ed.), pp. 135-150, NATO-ASI E330, Kluwer Acad. Publ., Dordrecht, The Netherlands.

Rajasekaran, E., Zeng, X. C. and Diestler, D. J. (1997), "Frictional Anisotropy and the Role of Lattice Relaxation in Molecular Tribology of Crystalline Interfaces," in *Micro/Nanotribology and its Applications* (B. Bhushan, ed.), pp. 371-377, NATO-ASI E330, Kluwer Acad. Publ., Dordrecht, The Netherlands.

Schimizu, J., Eda, H., Yoritsune, M. and Etsuji, O. (1998), "Molecular Dynamics Simulation of Friction on the Atomic Scale," *Nanotechnology* **9**, 118-123.

Schmitt, M., Paulmier, D. and Huu, T. L. (1999), "Influence of Diamond Crystal Orientation on Their Tribological Behavior under Various Environments," *Thin Solid Films* **343-344**, 226-229.

Shackelford, J. F. (1996), *Introduction to Materials Science for Engineers, 4th ed.*, Prentice Hall, New Jersey.

Sheehan, P. E. and Lieber, C. M. (1996), "Nanotribology and Nanofabrication of MoO$_3$ Structures by Atomic Force Microscopy," *Science* **272**, 1158-1161.

Sundararajan, S. and Bhushan, B. (1998), "Micro/Nanotribological Studies of Polysilicon and SiC Films for MEMS Applications," *Wear* **217**, 251-261.

Tabor, D. and Winterton, R. H. S. (1969), "The Direct Measurement of Normal and Retarded Van Der Waals Forces," *Proc. R. Soc. Lond.* **A312**, 435-450.

Tas, N., Vogelzang, B., Elwenspoek, M. and Legtenberg, R. (1997), "Adhesion and Friction in MEMS," in *Micro/Nanotribology and its Applications* (B. Bhushan, ed.), pp. 621-628, NATO-ASI E330, Kluwer Acad. Publ., Dordrecht, The Netherlands.

Tomlinson, G. A. (1929), "A Molecular Theory of Friction," *Philo. Mag.*, 7, 905-939.

Venkatesan, S. and Bhushan, B. (1993), "The Role of Environment in the Friction and Wear of Single-crystal Silicon in Sliding Contact with Thin-film Magnetic Rigid Disks," *Adv. Info. Storage Syst.*, **5**, 241-257.

Venkatesan, S. and Bhushan, B. (1994), "The Sliding Friction and Wear Behavior of Single-crystal Polycrystalline and Oxidized Silicon," *Wear* **171**, 25-32.

Weick, B. and Bhushan, B. (2000), "Grain Boundary and Crystallographic Orientation Effects on Friction," *Tribol. Trans.* **43**, 33-38.

Yoshizawa, H., Chen, Y. L. and Israelachvili, J. (1993), "Fundamental Mechanisms of Interfacial Friction. 1. Relation Between Adhesion and Friction," *J. Phys. Chem.* **97**, 4128-4140.

RELATIONSHIP BETWEEN STRUCTURE AND INTERNAL FRICTION IN CoPt AND FePd ALLOYS

E. KLUGMANN
Technical University of Gdańsk, Department of Solid - State Electronics
80-952 Gdańsk, Poland

1. Introduction

It is known that atomic ordering can affect the anelastic and magnetic properties of metals and alloys. The effects of atomic and magnetic ordering in "as-quenched" equi-atomic CoPt and FePd samples are presented as illustrative examples.

The internal friction (IF) level, the initial magnetic susceptibility and the magnetic after-effect (MAE) can be used as an indication of the degree of order.

In the disordered A1 (cubic) phase (Fig.1a) CoPt exhibits a high level of magneto-elastic damping and soft ferromagnetic properties, while in the ordered $L1_0$ (tetragonal) phase (Fig.1b) it show a very low IF level and hard magnetic properties. The state of ordered phase is characterized by lower symmetry and bigger size of unit cell called superstructure as seen in Fig.1.

The aim of the present paper is to investigate the ordering process and point the correlation between the ordering and change of IF, elastic modulus defect and magnetic susceptibility during continuous or successive annealing of disordered samples.

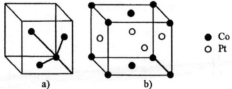

a)   b)

Figure 1. Crystal structure of the disordered Al (a) and ordered $L1_0$ (b) phase of the CoPt equi-atomic alloy.

2. Experimental methods

2.1. SAMPLE PREPARATION

High purity, equi-atomic CoPt and FePd samples were prepared in the form of wires with diameter of 1mm; these were then sealed off under a vacuum of 10^{-5} Pa in quartz tubes, annealed for 5 hours at 1430 K and water quenched in order to obtain the disordered (fcc) phase with soft ferromagnetic properties.

B. Bhushan (ed.),
Fundamentals of Tribology and Bridging the Gap between the Macro- and Micro/Nanoscales, 299–304.
© 2001 *Kluwer Academic Publishers.*

2.2. THE INTERNAL FRICTION

The IF measurements were made using an inverted torsional pendulum at a frequency, f_0, of 60 Hz, a strain amplitude of 2×10^{-5} and two heating rates (Klugmann et al., 1987). The shear modulus, G, was determined by means of frequency measurements ($G \sim f^2$). Thus the shear modulus defect:

$$\frac{\Delta G}{G_0} = \left(\frac{f}{f_0}\right)^2 - 1 \tag{1}$$

2.3. MAGNETIC PROPERTIES

The real component of the initial magnetic susceptibility χ (t,T) and its MAE defined as the time-dependent change of the reluctivity, r(t), after a demagnetization:

$$\frac{\Delta r}{r_1} = \frac{r(t_2, T) - r(t_1, T)}{r(t_1, T)} \tag{2}$$

where $r = 1/\chi$, were measured using a differential ac technique based on a two-channel phasesensitive detector and an automated mutual inductance bridge (Klugmann et al., 1994). Measurements were made on disordered metals by quenching samples during heating up to the Curie point The rate of temperature change was 0.3 K min^{-1}.

3. Results and discussion

3.1. INTERNAL FRICTION

The temperature dependence of internal friction, Q^{-1}, in CoPt at two heating rates is shown in Fig. 2. The IF of "as-quenched" CoPt samples was measured at two heating rates: 2 K min^{-1} (curve 1) and 6 K min^{-1} (curve 2).

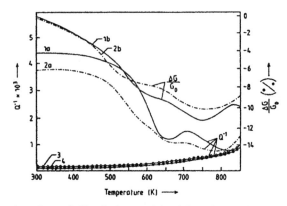

Figure 2. The temperature dependence of IF and shear modulus defect of CoPt for two different heating rates: curve 1: 2 K min^{-1}, 2: 6 K min^{-1}, curves 3 and 4 are damping data for saturated and for ordered samples respectively.

By applying a steady magnetic saturation field H_s=90 kA/m, it was possible to reduce the magneto-mechanical damping (see curve 3) to the level observed in the ordered CoPt alloy (curve 4).

This background IF increase with temperature according to (Chomka et al., 1977):

$$Q^{-1} = \frac{1}{\varpi \tau_0} \exp\left(-\frac{E}{kT}\right)$$ (3)

At T=530 K the background IF, as seen in Fig. 2 is very low: $Q^{-1} = 1.3 \times 10^{-4}$, and in this case at $\omega = 2\pi f_0 = 377$ Hz, and $\tau_0 = 10^{-13}$s, we get the activation energy E=1.5 eV.

Above the Curie point T_C= 816 K, the magneto-mechanical damping of CoPt disappears and as seen in Fig.2, Q^{-1} exhibit the same exponential temperature dependence and the same level as the background IF. This behavior is a consequence of the origin of magneto-mechanical damping.

Characteristically, ferromagnetic materials present a strong damping of their mechanical vibrations due to the irreversible motion of 90^0 domain walls (DW`s), which induces an increase of the magnetic damping. However, above the Curie temperature or in a strong magnetic saturation field, the domain structure disappear, and the magneto-mechanical damping too. This implies that, magneto-mechanical damping is the difference between the damping in the non magnetized state and under a magnetic saturation field H_s, which is high enough to cancel out any magnetic contribution:

$$Q^{-1}_{mag} = \frac{1}{\pi}\left(\delta_{H=0} - \delta_{H_s}\right) = \frac{1}{2\pi}\frac{\Delta W_{mag}}{W}$$ (4)

where δ is the logarithmic decrement of the mechanical vibration, W-is the vibrational elastic energy.

High internal friction level of disordered samples observed in fig.2 up to 450K results from the energy losses ΔW_{mag}, the origin of which is the irreversible DW`s motion.

A drastic (factor of three) decrease of IF level during the ordering process depends from: the vibration amplitude, the heating rates (see Fig.2) and the magnetic field. It is independent from the vibration frequency (Augustyniak et al., 1979).

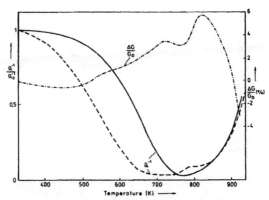

Figure. 3. Internal friction and shear modulus defect of a disordered FePd sample during anneal with $\dot{T} = 6$ K min^{-1} (continuous curve) and $\dot{T} = 2$ K min^{-1} (dashed and dot-dashed curves).

A low level of IF, observed at T>650K is due to the high magnetic anisotropy of the ordered phase and internal stresses produced during the ordering, which restrict the DW's motion and in consequence the magneto-elastic hysteresis losses. A maximum of IF seen in Fig.2 at 700K together with an inflection in the elastic module defect ($\Delta G/G_0$) are due to long-range order under the action of long-range ordering by point defect migration and dislocation pinning.

As seen in Fig.3, a monotonic temperature decrease of the heating-rate dependent magneto-mechanical damping in FePd alloy occurs up to the Curie point (~740K), reaching its minimum in the same region, where $\Delta G/G_0$ achieved a maximum. The anomalous increase of $\Delta G/G_0$ in FePd presumably reflects the change of the Voigt's elastic modulus, c, the origin of which is attributed to the increase of tetragonality and volume of regions where the magnetization vector lies along the c-axis, since the change in the lattice symmetry is accompanied by an increase in the magnetic anisotropy energy up to 1 J cm^{-3} (Klugmann et al., 1992).

3.2. MAGNETIC PROPERTIES

According to the order-disorder theory of alloys (Gomonaj et al., 1995), the crystal structure of a binary alloy like CoPt can be described by the long-range structural order parameter $\rho^{(i)}$ (i = 1,2,3,4) defined as:

$$\rho^{(i)} = P_1^{(i)} - P_2^{(i)} \tag{5}$$

where $P_\alpha^{(i)}$ is the probability of the α - type atom (α=1 for Co and 2 for Pt) to occupy an i-th site in the unit cell. Below 770 K, the ferromagnetic order in CoPt occurs, with the magnetization vector directed along the fourfold symmetry axis. The magnetization $\bar{M}.^{(i)}$ related to the structural order parameter is simply defined as :

$$\bar{M}.^{(i)} = 0.5 (1 + \rho^{(i)}) \bar{M}_1 + 0.5 (1 - \rho^{(i)}) \bar{M}_2 \tag{6}$$

where, $\bar{M}_1$ and $\bar{M}_2$ is the magnetization of the Co and Pt atom respectively.

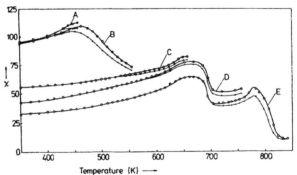

Figure 4. Variation of the initial susceptibility of disordered CoPt sample during successive annealing runs: A, B, C, D and E . o: $t_1 = 1$ s; $\bullet$: $t_2 = 30$ min.

The magnitude of Pt magnetization localized at the Pt atom is small ($|\bar{M}|_2/|\bar{M}|_1$ ~0.1) and depends mainly from the temperature. Assuming that the CoPt alloy contains effectively two sub-lattices, the Co and Pt magnetization can be parallel or anti-parallel. The parallel orientation as follow from experiments is realized in the fully- disordered ferromagnetic state, with $\rho = 0$. In the fully-ordered alloy, the anti-parallel orientation is realized, with $\rho = 1$.

In our experiment (Fig.4), we observe at 350 K a significant decrease of the magnetic susceptibility during step-wise ordering (successive annealing of a disordered sample).

Taking into account the relation between ρ and T in the form of a linear function, because of the constant heating rate during ordering (seen in Fig.4):

$$\rho(T) = 1.6 \times 10^{-3} \text{ K}^{-1} (T-350 \text{ K}), \quad \text{at T} \geq 350 \text{ K} \qquad (7)$$

we can find ρ after different annealing runs:

$\rho = 0$, run A, (disordered alloy);

$\rho = 0.32$, run B, annealing up to $T_{max.} = 550$ K;

$\rho = 0.48$, run C, annealing up to $T_{max.} = 650$ K;

$\rho = 0.64$, run D, annealing up to $T_{max.} = 750$ K (partially ordered alloy).

The results of MAE measurements together with IF measurements made on identically prepared CoPt samples are shown in Fig.5.

We see from Fig.5 that a significant MAE occurs between 430 K and 650 K, accompanied by a decrease of IF. The Richter type MAE peak, reversible magnetic relaxation peak, situated at 530 K, has an activation energy of 1.4 eV (Klugmann et al., 1990) and is ascribed to the reorientation of Co-Pt atomic pairs. This is superposed on a long - range, irreversible process which gives rise to a relaxation - like peak at 700 K and which is interpreted as a structural relaxation.

The MAE peak at 700 K is accompanied by an IF peak. An inflection in $\Delta G/G_0$ is also observed in this temperature range, as seen in Fig. 2.

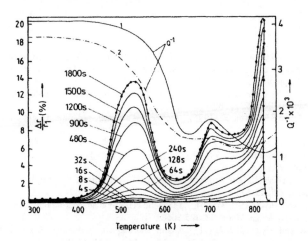

Figure 5. Isochronal relaxation spectrum of a disordered CoPt sample ($t_1 = 1$s, t_2 varied from 4 s to 1800 s).

Fig.6 gives the magnetic susceptibility χ of a disordered CoPt sample during heating from room temperature up to the Curie point and then cooled down with the effect of

304

lowering of the Curie temperature. It can be seen that the irreversible changes of the susceptibility (χ_1 -χ_3) in Fig. 6, correlate with the MAE peaks (Fig.5).

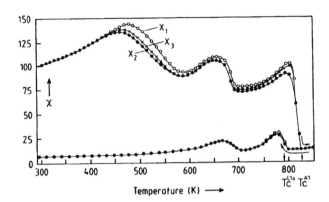

Figure 6. Spectra of initial susceptibility χ of a disordered CoPt sample measured during annealing : χ_3 at t_1 = 1s, and χ_2 at t_2 = 30 min after demagnetization; χ_1 at t_1= 1s after a second demagnetization.

We interpret this changes as being due to the long- time and long-range migration of point defects, leading to the establishment of long-range order, to the reduction of Co-Co atomic pairs with ferromagnetic bonding and to a lowering of the Curie temperature from 830 to 780 K in the partially-ordered sample. A lower value of T_C^{L1o} = 721 K was obtained (Klugmann et al., 1991) for the L1o superstructure according to the Inden model (Inden, 1982) based on the Bragg-Williams-Gorsky approximation.

4. References

Augustyniak, B. and Chomka, W. (1979), „Magnetic field and amplitude dependence of the internal friction in CoPt alloy", Papers of Commission of Metallurgy and Foundry Metallurgy, Polish Academy of Sciences, Kraków, No **26**, pp. 85-92.

Chomka, W. (1977), „The mechanisms of internal friction in metals", Scientific Papers of the University of Katowice (Poland), No **165** pp. 80-113.

Gomonaj,E.V. and Klugmann, E. (1995), „Atomic and magnetic ordering in CoPt alloy", J. Appl. Phys. **77**(5), 2160-2165.

Inden, G. (1982), „Magnetic and chemical ordering", in Proc. of Symp. Alloy Phase Diagrams, Boston, Massachusetts. Mat. Res. Soc. Vol.**19** (L.H. Bennet, et al. eds.) North-Holland, New York, Amsterdam, Oxford. pp.114-127.

Klugmann, E., Blythe H.J. and Augustyniak, B. (1987), „Kinetic of ordering in equi-atomic CoPt alloys", J. de Phys., Colloq. C8, **48**, 513-517.

Klugmann,E. and Blythe, H.J. (1990), „Magnetic relaxation in equi-atomic CoPt and FePd alloys during ordering", J. Magn. Magn. Mater. **83**, 305-306.

Klugmann, E. and Blythe, H.J. (1991), „Atomic and magnetic ordering in CoPt alloys studied by magnetic and internal friction measurements", J. Magn. Magn. Mater. **101**, 99-101.

Klugmann, E., Pastor, J. and Blythe H.J. (1992), „Magnetic and anelastic relaxation in equi-atomic FePd alloys during ordering", J, Magn. Magn. Mater. **112**, 389-381.

Klugmann, E. Blythe, H.J. and Walz, F. (1994), Investigation of thermomagnetic effects in monocrystalline cobalt near the martensitic phase transition", Phys. Stat. Sol. (a) **146**, 803-813.

DIRECT MEASUREMENT OF SURFACE AND INTERFACIAL ENERGIES OF GLASSY POLYMERS AND PDMS

LIHUA LI
Department of Chemical Engineering and Materials Science
University of Minnesota, Minneapolis, MN 55455

VENKATA SUBU MANGIPUDI
3M Austin Center, A142-3N-02, Austin, TX 78726

MATTHEW TIRRELL
Departments of Chemical Engineering and of Materials
Materials Research Laboratory
University of California Santa Barbara, Santa Barbara, CA 93106

ALPHONSUS V. POCIUS
3M Company, Adhesive Technologies Center, St Paul, MN 55144

1. Introduction

The determination of the surface energy of a solid is of interest in the fields of adhesion, adsorption, and wetting, where interactions at and across the interface operate. Adhesion is a phenomenon by which two materials, the adhesive and the adherend, form a bond. A fundamental issue in the science of adhesion is the measurement of the intrinsic strength of an adhesive bond between solids. The reversible work of adhesion, W, is a key parameter in understanding adhesion mechanisms. The determination of the thermodynamic work of adhesion between solid bodies in contact can impact materials design through prediction of adhesive and interfacial performance.

The intrinsic adhesion between polymeric systems is caused by any one or a combination of the following mechanisms: inter-molecular forces such as van der Waals interactions, interdiffusion and entanglements of polymer chains across the interface, acid-base interactions, donor-acceptor interaction or the formation of covalent bonds. It is now generally believed that the principal forces contributing to the work of adhesion between two phases are the Lifshitz-van der Waals forces and acid-base interactions. The contribution of these forces is quantified in terms of two material properties, namely, the surface and interfacial energies.

The intrinsic work of adhesion W_{adh} is an important quantity in determining whether an interface will be stable or will separate spontaneously. W_{adh} is the free energy difference between the separated surfaces and interface. If W_{adh} is positive, it represents the work needed to separate the surfaces. If it is negative, it represents the reduction in free energy by the spontaneous separation of the two surfaces.

Experimental measurements of surface free energy, in most cases, are themselves dependent upon specific models of interfacial energetics and therefore are indirect estimations of surface energy. Wetting methods, that probe solid surfaces by the

305

B. Bhushan (ed.),
Fundamentals of Tribology and Bridging the Gap between the Macro- and Micro/Nanoscales, 305-329.

deposition of liquid droplets of different liquid-vapor surface tensions, are performed to infer or estimate solid surface and interfacial energies by the application of various theories. They are easy to execute but difficult to analyze. Most important for the purpose of adhesion and tribology, wetting methods cannot be extended to the measurement of interfacial energies between two solid polymers. Interfacial energy between solids, unlike surface energy, is a property of two solids in contact, and there is no direct route to its experimental determination other than by contact mechanics (Tirrell, 1996).

Attempts have been made to estimate interfacial tension from surface tensions. Theories of interfacial tension were reviewed by Wu (1982). Wu also provided an extensive tabulation of experimental values for amorphous polymer-polymer interfaces, including both homopolymers and copolymers, at several temperatures. However, these data are all on polymer melts or liquids. We note that they are always much smaller than the surface tensions of polymers. The data on solid polymers are not available. Van Krevelen (1990) also examined the methods used to estimate the interfacial tension by calculations.

It has been of considerable interest to develop a reliable method to measure the intrinsic strength of an adhesive bond between two solid surfaces. Contact mechanics is a method by which one can measure, directly, the work of adhesion between two solid materials. This technique measures the deformation produced by contacting elastic hemispheres under the influence of surface forces and external loads. The mechanics of the system is described by a balance of the elasticity of the materials in contact and the surface forces between them. Load-deformation data can be translated directly to work of adhesion and surface energies.

Direct measurement of surface energies via contact mechanics methods based on the JKR theory has proven to be successful and accurate for surfaces and interfaces of elastic materials (Johnson et al., 1971). Fig. 1 defines the variables of the JKR contact mechanics problem.

According to the JKR theory, the contact radius a between contacting spherically symmetric bodies under an applied load P is given by

$$a^3 = \frac{R}{K}\left[P + 3\pi WR + \sqrt{6WRP + (3\pi WR)^2} \right] \quad (1)$$

In which:

$$\frac{1}{K} = \frac{3}{4}\left(\frac{1 - v_1^2}{E_1} + \frac{1 - v_2^2}{E_2} \right) \quad (2)$$

$$\frac{1}{R} = \frac{1}{R_1} + \frac{1}{R_2} \quad (3)$$

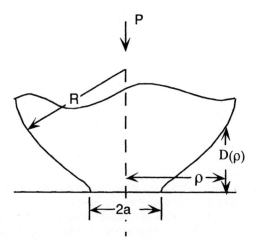

Fig. 1. Notation used to describe the deformed shape of a sphere in contact with a flat surface. Radius of curvature, R, contact radius, a and separation profile, D *versus* x $(= \rho/a)$ can be measured using the interference fringes in the SFA. ρ is the radial distance and P is the applied load.

The theory also calculates the surface separation profile outside the contact zone that is given by the following equation:

$$D(x) = \frac{a^2}{\pi R}\left[\left(x^2 - 1\right) + \left\{x^2 - 2 + \frac{4}{3}\left(\frac{a_o}{a}\right)^{3/2}\right\}\tan^{-1}\sqrt{x^2 - 1}\right]$$

(4)

where $x = r/a$, ρ is the radial distance, a_o is the contact radius at no applied load and D is the distance between the two surfaces.

The pull-off force is given by Eq. (5). The ratio of the pull-off force to the radius of curvature is termed the normalized pull-off energy, P_n, and is given by Eq. (6). P_n depends only on W, and Eq. (6) can be used to compare the data taken on samples of different radius of curvature. This theory also predicts that at the instant of separation the contact radius, a_s, is about 63% of the initial no load contact radius, a_o.

$$P_s = \frac{3}{2}\pi W R$$

(5)

$$P_n = \frac{3}{2}\pi W$$

(6)

In the case of self-adhesion, W_{adh}, the thermodynamic work of adhesion, or intrinsic work of adhesion, is twice the surface energy γ, as shown in Eq. (7). In the case of adhesion between dissimilar materials, the work of adhesion is described by Eq. (8), where γ_1, γ_2, and γ_{12} are respectively the surface energy of material 1 and 2, and the interfacial energy between 1 and 2.

$$W_{11} = 2\gamma_{11} \tag{7}$$

$$W_{12} = \gamma_1 + \gamma_2 - \gamma_{12} \tag{8}$$

The difference between JKR contact mechanics and Hertzian contact mechanics (Hertz, 1896) becomes apparent only when the deformation induced by interfacial forces is significant and detectable. Polymers, in general, are low surface energy materials, with their surface energies ranging from 20-60 mJ/m^2. For the surface and interfacial forces to cause significant deformations in the adhering bodies, the materials need to be relatively soft. Due to the action of the attractive forces, a finite tensile load is required to separate the surfaces from contact. This tensile load is called the pull-off force, P_s. It has been demonstrated that a material with a bulk modulus of about 10^6 Pa, would deform appreciably under the action of loads comparable to P_s. It is for this reason that JKR types of adhesion measurements are usually done on soft elastic materials such as cross-linked polyisoprene rubber or cross-linked poly (dimethyl siloxane) [PDMS].

We are interested in two types of interfaces: the case where the two polymers on either side of the interface are identical, sometimes called symmetric interfaces, and the asymmetric case where the two polymers on either side of the interface are dissimilar. Both the Surface Forces Apparatus [SFA] and the JKR apparatus will be used in this study. Materials such as glassy polymers are relatively hard, with bulk moduli of the order of 10^9 Pa. Increasing K by a factor of 10^3 decreases the contact radius by a factor of 10. For higher values of K, the contact radius a under low applied loads will be very small. It becomes difficult to measure the contact radius using an optical microscope. To get around this limitation, two methods have been developed to provide glassy polymer samples that can be utilized in a contact mechanics experiment. A thin layer is polymer (of thickness on the order of 200 nm) is attached to a previously prepared hemisphere of crosslinked PDMS or to a previously prepared cylinder of a highly crosslinked pressure sensitive adhesive-like network [PSA-LN] (Li et al., 2000). Since the polymer layer is relatively thin (thickness $\sim 0.2~\mu m$), the effective modulus of the composite is close to that of the subtending elastic network. The adhesion-induced deformation is appreciable, and the radius of the deformed contact zone can be measured using a simple optical microscope. This offers a new route for the application of JKR measurements to a variety of previously inaccessible polymers.

2. Experimental Methods

2.1. MATERIALS

In this work, a series of glassy polymers having a polyethylene backbone was studied to systematically vary the polarity of the polymer by substituents of various polarities. The polymers used in this study include poly (4-methyl 1-pentene) [TPX], poly (vinyl cyclohexane) [PVCH], poly (t-butyl styrene) [PtBS], poly (styrene) [PS], poly (vinyl benzyl chloride) [PVBC], poly (p-phenyl styrene) [PPPS], poly (methyl methacrylate) [PMMA], poly (acrylonitrile) [PAN], and poly (2-vinyl pyridine) [PVP]. The molecular structure and glass transition temperature of polymers used in this study are shown in Table 1. PS, PMMA and PVP were anionically synthesized. PVCH was synthesized by heterogeneous catalytic hydrogenation of polystyrene. The details of the synthesis were reported by Gehlsen and Bates (1993). TPX was obtained from a commercial supplier (Mitsui Chemical, Japan) and the rest of the polymers used in this study were obtained from Aldrich. All the polymers are glassy at room temperature. The molecular weights are at least twice of the entanglement molecular weight.

TABLE 1. Molecular structures and specifications of the glassy polymer series.

Polymer	TPX	PVCH	PtBS	PS	PVBC	PPPS	PMMA	PAN	PVP
M_W (kg/mol)	140	50~100	280	55	115	54	86	85	
Tg (°C)	30	120	130	100	~110	161	140	97	100
Elastic Foundation	Mica	PDMS	PDMS	PDMS	PSA-LN	PDMS	Mica	PSA-LN	Mica
Apparatus	SFA	JKR & SFA	JKR	JKR	JKR & SFA	JKR	SFA	JKR	SFA

2.2. SAMPLE PREPARATION USED IN THE SFA

For use in the SFA, the surfaces need to be very smooth and thin. Smoothness is a requirement to obtain well-defined contact area between the two surfaces and a reasonable interference pattern (Merrill et al., 1991). Due to these reasons, most studies carried out in the SFA employ molecularly smooth mica sheets as substrates and the molecules of interest are usually adsorbed or deposited in monolayers (~ thickness 10-50 Å) on mica. The forces of interaction between these modified mica surfaces are measured in a variety of media (Israelachvili et al., 1978). It has only been recently that mica has been replaced by other materials such as polymers, ceramics, and glass.

In our previous studies, we measured the adhesion between polymers films. In these studies mica was replaced by poly (ethylene terephthalate) [PET] and polyethylene [PE] (Good et al., 1960). Thin PET films were obtained by biaxial stretching, and PE films were coextruded with PET. Replacing mica with polymer films is desirable, since it would be possible to study surface energetics of polymers. However, it is not feasible to generate smooth and thin films of a variety of polymers by stretching or extrusion because of processing difficulties such as film tearing and cracking owing to brittleness or fragility. Due to the growing interest in the mechanical measurement of surface energy of polymers, it has been necessary to develop alternative techniques to generate very smooth and thin films that can be used in the SFA studies. In this work, we have made samples for adhesion measurements by solvent casting polymers on molecularly smooth mica. While this technique makes it possible to generate thin and smooth films of a wide variety of polymers for adhesion measurements, it is not without disadvantages of its own. Solvent casting results in a rather complicated five layer interferometer (mica-polymer-air-polymer-mica), shown schematically in Fig. 2. The analysis of five layer interferometry is described elsewhere (Mangipudi et al., 1995).

310

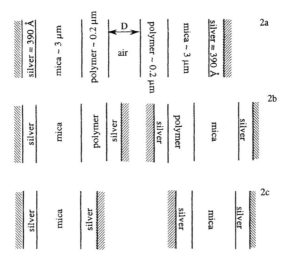

Fig. 2. Schematic of 5-layer interferometer in the SFA. Thickness of different layers can be determined by successive reduction of the 5-layer interferometer to simpler cases. Multilayer matrix method is used to analyze the 5-layer interferometer. In step (a) interference pattern for the 5- layer case is recorded. In step (b) one of the lenses is replaced by a silver layer, and the wavelength pattern for the 2-layer interferometer is noted. In step (c) polymer layer is dissolved, and the thickness of mica sheet is obtained from the wavelength pattern using the analytical solution for 1-layer interferometer. Once mica thickness is known, from step (b) polymer film thickness is determined. Steps (b) and (c) are repeated for the second lens to get thickness of mica and polymer film. Using the calculated thickness in step (a), air gap thickness can be determined.

Thin films of polymers were prepared by solvent casting from dilute solution. PS and PVCH were dissolved in toluene to make a 2 wt% solution. TPX was dissolved in cyclohexane at 50-60°C. PMMA was dissolved in methyl ethyl ketone [MEK], and PVP was dissolved in tetrahydrofuran [THF]. All the solutions were filtered using 0.2 μm Millipore filters. Polymer films of thickness 0.1-0.2 μm were cast on mica. TPX required special treatment due to its semicrystallinity. The polymer was dissolved in cyclohexane at 50 °C to make a 0.2 wt% solution. If the concentration was higher than 0.5 wt% or so, the polymer phase separated when solution was cooled to room temperature.

Freshly cleaved mica of thickness 2-3 μm was silvered on the backside and mounted on cylindrical quartz lenses of 2 cm radius of curvature. The mica sheets were adhered to the glass lenses using molten sugar. About 30 μl of the 2% polymer solution (~45 μl of 0.2% solution in the case of TPX) was placed on the mica surface, and the solvent was evaporated. The casting was carried out in a chamber that was saturated with vapor of the solvent. The solvent was allowed to evaporate at a very slow rate by controlling the vapor pressure of the solvent in the chamber. The polymer films obtained in this manner were extremely smooth. When the evaporation was carried out under ambient conditions the evaporation rates were high. The films obtained in that manner were not smooth enough and were not suitable for adhesion measurements. Spin coating of the polymer solution on mica was also attempted. However, due to the curvature of the surface we could not get uniformly smooth films. Also, contamination of the fresh mica surface by atmospheric impurities resulted in dewetting of the polymer. To minimize contact with impurities, and protect the cleanliness of the mica surface before the deposition of polymer film, it was necessary to carry out solvent casting in a laminar flow hood. It may be emphasized that it was very important to avoid contamination of the mica surface by low surface energy impurities that might be present in the atmosphere. Once the evaporation of solvent was complete, the samples were dried in vacuum for 24 hours. Later, the samples were annealed at 70-80 °C for 36-48 hours in vacuum. This would ensure the removal of any entrapped solvent.

2.3. SAMPLE PREPARATION USED IN THE JKR APPARATUS

Due to the difficulty in the analysis of five-layer interferometry, the JKR apparatus in place of SFA was developed for the contact mechanical studies. A schematic of the sample used in this study is shown in Fig. 3. A thin layer (around 140 to 200 nm) of a polymer is coated on the surface of an O_2-plasma modified cross-linked PDMS or PSA-LNs resulting in the formation of a composite. Contact between two cross cylinders of equal radii of R_c is identical to a sphere of $R_s = R_c$ in contact with a flat surface of $R_f = \infty$, i. e. (Johnson, 1985),

$$\frac{1}{R} = \frac{1}{R_s} + \frac{1}{R_f} = \frac{1}{R_c} \tag{9}$$

Cross-Linked PSA-LN-NoAA or PDMS

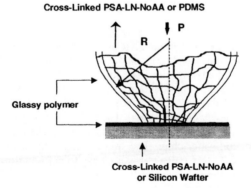

Fig. 3. The schematic of a polymer coated cross-linked PDMS cap or PSA-LN-NoAA in contact with a polymer coated flat surface. The PDMS cap or PSA-LN-NoAA is oxidized in O_2-plasma, and a thin polymer layer is then coated by solvent casting or spin coating. The flat surface is obtained using spun coated polymer film on a silicon wafer.

First, we generate the cross-linked PDMS caps or PSA-LN cylinders. Lenses of PDMS elastomers were prepared by placing droplets of addition-curing silicone using a syringe onto a pretreated glass slide. The glass slides are pretreated with a 10% solution of 1H, 1H, 2H, 2H-perfluorodecyltrichlorosilane (PCR, Inc.) in hexane. This treatment lowered the surface energy of the glass and caused the polymer solution to bead up into a hemispherical shape. The radius of curvature of these lenses was kept smaller than 1 mm to ensure that the shape was indeed spherical. These PDMS caps were left in a laminar flow hood for four hours before they were put into a vacuum oven for curing at 60 °C for two hours. Cross-linking occurred due to the presence of a suitable catalyst in the PDMS liquid solution. Following the cross-linking process, the sol fraction was extracted from the hemispheres by rinsing for two days with heptane in a Soxhlet extraction apparatus. Free radical polymerization was employed to synthesize acrylic PSA-LNs. PDMS flats can be generated by pouring the uncured mixture onto pretreated glass slides and allowing the PDMS to cure in place.

Our PSA-LNs samples were in the form of cylinders. This choice was made on the basis of the chemistry and composition control required of our samples. The detailed procedure is described in detail in another paper (Li et al., 2000). Briefly, degassed acrylic monomers were free-radically polymerized in quartz capillary tubes. After breaking the tube and extracting the resultant material in a suitable solvent, cylindrical samples of an elastic network were obtained.

Silicon wafers were cleaned in hot H_2SO_4/H_2O_2 solution at 90 °C for 30 minutes, rinsed with distilled water, and then dried under N_2. The flat surfaces are created by spin coating a dilute solution of the polymer onto a thus-cleaned pre-cleaned silicon wafer. For the polystyrene analogues, toluene is a good solvent. The solution was filtered using 0.2 μm filters (Millipore), and then reprecipitated. Resulting films were uniform in color, optically smooth, and free of pinholes. Surface roughness and film thickness were measured using AFM and Ellipsometry respectively. The higher the spin coating speed, the thinner the polymer film thickness. The optimum combination of polymer solution concentration and spin coating revolution speed gives the desired thickness and roughness. The polymer films generated are around 140-200 nm with roughness around 1.5 nm. Another flat used in the JKR measurements is silicon coated with addition-cured PDMS.

To prepare the samples to be used in the JKR apparatus, a thin layer of a polymer was coated on the surface of an O_2-plasma modified cross-linked PDMS spherical cap resulting in the formation of a composite. The function of O_2-plasma was to increase the surface energy of PDMS by the introduction of polar groups, such as carbonyl groups, into the surface regions of the foundation. We also used this method. A thin polymer film was spun coated on newly cleaved mica. The polymer film can be floated from the mica substrate in deionized water. The polymer film was picked up onto a PDMS cap mounted on a small silicon wafer chip. The spin coating method works well for PtBS, PVCH and PPPS. However, for PVBC, it is not easy to get a uniform film on top of a silicone hemisphere using the spin coating/float-off method. For PAN, it is impossible to float the film off the mica sheet. Thus, solvent casting was used for PVBC and PAN. The foundation for solvent casting was a PSA-LN. These samples were found to be free of contamination. When PDMS is used as the foundation, it is important to use the spin coating/float-off method instead of solvent casting in order to minimize the possibility of surface contamination by free silicone.

2.4. THE SURFACE FORCES APPARATUS [SFA] AND ADHESION MEASUREMENTS

The SFA has been traditionally used to measure the interactions between solid surfaces (Israelachvili et al., 1978; Israelachvili et al., 1972). We measured the pull-off force, P_S, radius of curvature, R, surface separation outside the contact zone, D versus x, and the contact radius as a function of applied load, a versus P. The radius of the contact zone, the surface separation profile outside the contact region and the radius of curvature of the geometry were measured from the interferometric fringes. All these measurements were carried out in dry nitrogen at 25°C. By measuring the pull-off force, the work of adhesion can be estimated.

The preparation of poly (4-methyl 1-pentene) [TPX] samples needs special mention here. TPX is semicrystalline in nature. The surfaces obtained were not uniformly smooth, as the polymer crystallizes on the substrate. However, it was possible locate some smooth regions using the criteria established in our work (Merrill et al., 1991). These criteria include the following: (i) The surfaces, when brought close enough, should jump into contact. (ii) The FECO fringes need to be sharp. (iii) The pull-off obtained from repetitive measurements at a given contact spot should be reproducible. The pull-off force and radius of curvature were measured in these regions.

2.5. THE JKR APPARATUS AND ADHESION MEASUREMENTS

A homemade automated JKR apparatus (Li et al., 2000) is used in this study to perform adhesion tests using a contact mechanical approach. The PDMS hemisphere and PDMS cap coated with glassy polymers adhere adequately to the glass slides. Cylindrical PSA-LN samples were glued to supporting surfaces by applying small amounts of thoroughly mixed fast setting epoxy. After reaching thermal equilibrium and complete curing of the epoxy, samples were crossed under close examination at lower magnification. A slight contact was made in order to align the video-zoom with the center of contact. The hemispheres or cylinders were then separated and the system was allowed to equilibrate mechanically and thermally for at least 3 hours. Samples were compressed stepwise allowing time for equilibration between each step. After a 10-step compression of 1 μm and equilibration at the maximum load for 30 minutes, samples were decompressed in a similar stepwise fashion. The displacement, load and contact area were recorded on a computer for analysis. The radius of the cylinders was measured sideways under a light microscope. By fitting a^3 versus P data to the JKR model, we can get W, thermodynamic work of adhesion and K, bulk modulus of the composite layer of polymer film on elastic foundation.

Using the displacement-controlled JKR apparatus, the adhesion measurements were taken in a quasi-static manner. There are two reasons that we restrict our load under 300 mg. One reason is that the application of JKR theory is more sensitive to the work of adhesion at lower load. At a very high load, the mechanics are Hertzian. The other reason is that at higher load, the linear elastic assumption is no longer valid.

2.6. CONTACT ANGLE MEASUREMENTS

Contact angle studies were conducted in order to compare the JKR method as a characterization tool for surface energy measurement of glassy polymers. The "recently advanced" and receding contact angles of several liquids, both polar and dispersive liquids, were measured. The surface tensions of testing liquids were measured using Fisher Surface Tensiomat (Model 21), as shown in Table 2.

TABLE 2. Surface tensions of testing liquids used in contact angle measurements.

Testing Liquids	Surface Tension γ_L (mJ/m^2)	Experimental γ_{LExp} (mJ/m^2)
Hexadecane	26.5	27.6
1,3-Butane Diol	37.7	38.3
Dimethyl Sulfoxide	41.2	41.0
Ethylene Glycol	47	47.7
Methylene Iodide	53.8	53.8
Formamide	57.8	58.8
Glycerol	63.0	64.9
Deionized Water	71.2	71.4

A liquid in contact with a solid can either completely wet it or exhibit a finite contact angle, as shown in Fig. 4. The contact angle can be static or dynamic. A static contact angle may be due to a stable equilibrium (the lowest energy state), or in a metastable equilibrium (an energy trough separated from the neighboring states by energy barriers). Contact angle measurements are dependent upon the direction in which the measurement is made. When a drop is laid down upon a surface and it advances over the surface as it spreads, the contact angle in this situation is known as the advancing contact angle. If liquid is withdrawn from a drop that has already come into equilibrium with the surface, the contact angle is known as the receding contact angle. In general, the advancing angle is larger than the receding angle.

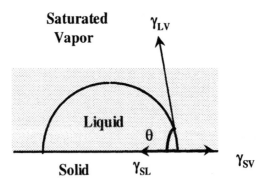

Fig. 4. The schematic of the contact angle of a liquid on a solid. By balancing components of interfacial free energies in the horizontal direction, we can obtain the Young's equation.

Contact angle measurements on glassy polymers coated on silicon wafers were done using a Rame-Hart goniometer with an environmental cell to keep the samples in the saturated vapor of the testing liquids. In contact angle measurements, a small drop of each liquid was placed on the surface using a clean syringe and allowed to equilibrate for about 60 seconds. The advancing and receding contact angles were measured. In the case of glycerol, due to its high viscosity, the equilibration time was more than 3 minutes. Three measurements were made on three drops that were placed in three different locations on the surface. In a sense, the advancing and receding contact angle correspond, respectively, to the loading curve and unloading curve in JKR measurements. The liquids used must not swell the polymer surface.

2.7. SURFACE COMPOSITION CHARACTERIZATION

The chemical analysis of the polymer surfaces used in adhesion measurements was done using X-ray photoelectron spectroscopy [XPS] with their standards reported in the literature to ensure that surfaces were free from contaminants. Along with each series of caps and flats, larger caps and also spin-coated flats were made for XPS analysis of the surface elements. The silicon concentration was less than 0.1%. Because PDMS has a low surface energy, PDMS may contaminate other polymers. Caps used in adhesion measurements are too small for standard XPS analysis, special larger caps were made for XPS analysis. The XPS analysis of polymer-coated silicon wafers showed clean surfaces without silicone contamination. Surface cleanliness of glassy polymer coated PSA-LN-NoAA was determined by the analysis of the surface chemistry using small-spot XPS. No silicone contamination was observed. The surface topography or smoothness of the region used for adhesion measurements can be qualitatively determined from the FECO fringes of multiple beam interferometry used in the SFA.

2.8. FILM THICKNESS CHARACTERIZATION

A Sopra ES4G Spectroscopic Ellipsometer is used to measure the thickness of the glassy polymer film spun coated on silicon wafer. The film thickness used in this study is around 140 ~ 200 nm. Atomic Force Microscopy (AFM) was also used to determine the surface topography of these samples. The AFM imaging indicated that the samples were generally smooth with root mean square (r. m. s.) roughness over the entire surface of about 1.5 nm.

3. Experimental Results

3.1. SURFACE ENERGIES OF GLASSY POLYMERS USING SFA

Adhesion measurements were made on several samples of each polymer. For each set of samples the measurements were made at several contact spots.

The separation profile outside the contact region, D versus x, is determined by measuring the FECO wavelengths. The details of calculating the distance from the FECO wavelengths are discussed elsewhere (Mangipudi, 1995). Fig. 5(a) shows the comparison of the measured profile with the predictions of the JKR theory under no applied load conditions. The theory also predicts the effect of applied load on D versus x profile, and Fig. 5(b) shows the comparison between experiment and the theory. It can be clearly seen that the agreement between the JKR theory and the experiment is good both under no applied load and finite applied load conditions.

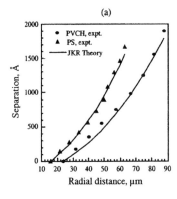
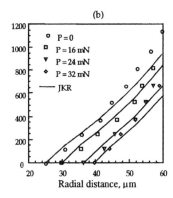

Fig. 5. Comparison of the measured surface separation profile outside the contact region with the prediction of the JKR theory. (a) shows the profile at no applied load for different polymers. (b) shows the separation profiles obtained at different applied loads in the case of PMMA-PMMA contact. The agreement between the experiments and theory is good. The separation at a given radial distance is determined from the interferometric fringes using the scheme illustrated in Fig. 2.

The surface energies of the polymers are determined from the pull-off force, P_S, and radius of curvature, R. Table 3 shows the average normalized pull-off energy (P_n) obtained from these measurements. The surface energies of the polymers are determined from the pull-off force, P_S, and radius of curvature, R, using Eq. (5) and (7), as listed in Table 3. The effects of contact time and rate of separation were found to be negligible. Such behavior indicates that the measurements are relatively dissipation free. Since all the polymers are glassy at the experimental temperature, the contribution of dissipation effects, usually related to viscoelastic properties and interdiffusion, is insignificant.

TABLE 3. Normalized pull-off energy and surface energy of glassy polymers using the SFA.

Polymer	Normalized pull-off energy (mJ/m^2)	Surface Energy (mJ/m^2)
TPX	250 ± 19	26.5 ± 2
PVCH	264 ± 9	28 ± 1
PS	415 ± 19	44 ± 2
PMMA	500 ± 33	53 ± 3.7
PVP	594 ± 38	63 ± 4

3.2. SURFACE ENERGIES OF GLASSY POLYMERS USING THE JKR METHOD

By simultaneously measuring the applied load, displacement between the two bodies, and contact area during cycles, we are able to employ a linear elastic fracture mechanics analysis to obtain the modulus, and the intrinsic work of adhesion W using Eq. (1). The loading, unloading and displacement history of a typical quasi-static step JKR experiment is shown in Fig. 6. Step quasi-static loading and step unloading were used in the measurements in an effort to closely approach the equilibrium state.

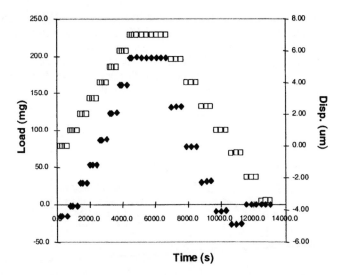

Fig. 6. The load (♦) and displacement (□) history used in a typical JKR measurement. Quasi-static step loading and step unloading were employed to approach equilibrium situation as close as possible.

As shown in Fig. 7, our measurement of the surface energy of PDMS is 21 mJ/m². This result agrees well with the known value for PDMS. There is no apparent adhesion hysteresis between loading and unloading curves. This indicates that the cross-linked PDMS is an ideal-JKR solid. The modulus of crosslinked PDMS was also determined using rheological measurements and results consistent with JKR measurements were obtained. Stress relaxation measurements show that the crosslinked PDMS samples are purely elastic.

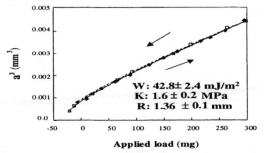

Fig. 7. The JKR plot of adhesion between a cross-linked PDMS hemisphere and a PDMS flat. There is no adhesion hysteresis between loading (♦) and unloading (□). From these data, W $_{PDMS-PDMS}$ = 42.8 mJ/m² and γ_{PDMS} = 21.4 mJ/m².

When a polymer coated PDMS cap is in contact with a PDMS flat, the same value of work of adhesion is obtained as when a PDMS cap is in contact with a polymer coated flat. This result validates our sample preparation and indicates that the value determined from the JKR method is the intrinsic adhesion energy since such a measurement should not be dependent upon sample orientation. Multiple loading and unloading data fall onto the same loading and unloading curves, further indicating that we are measuring thermodynamically reversible adhesion.

318

The JKR plot of self-adhesion of poly (4-tert-butyl-styrene) [PtBS] is shown in Fig. 8. PtBS-PtBS contact gives work of adhesion of 66 mJ/m^2. According to Eq. (7), the surface energy of PtBS is 33 mJ/m^2. A very small adhesion hysteresis exists between loading and unloading. Compared to the surface energy of PS (Mangipudi et al., 1996) of 38 mJ/m^2, we suggest that the tert-butyl on the phenyl ring decreases the surface energy due to the more hydrocarbon like character of PtBS in comparison with PS.

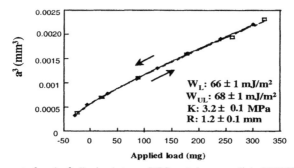

Fig. 8. Measurement of work of adhesion between PtBS coated on cross-linked PDMS hemisphere and PtBS-coated silicon wafer. There is little hysteresis between loading (♦) and unloading (#). From the loading data, we get $W_{PtBS-PtBS} = 66 \pm 1$ mJ/m^2 and $\gamma_{PtBS} = 33 \pm 1$ mJ/m^2.

As shown in Fig. 9, the surface energy of PVCH agrees well with the value obtained from the pull-off force using the SFA. Finite hysteresis exists between loading and unloading cycle for PVCH. However, repeating several cycles of loading and unloading processes, the data always fall onto the same hysteresis cycle. As shown in Table 4, there is also adhesion hysteresis in the case of self-adhesion of PPPS. The higher surface energy of PPPS than that of PS might be due to the substitution group of second phenyl ring on the phenyl ring. The aliphatic nature of PVCH leads to much lower surface energy than that of PS. Thus, our experimental results agree with the literature in that more aliphatic nature of the substitution group leads to lower surface energy, and more aromatic characteristic of the substitution group leads to higher surface energy. Aromatic groups can attract each other by pi-pi interactions that aliphatic hydrocarbons cannot do. As listed in Table 4, the surface energy of PVBC is a little higher than that of PS, which we assume is due to the presence of the more polar chloromethyl group.

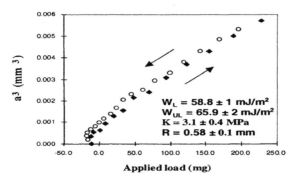

Fig. 9. The measurement of adhesion between PVCH-coated cross-linked PDMS hemisphere and PVCH-coated silicon wafer. There is a discernable hysteresis between loading (♦) and unloading (#). From the loading data, we get $W_{PVCH-PVCH} = 58.8 \pm 1$ mJ/m^2, and $\gamma_{PVCH} = 29.9 \pm 1$ mJ/m^2.

TABLE 4. The comparison between surface energy obtained from the JKR method (including the SFA and JKR Apparatus) and contact angle methods. Three different methods were used to estimate surface energy using the advancing contact angle data.

Polymer	Surface Energy Exp (mJ/m^2)	Surface Energy Cal-MP (mJ/m^2)	Surface Energy Cal-Fedors (mJ/m^2)	Surface Energy Cal-VK (mJ/m^2)
PDMS	21.4 ± 1.2	21.62	26.65	25.79
PVCH	29.9 ± 1	40.02	36.82	32.11
PtBS	33 ± 1	36.99	36.07	34.52
PS	38 ± 1	45.95	41	39.48
PPPS	41.7 ± 1	48.05	42	40.48
PVBC	43.3 ± 1.9	48.3	42.57	40.12
PMMA	53 ± 4	37.26	41.13	34.53
PE	33.3 ± 1.6	35.1	34.15	32.26
TPX	26.5 ± 2	32.52	32.17	30.39
PVP	63 ± 4	46.69	49.07	45.08
PET	61 ± 2	47.09	47.74	40.09
PAN	54 ± 1	61.14	62.33	53.53

The reproducibility of repeated loading and unloading experiments on PtBS-PtBS, PVCH-PVCH, PPPS-PPPS and PVBC-PVBC indicates that hysteresis is due to a reversible cause. It is reasonable to propose that there is a contact induced rearrangement of the interface when the surfaces are in contact under compressive stress due to some non-dispersive interaction. The surprising result is that the polymer reorients itself between the unloading and reloading cycles to provide reproducible measurements. The adhesion behavior of PAN is quite different from the PS analogues. As shown in Fig. 10, adhesion hysteresis increases with decreasing contact radius. This dependence might be due to the nitrile group intermolecular pairing (Andreeva et al., 2000).

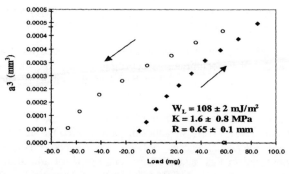

Applied load (mg)

Fig. 10. Measurement of work of adhesion between PAN-coated PSA-LN-NoAA cylinders. It can be seen that there was substantial adhesion hysteresis between loading (♦) and unloading (#). From the loading data, we get $W_{PAN-PAN} = 108 \pm 2$ mJ/m^2, and $\gamma_{PAN} = 54 \pm 1$ mJ/m^2.

In our present studies it was found that at a given load, the contact radius during loading was different from the contact radius during unloading. This is known as the contact hysteresis or adhesion hysteresis. The origins of this hysteresis are not completely understood. Adhesion hysteresis can occur due to variety of reasons. These include surface heterogeneity, stress induced surface reconstruction or rearrangement, surface roughness, interpenetration or interdigitation of molecular chains across the interface, bulk viscoelastic deformation of the adhering bodies etc. In many cases a combinations of some of these effects would play a role in causing the adhesion hysteresis.

The adhesion hysteresis between loading and unloading for the solid polymers studied is also summarized in Table 5. The extent of hysteresis is calculated using Eq. (10), where W_L and W_{UL} are the work of adhesion obtained from loading and unloading by curve fitting, respectively. The data in Table 5 seems to suggest that when more non-dispersive interactions are possible between mating surfaces, the more likely the possibility of adhesion hysteresis. The adhesion hysteresis carries signature of the non-equilibrium processes occurring in the materials (Ghatak et al., 2000).

TABLE 5. Surface energies of glassy polymers and interfacial energies between glassy polymers and PDMS using the JKR Apparatus.

Polymer	Solubility Parameter δ (J/cm^3)$^{1/2}$	γ_{JKR} (mJ/m^2)	Hysteresis γ (%)	$W_{PDMS\text{-}Polymer}$ (mJ/m^2)	Hysteresis W_{12} (%)	$\gamma_{PDMS\text{-}Polymer}$ (mJ/m^2)
PDMS	14.2	21.4 ± 1.2	0	42.8 ± 2.4	0	0 ± 0
PVCH	16.74	29.9 ± 1	12	48 ± 1	0	3.3 ± 1
PtBS	17.67	33 ± 1	3	45 ± 1	0	9.4 ± 1
PS	19.52	38 ± 2	6	49 ± 2	2	10.4 ± 2
PPPS	19.91	41.7 ± 1	14	52 ± 1	19	11.1 ± 1
PVBC	19.78	43.3 ± 1.9	19	53 ± 2	32	11.7 ± 2
PAN	24.55	54 ± 1	>100	55 ± 2	13	20.4 ± 2

$$\text{Hysteresis } \% = 100 \times \frac{W_{UL} - W_L}{W_L} \tag{10}$$

3.3. INTRINSIC WORK OF ADHESION AND INTERFACIAL ENERGY BETWEEN GLASSY POLYMERS AND POLY (DIMETHYL SILOXANE) [PDMS]

The intrinsic work of adhesion between a series of glassy polymers and PDMS were determined using the JKR technique. According to Eq. (8), the interfacial energy between these polymers can be determined based on the values of surface energies of each polymer and the work of adhesion between them.

As shown in Fig. 11, the work of adhesion between PtBS and PDMS was determined using the JKR method. The work of adhesion between PDMS and PtBS is 45 mJ/m^2. Using Eq. (8), the interfacial energy between PDMS and PtBS is 9 mJ/m^2. No hysteresis was observed in this measurement. The work of adhesion between PDMS and PVCH is very similar to that of PDMS and PtBS. No hysteresis was observed. The value of work of adhesion is 48 mJ/m^2, as shown in Fig. 12. As shown in Fig. 13, PDMS-PAN shows a finite adhesion hysteresis between loading and unloading. Similarly, there is finite adhesion hysteresis for PDMS-PPPS and PDMS-PBVC. In the cases of PPPS, PAN and PVBC in contact with PDMS, there is finite adhesion hysteresis.

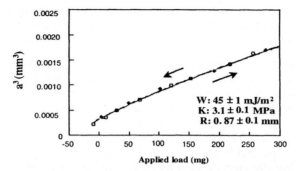

Fig. 11. The JKR plot of adhesion between PDMS cap and PtBS-coated silicon wafer. There was no adhesion hysteresis between loading (♦) and unloading (#). From these data, we get $W_{PtBS-PDMS}$ = 45 ± 1 mJ/m^2 and $\gamma_{PtBS-PDMS}$ = 9.4 ± 1 mJ/m^2.

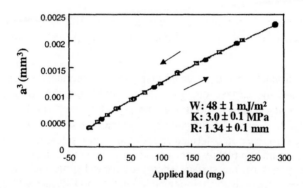

Fig. 12. Measurement of work of adhesion between PDMS cap and PVCH-coated silicon wafer. There was no adhesion hysteresis between loading (•) and unloading (□). From these data, we get $W_{PVCH-PDMS}$ = 48 ± 1 mJ/m^2, and $\gamma_{PVCH-PDMS}$ = 3.3 ± 1 mJ/m^2.

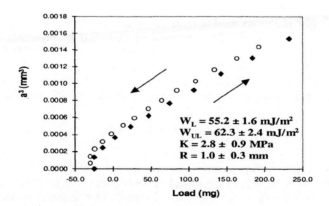

Fig. 13. Measurement of work of adhesion between PDMS cap and PAN-coated silicon wafer. It can be seen that there was some adhesion hysteresis between loading (♦) and unloading (#). From these data, we get $W_{PPPS-PDMS}$ = 55.2 ± 1.6 mJ/m^2 and $\gamma_{PPPS-PDMS}$ is 20.4 ± 2 mJ/m^2.

It seems that more polar and more aromatic nature of the glassy polymer leads to hysteresis. The extent of hysteresis depends on the chemical nature of the interface, for we didn't observe any hysteresis in the case of PtBS-PDMS, PVCH-PDMS.

The measured work of adhesion and the interfacial energy between glassy polymers and PDMS are tabulated in Table 5. The work of adhesion between PDMS and glassy polymers are quite low, suggesting weak interaction with these polymers. A much higher interfacial energy between PDMS and PAN might be due to the great polarity disparity. The series of interfacial energies range from 4 to 21 mJ/m^2 indicating a higher value than the interfacial energy between polymer melts. The large interfacial energy values should not necessarily be surprising. The interfacial energy between PE and PET is about 17 mJ/m^2 (Mangipudi et al., 1994). The interfacial energy between some common solvents and water can be as high as 53 mJ/m^2 (Israelachvili, 1991).

3.4. ESTIMATION OF SURFACE ENERGIES USING CONTACT ANGLE METHODS

Three different methods for analyzing contact angle data were performed. These include the Zisman's plot, Good-Girifalco-Fowkes's (GGF) equation, and Wu's equation of state.

The equilibrium contact angle θ is related to the various interfacial energies by Eq. (11), where γ_{LV}, γ_{SV} are the surface energies of the liquid and the solid in contact with saturated vapor of the liquid, respectively, and γ_{SL} is the interfacial tension between the solid and the liquid. γ_{SV} may be considerably less than the surface energy of the solid in vacuum, γ_S. The difference π_e is termed as the equilibrium spreading pressure, as shown in Eq. (12). If a liquid completely wets a solid, then the contact angle is zero. Since the contact angle depends on the surface energy of the substrate and the liquid, contact angle measurements are widely used for estimating the solid surface energy.

$$\gamma_{LV}\cos\theta = \gamma_{SV} - \gamma_{SL} \tag{11}$$

$$\pi_e = \gamma_S - \gamma_{SV} \tag{12}$$

The Zisman's plot method is used to measure the critical surface tension. The concept of critical surface tension was first proposed by Fox (Fox et al., 1952) and Zisman (1975). If the nature of surface interactions between the series of testing liquids is the same as with the substrate, then the cosine of the contact angle versus the liquid surface tension will fall on a straight line, which is known as Zisman plot. Using Young's equation and extrapolating to complete wetting, we obtain the critical wetting tension.

$$\gamma_C = \lim_{\theta \to 0} \gamma_{LV} = \gamma_s - \left(\gamma_{SL} + \pi_e\right) \tag{13}$$

As can be seen from above, γ_C underestimates the surface energy of the substrate by $(\gamma_{SL} + \pi_e)$. Obviously, γ_{SL} is not zero, though it can be minimized for those testing liquids that have similar polarity as the substrate. Zisman's plot is purely empirical. Truly linear variation of cosθ with γ_{lv} was observed only when contact angles on dispersive solids were measured using dispersive test liquids.

Due to the limitations of Zisman's approach, Girifalco and Good (1960) derived an expression for γ_{sl} in terms of γ_{sv} and γ_{lv} using the Berthelot relation for attractive constants. Using this in Young's equation results in the Good-Girifalco-Fowkes equation (GGF), as shown in Eq. (14).

$$\frac{(1 + \cos \theta)}{2} = \left(\frac{\gamma_{sv}}{\gamma_{lv}} \right)^{1/2} \qquad (14)$$

Using this relation, the critical surface tension of a solid, can be obtained from the plot of $(1+\cos\theta)/2$ versus. This method is valid only when the solid-liquid interactions are dominantly dispersive.

Wu (1979) derived an equation of state for γ_c that accounts for interactions between the solid surfaces and the testing liquid in a qualitative manner. γ_c is expressed in terms of ϕ to obtain a spectrum of critical surface tension of wetting given by

$$\gamma_{c,\phi} = \phi^2 \gamma_s - \pi_e = \frac{(1 + \cos \theta)^2 \gamma_{lv}}{4} \qquad (15)$$

where ϕ is the solid-liquid interaction parameter. ϕ depends on molecular parameters such as ionization potential, molecular polarizability, dipole moment, and molecular volume (Good et al., 1970). $\gamma_{c, \phi}$ is plotted against γ_{lv} to obtain a curve whose maximum, $\gamma_{c, \phi, max}$, equals the surface energy of the solid. In the limit of $\phi_{max} = 1$, the above equation gives $\gamma_{c, \phi, max} = \gamma_{c, Wu} \cong \gamma_s$. By matching the polarities of the testing liquids and the substrate, the interaction parameter approaches the maximum value of one. The maximum attainable value is a very close estimate for the surface free energy of the substrate. To ensure this, a wide range of testing liquids, from nonpolar hydrocarbons to highly polar liquids must be chosen.

Advancing and receding contact angles were measured using polar as well as dispersive liquids. There was a finite hysteresis in the contact angles measurement. The values of estimation of surface energy obtained from the advancing and receding angles are not very different, as these numbers are obtained using linear least square fits to the models described above. The phenomenon of having a different contact angle under advancing and receding conditions is known as contact angle hysteresis. Johnson and Dettre (Johnson et al., 1964) have described a number of reasons contact angle measurements are hysteretic, specifically: inhomogeneous surface chemistry, surface roughness, and molecular rearrangement in the solid induced by the liquid and vice versa. The hysteretic character of contact angle measurements raises some doubts as to their character as equilibrium measurements of surface energetics. They have also shown that advancing contact angles convey information about the dispersive part of the surface interactions, while receding angles are greatly affected by the polar contribution. Our results using contact mechanics seem to behave in a similar fashion in that adhesion hysteresis is observed when polar materials are placed in contact.

As shown in Table 4, three different methods commonly used for contact angle data analysis are compared with surface energies obtained from the JKR method. The results from the contact angle measurements indicate that it depends on not only the characteristics of the test liquids used in the measurements but also on the methods used to analyze the data.

4. Discussion

4.1. COMPARISON BETWEEN SURFACE ENERGY FROM THE JKR METHOD AND OTHER METHODS

The comparison between surface energy results from contact angle measurements and the JKR measurements is shown in Table 4. The critical surface tension obtained from the Zisman's plot is obviously lower than other estimation methods, which makes sense since it ignores the interfacial tensions between testing liquids and surfaces. More rigorous models like the Good-Girifalco-Fowkes equation, or Wu's equation of state might be a better estimate of surface energy. Similarly, the agreement between surface energy from the JKR method and contact angle method using Zisman plot was shown by Chaudhury for monolayer surfaces (Chaudhury et al., 1992).

A critical observation of the data listed in Table 4 reveals the following:
(i)   In the case of dispersive polymers like PE, PVCH, PtBS and TPX the values of surface energies obtained from JKR measurements agree well with the surface energy predicted from the contact angle data.
(ii) In the case of polymers like PMMA, PET and PVP, and PAN, where nondispersive surface interactions are dominant, the difference between the surface energy obtained from the JKR data and estimated from contact angle data is significant.
(iii) Surface energy predicted using Wu's equation seems to be in agreement with the JKR data for several polymers like PE, PVCH, TPX, PtBS and PS.

To understand the origins of these differences, we need to critically examine the assumptions associated with each of the models proposed for the interpretation of contact angle data. Zisman's method is purely empirical, and it does not account for the solid-liquid interactions. The linearity of $\cos\theta$ versus γ_{lv} hold only in the case of solids that are purely dispersive, e.g. contact angles of hydrocarbon liquids like alkanes on low energy surfaces like poly (tetrafluoroethylene). The deviation from this linear behavior becomes significant as the solid and/or the probe liquid becomes polar. In general, the Zisman plot method is shown to underestimate the surface energy of solid surfaces. Unlike Zisman's method, the Good-Girifalco equation (Good et al., 1960) takes into account the intermolecular interactions between the solid and the probe through the interaction parameter ϕ. For dispersive materials ϕ is unity, and the solid surface energy can be predicted from a plot of $\cos\theta$ versus $\gamma_{lv}^{-0.5}$, known as the GGF plot. This type of plot does indeed account for the solid-liquid interactions. However, in the case of solids and liquids where H-bonding effects become significant, this method may not work well as the original Good-Girifalco parameter, ϕ, does not consider H-bonding interactions (Good et al., 1970). It is for this reason that the GGF plot predictions for PE, PVCH, TPX, and PS agree fairly well with the JKR data, and the discrepancy becomes significant in the case of materials like PMMA, PVP, PAN and PET. It should also be emphasized that the effects of H-bonding interactions may become rather significant in the case of polymers like PET, PVP, and PAN, and high surface tension probe liquids like glycols, glycerol, formamide and water. It is necessary to account for the influence of these interactions in a more systematic manner, and develop better models to predict surface energy.

The comparison suggests that Wu's equation gives the closest estimate of surface energy to the JKR method. The Wu's equation of state (Wu et al., 1979) is an expansion of the Good-Girifalco equation in the Zisman's limit of $\cos\theta \rightarrow 1.0$. Since Wu's equation of state method accounts for solid-liquid interactions qualitatively, it might be a more reasonable method to use for contact angle analysis.

Burrell (1955) points out the interesting relationship between surface tension and solubility parameter of liquids, where δ is the solubility parameter and V is the molar volume. This relationship was extended to polymeric systems by substituting Zisman's critical surface tension γ_c in place of γ.

$$\delta = 4.1\left(\frac{\gamma}{V^{1/3}}\right)^{0.43} \tag{16}$$

Furthermore, computational results are obtained using the software program Synthia from Molecular Simulations, Inc., San Diego, CA. Based on the research of Bicerano (1993), Synthia property/structure calculations used topological/connectivitiy index methodologies essentially to extend traditional group additive methods into a formalism such that one does not need to compute sums of group additive contributions to plug into correlations to then calculate properties. The nice thing is that this eliminates the need to have a database of "QSPR" group contributions for quantities such as cohesive energy, Tg, etc. As shown in Table 6, the surface energy from the JKR method is compared to the estimation from the group contribution method using three different sources – Mumford and Phillips (MP) (1929), Fedors (1974) and Van Krevelen (VK) (1990). The agreement between calculations and experiment is fairly good.

TABLE 6. Comparison between surface energy from the JKR method and Synthia property/structure calculations.

Polymer	Surface Energy Exp (mJ/m^2)	Surface Energy Cal-MP (mJ/m^2)	Surface Energy Cal-Fedors (mJ/m^2)	Surface Energy Cal-VK (mJ/m^2)
PDMS	21.4 ± 1.2	21.62	26.65	25.79
PVCH	29.9 ± 1	40.02	36.82	32.11
PtBS	33 ± 1	36.99	36.07	34.52
PS	38 ± 1	45.95	41	39.48
PPPS	41.7 ± 1	48.05	42	40.48
PVBC	43.3 ± 1.9	48.3	42.57	40.12
PMMA	53 ± 4	37.26	41.13	34.53
PE	33	35.1	34.15	32.26
PAN	54 ± 1	61.14	62.33	53.53

From Table 5, it seems that both surface energy and interfacial energy increase with solubility parameter. Mean field theory predicts this proportionality for polymer melts. It would be interesting to see if it also works for solid polymers.

4.2. CORRELATION BETWEEN INTERFACIAL ENERGY AND SOLUBILITY PARAMETER

As shown in Fig. 14, the interfacial energy between PDMS and a series of glassy polymers increases linearly with the solubility parameter difference of the polymers in contact. This correlation could be useful in the design of polymer pairs with desired interfacial energy.

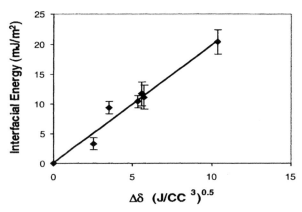

Fig. 14 The interfacial energy between cross-linked PDMS and glassy polymers is correlated with the solubility parameter difference between the polymer pairs in contact using a linear regression fit.

A similar correlation has been predicted from the mean field theory for polymer melts. Helfand and coworkers (Helfand et al., 1972) considered that polymer segments interdiffuse to lower the free energy and form the interface. By integrating the interfacial tension along some path leading from a state of known free energy to the final state with the lowest order density profile, we find that

$$\gamma = \left(\frac{\chi}{6}\right)^{1/2} \rho_0 bkT \tag{17}$$

Since the theory calls for a common value of ρ_0 and b for both polymers, we have (arbitrarily) chosen the geometric mean. A similar magnitude maybe expected for the surface tension of many other polymer pairs, since ρ_0 and b are not greatly variable, and interfacial tension varies only as $\chi^{1/2}$.

$$\chi = \frac{(\delta_1 - \delta_2)^2}{\rho_0 kT} \tag{18}$$

Thus the mean-field theory of polymer interfaces predicts that

$$\gamma_{12} \propto \chi^{0.5} \propto (\delta_1 - \delta_2) \tag{19}$$

The agreement between experimental results and mean field theory is intriguing. Although glassy polymers are not in their melt state, a condition approaching equilibrium might still be available since PDMS is far above its glass transition temperature and therefore the PDMS can reorient to accommodate being in contact with another material. This correlation can be understood by the effect of thermodynamic compatibility on interfacial behavior between polymer pairs. Abere (1963) derived the concept that good bonding between a substrate and an adhesive is attained when their cohesive energy densities [CED] or solubility parameters are matched from thermodynamic considerations. Iyengar and Erickson investigated the mechanism of adhesion of polar and nonpolar polymeric substrates (Iyengar et al., 1967). Thermodynamic compatibility of substrate and adhesive appears to be a key factor in promoting the bondability of PET. This type of compatibility is achieved when the cohesive energy density or solubility parameter of the polymeric substrate is matched with that of the adhesive.

The comparison of interfacial energy obtained from different experimental methods is not possible due to the unavailability of interfacial energy from methods other than the JKR method. Bicerano used the estimation of interfacial energy from surface tension according to Eq. (20) (Bicerano et al., 1993), where γ_{12} denotes the interfacial energy between materials 1 and 2, γ_d is the dispersive force component of γ, while γ_x combines the polar (dipole and induced dipole) and hydrogen bonding components. Reasonable results were obtained, however, the values of the components are not readily available without further experimental work.

$$\gamma_{12} \approx \left(\gamma_{d_1}^{0.5} - \gamma_{d_2}^{0.5}\right)^2 + \left(\gamma_{x_1}^{0.5} - \gamma_{x_2}^{0.5}\right)^2 \tag{20}$$

Mattice (Misra et al., 1995) studied surface energy of poly-(1,4-cis-butadiene) (PBD) using molecular simulation. The atomistic simulation on interfacial energy between polypropylene (PP) and PBD is in the range of 3-10 mJ/m^2 (Natarajan et al., 1998), which is in similar range to our experimental results. Note that the values calculated by Mattice et. al. are relatively high in comparison to the values measured for polymer melts in contact.

5. Conclusions

The JKR method has been used to determine the surface energy and interfacial energy of a series of glassy polymers with different solubility parameters. The reproducibility and in some cases, reversibility, of our measurements suggests that we are approaching the thermodynamic work of adhesion. The solubility parameter is shown to correlate with the surface energy and interfacial energy of solid polymers.

For a homologous series of polymers, the higher is the solubility parameter, the higher is the surface energy. Comparison between contact angle measurements and JKR measurements suggests that the JKR is a more reasonable method for measuring surface energy of solid polymers, because estimation of surface energy from contact angles depends on the theory used to analyze the data. Wu's equation of state method gives a good estimate of surface energy when used to analyze the contact angle data. The

328

Zisman critical surface tension is generally smaller than our measurements of surface energy. The simplicity and ease of contact angle measurements can be well exploited to predict the surface energy of solids only by understanding the solid-liquid interactions more thoroughly. Reasonable agreement is obtained between our results and calculated values based on the group contribution method.

In this study we also demonstrated that solvent casting and spin coating under well controlled conditions is an effective means to prepare smooth and thin polymer films for the JKR type of measurements. Since a variety of polymers can be cast or spun coated from their solutions, this technique makes it possible to study the interactions between a variety of polymers using the SFA which has otherwise been used predominantly to study the forces between model surfaces.

More importantly, the JKR method is the only means of measuring the interfacial energy between solid polymers. A linear correlation was found between the measured interfacial energy between various glassy polymers and PDMS and the solubility parameter difference for that pair of polymers. This finding also suggests that our results are in accordance with the mean field theory.

Acknowledgments: We gratefully acknowledge the financial support provided by 3M Company and the NSF-MRSEC at University of Minnesota. We thank J. Xu for donating the PVCH polymer, G. L. Korba for XPS analysis, R. B. Ross for providing Synthia calculations. We thank Mark Gehlsen, Mark Schultz and Frank Bates for providing PS, PVP and PVCH, and Philippe Guégan and Chris Macosko for providing PMMA used in this study.

References

Abere, J. F. (1963), *Conference of Polymers as Engineering Material,* Newark: New Jersey.

Andreeva, O. A. and Burkova, L. A. (2000), "Spectroscopic Manifestations of Nitrile Group Intermolecular Attraction in PAN", *J. Macro. Sci.-Phys.* **B39** (2), 225–234.

Bicerano, J. (1993), *Prediction of Polymer Properties,* Marcel Dekker, Inc.: New York.

Burrell, H. (1955), "Solubility Parameters", *Interchem, Rev.,* **14** (1), 3-&.

Chaudhury, M. K. and Whitesides, G. M. (1992), "Correlation between Surface Free-Energy and Surface Constitution", *Science,* **255** (5049), 1230–1232.

Fedors, R. F. (1974), "Method for Estimating both Solubility Parameters and Molar Volumes of Liquids", *Polym. Eng. Sci.,* **2** (14), 147–154.

Fox, H. W. and Zisman, W. A. (1952), "The Spreading of Liquids on Low-Energy Surfaces, .2. Modified Tetrafluoroethylene Polymers", *J. Colloid Interface Sci.,* **7** (2), 109-121.

Gehlsen, M. and Bates, F. S. (1993), "Heterogeneous Catalytic – Hydrogenation of Poly (Styrene) – Thermodynamics of Poly (Vinylcyclohexane) containing Diblock Copolymers", *Macromolecules,* **26** (16), 4122-4127.

Ghatak, A., Vorvolakos, K., She, H., Malotky, D. L. and Chaudhury, M. K. (2000), "Interfacial Rate Processes in Adhesion and Friction", *J. Phys. Chem. B,* **104** (17), 4018–4030.

Good, R. J., and Elbing, E. (1970), "Generalization of Theory for Estimation of Interfacial Energies", *Ind. Eng. Chem.,* **62** (3), 54–78.

Good, R. J., and Girifalco, L. A. (1960), "A Theory for Estimation on Surface and Interfacial Energies .3. Estimation of Surface Energies of Solids from contact Angle Data", *J. Phys. Chem.,* **64** (5), 561–565.

Helfand, E. and Tagami, Y. (1972), "Theory of Interfaces between Immiscible Polymers 2", *J. Chem. Phys.,* **56** (7), 3592–3601.

Hertz, H. (1896), *Miscellaneous Papers* (Jones and Schott, eds.), Macmillan, London.

Israelachvili, J. N. (1991), *Intermolecular and Surface Forces,* 2nd ed., Academic Press: London, Chap. 15, p. 313.

Israelachvili, J. N. and Adams, G. E. (1978), "Measurement of Forces between 2 MICA Surfaces in Aqueous – Electrolyte Solutions in Range 0-100 NM", *J. Chem. Soc.,* Faraday Trans. I, **74**, 975-&.

Israelachvili, J. N. and Tabor, D. (1972), "Measurement of Van der Waals Dispersion Forces in Range 1.5 to 130 NM", *Proc. R. Soc. Lond. A.*, **331** (1584), 19-&.

Iyengar, Y. and Erickson, D. E. (1967), "Role of Adhesive – Substrate Compatibility in Adhesion", *J. Appl. Poly. Sci.*, **11** (11), 2311–2324.

Johnson, K. L. (1985), *Contact Mechanics*, Cambridge University Press, Cambridge.

Johnson, Jr. R. E., Dettre, R. H. and Tucker, W. B. (1964), "Contact Angle Hysteresis .3. Study of an Idealized Heterogeneous Surface", *J. Phys. Chem.*, **68** (7), 1744–1750.

Johnson, K. L., Kendall, K. and Roberts, A. D. (1971), "Surface Energy and the contact of Elastic Solids", *Proc. R. Soc. Lond. A.*, **324** (1558), 301–313.

Li, L., Tirrell, M. and Pocius, A. V. (2000), *J. Adhesion*, to appear.

Mangipudi, V.S. (1995), "Intrinsic Adhesion between Polymer Films: Measurement of Surface and Interfacial Energies", *Ph. D. Thesis*, University of Minnesota.

Mangipudi, V. S., Huang, E., Tirrell, M. and Pocius, A. V. (1996), "Measurement of Interfacial Adhesion between Glassy Polymers using the JKR Method", *Macromol. Symp.*, **102**, 131–143.

Mangipudi, V. S., Tirrell, M. and Pocius, A. V. (1994), "Direct Measurement of Molecular-Level Adhesion between Poly(Ethylene-Terephthalate) and Polyethylene Films – Determination of Surface and Interfacial Energies", *J. Adhesion Sci. Technol.*, **8** (11), 1251–1270.

Merrill, W. W., Pocius, A. V., Thakkar, B. V. and Tirrell, M. (1991), "Direct Measurement of Molecular-Level Adhesion Forces between Biaxially Oriented Solid Polymer – Films", *Langmuir*, **7** (9), 1975-1980.

Misra, S., Fleming, P. D. and Mattice (1995), "Structure and Energy of Thin Films of Poly-(1,4-cis-butadiene): A New Atomistic Approach", *J. Computer-Aided Materials Design*, **2**, 101–112.

Mumford, S. A. and Phillips, J. W. C. (1929), "The Evaluation and Interpretation of Parachors", *J. Chem. Soc.*, **130**, 2112–2133.

Natarajan, U., Misra, S. and Mattice, W. L. (1998), "Atomistic Simulation of a Polymer – Polymer Interface: Interfacial Energy and work of Adhesion", *Computational and Theoretical Polymer Science*, **8** (3-4), 323–329.

Tirrell, M. (1996), "Measurement of Interfacial Energy at Solid Polymer Surfaces", *Langmuir*, **12** (19), 4548–4551.

Van Krevelen, D. W. (1990), *Properties of Polymers*, 3rd ed., Elsevier: Amsterdam, Chap. 8, 227-&.

Wu, S. (1979), "Surface-Tension of Solids – Equation of State Analysis", *J. Colloid Interface Sci.*, **71** (3), 605–609.

Wu, S. (1982), *Polymer Interface and Adhesion*, Dekker: New York, Chap. 3, 67-&.

Zisman, W. A. (1975), "Recent Advances in Wetting and Adhesion", *Adhesion Science and Technology* (L. H. Lee, ed.), Plenum Press: New York, **9A**, 5-&.

A MODEL FOR ADHESIVE FORCES IN MINIATURE SYSTEMS

ANDREAS A. POLYCARPOU and ALLISON SUH
Department of Mechanical and Industrial Engineering
University of Illinois at Urbana-Champaign, Urbana, IL 61801, USA

Abstract

Miniature devices including MEMS and the head disk interface (HDI) in magnetic storage often include very smooth surfaces, typically having root-mean-square roughness, R_q of the order of 10 nm or less. When such smooth surfaces contact, or come into proximity of each other, either in dry or wet environments, then strong intermolecular (adhesive) forces may arise. Such strong intermolecular forces may result in unacceptable and possibly catastrophic adhesion, stiction, friction and wear. In the present work, an adhesion model termed sub-boundary lubrication (SBL) model is used to calculate the adhesion forces at typical MEMS interfaces. The model uses the Lennard-Jones attractive potential to characterize the intermolecular forces, and a statistical surface roughness model. The normal separation between the MEMS surfaces that are investigated is of the order of 100 nm down to fully contacting interfaces. Several levels of surface roughness are investigated. The first case is composed of rougher interfaces with combined $R_q = 15.8$ nm. The second case is an intermediate rough case with $R_q = 6.8$ nm, and the last case is a super smooth interface with combined $R_q = 1.4$ nm. The latter is a typical roughness of polished polysilicon films, whereas the rougher interfaces are typical of as deposited polysilicon films. The model reveals the significance of the surface roughness on the adhesion forces as the surfaces become smoother, and suggests that a link between macrotribology and micro/nanotribology is adhesion.

1. Introduction

Advances in miniature systems require reduced wear, increased durability and finer tolerances (reduced separation between surfaces). In order to achieve these stringent requirements, to some extend, smoother surfaces, small normal loads (smaller masses), and molecularly thin lubricants are used. However, such interfaces are susceptible to strong attractive intermolecular forces causing severe adhesion and stiction problems, see for example Tian and Matsudaira (1993) and Komvopoulos (1996) for such problems with the HDI, and MEMS, respectively. In MEMS, high adhesion and stiction can either arise during manufacturing or during shipment and operation of the devices. During manufacturing wet etching and subsequent post-etching are involved. The use of liquid etchants can cause large capillary forces (meniscus) and the suspended masses can get stuck. During shipment and operation the devises may be exposed to vibration and shock causing the flexible masses to contact and again get stuck. An example for this case is the suspended accelerometer mass hitting a limit stop. The main explanation of the high adhesion and stiction in these cases is the formation of meniscus at the interface. The meniscus models share the common assumption that stiction results from a significant increase in adhesion due to the capillary forces of the adsorbed film at the interface and are valid for static conditions only. Meniscus models are useful when the lubricant at the interface is abundant, relatively thick and mobile. However, when the lubricant thickness is extremely small, of the order of few

B. Bhushan (ed.),
Fundamentals of Tribology and Bridging the Gap between the Macro- and Micro/Nanoscales, 331–338.
© 2001 *Kluwer Academic Publishers.*

332

monolayers that strongly adhere to the solids (immobile), the meniscus model breaks down (Israelachvili, 1991). For these cases, an alternative to the meniscus model has been suggested by Stanley et al. (1990). This alternative model, termed sub-boundary lubrication (SBL), is more likely for strong lubricant-solid bond and extremely thin lubricant thickness, and was used by Polycarpou and Etsion (1997, 1998) to develop the SBL static friction model that accurately predicted experimental stiction measurements on magnetic thin film disks. This is a fundamental model derived from first principles and can be used to explain many of the stiction phenomena in miniature systems. It is important to note that the Lennard-Jones potential is valid for both static and dynamic (sliding) conditions, whereas the capillary meniscus based models are strictly for static conditions-formation of meniscus bridges.

Some MEMS devices and actuators involve high-speed motion, e.g., the Sandia microengine (Tanner et al., 1998) with tight tolerances. Under high speed sliding, even though stiction may not be an issue, (excluding startup), high adhesive forces may be present. This case has been investigated for the HDI for ultra low flying heads (<10 nm) by Polycarpou (2000). It was found that with existing HDI interfaces having $R_q \approx$ 1 nm, and lube thickness of 1-2 nm, there are no significant intermolecular forces for fly heights (separations) greater than 10 nm. However, as the fly height decreases to 5-6 nm the intermolecular forces become significant, but not the contact forces.

Figure 1 shows a schematic of a typical MEMS interface and the relevant forces. Other attractive forces, Fo, include electrostatic, hydrodynamic and other forces.

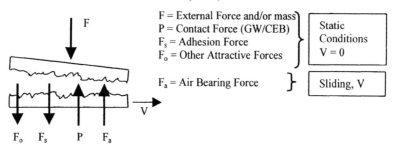

Figure 1. Schematic of the relevant forces at a typical MEMS interface.

During static contact, the air bearing force $F_a = 0$ and the external force, $F = P - F_s - F_o$. If $P > (F_s + F_o)$ then stiction exists in the usual sense. If $P < (F_s + F_o)$, then F is negative and it is the force that is required to separate the two surfaces, called pulled off force (related to the coefficient of adhesion). Polycarpou and Etsion (1997, 1998) showed that the adhesion force, Fs becomes significant and may dominate the forces at the interface for smooth surfaces and thicker lubricants at the surfaces (including humidity). During sliding (contact or no contact) all forces may be present and $F = P + F_s - F_s - F_o$ If there is no contact then $P = 0$ and when the adhesion forces are large, i.e., $(F_s + F_o) > F_a$, then the interface may collapse. For macrotribology, the intermolecular forces, F_s and F_o are typically small and can be ignored. However, this is not the case for miniature systems, and perhaps adhesion can be viewed as a link between macrotribology and micro/nanotribology.

2. Contact Asperity Model

A key element of the adhesion model is the surface topography of the interface. When two nominally flat surfaces touch, contact occurs at discrete contact spots due to surface roughness of both surfaces. Deformation occurs in the contacting region, and can be elastic, plastic or elastic-plastic, depending on the nominal pressure, surface roughness and material properties. In this work the Greenwood and Williamson (1966) statistical model (GW model) is used. Many researchers have proposed statistical models of surface roughness. They all share the common methodology of: (a) replacing the two rough surfaces by a smooth surface and an equivalent rough surface, (b) replacing asperities with simple geometrical shapes, and (c) assume a probability distribution for asperity parameters. The basic GW model assumes that each asperity summit has a spherical shape whose height above a reference plane has a normal (Gaussian) probability density function. It also assumes that the summits have a constant radius of curvature. The geometry of this model is shown in Fig. 2, where d is the separation of the surfaces, measured from the mean of the asperity heights; h is the separation of the surfaces measured from the mean of the surface heights; y_s is the distance between the means of asperity and surface heights; t is the lubricant layer thickness.

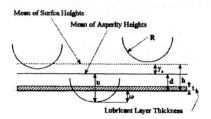

Figure 2. GW contacting rough surface model.

3. Contact Load

The contacting surfaces are modeled using the Chung et al. (1987) elastic-plastic model (CEB model), which is an extension of the GW. The deformation that occurs in the contacting regions can be elastic, plastic or elastic-plastic, depending on the nominal pressure, surface roughness and material properties. The dimensionless contact load $P^* = P/(AnE)$ is also given in Polycarpou and Etsion (1997) and consists of two integrals, the first for the contributions of the elastically and the second for the plastically deformed asperities. The parameters that enter into the calculation of P* are the roughness parameters of the standard deviation of surface heights, $\sigma = Rq$, the radius of curvature of asperity heights, R, and the areal density of asperities; η, the distribution of the asperity heights, $\varphi(u)$ which is assumed to be Gaussian, and the material properties of the hardness of the softer material, H; and the equivalent elastic modulus E^* (for more details see Polycarpou and Etsion, 1987).

4. Adhesion Force

Two basic adhesion models for the contact between an elastic sphere and a flat-as the assumption in the GW model- have been proposed. The first model due to Johnson et al. (JKR model) is suitable for soft materials with low surface energy, e.g., rubber. The second model suggested by Derjaguin et al. (DMT model), is more suitable for harder materials with high surface energy like metals. Both JKR and DMT models are based on the Lennard-Jones interactive potential and account for the strong intermolecular forces that arise as the two surfaces approach each other. These forces include the van der Waals attractive forces as well as short-range repulsive forces. Polycarpou and Etsion (1997) combined the DMT adhesion model with the GW asperity contact model to calculate the adhesion force in the static friction coefficient models for subboundary lubricated contacts. Such models work quite well for interfaces in the absence of thick lubricants and other attractive forces. There are two parameters in the Lennard-Jones potential. The first one is the inter-atomic spacing ε, which is in the range of 0.3-0.5 nm. The second parameter is the energy of adhesion $\Delta\gamma = 2\gamma$, where γ is the surface tension of the liquid and is measured experimentally. In some MEMS applications, in addition to the adhesion forces discussed above, other intermolecular forces may be present, such as electrostatic, and capillary forces, which are not considered in the current model.

The total adhesion force between the contacting surfaces in the presence of sub-boundary lubrication can be separated into three components. These components are from the contributions of the completely non-contacting asperities, the solid non-contacting-lubricant contacting asperities, and the elastic-plastic solid contact. The total dimensionless adhesion force $F_s^* = F_s / (AnE)$ consists of three integrals representing the aforementioned components of the adhesion force. In addition to the roughness and material parameters described above, the lubricant thickness and $\Delta\gamma$ also enter the calculation of the adhesion force (for more details see Polycarpou and Etsion, 1997).

5. Results and Discussion

5.1. SUERFACE CHARACTERIZATION

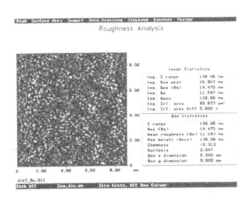

Figure 3. A typical AFM image for deposited polysilicon film.

Typical surface roughness measurements that are reported in the literature, give only amplitude descriptors, which are insufficient for modeling. Several roughness measurements were performed on typical MEMS samples (cantilever beams) using the AFM. The rougher samples are typical of as deposited polysilicon films with $\sigma = 14.3$ nm, and one such measurement is shown in Fig. 3. The second sample is a typical substrate where the deposited polysilicon structures are anchored, with an intermediate roughness of $R_q = 6.7$ nm. Finally, the smoother sample corresponds to a polished polysilicon film with $R_q = 1$ nm. The equivalent isotropic roughness parameters were then extracted, and are reported in Table 1. β is a roughness parameter ($\beta = \sigma R \eta$). All three type of surfaces considered are mostly

Table 1: Surface roughness parameters

Individual Surfaces	1. Deposited Polysilicon	2. Typical Substrate	3. Polished Polysilicon
σ (nm):	14.3	6.7	1.0
R (μm)	0.12	0.46	2.4
η (μm^{-2})	15	11	17
β	0.026	0.034	0.041
Surface Pairs	A. Rough: 1 on 2	B.Intermediate: 2 on 3	C. Smooth: 3 on 3
σ (nm):	15.8	6.8	1.4
R (μm)	0.116	0.45	1.7
η (μm^{-2})	14.7	11.1	17
β	0.027	0.034	0.041

isotropic with anisotropy indices $\gamma > 0.9$ ($\gamma = 1$ corresponds to a fully isotropic surface). The AFM samples were approximately 10 μm x 10 μm and care was taken to minimize scale effects that are inherent in such measurements.

Three simulations were performed by combining the surface parameters for the three samples in Table 1 as follows: Simulation A is a rough interface between surfaces 1 and 2 (as-deposited polysilicon on substrate), simulation B is an intermediate rough interface between substrate and the super smooth polished polysilicon surface, and the third case is the smoothest interface between 2 polished polysilicon surfaces. Table 1 also shows the equivalent rough surfaces in contact with an infinitely smooth surface for these 3 cases.

The interfacial combined Young's modulus, and hardness of these polysilicon materials are taken to be 95 GPa and 2 GPa, respectively (Bhushan, 1999). As far as the thickness of the lubricant is concerned, it is assumed that no additional lubricant is present at the interface other than humidity. For the simulations a small trace of water vapor (monolayer) is assumed to be present at the surface with a thickness of 0.1 nm. Also in the case of the smoothest interface the effect of water vapor thickness (corresponding to different levels of relative humidity) is investigated by assuming thickness ranging from 0.1 - 5 nm. The surface free energy for the water is = 140 mN/m, and $\varepsilon = 0.4$ nm.

5.2. SIMULATIONS

Fig. 4 shows the dimensionless contact load $P^*=P/AnE$ and the dimensionless adhesion force, $F_s^*=F_s/AnE$ versus the dimensionless separation, h/σ for the roughest interface (case A). Also shown in the figure is a vertical line indicating 3 times the value of the standard deviation of surface heights (3*Rq) and the contributions to the adhesive force from the non-contacting asperities, the lube contacting asperities and the elastic/plastic contacting asperities. The largest contribution is from the non contacting asperities. This is expected since the lube thickness is very small compared to the roughness. For separations $h^* > 4.74$ (or $h > 75$ nm), F_s^* is larger than the contact load but its value is small, $Fs^* \approx 10^{-9}$. For lower separations the contact load dominates and its value is about an order of magnitude larger than the adhesion force, clearly showing that adhesion is insignificant for this particularly rough interface. Furthermore, the contact force is significant over a large range of h^*. For example, at $h^*=3$, $P^*=2x10^{-5}$, indicating that if such surfaces contact, severe contact will occur that may lead to wear, even though the adhesion force will not be significant. Similar behavior is observed with the intermediate rough interface (case B, $\sigma =6.8$ nm) with insignificant adhesion

336

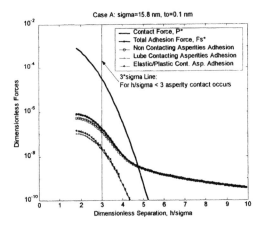

Figure 4. Case A: Roughest Interface: Dimensionless forces versus dimensionless separation.

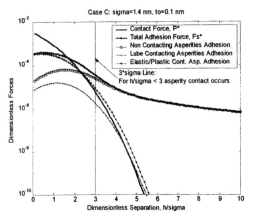

Figure 5. Case C: Smoothest Interface, Dimensionless forces versus dimensionless separation.

forces over a large range of separations. For $h^*>4.45$, $Fs^* \approx 10^{-8}$, which is about an order of magnitude larger than the roughest case A. For smaller separations, the adhesion force is less than the contact force, but their difference is less than in case A.

Different behavior is seen with the smoothest interface as shown in Fig. 5. In this case adhesion forces are very significant and may dominate the forces at the interface, and in some cases may even cause the collapse of the interface. For the practical range of interest of the dimensionless separation, i.e., $h^* > 3$ ($h>4.2$ nm) the adhesion force is always larger than the contact load. Only at very small separations, $h/\sigma <1.7$, $P^*>F_s^*$ which is under severe contact conditions and not of interest in this work. For larger separations say $h^*>6$ ($h>8.4$ nm) adhesion is lower than at lower separations, but still about 2 orders of magnitude larger that the intermediate rough case, showing the significant presence of adhesion forces.

To get a feeling for the differences in the adhesion forces for the cases described above, all 3 cases are plotted together in Fig. 6. Note that in this case the adhesion force per unit area is plotted versus separation. The onset of contact (3*Rq) occurs at 47.4 nm, 20.4 nm, and 4.2 nm with corresponding adhesion forces per unit area of 9×10^{-3}, 6×10^{-2}, and 1.9 $\mu N/\mu m^2$, for cases A, B, and C, respectively. Clearly the rougher interfaces can survive a contact condition but not the smoother interface, where adhesion forces dominate and may cause the interface to interlock.

The effect of humidity is studied for the smoother case, depicted in Fig. 7. As expected, increasing the water vapor thickness, increases the adhesion forces. For the larger lube thickness, the validity of the SBL adhesion model may be in question since the

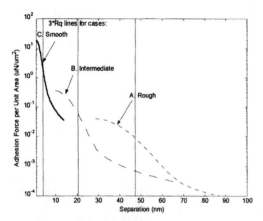

Figure 6. Adhesion Force Per Unit Area Versus Separation, cases 1, 2, 3.

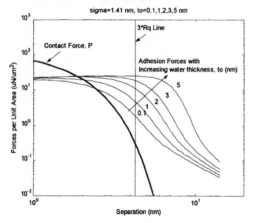

Figure 7. Effect of Humidity on adhesion, smooth interface, (A).

abundance of lubricant will probably form meniscus as well. Nevertheless, the model clearly shows what it has been observed in the literature, namely, increased adhesion with increased humidity. In addition, the model also explains that reducing humidity will decrease adhesion but not eliminate it.

Finally, the simulation results can also be used to study high speed MEMS and the effect of roughness on tolerances, and contact conditions. Fig. 8 shows the fly-height or clearance or stand-off distance (separation) versus forces in units of mN, for a nominal contact area of 100 μm x100 μm for cases B and C. As expected for the rougher interface (case B) at a clearance of 20 nm the contact force is 16 mN and the adhesion force is only 0.65 mN, thus the designer needs to be concerned more about contact and wear and not adhesion. In contrast for the smoother interface (case C), clearances can be much smaller before contact occurs.

More specifically at a clearance of 5 nm, the contact force is very small P=0.45 mN, but the adhesion force now becomes significant, Fs=8 mN.

6. Conclusions

A fundamental adhesion model based on the SBL model has been described and used to calculate the adhesion forces at typical MEMS interfaces. The model is valid for both static and dynamic contacts. The normal separation investigated is of the order of 100 nm down to contacting interfaces. Three different levels of surface roughness were investigated. The first two cases are rougher interfaces with combined standard deviation of surface heights, σ=15.8 nm, and 6.8 nm. In these cases, the contact force is larger than the adhesive force and most of the separations contact will first occur at

338

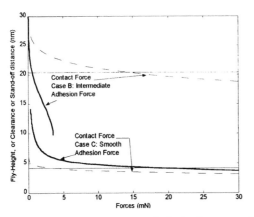

Figure 8. Clearance versus contact and adhesion forces, cases B, C.

relatively large separations. The third case is a super smooth interface with a combined $\sigma=1.4$ nm. In this case at separations of 10 nm or greater, there are no significant adhesive forces. For separations between 10 nm to 5 nm, adhesive forces become significant, and no significant asperity contact occurs. In this range, the possibility for collapsing the interface is very high. In summary, the adhesion model predicts that significant intermolecular forces exist at low separations/clearances of very smooth MEMS interfaces. For the rougher interfaces at relatively large clearances the contact force is also large but the adhesion force relatively small, and from a design point of view, one needs to be concerned more about contact and wear rather than adhesion. In contrast, for smoother interfaces, clearances can be much smaller before contact occurs, but the adhesion force becomes very significant. It is suggested that adhesion can be viewed as a link between macrotribology and micro/nanotribology.

7. References

Bhushan, B. (1999), *Handbook of Micro/Nanotribology,* second edition, CRC Press, 822. Chang, W. R., Etsion, I. and Bogy, D. B. (1987), "An Elastic-Plastic Model for the Contact of Rough Surfaces," *ASME J. of Trib.*,**109**, 257-263.

Greenwood, J. A. and Williamson, J. P. B. (1966), "Contact of Nominally Flat Surfaces," *Proc. of the Roy. Soc. of London,* **A295**, 300-319.

Israelachvili, J. N. (1991), *Intermolecular and Surface Forces,* Academic Press, second edition, San Diego.

Komvopoulos, K. (1996), "Surface Engineering and Microtribology for Microelectromechanical Systems," *Wear,* **200 (1-2)**, 305-327.

Polycarpou, A.A. (2000), "Modeling Adhesive Forces for Ultra Low Flying Head Disk Interfaces," in *Nanotribology: Critical Assessment and Research Needs,* S.M. Hsu, ed., Kluwer Academic Pub., in press.

Polycarpou, A. A. and Etsion, I. (1997), "Static Friction of Contacting Real Surfaces in the Presence of Sub-Boundary Lubrication," *ASME J. of Trib.,* **120**, 296-303.

Polycarpou, A.A., Etsion, I. (1998), "Comparison of the Static Friction Sub-Boundary Lubrication Model with Experimental Measurements on Thin Film Disks," *STLE Tribology Transactions,* **41(2)**, 217-224.

Stanley, H. M., Etsion, I. and Bogy, D. B. (1990), "Adhesion of Contacting Rough Surfaces in the Presence of Sub-Boundary Lubrication," *ASME J. of Trib.,* **112**, 98-104.

Tanner, D.M., Miller, W.M., Eaton, W.P., Irwin, L.W.; Peterson, K.A., Dugger, M.T., Senft, D.C., Smith, N.F., Tangyunyong, P., Miller, S.L. (1998), "The Effect of Frequency on the Lifetime of a Surface Micromachined Microengine Driving a Load," IEEE *36th Annual Reliability Physics Symposium Proceedings*, 26 –35.

Tian, H., and Matsudaira, T., (1993), "The Role of Relative Humidity, Surface Roughness and Liquid Build-up on Static Friction Behavior of the Head/Disk Interface," *ASME J. of Tribology,* **115**, 28-35.

SIMPLE MODEL FOR LOW FRICTION SYSTEMS

M. D'ACUNTO

Dipartimento di Ingegneria Meccanica e Nucleare.
Università di Pisa. Via Diotisalvi 2, I-56126 Pisa, Italy

Abstract. The paper presents a general model for low friction systems. The model describes a self-excited oscillator sliding along a periodic potential and it is subject to a linearly time force. The results show: first, low energy dissipation during the sliding process, and second, an intermittent behaviour between frictionless and non-zero friction sliding states.

1. Introduction

When two unlubricated solid bodies contact each other and one body subsequently slides against the other, a frictional force occurs. Several models have been proposed to explain the origin of this frictional force in macroscopic or in microscopic systems. Nevertheless, only the advent of Scanning Probe Microscopy has made possible the measure of frictional force at the microscopic scale, with high degree of precision (Bhushan, 1999).

At the microscopic scale, it is possible to consider conditions in which the energy dissipation is reduced until a nearly frictionless motion condition is achieved (Hirano and Shinjo,1990; McClelland and Glosli, 1992; Singer, 1994; Sokoloff, 1995). Starting from the pioneering model for wearless friction of Tomlinson (1929), more recent studies modeled the occurrence of friction as a consequence of the existence of unstable equilibrium positions during the translations of one body over another. Moving through these equilibrium positions leads to sudden displacement, as a consequence the sliding body vibrates and this vibration excites the bulk in an irreversible manner causing loss in energy.

The surface of the sliding body can be modeled as a chain of particles coupled in elastic way, the so called Frenkel-Kontorova (FK) model. Then a molecule of the upper body can be pinned at the lower body due to the roughness. It depins when the surface is moved further, after the depinning the molecule vibrates. If the molecule is coupled to an environment, then the vibration is damped because the vibrating molecule excites electronic or elastic waves into the bulks of the sliding bodies. Sokoloff (1995) has studied the interaction of a translating sinusoidal force with a layered crystalline lattice, modeled by a collection of springs and masses. In this investigation, the author

B. Bhushan (ed.),
Fundamentals of Tribology and Bridging the Gap between the Macro- and Micro/Nanoscales, 339–344.
© 2001 *Kluwer Academic Publishers.*

emphasized the importance of internal damping on the generation of friction force for any finite-sized crystal. The author has also compared frictional force between commensurate and incommensurate sliding surfaces and found the former values to be 12 orders of magnitude greater than the latter. An analogous mechanism describes the reduction of energy absorption by phonons and spin waves in a disordered solid, (Sokoloff 2000).

Similar effects, as found for the sliding chain model, should be observed in the Independent Oscillator (IO) model, Streator (1994). In this model, the atoms of one surface are modeled by masses which are independent of one another but are harmonically coupled to a rigid base.

In this paper, it is supposed that if a sliding body has a mechanism for which the loss energy can be reintroduced, the so called *self-excited oscillator*, then the energy dissipation would be much lower.

Dissipative process are described phenomenologically in classical equations of motions by a friction term. This term is usually assumed to be proportional to velocity. In a self-excited oscillator the friction term is dependent either to velocity and to position. A self-excited oscillator is a well known system which has some external source of energy upon which it can draw. Many example of such self-exciting oscillators in electrical, mechanical and biological systems can be found in literature (Jackson, 1989; Cartwrigth et al., 1997). The van der Pol oscillator is, probably, the most simple example of such system. As one of its most prominent features, stable limit cycles in phase space are established, emerging from a balance between energy gain and dissipation.

2. The model

A single self-excited oscillator tangential to a surface can be built in different ways, (Nayfeh and Mook, 1979; Chatterjee and Mallik, 1996).

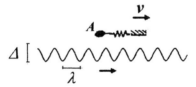

Figure 1. Schematic sketch of a self-excited oscillator A sliding on a periodic potential. The body of interest A is pulled by a linear spring with a force constant connected to a stage that moves with velocity v. The other body is described by a periodic potential and it moves with lower velocity respect to the stage. Δ is the strength of periodic potential, λ is the period of the potential. All parameters and variables are given in dimensionless units.

The oscillator A is elastically coupled to a large mass body travelling on a straight line with constant velocity v and moves along the direction of the force. A rigid body supports the whole system. The surface of the body below is rough, which is described by a periodic potential. The equation of motion for the self-excited oscillator for unity of mass can be described as follows:

$$\ddot{x} - \mu\left(1 - \beta x^2\right)\dot{x} + \Delta sin\left(2\pi x / \lambda\right) + \Omega^2 (x - vt) = 0 \qquad (1)$$

where μ is damping coefficient, Δ is the strength of periodic potential, Ω is the natural frequency of the oscillator, λ is the period of the potential, v is the external driving velocity. The coefficient $0 \leq \beta \leq 1$ is introduced in the equation of motion (1) in order to quantify the self-excitability of the system.

The potential $U(x)=1/2\Omega^2 x^2 - \lambda \Delta cos(2\pi x/\lambda)/2\pi$ presents an absolute minimum point in $x=0$, and a series of periodic minima. When $v=0$, the oscillator reaches to limit cycle (steady state) which are weakly or strongly deformed as a function of the strength of potential Δ. During the motion the dynamic friction force may either introduce energy into the system, if it has the same sign as the velocity of the block, or dissipate energy, if the signs of the friction force and the velocity of the block are different. Following the paper of Helman et al.(1994), the friction force is defined as the energy exchanged per traveled length.

During the motion, the energy gain and loss at the time τ can be expressed by the relation:

$$\Gamma = \mu \int_0^\tau \left(1 - \beta x^2 \right) \dot{x}^2 dt \qquad (2)$$

The expression (2) can assume negative values, it means that the system slides and absorbs energy from the external source. This is what happen for large values of the position. If $v=0$, $\Delta=0$, and $\beta=1$, equation (1) is reduced to a standard van der Pol oscillator and shows limit cycles. If $\Delta \neq 0$ and $v=0$, the body described equation (1) can drop between local minima. When the system is in a minimum the frictional coefficient gives an additional contribution to stabilize the motion. In fact, in the autonomous case, when the driving term is absent, the condition of steady-state is restored also for $\Delta \gg \mu$. If the non-autonomous case is bring into focus, i.e., $v \neq 0$, then the time needed to develop a sliding process is strongly dependent on the time necessary to jump between the first subsequent potential wells. Friction force is produced around the transition between potential wells.

3. Numerical Results

The main objective of this paper is to give some meaningful results via a phenomenological approach. The complex dynamics of equation (1) will be analysed in a separate report. Equation (1) has been integrated numerically using a fourth-order Runge-Kutta algorithm with adaptive step size, with parameter values fixed at $\mu=0.1$, and $\Omega=1.0$. The main interest is to study the dynamics of a sliding process, i.e., the transport of the body along a straight line. The efficiency of the transport is measured by the difference $vt-x$, or equivalently by the relation expressed as $\eta=x/vt$, namely the ratio between the travelled distance and the total possible sliding way. Figure 2 demonstrates a sliding process for two different initial conditions. First, the competition between potential wells and driven term produces an increasing instability, the sliding process presents characteristic jumps between potential wells. Initially, the oscillator experiments a *step by step* motion. The oscillation grows in intensity until a transition is performed. Figures 3 and 4 display this behaviour. When the time is increased, the jumps are reduced either in height and in frequency, and the friction force presents a characteristic friction-frictionless intermittent behaviour (Sokoloff, 1995). This intermittence can be explained in the following argumentation: significant energy is absorbed whenever the frequency of the driving force becomes equal to the resonant frequency of the oscillator. This condition is not stable because the resonant oscillator shifts off resonance. This is caused because the resonant frequency depends on the displacements of the oscillator. Consequently, the oscillator does not absorb energy from the driving force, until the resonance condition is restored.

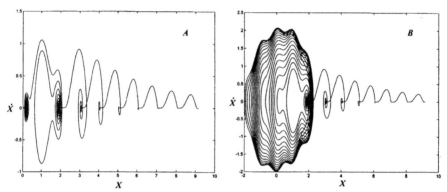

Figure 2. Phase space plots showing the sliding process as described by the equation (1) for two different initial conditions. In figure A the initial condition is (0,0), while in figure B the initial condition is (1.8,0).

In figure 3 we show a time series and the correspondent friction force plots. It is evident that the friction force manifests itself in occasion of the jump between potential wells, as shown in figure 4.

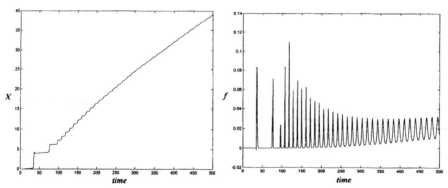

Figure 3 Time series (left) and the correspondent friction force plot (right); Δ=3.0, ν=0.1.

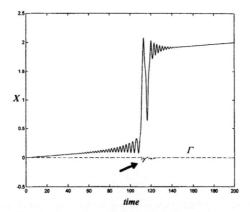

Fig.4 Plot showing the transition between two minima and the correspondent energy dissipation Γ (dashed line). The arrow indicates the non-zero oscillating value for Γ. $\Delta=3.0$, $v=0.01$.

Figure 5 shows that the efficiency of the sliding process seems to be independent by the strength of potential. The variation of the strength of potential does not change the value of η, nearly 80%. This result seems to show that the coupling with the periodic potential does not affect the sliding process, rather, it influence the initial dynamics. Moreover, the time series as a function of the parameter of self-excitability, β, is plotted. It is interesting to note that the better value for the efficiency is obtained for $\beta=0.5$, instead than $\beta=1$. Nevertheless, for both the values the efficiency does not change in a drastic manner. Figure 6, displays a comparison of the energy dissipation between a self-excited oscillator as described in this paper and a non self-excited oscillator, i.e., when $\beta=0$. We have considered the initial time region for which the discrepancy becomes significant. Moreover, the energy dissipation quantity is taken in absolute value for the case $\beta=1$. The result shows that a self-excited oscillator slides in condition of nearly frictionless motion.

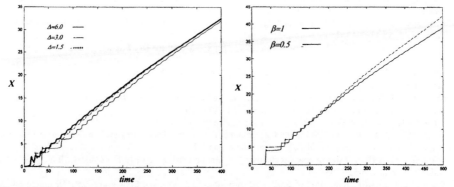

Figure 5 Time series comparison for several values of the strength of potential Δ (left). Time series comparison for two different value of the self-excitability parameter β (right).

344

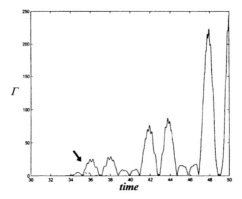

Figure 6 Comparison of the energy dissipation function Γ between β=1 (dashed line) and β=0 (continue line). The dashed line is zero except that in the points indicated by the arrow.

4. Conclusions

This work discusses a simple model of classical mechanics with the purpose of describing reduction of loss energy and friction during a solid-solid sliding process. The model describes a self-excited oscillator sliding along a periodic potential and subject to a linear time force. Figures 4 and 6 represent the main results of the present paper. During the sliding process, the system exchanges energy in correspondence of the transitions between potential wells, as prescribed by Tomlinson model. Moreover, the self-excitability of the system guarantees that the energy is lost and suddenly regained during the transition. Then, figure 6 shows that a self-excited oscillator, if compared to a Coulomb friction term, slides in condition of nearly frictionless motion.

5. References

Bhushan B., (1999) *Handbook of Micro/Nanotribology*, second edition, CRC Press, Boca Raton, FL.

Cartwright J.H.E., Hernandez-Garcia E., and Piro O., (1997), "Burridge-Knopoff Models as Elastic Excitable Media", *Phys. Rev Letters* **79** 527-530.

Chatterjee S., and Mallik A.K. (1996) "Three Kinds of Intermittency in a Nonlinear Mechanical System", *Phys. Rev E*, **53** 4362-4367.

Helman J.S., Baltensperger W., Holist J.A., (1994), "Simple Model for Dry Friction", *Phys Rev. B*, **49** 3831-3838.

Hirano M., and Shinjo K., (1990) "Atomistic Locking and Friction", *Phys. Rev. B* **41** 11837-11851.

Jackson E.A., (1989), *Perspectives of Nonlinear Dynamics*, Cambridge University Press, Cambridge.

McClelland G.M., and Glosli J.N., (1992) in *Fundamental of Friction: Macroscopic and Microscopic Processes*, Singer and Pollock eds., Nato-ASI Series E, Vol.220.

Nayfeh A.H., and Mook D.T., (1979) *Nonlinear Oscillations*. Wiley

Singer I.L., (1994), "Friction and Energy Dissipation at the Atomic Scale-a Review", in *Dissipative Processes in Tribology*, (D. Dowson et al. eds.), Elsevier Science, pp 3-20.

Sokoloff J.B., (1995), "Microscopic Mechanism for Kinetic Friction: Nearly Frictionless Sliding for Small Solids", *Phys. Rev. B* **52** 7205-7214.

Sokoloff J.B., (2000), "Reduction of Energy Absorption by Phonons and Spin Waves in a Disordered Solid Due to Localization", *Phys. Rev. B* **61** 9380-9386.

Streator J.L., (1994), "A Molecularly-Based Model of Sliding Friction" in *Dissipative Process in Tribology*, (D. Dowson et al. Editors), Elsevier Science, pp 173-183.

Tomlinson G.A., (1929) "A Molecular Theory of Friction", *Phil. Mag. Series* **7** 7 905-939.

ULTRA-LOW FRICTION BETWEEN WATER DROPLET AND HYDROPHOBIC SURFACE

K. HIRATSUKA, A. BOHNO and M. KUROSAWA
Dep. Precision Engineering
Chiba Institute of Technology
2-17-1, Tsudanuma, Narashino-shi, Chiba, 275-8588 JAPAN

1. Introduction

Recent advances in micro-machine fabrication have enabled production of machine parts under millimeter in size (Mehregany, 1988). In machine pairs having such small sizes, the surface force is more predominant than the body force. When water is present at the gap between the two parts, the meniscus force attracts two surfaces and increases friction (Bhushan, 1996). In order to avoid stiction between hard disk (media) and head, both the quantity as well as quality of lubricant must be carefully chosen.

The PTFE (polytetrafluoroethylene) surface is known to have an excellent water-repellent property. Especially when the surface has some roughness, the contact angle of water can increase up to 150° (Yamauchi, et al., 1996) and 170° (Shibuichi, et al., 1996). On such a surface, the water droplet can move around as if it is frictionless. In this paper, the authors have demonstrated that the water droplet supports the load and exhibits ultra-low coefficient of friction of the order of 10^{-4}.

2. Experimental Procedure

Fig.1 shows the schematic of test rig. The upper glass plate had 50mm square shape, and the lower had 50mm x 76mm rectangular shape. The thickness of both the plates was 2mm. The upper glass plate moved vertically for loading. The lower one was subjected to horizontal reciprocating motion. Each surface was coated by PTFE powder by spray coating

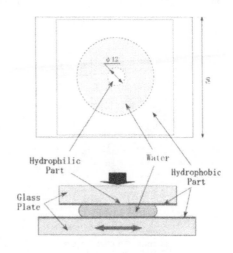

Figure 1. Schematic of test rig

345

B. Bhushan (ed.),
Fundamentals of Tribology and Bridging the Gap between the Macro- and Micro/Nanoscales, 345–348.
© 2001 *Kluwer Academic Publishers.*

method. The spray contains PTFE powder of several micrometers in diameter with butyl-acetate as a solvent. Only the central region (12mm in diameter) of the upper plate was not coated, so that it remained hydrophilic and water droplet can stick on to this surface. These two plates were then brought in to contact with each other.

Water (0.1mL) was applied to the central part of the upper plate. As the upper plate approached the lower one, water spread on to the contacting surfaces of two plates and supported the applied load. The water did not squeeze out from the gap because the hydrophilic area on the upper plate did not release it. The gap between the specimens became smaller for higher load. In the present work, loads up to 100mN were successfully applied without rupturing the water film.

After the application of load, the lower specimen was set in to reciprocating motion. Loads and friction forces were measured by the displacement of the double canti-lever with the help of laser displacement gauges, independently.

3. Results

Fig.2 shows the variation of friction force at a load of 101 mN and sliding velocity of 0.1mm/s in presence of water. The lower glass plate was reciprocated thrice. The average friction force for both directions was 0.083mN and the coefficient of friction obtained was 0.00083.

Fig.3 shows the effect of sliding velocity on the coefficient of friction under a load of 83mN. In this experiment, the sliding velocity was increased stepwise from 0.1mm/s to 0.3mm/s, then decreased to 0.03mm/s.

The following observations can be made from this graph:
1) The coefficient of friction is in the range of 0.0006 to 0.0011.

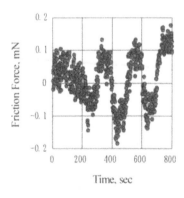

Figure 2. Variation of friction force as a
function of time, Load=101mN,
Sliding Velocity=0.1mm/s

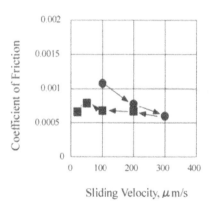

Figure 3. Effect of sliding velocity on the
coefficient of friction, Load=83mN

2) As the sliding velocity increases, the coefficient of friction decreases.

3) The coefficients of friction are lower during the decreasing process of sliding velocity as compared to that of increasing process of sliding velocity.

4) The effect of velocity change is irreversible.

   The coefficient of friction remained almost the same at 58mN load.

   In Fig.4, the effect of load on the coefficient of friction is collated. It is clear that the coefficient of friction is almost independent of the applied load. The coefficient of friction is lower when the sliding velocity is high under all loading conditions.

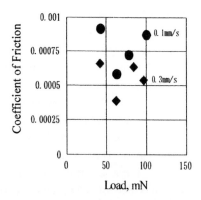

Figure 4. Effect of load on the coefficient of friction

The contact angle of water droplet on the hydrophobic surface is about 130°.

4. Discussion

It is remarkable to note that the coefficient of friction between the interposed water and the hydrophobic surface is less than 10^{-3}. Liquid water plays the role of a lubricant. However in this case it is one element of the friction pair.

   Let us consider the force balance of the water droplet when it is pressed and spread between the two plates. By equating the pressure by the applied load and the normal component of the surface tension, we get the following equation:

$$\frac{W+2\pi R\gamma\sin\theta}{\pi R^2} = \gamma\left(\frac{1}{R}+\frac{1}{r}\right)$$

where γ is the surface tension of water, θ is the contact angle, W is the applied load, R and r are the radii of the droplet. As long as the water film is not broken and squeezed out, it supports the load by the surface tension.

   The origin of the friction force is interesting because the friction pair in this system is liquid vs. solid. The authors propose the following possible mechanisms:

1) Contact angle hysteresis

   When the receding and advancing contact angle is different, the difference in the lateral component of surface tension contributes the friction force. At the rear part of the contact area, the adhesion occurs. This adhesion hysteresis should be the cause of friction force. It is implied that this is another expression of contact angle hysteresis (Israelachvili, 1999). The difference in contact angles is calculated to be very small.

2) Viscosity of water

The inside movement of water film should be the resistance against the sliding action. The viscosity of the liquids, therefore, is another source of the friction force.

In the present system, as the sliding velocity was increased, the coefficient of friction went down. It is deduced that the contribution of viscosity is small.

From Hertz theory, the coefficient of friction is in proportion to $W^{-1/3}$. However the friction coefficient vs. load plot in Fig.4 does follow this relationship. In this system, the specimen shape changes elastically. However, the deformation is not limited to the top region. Therefore, the coefficient of friction need not follow the Hertz theory.

It is also pertinent to discuss here whether this film would "Slide or "Roll". It is likely to be determined by the competitive action of the adhesive force and viscous resistance. If the viscous force is less than the adhesive force, the movement could be rolling.

5. Conclusions

Water film confined between hydrophilic and hydrophobic surfaces can support loads up to about 100mN and exhibits coefficients of friction as low as 0.0005. This phenomenon enables us to use the water as a sliding material when the loads are not too high.

Acknowledgment

The authors are grateful to Dr. K.Nakamura in NTT Advanced Technology Cooperation for supplying the hydrophobic material and helpful discussions.

References

Bhushan, B., (1996) *Tribology and Mechanics of Magnetic Storage Devices*, 2nd ed., Springer-verlag, New York.

Israelachvili, J.N., and Berman, A.D., (1999) Bhushan, B. ed., *Handbook of Micro/Nano Tribology,* 2nd ed. CRC Press, Boca Raton, Florida.

Mehregany, M., Gabriel, K.J., and Trimmer, W.S.N. (1988), "Integrated fabrication of polysilicon mechanisms", *IEEE Trans. Elec. Dev.* 35, 6, 719-723.

Shibuichi, S., Onda, T., Satoh, N., and Tsujii, K., (1996), "Super water-repellent surfaces resulting from fractal structure", *J. Phys. Chem.* 100, 50, 19512-19517.

Yamauchi, G., Miller, J.D., Saito, H., Takai, K., Ueda, T., Takazawa, H., Yamamoto, H., and Nishi, S., (1996), "Wetting characteristics of newly developed water-repellent material", *Colloids and Surfaces A: Physicochem. Eng. Aspects* 116, 125-134.

AFM AS A NEW TOOL IN CHARACTERISATION OF MESOPOROUS CERAMICS AS MATERIALS TO TRIBOLOGICAL APPLICATIONS

I. PIWOŃSKI,
University of Lodz,
Department of Chemical Technology and Environmental Protection
163 Pomorska St., 90-236 Lodz, Poland

J. GROBELNY
University of Lodz,
Department of Solid State Physics
149/153 Pomorska St., 90-236 Lodz, Poland

Abstract

Surface topography and the relationship of friction forces versus load forces between silicon nitride tips and mesoporous silica were investigated. Atomic Force Microscopy (AFM) was used in our investigation. Mesoporous silicas of different porosity were used. The relative friction coefficients were determined for a silicon nitride tip against both dry and lubricated silicas. Relative friction coefficient values for lubricated surfaces are about two times lower than for dry surfaces.

1. Introduction

Use of compliant layers in cushion form bearings for total artificial joints is generating considerable interest. Porous elastic layers under a mixed lubricating regime show significantly lower friction coefficients than do non-porous elastic layers (Caravia at al. 1994). It is known that several microstructural parameters influence the mechanical property-porosity correlation, including pore shape, orientation and distribution. Generally, porosity negatively effects the mechanical strength and stiffness of a ceramic material. Pore-stress concentration as a residual thermal stress, develops from the effects of fabrication temperature on the mechanical property-porosity relationship (Rice, 1993; Boccaccini, 1998). More important is the minimum-solid area. This is particularly true for the tensile fracture strength of ceramics failing from large pores containing a moderate level of porosity (<40 vol%) with no other microstructural defects contributing to the failure's origin (Boccaccini, 1998). Pores in a ceramic surface act as lubricant reservoirs.

Mesoporous silicates and aluminosilicates discovered in Mobil Oil Corporation's laboratories in the early 'nineties attracted the interest of scientists around the world.

B. Bhushan (ed.),
Fundamentals of Tribology and Bridging the Gap between the Macro- and Micro/Nanoscales, 349–354.
© 2001 *Kluwer Academic Publishers.*

High surface area, highly regular pore system and sharp adjustable pore size distribution are characteristic features of these materials. Mesoporous ceramic formation is caused by the interaction between an organic matrix „template," and an inorganic silicate species. This interaction is called the „liquid crystal templating" (LCT) mechanism. Many types of surfactants can be applied as „templates". Different sources of silica also can be used. It seems that between the many available directing agents, the most popular is cationic CTA^+ cetyltrimethylammonium. Other surfactants can also be used during the preparation of porous ceramics. Decreasing the amount of surfactant generally leads to decreases in surface area and pore volume. Increasing wall thickness, on the other hand, makes the sample more resistant to mechanical force. This class of materials can be used to support catalysts, as adsorbates, and as hosts to confine guest molecules. They also can be used as new materials in tribology. The possibility was investigated in this study.

Atomic Force Microscopy (AFM) techniques are increasingly used for tribological studies of engineering surfaces, at scale ranges from atomic through molecular, to micro. The technique has been used to study surface roughness, adhesion, friction, scratching/wear, indentation, material transfer detection and boundary lubrication, and for nanofabrication/nanomachining purposes. Friction directionality resulting from surface preparation is observed at both micro and macro scales. Anisotropy is surface roughness. As there is a lower ploughing contribution at microscale measurements, microscale friction is generally found to be smaller than macroscale friction. Microscale friction is load dependent. The friction value, therefore, increases with an increase in the normal load at contact stresses higher than the hardness of the softer material, up to the macrofriction level. AFM was also used to evaluate wear. Ceramics exhibit both significant plasticity and creep on a nanoscale. It was found the wear to be initiated as nano scratches.

In this paper, both surface topography and friction of porous silica with amounts of pores having an average pore diameter of 2-3nm and more, synthesised using various amounts of surfactant, are studied. The topographic results and the relationship of friction to load force are presented and discussed.

2. Experimental

2.1. PREPARATION OF POROUS SILICA

In the preparation as silica source tetraethoxysilane (TEOS) was used, and non-ionic alkylpoliethylene-glycol ether surfactants. Here $C_{15}H_{31}O(CH_2CH_2O)_9H$ is used as a directing agent with surfactant/ TEOS weight ratios of 0,26 and 0,09 respectively. The final product forms after combining the surfactant and silica source with water in acidic pH, after three to four days at ambient temperature. After synthesis, the organic matrix is removed via extraction in ethanol, using a Soxhlet apparatus.

Another samples were prepared using not only non-ionic surfactants but also polymers such as polypropoxyglycol. CTAB – cetyltrimethylammonium bromide was used successfully as a directing agent pure and in the mixture with swelling agent –

mezytylen, to increase pore diameter. Mesoporous silica was lubricated with
n-hexadacane.

2.2. AFM STUDIES

A home made AFM was used to measure relative coefficients of friction between the
silicon nitride tip, which has pyramidal shape with an average radius of 10 nm, and
silica surfaces in an ambient atmosphere at room temperature. The four-quadrant
photodiode on the conventional AFM allows it to measure the torsional movement of
the cantilever. While scanning the sample perpendicular to the long axis of the
cantilever beam, the cantilever twists from friction between the tip and the surface.
The amount of twist is proportional to the friction force between the sliding tip and the
surface. The AFM can simultaneously image topography and torsional deflection.
Friction forces in arbitrary units were measured on 15x15 nm sample surfaces by sliding
the tip, with the scan disabled, repeatedly across the sample surface in the x direction.
Slopes of the relationship of friction versus load forces, expressed in arbitrary units of
relative friction coefficients, were calculated from the curves. All measurements were
repeated several times at randomly selected location, and show good reproducibility.

3. Results and discussion

Solid surfaces contain different amounts of pores and, irrespective of the method of
formation, contain deviations from the surface's prescribed geometrical form. The area
of contact generally must be minimised to minimise adhesion friction and wear.
Characterising surface roughness is, therefore, important for predicting and
understanding the tribological properties of the solids in contact. Figures 1 present AFM
topography, 3-D images, and profiles of cross-sections of the tested surfaces.
Samples before extraction (not shown) had smooth surfaces, with islands of different
dimensions in the range of 300÷600 nm. The section along the line shows vertical
differences up to 70 nm. After extraction, the surface is rough (*Figure 1 a-c*), with
unevenly distributed holes, and with grains which resemble hills.
The cross-section along the line (*Figure 1c*) shows roughness with heights up to 18 nm,
in a range in width between 20 and about 200 nm. At very high magnifications (*Figures
1 d-f*), pores with diameters about 2 nm are evidently seen on the surface's cross-section
profile. Morphology of a silica surface with lower porosity before extraction was also
investigated. There are a few islands. Their height is low (below 3 nm) when compared
to a surface of silica with a higher porosity before extraction.
Variations of the friction forces for unlubricated and lubricated silica surfaces with a
normal load of F_f (F_n), are presented in Figures 2 and 3.
In both unlubricated and lubricated materials with both more and less porosity, there is
an increase of friction forces in a range of normal loads, with some deviations from
linearity at lower loads for unlubricated surfaces with higher porosity (*Figure 2a*).
This F_f (F_n) curve is characterized by a linear region at the lowest load up to 75 arbitrary
units of load, and the common linear growth for normal forces larger than 100 arbitrary
units. In this case, the relative friction coefficient is very low, 0,086 at the lowest load.

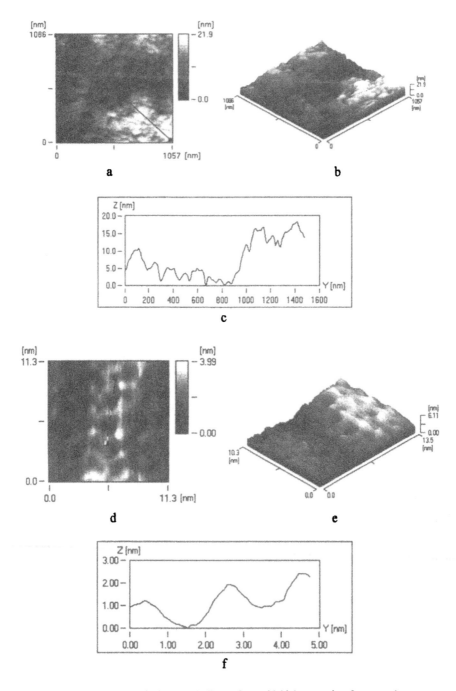

Figure 1. AFM topography images of silica surfaces with higher porosity after extraction.

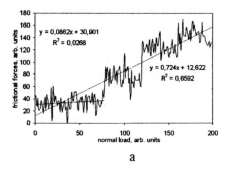
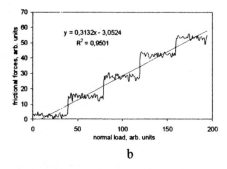

a   b

Figure 2. Friction vs. applied load for a silicon nitride tip sliding against silica with higher porosity: a) for unlubricated surfaces, b) for lubricated surfaces.

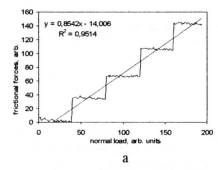
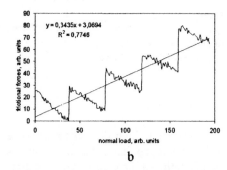

a   b

Figure 3. Friction vs. applied load for a silicon nitride tip sliding against silica with lower porosity: a) for unlubricated surfaces, b) for lubricated surfaces

Extrapolation of F_f (F_n) of the portion of the loading curves at the lowest normal loads to zero normal load, gives small (but consistent) negative interceptions with the horizontal axis associated with the presence of adhesive surface forces. Lower relative μ's are generally found for lubricated surfaces. Deviations from linear F_f (F_n) behaviour, known as Amonton's law, are observed when the actual area of contact does not increase proportionally with the normal load applied. In this case friction decreases, usually with the increasing load. Relative friction coefficients of μ at generally all loads, and especially for higher loads of unlubricating silica with higher porosity, are approximately doubled for unlubricated surfaces from lubricated. Friction coefficients after scanning the surface area are relatively stable and reproducible. Local variations are significant for the unlubricated, more porous silica (*Figure 2a*). They seem to depend more on the local surface slope being connected to the presence of pores than to surface height distribution.

AFM studies are still continued in our laboratory. The measurements described above have been completed recently. In tribological test of silica two mesoporous samples and one non-porous were used. One of them, which was synthesised using 1 g of non-ionic polyoxyethylene(4) lauryl ether (Brij) was used twice and the second one, synthesised in the same conditions but without surfactant, (in this way non-porous sample was obtained) was treated as a reference. The measurements were performed for "dry" mesoporous surface and with the surface covered with lubricant (n-hexadecane). The results were compared with frictional measurements performed for non-porous dry silica. This time frictional forces in arbitrary units were measured not in mesoporous scale range, but in bigger area 840x840 nm. The sample surfaces was scaned repeatedly in the x direction in the range of normal loads 0.018 – 0,009 a.u. The relative coefficients of friction were calculated from the slopes of the relationship of friction versus load forces. Measurements were repeated several times at randomly selected location and showed good reproducibility. From the obtained charts calculated relative coefficient of friction for "dry" mesoporous sample (0,3092 a.u.) is higher than for lubricated mesoporous sample (0,2367 a.u.) and non-porous "dry" silica (0,2838 a.u.). The differences in relative coefficients of friction for lubricated and unlubricated mesoporous surfaces are not as big as in the case where the measurement was performed in the range 15x15 nm., but the main trend of frictional forces changes is the same.

Water and other surface contaminants can affect the μ dramatically. Since the AFM was operated in a repulsive mode, attractive forces due to water adsorbed on the surfaces are not observed in the image.

4. Conclusions

Experimentally, high differences are observed in AFM images of two silica surfaces before extraction that differ in porosity. After extraction, surface morphology changes significantly. At the highest magnification, pore holes with expected diameters of 2 nm and more are seen.

From this study, it is also evident that relative friction coefficients are more than doubled for unlubricated surfaces over lubricated. At the lowest loads, very low relative friction coefficients for unlubricated silica with higher porosity are observed. The reason for this is not entirely clear. The relative coefficient of friction of "dry" mesoporous sample is higher than for lubricated mesoporous sample and for non-porous "dry"silica.

5. References

Boccaccini, A. R. (1998), Influence Of Stress Concentrations On The Mechanical Property - Porosity Correlation In Porous Materials. *J. Mat. Sci. Lett.*, **17**, 1273-1275.
Caravia, L., Dowson, D., Fisher, J., Corkhill, P.H., Tighe, B. J. (1994) Sliding Friction In Porous And Non-Porous Elastic Layers: The Effect Of Translation Of The Contact Zone Over The Porous Material, Dowson, D at al. (Editors), „Dissipative Processes in Tribology", *Elsevier Science B. V.*, 261-266.
Rice, R.W. (1993) Comparison Of Stress Concentrations Versus Minimum Solid Area Based Mechanical Property-Porosity Relations, *J. Mat. Sci.* **28**, 2187-2190.

This work was supported by the State Committee for Scientific Research of Poland (KBN) under Grant number 7 TO8C 020 10.

DISCUSSION FORUM REPORT: BRIDGING THE GAP BETWEEN MACRO- AND MICRO/NANOSCALE ADHESION AND FRICTION

Matthew Tirrell
University of California at Santa Barbara
Office of the Dean
College of Engineering
Santa Barbara, CA 93106-5130
USA

Ernst Meyer
Institute of Physics, University of Basel
Klingelbergstr. 82
4056 Basel
Switzerland

1. Introduction

A discussion forum was held to discuss issues and needs in bridging the gap between macro- and micro/adhesion and friction. The following summarizes the observations.

2. Issues

- The technological fields of magnetic storage, MEMS (micro-electro-mechanical systems) and biomedicine will provide on-going stimulus for bridging the length-scale gap in friction and adhesion; at the same time, many longstanding problems in other important technologies (transportation, defense, manufacturing) derive from lack of adequate understanding of the inter-relationship between smaller and larger scale frictional processes.

- Friction and adhesion encompass a class of problems; they are not uniform or unified scientific fields. Therefore, any successful efforts or progress will arise from multidisciplinary approaches. The problems arising at different length scales, such as interpreting a frictional force measurement, or understanding the origins of stick-slip phenomena, may have parallels, but are not necessarily identical, even if the phenomenologies of the effects are similar. That is, while small-scale measurement can give considerable useful information and insight in its own right, it is not accurate to say that the emergent properties of friction between larger objects can be totally understood from nanoscale measurements. This is not to say, however, that extensive comparisons between nanoscale and macroscale measurements that are putatively the same would not be very informative. This could usefully be applied to measurements of rate and load dependence of friction and several other observables.

B. Bhushan (ed.),
Fundamentals of Tribology and Bridging the Gap between the Macro- and Micro/Nanoscales, 355–357.
© 2001 *Kluwer Academic Publishers.*

- There is a need to understand contact between materials surfaces better, from small groups of atoms and molecules to interactions over larger surfaces. There are models for macroscopic contact that seem to describe successfully on a macro-level the extent and nature of interface between materials, depending on bulk material properties, surface and interfacial energies, environmental conditions such as humidity and the size of the objects in contact. The nature and quality of contact on smaller scales in less well-understood and less well-characterized. Several aspects deriving from differences in small- and larger scale contacts differentiate frictional processes occurring at different scales (heat generation and removal, debris ejection, etc.) Likewise, adhesion processes, such as separation mechanisms (peeling, cracking, fingering, etc.) are scale-dependent. Surface films, sometimes resulting from contamination even in relatively clean environments, often dominate the behavior of friction in a contact zone. Adhesion and friction can, in fact often do, depend on the history, sequence and time-dependence of contact between surfaces.

- It should be recognized that friction between two surfaces in relative motion can be an asymmetric process for the surfaces involved. For example, if an object of smaller extent is driven to slide over a surface of larger (let's say, indefinite) extent, the contact area of the smaller surface will experience continual friction while the contact area of the larger surface will be loaded and unloaded in a time-dependent fashion. For friction between two different materials, this means that, even in principle, if we interchange the two materials composing the smaller and larger surfaces, we should not expect the same results. This is particular germane to the use of some nanoscale friction probes. The averaging inherent in larger areas of contact must be appreciated; nanocontacts will be "site-dependent" and inherently affected by fluctuations.

- Friction, and the role of adhesion therein, comprise processes that require multiscale modeling. Theory and modeling of friction are underdeveloped. There are relationships known as "laws" that are rarely obeyed. Part of the deficiency in this regard is related to insufficient precision in defining objectives; for example, modeling an empirically, directly observable friction force may be more straightforward than modeling a friction coefficient, the very definition of which come from a conjectured constitutive equation that may not apply. Development of simulations, models and, where possible, predictive theories, at scales from atomic (quantum mechanical, electronic structure-based), through continuum models to useful engineering models, is an important endeavor. "Bridging the gap" in theory and modeling will require a set of models; it is unwise to expect quantum-based engineering models in any foreseeable future. However, further work on the inter-relationships among models at different scales should be useful in understanding emergent, collective frictional properties. In the near term, the modeling approach should be well-adapted to producing the desired kind of result. A desirable and realistic goal is to identify length scales at which fundamentally new behavior becomes important.

3. Needs

- There is a major need for better and more complete characterization of surfaces involved in friction experiments. This lack has held back the field of tribology. Similarly, techniques used for producing friction data must be better defined and, at least, thoroughly explained in order for any publication to be broadly useful.

- New methods are necessary to carry out the required full characterization and to interpret friction measurements more insightfully. These include: measurements of local frictional heating and subsequent heat dissipation (micro- or nano-tribothermography); better techniques to measure mechanical properties of thin films and surface layers; methods for determining the nature and quality of material contacts; ways for characterizing subsurface properties of materials (i.e., material at some depth away from the surface but not yet in bulk region). A possible realization might be the magnetic resonance force microscopy (MRFM). This technique would combine the subsurface information with chemical information and high resolution. Due to its rather early stage of development, MRFM has not yet been applied so far in tribology. Moreover, combinations of methods must be brought together for fuller characterization. New imaging methods and fast characterization tools will be especially helpful.

- Development of miniaturized tribometric sensors (for imbedded, *in situ*, real-time friction measurement) would be a revolutionary asset. The combination of micro-(nano-)tribometers with large-scale machinery appears to be a promising approach to strengthen the bridge between macro- micro- and nano-tribology.

MODELING (and) WEAR MECHANISMS °

Ken Ludema, Professor Emeritus
Department of Mechanical Engineering and Applied Mechanics
University of Michigan,
Ann Arbor, MI, USA 48109-2125

1. Introduction

In all technical fields the highest accomplishment is to formalize knowledge into mathematical (equation) format. This action serves two purposes, namely,

　　a. To provide equations that engineers can use in product design, and,

　　b. To add purpose and discipline to research.

Overall the state of developing equations (modeling) in tribology ranges from very accomplished to primitive. Following is a list of tribology topics, covering that range, in the same order:

　　a. Thick fluid film lubrication (hydrodynamic and elastohydrodynamic),

　　b. Solid mechanics (elastic and plastic) of contact,

　　c. Temperature rise in sliding surfaces,

　　d. Surface topography,

　　e. Frictional vibrations,

　　f. Solid mechanics of visco-elastic contact,

　　g. "Wear" by rolling contact fatigue,

　　h. Wear by erosion,

　　i. Wear by abrasion,

　　j. Wear by sliding,

　　k. Wear in the "boundary lubricated" state,

　　l. Wear protection by chemical boundary film formation,

　　m. Criteria for galling (scuffing, scoring, etc.) in lubricated sliding,

　　n. Friction.

The reason for the primitive state in the latter areas is that the governing mechanisms are not known. The lack of equations has not totally thwarted developments in tribology, however, because we see:

° Based on Chapter 10 in the book, "Friction, Wear and Lubrication: a Textbook in Tribology", by K.C Ludema; CRC Press, ISBN 0-8493-2685-0, 1996

B. Bhushan (ed.),
Fundamentals of Tribology and Bridging the Gap between the Macro- and Micro/Nanoscales, 359–375.
© *2001 Kluwer Academic Publishers.*

a. Much good research, though uncoordinated overall,

b. Many good technical solutions generated from sound technical intuition, but the lack of inclination to develop equations has allowed:

a. Uncoordinated and sloppy investigations, filling the literature with junk,

b. Technical stagnation of some tribology intensive products because of confusion on the basic principles of tribology.

The rate of progress in the development of general or global equations of the latter categories is so slow that most design engineers will likely not benefit from them during their careers. We in the research community therefore have the dual duties to both develop useful wear equations and to develop rational methods for designers to use in their work while waiting for good tribology models. In my view that will involve education and outlining the rationale for simulative testing. But first, we must acquire an overall perspective in the field.

Without models, the easiest thing we can do has already been done. Various handbooks on tribology offer the following sage advice for assuring long life of sliding members in mechanical components:

a. Maintain low contact pressure,

b. Maintain low sliding speed,

c. Maintain smooth bearing surfaces,

d. Prevent high temperature,

e. Use hard materials,

f. Insure a low coefficient of friction (μ),

g. Use a lubricant.

These points may assure the long life of sliding members but will also assure failure in business. For competitive purposes the designers need to know how to achieve some specific target in product life, which life is usually shorter than ideal from the consumer's point of view. Thus the designers need to know where the margins are between "low contact pressure" and dangerous contact pressures, and so with the other variables.

2. Models and Equations

The terms, "models" and "equations" have very specific meaning in some sciences and technologies. In this paper the term "model" means an equation describing the relationship between relevant variables, in mathematical form. The "equation model" is distinct from a "word model" or a "pictorial model." Word models are descriptions in words only, of phenomena or behavior of materials, etc. Pictorial models are sketches (etc.) and/or a series of sketches describing the functioning of some device or phenomenon.

Mathematical models are seen in several forms. In several fields, models are expressions that simulate, or describe the output response of an entity of unknown internal composition ('black box') to some input variable. There are also "molecular models", which are descriptions in mathematical form of the physical response of groups of molecules to some input, such as shearing, heating, adsorption etc. Whatever the form, a model can never be a complete description of a system: some "best" assumptions are used to begin the model and the model is slowly improved over time, but never (hardly ever) perfected.

2.1 FUNDAMENTAL EQUATIONS

The most useful type of equation is derived strictly from a knowledge of the controlling variables. The final form of such equations can be very satisfying in that all variables are readily measured, none is omitted, and each is independent of the others and no "constant of proportionality" is needed. They are (or should be) complete in themselves without constants of proportionality or adjustable constants that represent some here-to-fore unknown phenomena. But an additional and vital criterion is that the equation must accurately describe nature. There likely are very few such equations for wearing of materials, or at least very few that have been validated by researchers other than the author of the equations.

One such equation is due to Evans et al. (1981) for ceramic wear by deep scratching with a hard indenter. They assumed that material removal begins as a loosening of material by linking of the two types of cracks that develop under a sharp indenter, the one during indentation and the other during release. A sharp indenter extends the crack system over a sliding distance, S, to produce a wear volume, X:

$$X = \frac{W_n^{9/8}(E/H)^{4/5}S}{K_c^{1/2}H^{5/8}}$$

where W_n = normal contact force
   K_c = fracture toughness   H = hardness
   E = elastic modulus   S = sliding distance

Evans et al. reported a qualitative correlation of this equation with the wear rate of glass, but very poor correlation with polycrystalline structural ceramics, which conclusions have been verified by others in slow speed sliding. Likely this equation works because surface effects are not significantly involved in deep scratching.

2.2 EMPIRICAL EQUATIONS

At the opposite end of the scale from fundamental equations are empirical equations. These are developed from experimental data by 'fitting' equations to the plotted data or by estimating equations by "least squares" or other methods. For most wear phenomena, empirical equations constitute the "state of the art". For example, the life, "T", of a cutting tool in a lathe has been found by experiment to depend on the cutting speed, "V", in a logarithmic manner. This relationship was not previously known. The best relationship between these variables is:

$$VT^n = C$$

where n and C are taken from experiments. Equations of this type are valid only over the range of tests, which, for tool life, is usually selected to cover some part of the practical range of cutting conditions.

Any one set of constants, n and C applies to very specific conditions of depth of cut, feed rate, tool shape, tool material, sharpness of tool, material being cut, type of coolant, and perhaps machine related variables such as vibration characteristics. Though tool life testing has been under way for 75 years or so, few of the variables beyond the depth of cut and feed rate have been formally "modeled". By general agreement in the community of users of cutting tools, one ceases to make the equation more complicated beyond some point.

Tool life equations are not actually wear rate equations. Rather they express a cutting time after which the tool is useless, ie, its cutting edge has rounded off and no longer cuts. The equation could be called a performance model, and most wear equations in the literature are of the same type. The actual wear (material loss) rate of tool material is non-linear in time, increasing toward the end of the test. Even then, a simple expression of persistence of the tool in cutting is inadequate in engineering practice. Often the more relevant condition for stopping a cut is the deteriorating condition of the surface of the part being cut: surface roughness and residual stresses in the part usually increase somewhere in the last half of tool life. Thus tool life should be evaluated according to its useful cutting life rather than the time over which it continues to remove material, and the same with other forms of wear.

2.3 SEMI-EMPIRICAL EQUATIONS

Most often authors appear to develop equations that begin with a (usually short) list of variables that may be relevant. Wear rate is then taken as the product of all of the named variables, perhaps with each variable raised to some exponent. Each of the variables and exponents is assumed to be independent of all others. Numerical relevance is added in the form of a "constant" derived from experiment.

3. What Equations Are Available for Wear

3.1 TWO CLASSIC EQUATIONS

There are very many equations for predicting wear life and wear rates but two will be cited first because of their wide use. The two are the Archard equation and the Kruschov/Babichev equation.

a. Archard (1953) published an equation for the time rate of wear of material, Ψ, in the form:

$$\Psi = kWV/H$$

where W is the applied load, H is the hardness of the sliding materials, V is the sliding speed, and k is a constant, referred to as a wear coefficient.

The selection of variables was not a blind choice of what "might" be relevant to wearing. Archard had a specific physical model in mind: he assumed that when any two asperities come into contact with each other they bond, weld or adhere together. Wear loss is then due to adhesive "tearing out" of one or both asperities from their locations. The area of adhered contact, "A_r", was taken to be W/H. "V" is a measure of the number of these real contact areas that will come into and move out of contact in unit time of sliding. This equation is widely known as an equation for ADHESIVE WEAR.

Archard could not, however, predict how big each torn volume would be, so an experimental "constant of proportionality", "k" accounted for this uncertainty. Several authors had attempted to refine the meaning of "k" as a probability factor for wear particle formation from the system, but without much success. The reason lies in the undefined nature of oxides, adsorbed substances and other adhesion preventing substances on virtually all surfaces. Simple substrate-to-substrate adhesion as the initiator of wear is very uncommon in long lasting systems. Further, the common assumption is that loosened particles disappear from the system, whereas in practice loosened particles remain in the contact region for a long time, serving as "solid lubricants"?.

b. Kruschov et al. (1956) published the results of a large testing program in relatively deep ABRASIVE WEAR. A curve fit to their data resulted in the equation:

$$\Psi \propto WV/H \text{ (same variables as above)}$$

at least for single phase microstructures. They, and later authors found more complicated behavior for two phase microstructures. These authors made no assumptions about contact areas at all.

In the 1950's and 1960's there were some spirited discussions as to whether the wearing of practical machinery is due primarily to "adhesion" or "abrasion". Many papers were published in which one mechanism or other was "proven" from measurements of wear of functioning machinery. The proof to the author lay in the linearity of wear rate with changes in the variables W, V and H and each equation had partisan support. The proponents of each mechanism claimed a high fraction of causes for wear in practice for their favorite mechanism, such that the total exceeded 100% by a large margin! One author had even counted papers he found in the literature, finding that abrasive wear papers were in the (slight) majority and therefore abrasive wear was the most likely cause of wear in all of technology! (It seems unnecessary to state that abrasive wear means wear by abrasive substances. Many papers attribute wear to abrasion if the authors are not confident that adhesion has occurred, whether abrasive substances are present or not! The evidence for adhesive wear is usually even more elusive and wear is often deemed to be adhesive if there is no visible lubricant present. Likely authors need the comfort of familiar terms and frequent use increases comfort with a particular term.)

A more reasoned view is that neither of the above equations is adequate for wear life prediction in practical situations. Neither one is sufficiently general or global for use in engineering design. Neither one incorporates more than 3 of the likely 30++ parameters needed for completeness (a point to be made later). The imprecision of the Archard equation may be seen in the very large range of values of 'k' inferred from measurements of wear of practical machinery, extending from 10^{-4} to 10^{-9}. No one is able to predict 'k' for any particular application to better than one order of ten accuracy without considerable experience with their products. Designers need predictions in the range of accuracy of $\pm 10\%$.

3.2 MORE EQUATIONS

Though equations other than those above are rarely cited, most of them likely reflect some reality somewhere. Perhaps a composite equation could be assembled from them, or perhaps the preponderance of use of some parameters in all equations might indicate the importance of one material property or operating condition over another. It was to answer these questions that a search was conducted for common themes in wear equations, Meng (1994).

3.2.1 The Search

A search was done in the 4706 papers in Wear Journal from 1957 to 1990, and 751 papers in the proceedings of the conferences on Wear of Materials from 1977 to 1991. Many more journals could have been scanned but these two sources are specifically devoted to wear and they are well reviewed. Papers from other journals were also analyzed if they were referenced in one of the primary papers.

The great majority of papers contained discussions and logical descriptions of how wear progresses, which could be called word models. Most of these are accompanied by micrographs, electron/x-ray spectra and other evidence of wear damage.

Over 300 equations were found for friction and wear. The exact number is difficult to state because some equations are very small revisions of previous equations. The equations were scanned to determine whether they converged upon a few most desirable variables. In principal, it would seem that one (or a few) good equation(s) could be condensed from the great number of published equations, provided the true importance of a few central variable could be determined.

One of the first discoveries was that many of the equations appeared to contradict each other and very few equations incorporated the same array of variables. It is common to find, for example, Young's modulus in the numerator of some equations and in the denominator of others. It therefore seemed obvious that a simple tabulation of the uses of each variable would not indicate the importance of that variable in the wearing process. A method was necessary to separate out those equations that properly represent variables from those that do not. Perhaps from these equations a reasonably authoritative hierarchy of factors could be established to help develop fundamental equations.

3.2.2 Analysis of equations:

The 300+ equations were evaluated against four criteria for the purpose of finding the most useful of them. The following criteria were applied:

> i. Historical significance
> ii. Applicability
> iii. Logical structure
> iv. Nature of supporting information, especially from experiments.

3.2.2.1 Historical Significance.

The equations of authors who published a progression of thought in the same general topic were given higher credence than those of authors who published only once. For example, of the 5137 authors named on the 5467 papers, 3257 were named on papers only once, 810 were named on two papers and 362 on three papers. Fewer than 100 authors published more than 10 papers and only 291 have remained in the field more than 5 years.

A second criterion in this category was to see how peers regarded a published work by analysis of reference lists. This procedure favors older papers, which was not intended, but it adds the opinions of a broad range of authors. (It is particularly instructive to see how an author responds to commentary on his work, and to note how often an author quotes his own work beside the work of others.)

It was also noted that papers older than 15 years old are seldom referenced, except sometimes in a group of references taken from another, fairly recent paper. Also, mathematical papers are referenced many more times than are those containing micrographs and other tedious information to comprehend. An analysis of significance based on these criteria requires some judgment.

3.2.2.2 Applicability.

Equations contain many different variables. Most often the familiar variables are used, such as hardness, Young's modulus, etc. However, many equations contain variables that are not readily definable, or are only available from experiment. Examples are 'grain boundary strength' or 'atomic damping factor' or 'surface stiffness'. Rarely do the authors of such variables follow up and measure these quantities themselves, and neither does anyone else. Such equations have limited use. Many equations have limited use also because they only show relationships between variables without providing information on the resulting friction or wear rate.

3.2.2.3 Logical structure.

Some equations are built on strings of poorly rationalized assumptions. It is difficult to show precisely where, in such equations, the overall argument departs irretrievably from reality. As mentioned before, adhesion is often invoked but the nature of the assumed adhesion is never described, nor is evidence for adhesion shown in connection with the equations found in this study.

3.2.2.4 Nature of supporting information, especially from experiments.

The most helpful papers include data from experiments covering a wide range of variables such as sliding speed, surface roughness, etc. Furthermore, papers that include a lengthy analysis of previous work are much more likely to place the new results in the proper context than are those in which references are merely cited. Finally, those papers in which wear rates and transitions in wear rates are identified with observations (with and without microscopes) of surfaces appearance, nature of wear particles and other important features of surfaces are more useful than others.

3.2.3 Results of applying the above criteria to equations in erosion:

Erosion by solid particle impingement is one of the less (?) complicated types of wear. Ninety eight equations were found in this topic, but not all of them survived careful scrutiny. When the above four criteria were applied, 28 equations appeared useful for further analysis. Few of these equations contain the same array of factors: 28 equations contained some 33 factors, counting a few combined and adjustable coefficients. These are shown in Table 1.

It is interesting to speculate on the reasons authors choose particular factors for their equations. Academic specialty is clearly one reason. There is really no sure way to discern whether all of the necessary variables and parameters were included or only the convenient ones. Dimensional analysis has been used by some authors to choose factors, but inherent in this method is the assumption that all relevant factors are known and happen to have useful units.

One logical indicator of completeness in, and relevance of erosion equations might be consistency in the exponent on velocity, V. These range from about 1.5 to 6. Ordinarily one would expect that this exponent should be 2, reflecting the idea that particle energy would be the operative measure of impact severity. Several authors note that the larger exponents are found when ceramic materials are the target. Perhaps the higher exponents for ceramics actually reflects the sensitivity of ceramic materials to the strain rate differences inherent in differences in V. Thus an exponent of other than 2 may indicate that dynamic material properties should be used in equations rather than static properties. One candidate property is fracture toughness, a property that is measured in a quasi-static

Table 1. Parameters chosen in each of the 28 equations for erosion by solid particles

Parameter	1	2	3	4	5	6	7	8	9	10	11	12	13	14	15	16	17	18	19	20	21	22	23	24	25	26	27	28
Density	X																						X	X			X	X
Hardness						X	X																				X	X
Moment of Inertia								X																				
Roundness						X	X				X			X														
Single mass			X						X	X										X								
Size		X	X		X	X	X	X	X	X		X	X		X	X	X	X	X	X	X	X	X	X			X	X
Velocity	X	X	X	X	X	X	X	X	X	X	X	X	X		X	X	X	X	X	X	X	X	X	X			X	X
Rebound Velocity				X																								
Kinetic Energy of Particle																									X			
Density	X					X	X		X		X			X		X	X	X		X	X	X	X	X			X	X
Hardness						X	X		X		X					X	X	X			X	X	X	X			X	X
Flow Stress	X																				X							
Young Modulus																									X		X	X
Fracture Toughness								X							X	X	X	X			X			X	X		X	X
Critical Strain																					X					X		
Depth of Deformation																										X		
Incremental strain per impact																										X		
Thermal Conductivity														X														
Melting Temperature														X								X						
Enthalpy of melting														X														
Cutting Energy		X		X																								
Deformation Energy		X		X																								
Erosion resistance						X	X				X																	
Heat Capacity														X								X	X					
Grain Molecular weight																												
Weibull Flaw Parameter			X				X																					
Lame Constant																	X	X										
Grain Diameter																									X			
Impact Angle	X	90		X		X	90	X			X		X							X	90	90	90					
Impact Angle Max. Wear												X																
K.E. Transfer from P to T														X														X
Temperature																						X						
Constant	2	3	3	1	2	10	1	3	2	1	6	1	4	4	1	1	1	1	4	1	8	1	1	1	1	1	3	3

test. Alternatively an exponent other than 2 may indicate the absence of one or more other important variables. To pursue these possibilities it is necessary to acquire a large body of data from tests using wide ranges of the many variables.

3.2.4 Observations:

The primitive state of equation development in friction, scuffing and wear is clearly the result of the complexity of the topics but there are other reasons as well. Some are:

    a. The efforts of the several disciplines in the field are poorly coordinated,

    b. There is insufficient consensus on coherent methods of constructing equations. This may be seen in the continued use of a limited set of material parameters in particular, and omission of other important ones. For example:

        1. Few of the mechanical properties used in tribology equations are unique, ie., many of them are the result of the same basic behavior of atoms., e.g., hardness, Young's modulus and melting point, and yet several are found together in most equations.

        2. Some of the properties are not intrinsic material properties, such as hardness or stress intensity factors.

        3. Some variables should not be found in first-principles wear equations, such as temperature or the coefficient of friction. Temperature does not cause wear, but does influence the material properties that control wear (among other things). The basic mechanisms of friction and wear are probably the same so one can not be used to describe the other.

        4 Few of the properties are related to the mechanisms whereby wear particles are generated or expelled.

        5. Some properties are rarely seen in equations though sliding of one surface over another certainly calls for these properties, e.g., fatigue properties, oxide properties, debris content and stickiness, strain rate sensitive mechanical properties, etc.

        6. The influence of contact geometry, duty cycling and other factors on wear particle retention are not often considered.

        7. The influence of vibration, slight deviation from repeat pass paths, the difference between cyclic sliding, repeat pass and single pass sliding is not usually considered.

        8. Most researchers select variables for study exclusively from their own discipline, as if no other properties are relevant. Alternatively, most authors may assume that more variables can be added as they arise, following the principle of linear superposition.

        9. Wearing usually takes place by a combination of mechanisms, which combination changes with time. There are few studies on transitions between the balance of mechanisms and on partitioning between mechanisms.

4. Inferring Wear Mechanisms From the Nature of Data Curves

Most equations in the literature suggest a monotonic influence of variables on wear rate, probably derived from the assumption that only a single wear mechanism is operative over the whole useful range of the variables. Two examples of multiple mechanisms are given, where varying rates of oxidation control the wear processes on the surfaces. These are by J.K. Lancaster and by N.C. Welsh.

a. Lancaster (1963) measured the wear rate in dry sliding of 60Cu-40Zn brass pins on High Speed Steel (HSS) rings, over a very wide range of sliding speed and temperature, and got the results for the brass shown in Figure 1. He classified wear in relative terms, mild and severe, without implying mechanisms of wear.

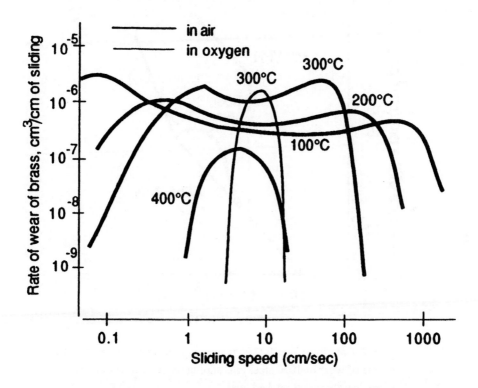

Figure 1. Data from Lancaster (1963) showing the variation in the distance rate of wear of brass versus sliding speed, for 60Cu-40Zn brass pins sliding on high-speed steel, at various temperatures and in two atmospheres.

The transition between severe wear and mild wear is influenced by atmosphere, as well as by sliding speed and ambient temperature, which, according to Lancaster, influences the thickness of oxide. Where oxides are thick the wear rate is low and the differences are in orders of ten. The oxide thickness is a function of two factors, namely, the time available to reoxidize a denuded region following a sliding cycle (on the steel ring) and by the rate of formation of the oxide as influenced by the temperature due to sliding.

b. Welsh (1965) worked with 0.5%C steel on steel, used a pin-on-ring configuration and found some unexpected transitions between severe wear and mild wear, as shown in Figure 3. The large transitions ($\approx$ 2.5 orders of ten) in the data for the softest steel seems impossible and yet it is real: these data for 1050 steel as well as for other steels have been verified by research students many times. Welsh also attributed these transitions to the thickness of oxide. Load influences the relative amount of oxide left after a slider "passes by" but also influences the temperature rise on sliding surfaces. The thickest oxides exist where wear rates are low.

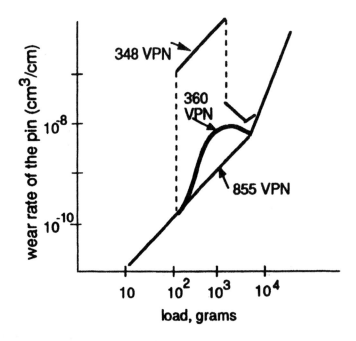

Figure 2. Data from Welsh (1965) showing the variation in the time rate of wear of 1050 steel pins sliding on 1050 steels of three hardnesses versus applied load.

Hardness has a major effect as suggested in the Archard equation but in only one range of load, and mostly to diminish the extent of transition to severe wear. It may be speculated that the critical oxide thickness is less for hard substrates than for the soft substrate.

A most unusual feature of the data of both Lancaster and Welsh is the very wide range of sliding speed and load, covering several orders of ten. The great preponderance of data in the literature cover quite narrow ranges variables. Perhaps it is easier to write equations from limited data.

Equations for the curves of Lancaster and Welsh could be written but would have very little value for any condition outside those of the conducted tests. But, since we know that oxides are prominent in the limiting of wear in some regimes of load and sliding speed, it may be possible to write more general equations. These equations could incorporate variables that relate to oxide formation rates (oxygen partial pressure, contact area shape, duty cycle, etc.) and oxide loss rate (contact pressure, length of pathway for expulsion of oxide from the contact region, etc.). Such equations would contain many more variables than those connected only with the shape of curves in Figures 1 and 2 and would therefore be useful to predict wear rates over a wider range of operating conditions.

5. The Influence of Test Specimen Shape

Whereas the work of Lancaster and of Welsh demonstrated the importance of oxide film thickness, it is not likely that universal conclusions can be drawn from their work. They used the pin-on-ring test geometries. Other geometries will produce different results as may be seen in the comparison of three common types of bench test machines. These are:

  a. Pin-V test in which a 1/4-inch diameter pin of 3135 steel rotates at 200 RPM with 4-line contact provided by two V blocks made of 1137 steel. (The 1137 steel is chosen for the ease of specimen making!)

  b. Block on ring test where a rectangular block of a chosen material slides on the outer (OD) surface of a ring of hard, case carburized 4620 steel.

  c. The 4-ball test where a ball rotates in contact with three stationary balls, all of hardened 52100 steel.

The 4-ball test and the ring on block test were run over a range of applied load and speed. The pin-V test was run over a range of load only. All tests were run continuously, i.e., not in oscillating or stop-start sequence mode. All tests were run with several lubricants, which is very different from the conditions in the tests of Lancaster and of Welsh. However, it is highly unlikely that dry sliding would produce different conclusions than those that follow.

Results from the ring-block test were not sufficiently reproducible or consistent for reliable analysis. Results from the other two tests were adequate for the formulation of a wear equation from each, as follows:

$$\text{Pin-V test: Wear rate} \propto (\text{Load})^2$$
$$\text{4-ball test : Wear rate} \propto (\text{Load})^{4.75} \times (\text{Speed})^{2.5}$$

These results may be compared with "linear laws" of wear discussed frequently in the literature, which would be of the form:

$$\text{Linear law: Wear rate} \propto (\text{Load})^{1.0} \times (\text{Speed})^{1.0}$$

6. Toward Better Descriptions of Types, Terms and Processes of WEAR

The field of wear is very unsettled as to descriptions of wearing processes, so much so that the modeling exercise can hardly begin. Modeling begins with accurate mechanisms of wear, and makes little progress with mere descriptions of wear. Following is a listing of 24 common terms discussed in the literature, to describe wear (and there are many more). Some terms communicate marginally more than others on the actual causes of loss (wear) of material from a surface, some are very subjective in nature and communicate only between people who have observed the particular wearing process together. Following are six categories of terms, progressing from the more subjective to the more 'basic'. The grouping of terms:

a. The first group could be classified as <u>subjective</u> or <u>descriptive</u> terms in that they describe what <u>appears</u> to be happening in the vicinity of the wearing surfaces:

blasting, hot gas corrosion, percussive, hot, mild, impact, deformation, pitting, frictional, mechanical, seizing, welding

b. The second group contains terms that appear to have more meaning than those in group "a", in that some mechanisms are often implied when the terms are used. These types of wear do not necessarily involve loss of material but some change in the sliding or contacting function of the machine.

galling (may relate to surface roughening due to high local shear stress)

scuffing: scoring (often taken as synonymous terms) probably relate to some stage of severe surface roughening that appears suddenly in lubricated systems

c. **Adhesive** wear is the most difficult term to define. It may denote a particular type of material loss due to high local friction (which is often attributed to adhesion) and is a tempting term to use because high local friction produces tearing and fragmentation whereas lubricants diminish tearing. Often **lubricated** wear is taken to be the opposite of adhesive wear. An often overlooked aspect of "adhesive wear" is that if adhesion is effective then the particles torn loose by adhesion will readhere and not be expelled as debris!

d. Terms that derive from cyclic stressing, implying **fatigue** of materials:

fretting, a small (few microns?) amplitude cyclic sliding that displaces surface substances (e.g., oxides) in microscopic contact regions and may induce failure into the substrate, sometimes generating debris from the substrate and/or cracks that emanate into the substrate)

delamination describes a type of wear debris that develops by low cycle fatigue when surfaces are rubbed repeatedly by a small (often spherical) shaped slider.

e. The fifth group can probably be placed in an orderly form but individual terms may not have originated with this intent. These relate to the types of wear known as **abrasive** wear. In general abrasive wear consists in the scraping or cutting off bits of a surface (oxides, coatings, substrate) by particles, edges or other entities that are hard enough to produce more damage to a solid other than to itself. Abrasive wear does not necessarily occur if substances are present that 'feel' abrasive to the fingers! The abrasive processes may be described according to size scale as follows:

> **polishing** which likely is mostly a process of removal of oxides and products of other chemical conversion, followed by renewal of the film, and removal, etc.

> **scouring** which may involve some removal of substrate materials as well as oxides, et.al.,

> **scratching** which involves more removal of substrate material and relatively less oxide,

> **grinding** which removes substrate material predominantly,

> **gouging** which is strictly not an abrasive process but does remove material as ordinary abrasive processes do,

f. Wear by impingement, over angles ranging from near 0° (parallel flow) to 90°.

7. Mechanisms of Wear

In order to model the wearing process we need to properly describe wear mechanisms. The following definition is offered:

> MECHANISMS OF WEAR -- the succession of events whereby atoms, products of chemical conversion, fragments, et al, are induced to leave the system (very likely after some circulation) and identified in a manner that embody or immediately suggest solutions.

These solutions may include choice of materials, choice of lubricants, choice of contact condition, choice of the manner of operation of the mechanical system, etc. This definition encompasses the following (not a mutually exclusive list):

a. The conditions imposed upon surfaces: for example, contact stress, shear stress and number of cycles of these stress applications,

b. The changing composition of substances in the sliding interfaces due to both sliding and due to standing still: these will likely influence the friction (shear stresses) imposed in sliding contact,

c. The processes whereby particles (atoms, etc.) are separated from a surface,

d. The circulation (or hindrance to circulation offered by "stickiness") of those particles before they are expelled as debris,

e. And doubtless more.

I have not made much progress in many aspects of the overall sequence, but I offer a perspective on how models might express material separation processes in Table 2:

8. What To Do Until Complete Models Are Available

Designing of consumer products with predicted life is complicated and simply cannot be done today for most products. Design managers find this fact to be incredible in this era of the invincibility of the computer. There appears to be the general confidence that even the most complicated problems can be manipulated by computers, but that applies only to simple phenomena. Finite element methods are very helpful in calculating the stresses and strains in very complicated structures, but the complexity arises in the modeling of the structure in terms of the millions of interacting elements, not in the basic equations of elasticity. The equations for elastic bending of beams contains at most two material properties, and 2 or 3 geometric variables. Wear involves upwards of 30 variables, and there may be as many as 100. Thus alternative methods of making progress in designing for desired wear life must be adopted because computer methods simply will not be available for many years. It seems that with each passing year any possibility of computer methods to predict wear becomes more remote because of the high rate of introduction of new materials, lubricants and sliding conditions. New information does not come automatically but rather must be sought out by people with the need for the new information.

The elements in practical tribological design include the following:
   a. Acquire broad knowledge of and experience with the product under development, from design to field use. This includes observing and recording the progression of wear and surface changes of products in field use.
   b. To aid in product development, begin simulative testing of production products. Avoid the temptation to resort directly to simple bench tests since these require more knowledge of tribology than do simulative tests. In the tests, cover a wider range of all variables than expected in the prototype. Prove simulation by examination of surfaces, that is, "ask the material" , on the surface and below what is happening because we humans are not yet adequately informed to know how material respond to the imposition of sliding, etc.
   c. Write equations, where ever possible, incorporating all variables. A frequent problem is to ignore some variables, only to discover later that some particular variable was more important than supposed.
   d. Stay with the product and the equations for a career.

9. Short Summary

Designing of products for desired performance and life needs useful models, and these will develop only from knowledge of the basic mechanisms of WEAR. The field needs much careful and coordinated work, which is lacking at this time. Surely diligent scientists could infuse the field with the scientific method and propel the development of good wear models.

References

Archard, J. F., (1953), "Contact and Rubbing of Flat Surfaces," *Journal of Applied Physics* **24**, 981 - 986.

Evans, A. G., and Marshall, D.B., (1981) "Wear Mechanisms in Ceramics," in *Fundamentals of Friction and Wear of Materials* ASM, 439 - 450, Ed., D.A. Rigney.

Kruschov, M. M., and Babichev, M.A., (1956) *Friction and Wear in Machinery* **11**, a translation by the American Society of Mechanical Engineers from the Russian, Machinistrinia **11**.

Lancaster, J. K., (1963), "The Formation of Surface Films at the Transition Between Mild and Severe Metallic Wear," *Proceedings of the Royal Society of London* **273**, 68 - 89.

Meng, H-C., (1994), *Wear Modeling: Evaluation and Categorization of Wear Models*, PhD thesis, University of Michigan

Welsh, N. C., (1965), "The Dry Wear of Steels I: General Patterns of Behavior," *Philosophical Transactions of the Royal Society of London* **257**, part 1, 31 - 70.

SURFACE DAMAGE UNDER RECIPROCATING SLIDING

S. FOUVRY, Ph. KAPSA
Laboratoire de Tribologie et Dynamique des Systèmes,
UMR 5513, CNRS, Ecole Centrale de Lyon
36, avenue Guy de Collongue
69131 ECULLY cedex – FRANCE
philippe.kapsa@ec-lyon.fr

1. Introduction

Reciprocating sliding is often present in mechanical systems. This particular sliding condition is called fretting when sliding amplitude is small compared to the contact area. Fretting also commonly refers to the degradation of the fatigue properties of a material due to repeated sliding of two contacting surfaces. The small relative displacement amplitudes range typically between 10-100 μm. Examples of practical situations where cracking or wear induced by fretting influence mechanical integrity include such various applications as bolted and riveted joints, key-way-shaft coupling and shrink-fitted couplings. To prevent fretting damage, fretting wear and fretting fatigue have been widely studied for more than thirty years considering macroscopic and microscopic phenomena.

The aim of this paper is to show that surface damage under reciprocating sliding present at least two very important aspects : mechanical and physicochemical. These two aspects have to be considered to understand the elementary phenomena which lead to surface damage. Various experimental results will be presented in order to illustrate these aspects.

2. Fretting

Fretting is a contact loading associated with very small alternated displacement amplitudes. This relative displacement induces a tangential loading, which can be described by the fretting loop : $Q(t)=f(\delta^*(t))$ as shown in Figure 1.

Two fretting condition are identified : partial and gross slip. Partial slip, which is characterised by a closed elliptical fretting loop, is associated with a composite contact of sliding and sticking zones (Figure 1). The gross slip condition which is identified by a quadratic dissipative fretting loop, is related to full sliding occurring over the entire interface.

The friction coefficient changes during the test and this can lead to a change in sliding condition. It is then possible to define various fretting regimes as shown in

B. Bhushan (ed.),
Fundamentals of Tribology and Bridging the Gap between the Macro- and Micro/Nanoscales, 377–391.
© 2001 *Kluwer Academic Publishers.*

Figure 2. These regimes are illustrated in the form of fretting log where the cycles are plotted on a logarithmic scale for the number of cycles. Fretting maps were introduced by Vingsbo et al (1988); Vincent et al (1992) have shown that the damage progression strongly depends on the sliding regimes (Figure 3).

Cracking is mainly encountered in the partial and mixed fretting regimes whereas wear is observed for larger amplitudes in gross slip.

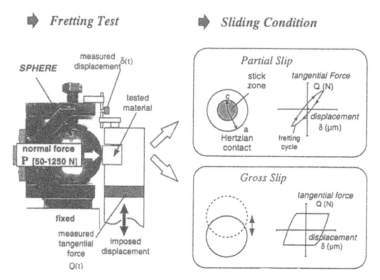

Figure 1. Fretting test with a sphere on flat contact and sliding conditions : partial and gross slip.

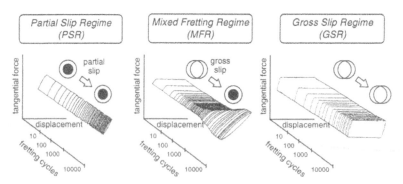

Figure 2. Fretting regimes. The evolution of loop shape indicates the change in the sliding condition during the test duration.

Considering the Mindlin (1949) formalism, the tangential force amplitude which controls the cyclic loading and then the fatigue cracking, strongly increases under partial slip until a maximum value at the transition from partial to gross slip. On the other hand, the interfacial dissipated energy which activates the debris creation and ejection is smaller under partial slip whereas it presents a fast straight rise under gross slip condition.

Therefore to formalise fretting damage, the following aspects must be identified :
- The sliding condition,
- Characterisation of the crack nucleation under Partial and Mixed fretting regimes,
- Quantification of the wear kinetics under gross slip situations.

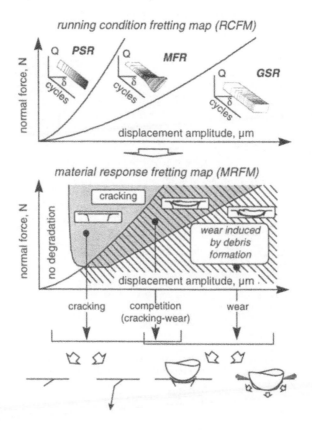

Figure 3. Fretting charts : from the sliding conditions, various material responses can be defined leading to different surface damage.

3. Some mechanical aspects of surface damage

Two aspects of this mechanical processes are considered here : the damage associated to the type of behavior of the material (elastic, plastic, ...) and the damage associated to cracks.

380

3.1 ELASTIC OR PLASTIC BEHAVIOR

Fretting experiments performed with an alumina ball sliding on a sintered steel with various normal load show that the wear behavior is related to the material behavior under friction (Figure 4). The lowest wear coefficient is obtained for elastic behavior. For sliding condition leading to elastic shakedown, the wear coefficient is increased by a factor of 2. For rachetting (or plastic shakedown), the increasing factor is about 10. This indicates the important effect of plasticity on the wear process. Elementary phenomena responsible of loss of matter are nevertheless to be understood in more details.

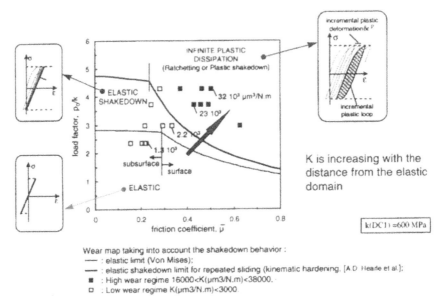

Wear map taking into account the shakedown behavior :
— : elastic limit (Von Mises);
— : elastic shakedown limit for repeated sliding (kinematic hardening, [A.D. Hearle et al.];
■ : High wear regime 16000<K(μm3/N.m)<38000.
□ : Low wear regime K(μm3/N.m)<3000.

Figure 4. Experimental results showing the increase of the wear coefficient K with the material behaviour. Alumina sphere sliding on sintered steel flat in a fretting experiment.

3.2 CRACKING PROCESS

To quantity crack nucleation under partial slip condition, different multiaxial approaches have been proposed, Nowell (1990), Szolwinski (1996), Fouvry (1996). The Dang Van crack nucleation model permits a good estimation of the location of the first crack observed at the contact border (Figures 5, 6) and predicts the cracking risk in small contacts from macroscopic variables such as the bending and shear fatigue limits (σ_d, τ_d). It appears essential to consider the size effect (Figure 6). The stress analysis must be based on the elementary volume representative of the microstructure. The analysis of several aeronautical alloys displaying a similar tempered microstructure (type : 30NiCrMo) concludes that the crack nucleation can be predicted independently of the contact dimension or the fretting situation if the stress imposed to the contact is averaged on a 5 μm edge cubic volume, Fouvry (1999) (Figure 6).

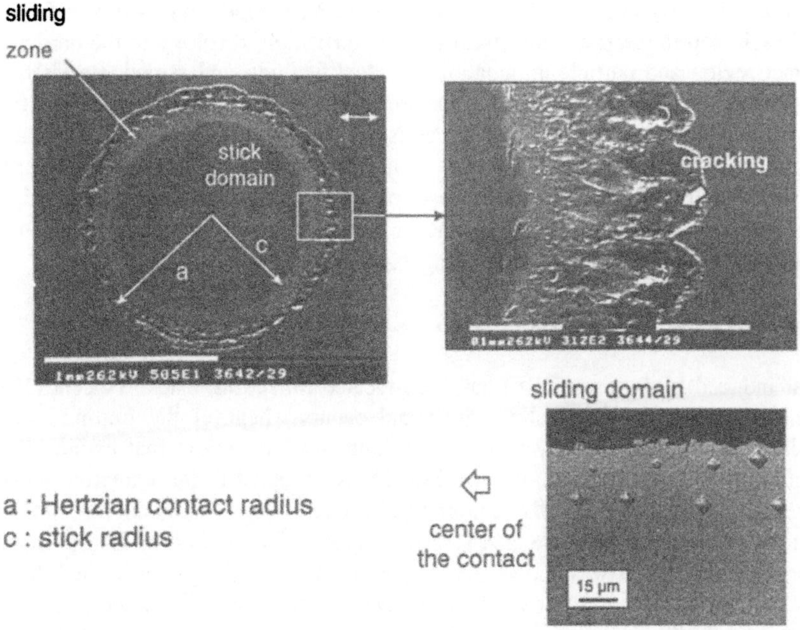

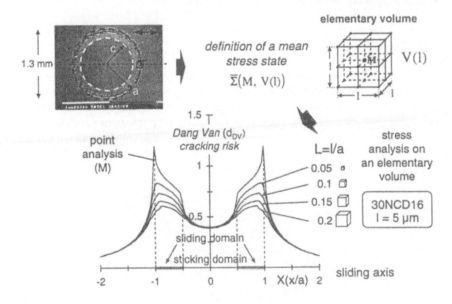

Figure 5. Cracking under partial slip regimes (30NiCrMo, 10^6 cycles), first crack on the contact borders.

Figure 6. Size effect in fretting contacts. The averaging of Dang Van craking risk in an elementary volume modifies its distribution.

4. Physicochemical aspects of damage

Wear is often associated with a loss of matter. Wear modelling will remain a utopia until each step in the process of wear particle formation, rheology of the particle in the contact region and particle elimination are identified and understood, Berthier (1988). Several steps have been identified in the creation of a wear particle so far. The first step is often related to one of the well established wear mechanisms (adhesion, abrasion, fatigue...) which can be described from an accurate analysis of stresses and temperature developed within the contact. In the cases of both fretting-wear and fretting-fatigue, the third body approach is a basic need since a particle must stay for a certain period in the contact after being detached before being expelled.

4.1 WEAR INDUCED BY FRETTING

Wear induced by fretting (WIF) has been located in fretting maps, Vincent (1992). It mainly concerns the gross slip regime and occurs whenever the fretting oscillatory displacement is controlled by vibration (fretting-wear) or by external loading (fretting-fatigue). WIF was analysed for the last 15 years at ECL for numerous coated or uncoated materials. Typically aluminium, copper, iron and titanium alloys were extensively studied, Blanchard (1991), Fouvry (1996), Martin (1996). Except in the case of brittle materials, it was shown that a specific superficial layer is formed during the very first cycles of fretting loading. This layer is called the tribologically transformed structure or TTS. TTS formation is considered as the first step to establish the third-body layer (powder bed) which usually separates the two contacting surfaces and in which the displacement can be accommodated (Figure 7).

4.2 GENERAL FEATURES OF THE TTS

After a fretting test the contact surface was usually covered with oxide debris. After elimination of the debris bed, a chemical etching may reveal zones which react differently to the etching as compared to regions located outside the contact. This was clearly observed after short duration tests for which very few particles detached from the specimens. This transformed structure was called the tribologically transformed structure or TTS. A better observation of TTS was obtained for longer test duration samples after cross - section, polishing and etching.

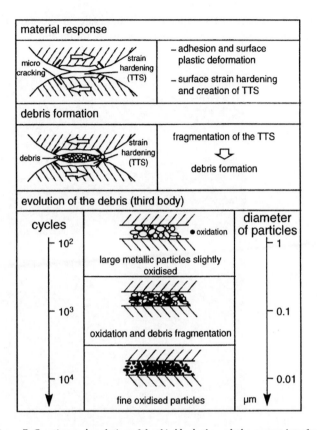

Figure 7. Creation and evolution of the third body through the contact interface.

The general feature of a cross-section observation is presented Figure 8: Whatever the alloy, compacted debris were seen on the top of the wear crack. Below or beside the debris layers, a localized TTS was revealed. Some times a plastically deformed area was noted beneath the TTS. TTS was always revealed (by chemical etching) in a different contrast, light or dark, with respect to the bulk structure. Light or dark contrasts depends merely on etching time but usually not on different chemical reactions of etching.

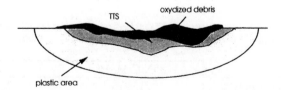

Figure 8 : Scheme of a metallographic cross-section of a fretting scar.

Examples of TTS depending on the bulk material are shown in Figure 9. Cracks could be observed within the TTS: their shape and direction differed for the

384

several alloys. For instance TTS seemed to desintegrate in the case of steels while longer cracks parallel to the surface were observed for titanium alloys.

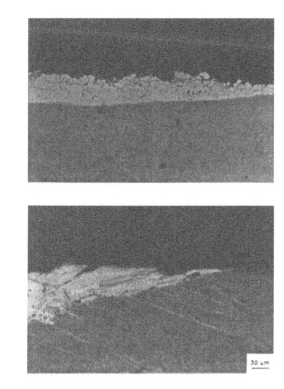

a

b

Figure 9. Optical microscopy observations of the TTS of : a - Maraging M250, b - Ti-15V-3Al-3Cr-3Sn (α).

These differences in crack morphology were related to the crystallography of the TTS. Micro-Vickers hardness measurements always indicated a great increase in hardness in the TTS as compared to bulk material whatever the initial microstructure. Hardness values were very high and much higher than hardness obtained after quenching in the case of steels.

From microindentation measurements, it appears that the Young modulus remained the same as for the bulk material (Table 1).

Table 1. Example of mechanical properties of a TTS in the case of tempered low alloy steel. Young's modulus values were calculated from micro – indentation test results.

	35NiCrMo16	TTS
Young modulus in GPa	215±15	200±20
Hardness in daN/mm^2	670±40	950±120

4.3 TTS PARAMETERS

The analysis of the effect of testing or structural parameters on the formation of TTS were conducted by the measurement of the TTS thickness. Referring to the initial surface and regarding cross-sections, several quantities were considered (Figure 10) i.e. the limit depth, Z_{tot}; the thickness of the remaining TTS, Z_{TTS}; the thickness of the removed TTS, Z_u. Of course $Z_{tot} = Z_{TTS} + Z_u$

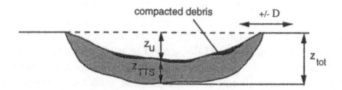

Figure 10. Parameters to be considered in the study of TTS amount variations.

The number of cycles, the crystallographic nature of the bulk material, and fretting parameters, normal load and displacement were considered. The main results are summarised in the following figures. TTS formed very rapidly : fewer than 10 cycles are sufficient to note the first localised TTS. For the wide range of testing conditions, fewer than 1 000 cycles were required to form a well established and stabilised TTS layer. At this stage debris formation rate is diminished (Figure 11).

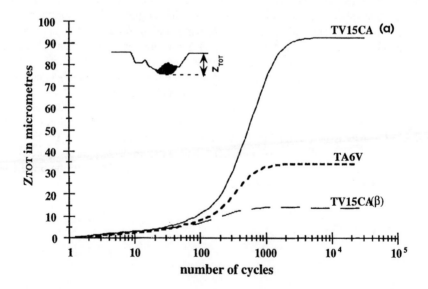

Figure 11.: Variation of Z_{tot} versus number of cycles for various titanium alloys (± 50 μm, 300 N, 1 Hz).

Following Archard's calculations, Vizintin (1995) claimed that flash-temperatures occurring during friction may produce a martensitic structural change in the material, which generates TTS. However, numerical simulations of real contact surfaces, Colombié (1984), as well as experiments, Sproles (1978), gave evidence of a temperature rise not higher than 100°C, for test frequencies lower than 1 Hz. Still, TTS structures were observed even at very low test frequencies. Hence, TTS nucleation may originate from mechanisms different from high temperature rises in fretting.

From their numerous observations and analyses, Rigney et al. concluded that the creation of such structures was linked to the presence of high plastic strains together with transfer phenomena, Rigney (1984), Sawa (1987), Heilman (1983). The authors called such a structure «mechanically mixed layer», thanks to its analogous origin with mechanical alloying, Benjamin (1974). According to this model a second phase is introduced which is absolutely necessary to generate very fine grains. Furthermore, this transformation occurs only if both a mechanical and a chemical process are present. When a counter-face material does not provide this second phase (i.e., in the case of homogeneous contacts), the contact's atmosphere, especially oxygen, enables the chemical process mentioned previously. However, no second phase from a counter-face was ever detected in the TTS analyzed after fretting tests even in heterogeneous contacts. The part played by oxygen remains, nonetheless, uncertain.

4.4 QUANTIFICATION OF THE TTS FORMATION THROUGH AN ENERGY APPROACH

Comparison between experiments on various microstructure and TEM observations indicates that plastic deformation is seen at the origin of TTS formation. Such conclusions require long and fastidious FEM computation. An alternative approach is to consider the interfacial shear work dissipated during the test. Such approach was first proposed to quantify the wear volume. Linear changes in the microstructure with the cumulated dissipated energy were observed which suggest a connection between nenrgy accumulation with wear rate, Fouvry (1995), Mohrbacher (1995).

Such approach that relates the total wear volume with the total cumulated energy is not directly transposable to the present analysis of TTS. A local description of dissipated energy is here required. This analysis is conducted assuming Hertzian shear and pressure field distributions (Figure 12). The maximum density of dissipated energy is observed at the centre of the fretting scar which also corresponds to the point where the wear and TTS analysis was performed.

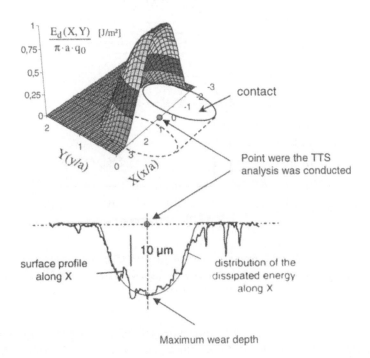

Figure 12. Comparison between the local distribution of interfacial shear work and the surface wear.

The local dissipated energy at the center of the contact is calculated for each fretting cycle and cumulated for the whole test duration taking into account the contact extension. An analysis was done to quantify the TTS formation on a tempered low alloyed steel (35NiCrMo). A 11.5 mm chromium steel ball radius was used as counterbody. At least 40 fretting tests have been performed for fretting cycles between 20 and 10000, normal force between 150 and 500 N and displacement amplitude between ±15 to ±50 µm.

Figure 13 compares the thickness of the TTS thickness with the local cumulated dissipated energy. It indicates that an energy threshold is required to first transform a 40 µm thick TTS layer. The thickness remains constant independent on the successive amount of energy dissipated through the interface.

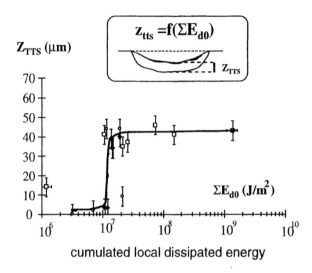

Figure 13. Evolution of the TTS thickness of a low alloyed steel (35NiCrMo) versus the cumulated local dissipated energy : □ : different number of cycles, O : different displacements, ◆ : different normal forces.

An energy threshold of TTS formation is estimated around 10^7 J/m^2 for a 40μm TTS thickness. It corresponds to a critical energy of TTS formation equivalent to 2.5 10^{-7} J to transform 1μm^3 to TTS then : $E_{TTS} = 2.5 10^{-7} J/\mu m^3$.

Figure 14 shows rate of energy dissipation as Z_{tot} increases. During the first stage, the total depth evolution is similar to TTS, no wear occurs. The dissipated energy is associated only with plastic transformation. At the transition of energy, the TTS layer is suddenly generated. Added energy contributes to damaging the TTS surface layer, generating debris and initiating wear. The local energy approach identifies different wear processes beginning with initiation of the TTS layer by plastic strain accumulation. After the TTS stabilization, wear extends in depth through several layers of progression defined first by a subsurface plastic region, a constant TTS layer and then surface wear. The TTS generation kinetics next to the plastic region is assumed to be similar to TTS surface degradation in order to maintain a constant TTS thickness. Reaching such a steady state wear process, a linear progression is usually observed between the wear volume and the cumulated dissipated energy. An energy wear coefficient can then be extrapolated (α_V). Considering that a specific quantity of energy is required to first transform the TTS before the wear process begins, an energy shift must be defined from the linear approximation. This is effectively observed for a different material in Figure 15 which confirms the presence of an energy shift along the energy axis and allows the evaluation of a specific energy of incubation related to the contact configuration (ΔE_{dV}).

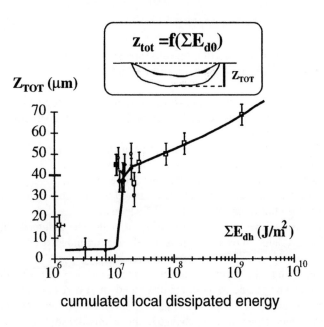

Figure 14. Evolution of the total thickness of a low alloyed steel (35NiCrMo) versus the cumulated local dissipated energy.

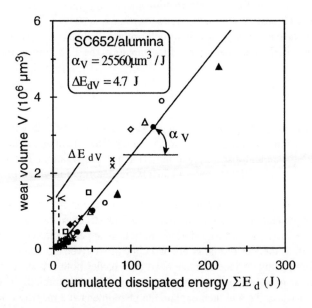

Figure 15. Wear energy analysis of the SC652/alumina (R=12.7mm) contact (P : 50 – 200 N, δ^* : ±25 - ± 200 µm, N : 500 – 10000 cycles); Identification of the wear energy factor (α_V) and incubation energy required by the TTS transformation (ΔE_{dV}).

4.5 SYNTHESIS : TTS PHENOMENON

During reciprocating sliding, a Tribologically Transformed Structure (TTS), from which debris are made, is formed within a very few number of cycles. Some features can be summarized.

TTS appears as a nanocrystalline structure, corresponding to the chemical composition of the initial material and made of the more stable structure regarding the phase diagram of the material.

TTS has similar elastic Young modulus as the original metallic structure but its hardness is significantly higher and can be greater than 1000 Hv.

Contrary to a compacted debris layer, TTS contains a very small amount of oxygen, in the same order of magnitude as observed in the initial bulk material. Oxygen is therefore not the controlling factor of the TTS transformation.

Considering the previous remarks, it can be assumed that the TTS transformation is related to excessive plastic deformation induced by the contact which generates a recrystallization of the microstructure.

Initiated by the cyclic contact loading, it was shown that the TTS transformation can be related to a critical dissipated energy. A threshold energy of TTS formation can be calculated which is related to a critical plastic work density. Below this energy the microstructure is unchanged, above this density, the microstructure is transformed to TTS.

Under the steady state of wear, the TTS layer remains constant which implies that the kinetics of the TTS transformation next to the inner plastic domain must be equivalent the TTS destruction on the fretted surface.

A global energy approach is directly related to wear energy and confirms the necessity of an initial input of energy to first transform the TTS before wear begins.

5. Conclusion

Damage under reciprocating sliding or fretting are related to the following :

The mechanical aspects of damage are related to material behavior (elastic, plastic, racheting, ...) and also to crack initiation and propagation. The wear coefficient is lower when the materials are only under elastic deformation while plasticity induces high wear.

Physicochemical aspects mainly concern surface transformation. Among them, structural changes leading to the formation of a new structure called TTS is important. Considering the dissipated energy during friction, it appears that the formation of this structure as well as the loss of matter are associated to a given amount of energy.

Understanding surface damage need to consider both aspects at various scales. The microscopic or nanoscopic phenomena have to be "assembled" in order to make some scenario to describe and then explain the phenomena at a macroscopic scale.

6. References

Benjamin, J.S. and Volin, T.E. (1974) "The Mechanism of Mechanical Alloying", *Metal. Trans.* 5, 1929-1934.

Berthier, Y., Vincent, L. and Godet, M. (1988) "Velocity accommodation in fretting", *Wear* **125**, 25-38.

Blanchard, P. (1991) "Usure induite en petits débattements : Transformation Tribologique Superficielle d'alliages de titane", PhD Thesis 91-32, Ecole Centrale de Lyon, Ecully, France.

Colombié, C., Berthier, Y., Floquet, A., Vincent, L. and Godet M. (1984), "Fretting: Load-Carrying Capacity of Wear Debris", *J. of Tribology ASME* **106**, 194-201.

Fouvry, S., Kapsa, Ph. and Vincent, L. (1995) "Wear phenomena quantification of Hard coating under Fretting situation", *International Tribology Conference ITC - Yokohama, Oct. 29 - Nov. 2 1995*, 277-282.

Fouvry, S., Kapsa, Ph. and Vincent L. (1996) "Quantification of fretting damage", *Wear* **200**, 186-205.

Fouvry, S., Kapsa, Ph. and Vincent L. (1997) "Tenue et performances d'un dépôt TiN sollicité en Fretting : Quantification de la fissuration et des volumes usés", *Tenue et performances mécaniques des dépôts durs, Rapport CETIM Tome 2*.

Fouvry, S., Kapsa, Ph. and Vincent, L. (1999), "A multiaxial fatigue analysis of fretting contact taking into account the size effect", *ASTM STP* **1367**, 167-182.

Heilmann, P., Don, J., Sun, T.C., Rigney, D.A. and Glaeser, W.A. (1983) "Sliding Wear and Transfer", *Wear*.

Johnson, K.L., (1985) *Contact Mechanics*, Cambridge University Press, Cambridge, UK.

Kalker, J.J. (1990) "Three-dimensional elastic bodies in rolling contact" *Kluwer Academic Publishers*, Dordrecht, The Netherlands.

Martin, B., Vincent, L., Wright, C.S., Eagles, A.E. and Wronski, A.S. (1996) "Wear and cracking of sintered high speed steel matrix composites under fretting conditions", *Powder Metallurgy and Particulate Materials*, Ed. MPIF.

Mindlin, R.D. (1949) Trans. ASME, Series E, *Journal of Applied Mechanics* **16**, 259-268

Mohrbacher, H., Blanpain, B., Celis, J.P., Roos, J.R., Stals, L. and Van Stappen, M. (1995) "Oxidational wear of TiN coating on tool steel and nitrided tool steel in unlubricated fretting", *Wear* **188**, 130-137.

Nowell, D. and Hills, D.A. (1990), "Crack Initiation criteria in fretting fatigue"; *Wear* **136**, 329 - 343.

Rigney, D.A. (1984), "Wear Processes in Sliding Systems", *Wear* **100**, 195-219.

Sawa, M. and Rigney, D.A. (1987), "Sliding Behaviour of Dual-Phase Steels in Vacuum and in Air", *Wear* **119**, 369-391.

Sproles, E.S. and Duquette, D.J. (1978), "Interface Temperature Measurements in the Fretting of a Medium Carbon Steel", *Wear* **47**, 387-396.

Szolwinski, M.P. and Farris, T.N. (1996), "Mechanics of fretting fatigue crack formation", *Wear* **198**, 93-107.

Vingsbo, O. and Soderberg S. (1988), "On fretting maps", *Wear* **126**, 131-147.

Vincent, L., Berthier, Y. and Godet M. (1992), "Testing methods in fretting fatigue: a critical appraisal", *ASTM STP* **1159**, 33-48

Vizintin, J., Podgornik, B., Kalin, M., Pezdirnik, J., and Vodopivec J. (1995), "Three-Body Contact Temperature in Fretting Conditions", *Proc. of the 22th Leeds-Lyon Symposium on Tribology*, Lyon.

WEAR PARTICLE LIFE IN A SLIDING CONTACT UNDER DRY CONDITIONS : THIRD BODY APPROACH

J. DENAPE*, Y. BERTHIER** and L. VINCENT***

*Laboratoire Génie de Production, Ecole Nationale d'Ingénieurs
B.P. 1629, 65016 TARBES cedex – FRANCE
** Laboratoire de Mécanique des Contacts, INSA de Lyon,
20 av A. Einstein, 69621 VILLEURBANNE cedex - FRANCE
*** Ecole Centrale de Lyon, BP 163 ECULLY cedex - FRANCE

The third body approach is a very useful and pragmatic tool for analysing and understanding the friction and wear behaviour of sliding materials. This approach, introduced by M. Godet in the middle of the 1970's, is based on the concept of dynamic screen played by the whole debris (« third body ») detached from the rubbing surfaces (« first bodies ») and trapped in the contact zone. More recently, Y. Berthier proposed additional concepts for a systematic structuration of friction and wear analyses including scale factors and dynamic interactions. The speed accommodation occuring within the contact zone, is then described in terms of sites and modes which migrate and change during the friction process.

The aim of this paper is to give a review of such phenomenological models which involve mechanical, material and physico-chemical disciplines and extend over macroscopic to microscopic scales. The first part of the paper highlights the interactive contributions of each element constituting any tribological system. The second part is devoted to reconstructing the contact history through the evolution of each element of the tribological system. The validity and the efficiency of such an approach is illustrated by varied test results and examples of practical applications.

1. Interactive Contributions of Triboelements

1.1. TRIBOLOGICAL SYSTEM AND THIRD BODY CONCEPT

The contact behaviour results from the interaction of various parameters pertaining to macroscopic and microscopic scales. The analysis of any tribological process requires a structured approach considering such scale factors so as to identify their effective contribution. Therefore a *tribological system* must be defined considering three scale levels (figure 1) :

B. Bhushan (ed.),
Fundamentals of Tribology and Bridging the Gap between the Macro- and Micro/Nanoscales, 393–411.
© 2001 *Kluwer Academic Publishers*.

- The *working device* (mechanism) determines the contact operating conditions by transmitting static and dynamic loads and by imposing a kinematic and environment,
- The *contacting elements* (called « *first bodies* ») react to the load imposed by the working device by bulk deformation and degradation (particle detachments, cracks, surface transformations) and accommodate a part of the velocity difference between the contacting materials,
- The *interface elements* (called « *third body* ») transmit the load imposed by the working device from one solid to the other, separate the frictional surfaces, control their degradation, and accommodate the major part of the velocity difference.

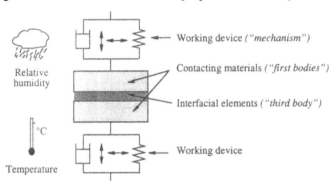

Figure 1. Synthetic view of a tribological system showing three different scales of analysis.

Third bodies are present between virtually all practical contacting and sliding surfaces. They include *static screens* covering the surfaces of first bodies (adsorbed gases, organic contamination, oxide films…) as a consequence of environmental interactions (typical size : 0.3 – 100 nm). They also refer to *dynamic screens* (typical size : 0.1 to 100 μm) induced either by the damage process of first bodies (wear particles) or by being intentionally introd uced as a lubricant (solid, liquid or gas). Such an approach is a development from the lubrication concepts applied to dry friction.

1.2. STATIC AND DYNAMIC SCREENS

The existence of *static screens* controls adhesion phenomena between two solids pressed together. In air and even in a vacuum, no adhesion occurs between two surfaces because of the presence of such contamination films. In contrast, carefully cleaned surfaces (ionic etching) lead to strong adhesion and damage surfaces in vacuum as a result of physico-chemical bonding possibilities (figure 2). Furthermore, these screens can be destroyed by friction.

The action of *dynamic screens* have been visualised by an original study using a piece of brittle common blackboard chalk (calcium carbonate) rubbing at low speed against a flat frosted glass disc (Play et al. 1977). The process of third body formation, motion and elimination was observed by transparency. Two regimes were noted, the first during the initial rotation when the disc was clean, the second during the following rotations when the wear trace was formed.

Observations during the initial pass have shown that (figure 2) :
— a part of the detached particles runs across the whole pin section,
— particle velocities are not equal among themselves and are lower than those of the disc,
— the rest of the particles is immediately eliminated from the contact area,
— two wear mechanisms are observed on the pin : a smooth frontal area resulting from a direct abrasion between the bulk chalk and glass, while at the contact exit, a rougher zone made out of compacted particles separates the glass from the chalk bulk.

During the following passes, it has been observed that :
— particles are reintroduced in the contact zone,
— their accumulation leads to the separation of the sliding surfaces by formation of a chalk powder bed which acts like a solid lubricant,
— the wear rate of chalk is consequently highly reduced.

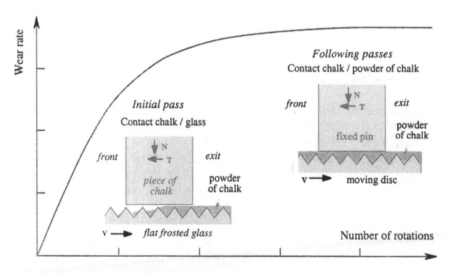

Figure 2. The wear of chalk : separation of the sliding surfaces by a dynamic screen of debris, reduction of the wear rate by a load-carrying phenomenon generated by the particle motion (Play et al. 1977). N : normal load, T : tangential force, V : disc velocity.

These observations suggest that motion of a sufficient quantity of chalk particles can generate a mechanical *load-carrying phenomenon* which protects the sliding surfaces as occurs in fluid film lubrication : it's a matter of surface autoprotection by the debris of the surface itself.

The quantitative wear rate of chalk is highly determined by the contact conformity and the device stiffness (figure 3). Actually, the wear rate is related to how long debris remains in the contact zone : for a same contact area, the wear rate is significally reduced if the higher length of contact is turned facing the moving direction.

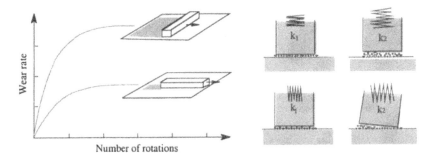

Figure 3. Contribution of the contact configuration (first bodies level) and of the device stiffness (mechanism level) : a) the wear rate is determined by the stay duration of debris in the contact zone, b) Each test displays the signature of the tribometer.

Furthermore, with a machine of weak rigidity, the entry conditions are modified and the geometric conformity is lost. The load is then continuously redistributed and the contact conditions, compared to rigid tribometer, are totally different, yielding, as expected, dissimilar results. Finally, the machine dynamics, in a vibrating environment as is often found in industrial practice, can strongly influence wear by favoring the elimination of particle powder.

1.3. DEBRIS BEHAVIOUR AND WEAR MODELLING

Wear mechanisms are relevant to a linking of stages with their own kinetics and high mutual interactions. Wear modelling can be described in three stages focused on the emission and flow of debris (Godet 1984, 1990) :

- The first stage is the *formation of debris* : surface damage produces detachment of particles by abrasion, adhesion, delamination... Such material loss, that comes from the surfaces, is referred to an « apparent » wear rate which actually corresponds to a *source flow* of third body.
- The second stage corresponds to the *circulation of debris* : particles are trapped and circulate within the contact zone (*internal flow* of third body between first bodies). Their accumulation generates a mechanical load-carrying effect which leads to the separation (complete or partial) of the contacting surfaces.
- The third stage concerns the *elimination of debris* : debris are finally, either temporarily or definitively, ejected outside the contact zone. The « real » wear rate is specific to this latter case. It corresponds to the material loss that comes from the contact zone, i.e. the *wear flow* of third body.

This approach highlights that a wear rate results from an equilibrium between the detachment and elimination of debris : « *A good friction pair is one which is willing to sacrifice its surface to save its volume* (Godet 1984) ». Therefore, the tribological performances of a material pair are mostly related to their aptitude for generating a tough third body, an aptitude strongly relevant to environmental parameters (load, speed, temperature, contact shape, device design...).

This approach naturally explains why the coefficient of friction and wear resistance are not intrinsic properties but only use parameters which are strongly dependant on the whole tribological system. In the same way, extensive variations in wear rates recorded in apparently similar conditions, can be attributed to minor differences in machine design. Such differences cause, for example, changes in natural frequencies which considerably modify the debris flow.

Similary, the multilaboratory tribotesting round, assumed by the VAMAS program in 1982, comes from the basis that the absolute values of friction and wear may differ considerably from test equipment to test equipment and from laboratory to laboratory. Alternatively, good reproducivity has often been reported for tests performed in individual laboratories. Thus, similar tribological tests, performed under well defined conditions (test system, operating parameters, atmosphere, temperature, surface preparation...), have been conducted with αAl_2O_3 ceramic and AISI 52100 steel (100Cr6) combinations by a large number of laboratories from different countries (Canada, France, Germany, Italy, Japan, the United Kingdom and the United States of America). Friction coefficients obtained with steel/steel pairs varied from 0.4 to 0.9 corresponding to a reproducivity (in terms of relative standard deviation) of about 20 % (109 data from 26 laboratories) while reproducivity of wear data reached 19 to 38 % (47 data from 11 laboratories). However, the repeatativity within one laboratory was respectively 9 to 13 % for friction coefficients and 14 % for wear data (Table 1, Czichos et al, 1987, 1989).

TABLE 1. Reproducibility and comparability of friction and wear data. Results of the multilaboratories tribotesting (VAMAS). Tested materials : AISI 52100 steel and Al_2O_3 ceramic. Testing device : pin-on-disc configuration (ball diameter : 10 mm, disc diameter : 40 mm, track diameter : 32 mm). Testing conditions speed : 0.1 m/s, applied load : 10 N, sliding distance : 1 km, relative humidity : 50 ± 10 %, room temperature : 23 ± 1°C under dry conditions (Czichos et al, 1987, 1989).

Ball Disc	Steel Steel	Ceramic Steel	Steel Ceramic	Ceramic Ceramic	Steel Steel
Friction coefficient	0.60 ± 0.11	0.76 ± 0.14	0.60 ± 0.12	0.41 ± 0.08	0.59 ± 0.25
Number of data	109	75	64	76	83
Number of lab.	26	26	23	26	
Wear rate (µm/km)	0.70 ± 20	very small	81 ± 29	very small	101 ± 40
Number of data	47		29		60
Number of lab.	11		11		

	Tribological quantity	Repeatatility (%) within one lab.	Reproducibility (%) between laboratories
D = 1 km T = 23°C RH = 50±10% F_N = 10 N v = 0.1 m/s	Friction coefficient	± 9 to ± 13	± 18 to ± 20
	System wear rate	± 14	± 29 to ± 38

1.4. APPLICATION TO FRICTION AND WEAR OF CERAMICS

The friction and wear behaviour of four structural ceramic pairs has been investigated in dry condition at room temperature using a rotating tribometer. This tribometer rotates a ceramic roller (with a transverse radius of curvature) against a flat beam of the same ceramic. This contact geometry favours the concentration of the wear debris in the sliding interface (Denape et al. 1990).

398

Microscopic examination revealed an important quantity of debris within and also outside the wear track, indicating that it is either eliminated from the sliding interface or recycled. Two types of debris were identified :

— Free individual rounded particles, with a size of 0.1 to 0.5 μm (figure 4a),

— Compacted debris in more or less large dense films adherent to the worn surfaces whose examination often requires removal of the layer of fine particles which masked the subjacent material (figure 4b).

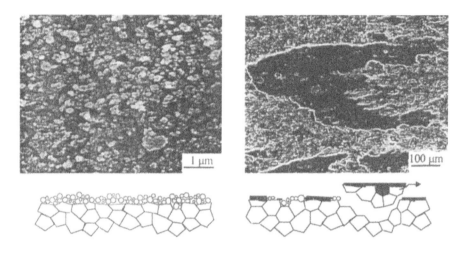

Figure 4. Friction and wear of ceramics : a) Bed of individual particles : low wear by load carrying effect of debris, b) Adhesion on the surfaces : high wear as a result of a lower load-carrying capacity of debris.

Friction coefficient and wear rates (expressed in $m^3.N^{-1}.m^{-1}$, i.e. in Pa^{-1}) do not imply significative influence on applied load for a constant sliding speed. In contrast, the sliding speed is, with this contact configuration, a preponderant factor which reveals fundamental features, and highlights the dynamic aspect of wear mechanisms. Plots of wear rate and friction coefficient against the sliding speed have shown that (figure 5) :

— The wear rate and the friction coefficient always exhibit opposite changes with increasing sliding speed,

— There is a critical sliding speed leading to a minimum wear rate and a maximum friction coefficient at about 0.5 m/s for all four ceramics so that the minimum wear rate corresponds paradoxically to the maximum friction coefficient.

— At increasing sliding speeds, below the critical speed the wear rate decreases, and then increases beyond it while the opposite trend is observed with the friction coefficient.

These unexpected results can be clearly explained on the basis of the third body approach. The influence of sliding speed on the friction coefficient and wear rate can be explained in terms of accumulation and elimination of debris :

— Below the critical speed, increasing speed causes wear rate reduction, indicative of a rise in load-carrying capacity of debris. This phase confirms an accumulation of debris inside the contact zone which corresponds to a rise in the friction coefficient.

— Beyond the critical speed, the friction coefficient decrease and the wear rate increase indicate a reduction of the quantity of particles in the contact zone which leads to a reduction of load-carrying capacity of debris.

— The critical speed marks the limit of both phases of accumulation and elimination. This speed is related to the centrifugal action of the rotating roller which becomes sufficient to expulse debris from the track definitively and prevents their recycling. This is a characteristic of the device kinetic : it is logically independent of the ceramic.

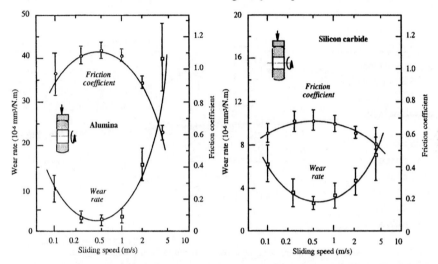

Figure 5. Friction and wear of ceramics : Variations of wear rate and coefficient of friction with sliding speed (5 N). The critical sliding speed at 0.5 m/s is related to a maximum of friction coefficient caused by a significative amount of debris trapped inside the contact zone which leads to a minimum of wear rate by an optimal load-carrying effect (Denape et al. 1990).

These results establish that, in dry friction, variations in the friction coefficient reflect the evolution of the quantity of debris in the sliding interface while variations in the wear rate are determined by the dynamic equilibrium between production and elimination from the wear track.

Similar experiments, performed on the same tribometer but in the presence of distilled water to enhance the escape of wear debris, induced a drop in the friction coefficient and, in parallel, a marked wear rate increase when compared to dry friction conditions. As expected, microscopic observations confirmed that all the debris was removed from the sliding interface. This indicates that the elimination of debris causes wear rate increase and therefore that wear particles present an effective load-carrying effect.

However, wear debris has a dual action, depending on their quantity inside the interface :

— When isolated, it interacts with the sliding surfaces by plowing and abrasion, which increases the friction coefficient (aggressive aspect),

— When numerous enough, it presents load-carrying capacity by forming layers separating the sliding surfaces, which decreases the wear rate (protective aspect).

2. Tribological life of a contact

2.1. MECHANICAL RESPONSE OF A TRIBOLOGICAL SYSTEM

The relative displacement of two contacting solids implies a velocity gradient through the interfacial elements edged by the solids. More generally, the velocity accommodation can be localised at different sites and produced by different modes (Berthier et al. 1990, 1992).

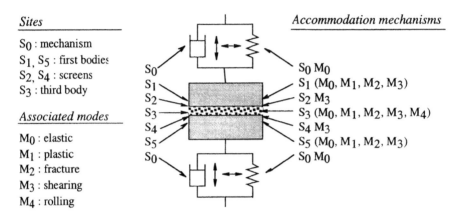

Figure 6. Mechanisms of speed accommodation in the tribological system.

In a basic contact, the sites of accommodation, referred to as S_i, should be (figure 6) :
— S_0 in the working device (mechanism),
— S_1 and S_5 in the skin of the sliding materials (first bodies),
— S_2 and S_4 in the natural screens (interface separating first and third bodies)
— S_3 in the volumic part of the third body.
The modes, referred to as M_j, cover the basic concepts of fracture mechanics and material behaviour, that is[1] :
— M_0 the elastic mode,
— M_1 the plastic mode,
— M_2 the fracture (normal cracking) mode,
— M_3 the shearing mode,
— M_4 the rolling mode.
The combination of one site and one mode, referred to as S_iM_j, describes the velocity accommodation mechanism occuring on the observed spot. That leads theoritically to 24 possibilities of accommodation mechanisms but, for symmetry reasons, only 8 of them are usually observed in practice (figures 7 and 8).
— $(S_0+S_1) M_0$: elastic deformation in the solids (mechanism, first bodies and possibly third body). This mechanism plays a significant role in the quasi static

[1] The plastic mode has been added to the present description with regard to the original approach proposed by Berthier et al. (1990).

contacts. It is observed in fretting conditions when total adhesion occurs between contacting surfaces. The imposed displacement is then accommodated in the samples and their supports.

— S_1M_1 : shallow or deep deformation without material loss (superficial creep) essentially governed by plastic deformation.

— S_1M_2 : superficial cracking of the surface frequently observed on brittle solids (Hertzian cracks). Displacement is then taken in account by the elastic deformation in the first body whose stiffness is reduced by the presence of cracks.

— S_1M_3 : superficial shearing leading to the detachment of particles.

— S_2M_3 : shearing in partition screens, characterises a wall sliding mechanism.

— S_3M_0 : elastic deformation of interfacial zone occurs for extremely short displacements (some micrometers) when an elastic comeback at discharge can be observed.

— S_3M_1 : plastic deformation of particles or of transfer layers.

— S_3M_2 : cracking phenomenon on transfer films or grinding of particles.

— S_3M_3 : shearing of third body commonly observed in numerous situations (main mechanism in fluid lubrication).

— S_3M_4 : roll formation or spherical agglomerates of debris inside the contact area.

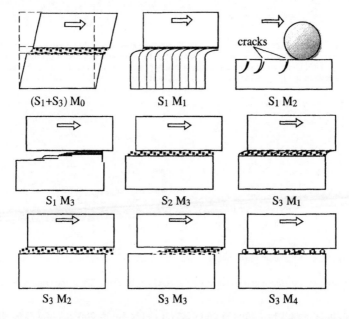

Figure 7. Identification of elementary accommodation mechanisms in a sliding system : a) Elastic deformation (S_1+S_3) M_0 of contacting elements and third body (fretting conditions), b) Superficial creep S_1M_1, c) Hertzian cracks S_1M_2 (brittle materials), d) Shearing of first body S_1M_3, e) Sliding in separating screens S_2M_3, f) Plastic deformation of third body, g) Normal cracking in the third body S_3M_2, h) Shearing of third body S_3M_3, i) Rotation of third body S_3M_4.

402

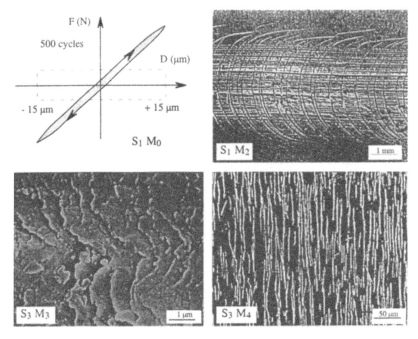

Figure 8. Identification of elementary accommodation mechanisms in a sliding system : a) S_1M_0 associated to the stick regime in fretting conditions (F(N) : tangential force versus number of cycles, D : displacement amplitude), b) S_1M_2 Hertzian cracks on glass in a contact steel (ball) / glass (disc), c) S_3M_3 shearing of debris in a Al_2O_3 / Al_2O_3 contact, d) S_3M_4 roll formation in a SiC / SiC contact.

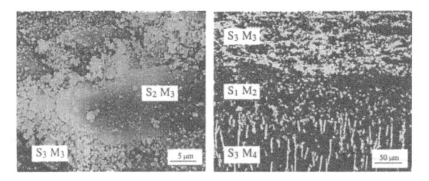

Figure 9. Coexistence and relays among several mechanisms within the contact zone.

In dry friction, several mechanisms can coexist in the same contact and take over during the displacement : at a given time, the contact is subdivided into active zones, disactive zones or re-active zones. The materials « choose » to locally activate the suitable accommodation mechanism capable of side stepping the locking effect imposed by the surface adhesion. The active site is the one for which the stress field exceeds the admissible stresses. Surface roughness, screens composition can favour the release of a mechanism to another (figure 9).

2.2. APPLICATION TO AN ARTIFICIAL JOINT

The validation of a hip prosthesis needs accurate data with regards to the tribological behaviour of the material pairs for a geometric and kinetic configuration as close as possible to the physiological reality. Simulations are usually carried out with working-out tribometers using a flexion-extension movement with a specific load cycle. Results reported here, were performed using femoral heads (referred to as S_5) made of alumina (99.7 Al_2O_3) or zirconia (TZP) ceramics and sockets (referred to as S_1) made of polyethylene (UHMWPE). Tests were conducted in distilled water at 37°C for a rotation angle of ± 15°. The applied load during flexion was 350 daN (6.4 MPa) while 80 daN during extension (stairs ascent simulation) at a frequency of 0.33 Hz (3 cycles per seconde) during 2.10^6 cycles (7.7 days). The surface degradation on the polyethylene sockets directly accounts for the imposed kinetic and the load cycle. Three distinct zones were observed (figure 10) :
— An equatorial back zone (zone 1), located at the beginning of the flexion (back edge of the socket) where high friction coefficients ($\mu \approx 0.3$) were recorded, was characterized by abrasive damage associated with permanent deformation (creep). This zone corresponds to the debris formation (source zone) and the accommodation mechanism was identified as $S_1(M_2+M_3)$.
— A polar zone (zone 2) of a low friction coefficient ($\mu \approx 0.05$), has shown thin superposed layers of debris beneath which the initial surface with machining striations still remained (protective effect of the debris). This zone, mechanically never released (permanent contact), corresponds to the debris retention (relay zone) associated to an S_3M_3 accommodation mechanism.
— An equatorial front zone (zone 3), located at the beginning of the extention (front edge of the socket) where high friction coefficients ($\mu \approx 0.35$) were recorded again, showed decohesion and fracture of the layers. This zone, totally released periodically, corresponds to the debris ejection (well zone) now associated to a $S_3(M_2+M_3)$ mechanism.
The originality of this particular configuration lies in the fact that, in a classic tribological situation, such zones are usually spread over the whole contacting surface and then remain undistinctible.

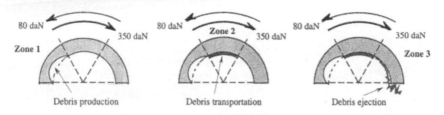

Figure 10. Hip prosthesis simulation (Boher 1992) showing the location of a) debris formation of UHMWPE (S_1M_2, source flow, equatorial back zone), b) debris transportation (S_3M_3, internal flow, polar zone) and c) debris ejection (S_3M_2, wear flow, equatorial front zone).

2.3. RESPONSE OF THE CONTACTING ELEMENTS AND DEBRIS EVOLUTION

Besides surface geometrical evolution (roughness change), the material surface may be subjected to mechanical and microstructural evolutions during friction. High local pressure may produce metal working (hardening mechanism) while high contact temperature usually leads to a drop in hardness with quenched steels (softening mechanism) equivalent to an over tempering effect (figure 11). Moreover, these mechanical and thermal conditions favour the formation of new superficial phases, called *Tribologically Transformed Structures* (TTS).

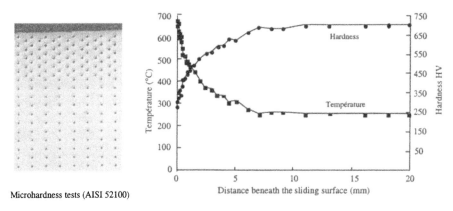

Microhardness tests (AISI 52100)

Figure 11. Cross section of a AISI 52100 steel pin (100Cr6) after sliding on alumina at 1.5 m/s, 360 N showing a hardness drop caused by heat generation during friction. The temperature curve is evaluated by a hardness-time-temperature equivalence parameter and the tempering curve of the steel (Dalverny 1998).

Such superficial transformed layers have been identified and analysed on various kinds of metallic alloys in fretting conditions (Fayeulle et al. 1991, Zhou et al 1992). Some general features have been deduced from these studies (figure 12) :

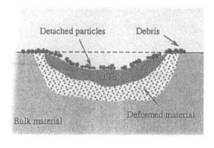

Figure 12. Tribological transformed structures observed in fretting conditions : a) Diagram of a superficial transformed layer, b) Cross section of maraging steel (Fayeulle et al., in Dowson 1992).

— The transformed layer is formed from the most stable phase, whatever the initial structure : TTS of steels is always ferrite while the one of titanium alloys is the α phase, even with β alloys.

— The chemical composition of the transformed layer is similar to that of the bulk alloy. That is not an oxide layer produced during friction (no increase in the oxygen content).

— The structure of the transformed layer is made of small crystallites (TEM revealed grain size ranging from 20 to 50 nm).

— The hardness of the transformed layer is highly increased compared to that of the initial alloy. This effect is probably related to the small grain size.

— The thickness of the transformed layer (20 to 100 μm) depends on the alloys.

— Below the transformed layer, deformation features are often detected : deformed grain boundaries, twins or shear bands.

Any TTS occuring on a tribological surface, involves a modification of the critical stress and strain field in the vicinity of the contact zone. Over-stresses are thus asociated to superficial cracking phenomenon while over-strains are related to superficial creep (figure 13). The dominating response is, of course, specific to the material nature and predetermines the velocity accommodation mode. The TTS often leads to the detachment of particles, therefore to the formation of the third body, i.e. the S_3 site.

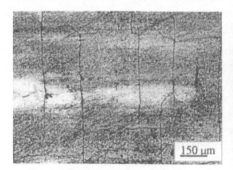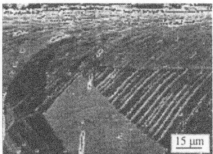

Figure 13. Modification of the material critical stress and strain field as a result of tribological transformed structures (TTS) : a) superficial cracking by over-stress, b) superficial creep by over-strain.

The detached particles are also subjected to further physico-chemical and rheological changes. A physical and chemical analysis (XRD) of a steel / steel contact in vibratory regime showed distinct features of debris trapped in the contact area with time (Colombié et al. 1984).

When just detached, particles have an average size of about 1 μm and showed a metallic coloration corresponding to a major αFe composition. Then, they are grinded and their average size decreases quickly. The small size of such debris, characterized by a high specific surface typically ranging between 20 and 600 m²/g, favours a high chemical reactivity with the atmosphere. Final particles exhibited a size of approximately 0.01 μm with a reddish color attributed to a complex mixture of αFe, Fe_2O_3 and Fe_3O_4.

406

However, this chemical interactivity happens through contact openings controlled by clearences, roughness, unloading spots between touching asperities. Therefore, the physico-chemical action is adjusted by the mechanism dynamics.

Finally, the competition between physico-chemical and mechanical processes controls the debris rheology and the wear flow. The microstructural evolution of both contacting surfaces and debris also modifies their surface energy and therefore the adhesion processes.

2.4. RECONSTRUCTION OF THE DYNAMIC HISTORY OF A CONTACT

The *tribological life of the contact* can be reconstructed according to the interactive contributions of each element of the tribological system. The life of a contact is divided into three stages (Noll 1998, figure 14) :

- A *conception stage* corresponding to the initial mechanical adaptation of both the working device and the contacting solids and where the static screens play a main role.
- A *birth stage* corresponding to the third body formation by particle detachment and the first bodies response including superficial structural transformations of the contacting materials.
- An *actual life stage* corresponding to a dynamic equilibrium between the trapping of the debris inside the contact zone and their ejection outside the contact zone. The load-carrying effect is controlled by rheology and cohesion of the wear particles.

High adhesion mechanism favours the formation of film transfers stuck on the first bodies which traps debris but also reduces the third body flow : the stress field is insufficiently decreased and cracks can form in the contacting bodies. In contrast, low adhesion mechanism prevents debris agglomeration and a bed of powder can be come established between the contacting surfaces. Under favourable trapping conditions (contact dynamic and component shape), the velocity difference is mainly accommodated in the third body which prevents the contacting surfaces from further damage.

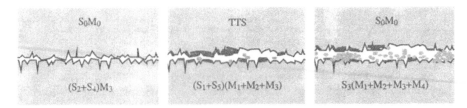

Figure 14. Reconstruction of the dynamic history of a contact : a) conception stage : elimination of natural screens and effective contact of asperities, b) birth stage : first bodies response (TTS) and third body formation (particle removal) and c) actual life stage : third body dynamic (particle transformations, evacuation and elimination of particles.

The two first stages usually occur very quickly. In fretting conditions, the three stages can be quite easily separated by plotting the successive cycles defined by the tangential force as a function of the displacement amplitude along a time axis (3D representation referred to as « friction log », figure 15). For a steel/steel contact in a fully slip regime where the cycles remain nearly rectangular, the variation of friction coefficients indicates the stage transitions.

— The first cycles constitute a run-in stage corresponding to the elimination of the natural screens by an accommodation mechanism of an (S_2+S_4) M_3 type. This conception stage is associated with a rise in the friction coefficient as a consequence of an increase in effective metal / metal contacts.

— The following cycles correspond to the first bodies response that is the birth stage. Plastic deformation, transformed layers formation and debris detachment occur, associated to a decrease in the friction coefficient. The accommodation mechanism is now essentially identified by $(S_1+S_5)M_1$.

— The steady state regime is reached when the debris trapped in the contact zone is numerous enough to form a bed of powder. This regime represents the own life stage : the friction coefficient is stationary and the accommodation mechanism is now located within the third body as S_3M_3.

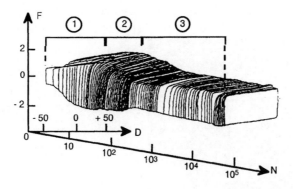

Figure 15. Friction log showing the three stages of contact life in fretting conditions (Fouvry et al. 1995).
Zone 1 : conception stage corresponding to the elimination of natural screens $(S_2+S_4)M_3$.
Zone 2 : birth stage corresponding to the debris detachment $(S_1+S_5)M_3$.
Zone 3 : actual life stage corresponding to the formation of the bed of powder (S_3M_3).

2.5. APPLICATION TO COATINGS FOR LENSES

Transparent plastic is widely used for optical applications such as corrective lenses. But this material is unfortunately not scratch resistant. Its damage leads to a loss of optical properties by reducing light diffusion. Coating of these plastics by a harder material seems to be a means of overcoming this disadvantage. The coatings presented here are thin films (thickness of about 5 µm) containing colloidal particles of silica embedded in an organic compound deposited by a dipping technique. Tribological experiments were performed on an alternating pin-on-plate tribometer using an hemispheric pin made of polished 52100 steel (Etienne et al. 1996).

The analysis of friction damage can be successfully interpreted again using the third body concept and the velocity accommodation mechanisms approach. Three different stages can be discussed in terms of debris behaviour in the contact zone :

— The conception stage can be linked to the damage initiation by cracking without material loss. This first stage is characterized by semi-circular cracks (Hertzian cracks of both substrate and coating) corresponding to an elastic response of the sample (figure 16a). This crack pattern originates from the stress distribution in the material which is induced by the pin displacement. The cracks occur in the tension zone (mode I) behind the pin and show a well-known semi-circular shape whose concave curvature is oriented with the direction of the pin motion. No structural change was registered on the contact surface. By referring to the site S_{1S} for the substrate and the site S_{1C} for the coating, the appropriate accommodation mechanism is $(S_{1S}+S_{1C})(M_1+M_2)$.

— The birth stage starts with the particle emission by crack interaction. With a reciprocating displacement of the pin, the surface is subjected to two arrays of cracks oriented according to the two opposite sliding directions (figure 16b). This second stage involves the destruction of the entire coating by the removal of large particles of coating. The film adhesion plays an important role as it favours or delays the occurence of this stage. The associated accommodation mechanism can be labelled as $S_{1C}M_2+S_{1i}M_3$ where S_{1i} is referring to the substrate / coating interface.

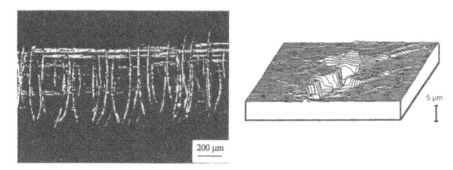

Figure 16. a) Damage initiation by cracking of both substrate and coating : $(S_{1S}+S_{1C})\ M_2$, b) Coating destruction by shearing of the interface substrate / coating : $S_{1C}M_2 + S_{1i}M_3$.

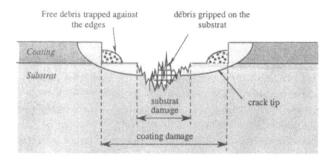

Figure 17. Schematic picture of the wear profile on coated plastic (Etienne et al. 1996).

— The actual life of the contact is mainly related to the **debris behaviour** (accumulation and elimination) inside the friction track. During this final stage, the detached particles are finely ground inside the contact zone that is a S_3M_2 mechanism. As a consequence of the hemispheric shape of the pin, the ground particles are drained off the borders of the wear track where the pressure is lower. A part cannot escape and pile up against the steep lateral sides of the track where they act as a protective screen by assuming a load-carrying effect by an S_3M_3 mechanism (figure 17). In the center of the track, where the pressure is the highest, the particles are not numerous enough to assume their role as a screen. On the contrary, they stick to the sliding surfaces and contribute to the wear damage by abrading the opposite material. The substrate is then affected by an $S_{1S}M_3$ mechanism.

3. Conclusion

A realistic analysis of a practical tribological application requires one to consider all the contributions of the *tribological system* with respect to different scales including the *working device* (mechanism), *contacting materials* (first bodies) and *interfacial elements* (third body).

- At the *mechanism* level, friction and wear are mainly influenced by the *stiffness* of the working device in which the contact is located. This is a mechanical effect acting at the scale from the meter down to the centimeter.
- At the *first bodies* level, the *contact geometry* is a significant feature. This is also a mechanical characteristic but at around the millimeter scale.
- At the *third body* level, *adhesion* of particles on the first body controls interactions between the contacting surfaces. This is a physico-chemical action related to the material nature which acts on the scale of some nanometers.

The global contact behaviour results from dynamic interactions between each element of the tribological system. Neglecting one of these contributions can lead to erroneous conclusions or hazardous predictions. Engineers must be convinced that tribology cannot always save a contact already compromised by a weak mechanical design and that every test result includes the tribometer characteristics. In the same way, a third body always separates the contacting materials and imposes a dominating action. The presence of debris inside the contact zone does not necessarily increase wear : generally in a dry contact, their protective action can be higher than their possible noxious action.

During friction, the velocity gradient between contacting bodies can be accommodated in various sites and different modes. These sites and modes migrate and change inside the contact zone during the sliding process (relay mechanisms) depending on the mechanical and physico-chemical local conditions.

- *Friction* is relevant to the instantaneous changes in sites and modes of velocity accommodation within the contact zone. Transitions in friction coefficients give information for separating the successive stages which define the contact life.
- *Wear* must be analysed as a dynamic accommodation mechanism considering emission, retention and ejection of debris (particle flow). Adhesion, abrasion, delamination, tribocorrosion... are only mechanisms of particle detachment which are only one stage of the overall wear process.

However, the third body approach is not a predictive model but it provides a useful and structuring tool, a general and unifying method based on experience for the mechanical engineer and it also identifies the multiscale interactions occuring in a tribological process.

References

Berthier Y., Godet M. and Brendle M. (1989), "Velocity Accommodation in Friction", *Tribology Trans.* **32**, 4, 490-496.

Berthier Y. (1990), "Experimental Evidence for Friction and Wear Modelling", *Wear* **139**, 1, 77-92.

Berthier Y., Vincent L. and Godet M. (1992), "Velocity Accommodation Sites and Modes in Tribology", *Eur. J. Mech. A/solids*, vol **11**, 1, 35-47.

Boher C. (1992), "Experimental simulation of the tribological behaviour of hip prosthesis materials", Thesis of the Polytechnic Institut of Toulouse (France), n°561, 156 pages (in french).

Colombie C., Berthier Y., Floquet A., Vincent L. and Godet M. (1984), "Fretting : Load Carrying Capacity of Wear Debris", *J. of Tribology*, vol **106**, 194-201.

Czichos H., Becker S. and Bexow J. (1987), "International Multilaboratory Sliding Wear Tests With Ceramics and Steel", *Wear* **135**, 171-191.

Czichos H., Becker S. and Bexow J. (1989), "Multilaboratory Tribotesting : Results From the Versailles Advanced Materials and Standards Programme on Wear Test Methods", *Wear* **114**, 109-130.

Dalverny O. (1998), "Tribological Life at High Temperature and Evolution of Surface Temperature Rise in Ceramic Contacts", Thesis of the University of Bordeaux I (France), n° 1841, 232 pages (in french).

Denape J. and Lamon J. (1990), "Sliding Friction of Ceramics : Mechanical Action of the Wear Debris", *J. of Mat. Sci.* **25**, 3592-3604.

Dowson D. et al. editors (1992), "Wear Particles : From the Cradle to the Grave", Proc. 18th Leeds-Lyon Symp. on Tribology (sept. 91) *Tribology Series* 21, 550 pages, Elsevier Sci. Publi. 550 pages (ISBN 0-444-89336-9).

Dowson D. et al. editors (1996), "The Third Body Concept : Interpretation of the Tribological Phenomena", Proc. 22th Leeds-Lyon Symp. on Tribology (sept. 95) *Tribology Series* 31, Elsevier Sci. Publi. 764 pages (ISBN 0-444-82502-9).

Etienne P., Denape J., Paris J.-Y., Phalippou J. and Sempere R. (1996), "Tribological Properties of Osmosil Coatings", *J. of Sol-Gel Sci. and Techno.* **6**, 287-297.

Fayeulle S., Vannes A.B. and Vincent L. (1991), "Fretting Behaviour of Titanium Alloys", *Trib. Trans.*, **36**, 267-275.

Fouvry S., Kapsa P., Vincent L. and Dang Van K. (1995), " Theoretical Analysis of Fatigue Cracking Under Dry Friction", *Wear of Materials*, Boston.

Godet M. (1982), "Extrapolation in Tribology", *Wear* **77**, 29-44.

Godet M. (1984), "The Third Body Approach : A Mechanical View of Wear", *Wear* **100**, 437-452.

Godet M. (1984), "Mechanics Versus or With Materials in the Understanding of Tribology", *Journ. of Lubrication Engineering*, vol **40**, 7, 410-414.

Godet M. (1990), "Third Bodies in Tribology", *Wear* **136**, 1, 29-45.

Godet M., Berthier Y., Lancaster J. et al (1988), "Wear Modelling : How Far Can Give Get First Principles ?", In *Tribological Modelling for Mechanicals Designers* ASTM STP 1105 ed. Ludema, Bayer, 173-179.

Godet M., Berthier Y., Lancaster J. and Vincent L. (1991), "Wear Modelling Using Fundamental Undrerstanding on Practical Experience", *Wear* **149**, 325-340.

Oktay S.T. and Suh N.P. (1992), "Wear Debris Formation and Agglomeration", *J. of Tribology* **114**, 379-393.

Godet M, Play D. and Berthe D. (1980), "An Attempt to Uniform Theory of Tribology Through Load Carrying Capacity, Transport and Contiuum Mechanics", *J. Lub. Tech.* **102**, 153-164.

Play D. and Godet M. (1977), "Visualisation of Chalk Wear", The Wear of Non-Metallic materials, 3rd Leeds-Lyon Symp., *Institution of Mechanical Engineers*, 221-229.

Meng H.S. and Ludema K.C. (1993), "Wear Life Equations for Mechanical Designers : State of the Art", WOM-93, 9th Intern. Conf. on *Wear of Mat.* (San Francisco, USA), 2-13.

Noll N. (1997), "Conception and birth of a tribological contact", Thesis of the National Institut of Applied Sciences of Lyon (France), n°97 ISAL (in french).

Sauger E., Ponsonnet L., Martin J.M. and Vincent L. (2000), "Study of the Tribologically Transformed Structure Created During Fretting Tests", *Trib. Intern.* to be published.

Zhou Z.R., Fayeulle S., and Vincent L. (1992), "Cracking Behaviour of Various Aluminium Alloys During Fretting Wear", *Wear*, **155**, 317-330.

FRETTING WEAR BEHAVIOUR OF A TITANIUM ALLOY

V. FRIDRICI, S. FOUVRY, Ph. KAPSA
Laboratoire de Tribologie et Dynamique des Systèmes
UMR CNRS 5513
Ecole Centrale de Lyon, Bât H10
36, avenue Guy de Collongue
69131 ECULLY cedex – FRANCE

1. Introduction

When two contacting surfaces are submitted to tangential loading due either to vibrations or to a fatigue loading in one of the contacting component, small amplitude displacements (from one tenth of a micron to hundreds of microns) are induced at the interface. Depending on the materials properties and the normal and tangential loadings, this phenomenon, known as fretting (Waterhouse 1972), can lead to crack nucleation (and possible propagation if one of the component is subjected to a fatigue loading) or to wear by debris formation (Fouvry et al. 1996).

Depending on the tangential displacement amplitude, two different fretting conditions are observed experimentally : if the fretting loop (tangential force Q versus displacement δ) is elliptical (Figure 1 (a)), the condition is called partial slip (indicating that the contact is divided into a central stick domain and an external domain where sliding occurs). The condition of gross slip is observed for higher values of displacement amplitudes, when the central stick domain has disappeared. The fretting loop is then quadrilateral (Figure 1 (b)).

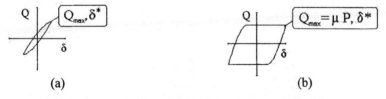

Figure 1. Fretting loops : (a) partial slip condition; (b) : gross slip condition.

Three fretting regimes are defined : the partial slip regime (partial slip condition is maintained during the test), the mixed slip regime (both partial and gross slip conditions appear during the test) and the gross slip regime (gross slip condition is maintained during the test). These three regimes induce three different types of friction logs (three-dimensional representations of the $Q - \delta$ cycles for a given test) that are represented in Figure 2. Debris formation is more likely to occur in the gross slip regime.

413

B. Bhushan (ed.),
Fundamentals of Tribology and Bridging the Gap between the Macro- and Micro/Nanoscales, 413–421.
© 2001 *Kluwer Academic Publishers.*

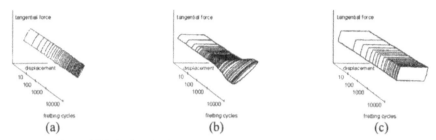

Figure 2. Friction logs for the (a) partial, (b) mixed, (c) gross slip regimes.

In this study, we focus on the behaviour under fretting-wear conditions of a titanium alloy. As loss of matter is the main interest, most of the tests have been conducted in the gross slip regime. The contact geometry is a cylinder on plane. The influence of the following parameters: tangential displacement amplitude, number of cycles and surface properties (with polished or shot peened specimens), has been investigated.

2. Testing apparatus and materials

2.1. FRETTING RIG

A schematic of the fretting rig used in this study is shown in Figure 3 (Blanchard et al. 1991). A tension – compression hydraulic machine is used to impose the surface displacement between the plane and the cylinder. During a fretting cycle, the displacement δ, the normal force P and the tangential force Q are recorded. This allows one to draw the fretting loop $Q - \delta$ and, at the end of the test, the friction log.

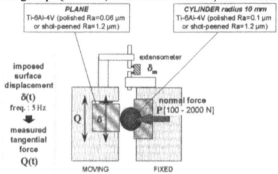

Figure 3. Schematic of the fretting rig.

2.2. MATERIALS

The material under investigation in this study is an alpha/beta titanium alloy (Ti-6Al-4V) broadly used in aeronautics.

Its chemical composition and mechanical properties are given in Table 1 and Table 2 respectively.

TABLE 1. Chemical composition (wt %)

Ti	90
Al	6
V	4
O	< 0.2
Fe	< 0.25

TABLE 2. Mechanical properties

Young's Modulus (GPa)	115
Poisson ratio	0.29
Yield stress $R_{0.2}$ (MPa)	970
HR_C	41
Density (g/cm^3)	4.4

Two types of contact have been studied : polished (Ra = 0.1 μm) cylinder on polished (Ra = 0.06 μm) plane and shot peened (Ra = 1.2 μm) cylinder on shot peened (Ra = 1.2 μm) plane, all made of Ti-6Al-4V. In both cases, the diameter of the cylinders was 20 mm and the length of the contact area was 3 mm (obtained by manufacturing a 3 mm wide strip in the plane). All the surfaces were cleaned with acetone and ethanol before testing.

Concerning the shot peened specimens, planes and cylinders were treated using conventional shot peening (balls of 0.315 mm in diameter) leading to an ALMEN intensity (French standard : AFNOR NFL 06832) of 0.15 mm. The shot peening conditions used give an overlapping rate of 125%. Shot peening results in a maximum compression stress of about 1 GPa at the surface.

2.3. EXPERIMENTAL CONDITIONS

One million cycle fretting tests have been run for both polished and shot peened specimens. The displacement was varied from +/- 5 μm to +/- 25 μm with intervals of 5 μm. In these tests, the initial normal force was 400 N, corresponding to a maximal hertzian pressure of 515 MPa and an initial contact width of 330 μm.

Furthermore, to determine the wear kinetic of the two kinds of specimens, other tests have been conducted with different numbers of cycles (ranging from 100000 up to one million) with a displacement amplitude of +/- 25 μm. For all the tests, the frequency of the stroke was 5 Hz. The tests were performed in air at room temperature.

3. Determination of the transition amplitude

With a view to determining the displacement amplitudes for which the gross slip regime occurs, the displacement amplitude at the transition between partial and gross slip has to be known. The incremental displacement method (Voisin 1992) allows one to determine this transition amplitude with one test, using only a single pair of specimens (one plane and one cylinder). In this method, 500 cycles are run at a given displacement amplitude (with partial slip cycles) after which the displacement amplitude is increased and maintained constant during 500 cycles... until gross slip fretting cycles occur. Figure 4 shows the variation of the stabilised Q_{max} / P ratio (at the end of each set of 500 cycles) for polished and shot peened specimens. In the partial slip condition, the Q_{max} / P ratio varies linearly with the displacement amplitude, as calculated for a sphere on flat contact (Mindlin 1949). After the transition, this Q_{max} / P ratio (i.e. the friction coefficient μ, in the gross slip condition) does not depend anymore on the displacement amplitude.

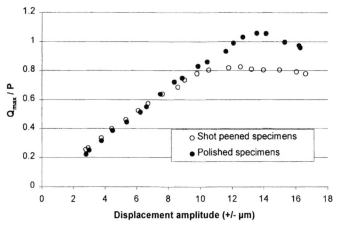

Figure 4. Variation of the Q_{max} / P ratio with displacement amplitude during an incremental displacement test for polished and shot peened specimens.

The transition amplitude is found to be about 14 μm for polished specimens and 11 μm for shot peened specimens. Furthermore, with polished specimens, a friction coefficient greater than 1 is observed at the transition. For shot peened specimens, the coefficient of friction at the transition is less (about 0.8).

The difference in transition amplitude and friction coefficient between polished and shot peened specimens is the influence of shot peening on the initial surface roughness. Indeed, Rabinowicz (1992) pointed out that the area of contact is smaller for shot peened specimens than for polished ones. Moreover, Bowden et al. (1968) showed that the friction coefficient is greater with polished specimens. Furthermore, the greater the friction coefficient, the greater the transition displacement amplitude.

To study the fretting-wear behaviour of Ti-6Al-4V, one million cycle tests have been conducted at different displacement amplitudes in the gross slip regime (up to +/- 25 μm). To confirm that wear is not a predominant damage in partial slip, few tests have been conducted between +/- 5 and +/- 10 μm. After the fretting tests, tactile 3D profilometry measurements (Figure 6) are realised on the ultrasonically cleaned wear scars to determine the wear volume on the plane and on the cylinder.

The results of all these tests are presented in the following paragraph.

4. Effect of the displacement amplitude

4.1. TRIBOLOGICAL BEHAVIOUR

Figure 5 shows a typical evolution of the friction coefficient during a long-term fretting test (δ* = +/- 25 μm), in the gross slip regime. Two stages can be seen. During the first hundreds of cycles, the friction coefficient increases, corresponding to the establishment of metal/metal contact and the formation of a tribologically transformed structure (Blanchard 1991). Then, the coefficient of friction reaches a constant value of about 1.0 (approximately the value determined by the incremental displacement method for polished specimens).

As noted above, the friction coefficient is greater with polished specimens than with shot peened ones because of the asperities induced by shot peening. But, these asperities are rapidly worn off and, during the second stage, the friction coefficient is the same for both types of specimens. This indicates that three-body contact is similar in the two cases and that the residual stresses induced by shot peening have no influence on the contact.

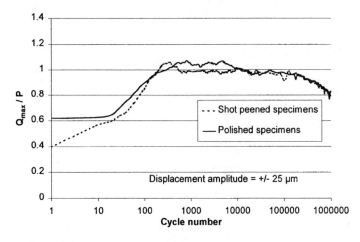

Figure 5. Variation of the friction coefficient during a fretting test (δ = +/- 25 μm).

4.2.   WEAR VOLUME

Figure 7 shows the evolution of the wear volumes (corresponding to the plane, the cylinder and the total) as a function of the displacement amplitude, at the end of the one million cycle tests. As expected, the total wear volume is almost zero if the displacement amplitude is less than 12.5 μm approximately, and, for higher values of the displacement amplitudes, the higher the amplitude, the higher the wear volume. The results indicate that shot peening has no significant effect on the long term wear behaviour of Ti-6Al-4V since, under the same experimental conditions, similar wear volumes are measured with shot peened and polished specimens. This confirms that shot peening, through the modification of surface roughness, has only an influence during the first hundreds of cycles of the tests, as long as the asperities are not worn. Thus, it seems that the compression residual stresses induced by shot peening have no impact on the fretting wear of Ti-6Al-4V.

418

Figure 6. 3D wear scar profilometry ($\delta^* = +/- 12.5$ μm, half of the polished plane).

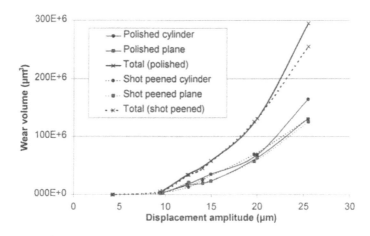

Figure 7. Evolution of the wear volumes as a function of the displacement amplitude.

An energy-based approach, developed for steels and hard coatings (Fouvry et al. 1996; Fouvry et al. 1997), can be used to analyse the wear process. The total cumulated dissipated energy (which corresponds to the volume contained in the friction log) is determined for each experiment : it is linked to the normal force, the coefficient of friction, the displacement amplitude and the number of cycles. Then, the total wear volume is plotted against this total cumulated dissipated energy. Contrary to the case of steels, the displacement amplitude has a non-linear influence on the total wear volume, for both polished and shot peened specimens (Figure 8). This indicates a particular impact of the displacement amplitude on the wear process and further works have to be done to clearly understand the physical phenomenon that leads to this variation. One way is to study the evolution of wear versus the number of cycles in order to determine the wear kinetic for given experimental conditions (in terms of displacement amplitude and normal force). Results concerning wear kinetic are given in the following paragraph.

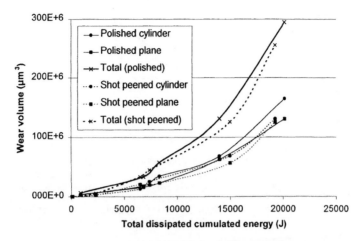

Figure 8. Variation of the wear volumes with the total dissipated cumulated energy, for different values of displacement amplitudes.

5.   Effect of shot peening on the wear kinetics

Tests have been conducted at a displacement amplitude $\delta^* = +/- 25$ μm for different numbers of cycles (ranging from 100000 to one million) with polished and shot peened specimens. The results show a linear variation between the wear volume and the number of cycles (Figure 9) and between the wear volume and the dissipated energy as a function of the number of cycles (Figure 10). Consequently, the energy-based approach can be applied to define an energy wear coefficient using the following equation :

$$V_W = a\,E_d + b$$

where V_W is the wear volume (μm^3), E_d the total dissipated cumulated energy (J), a the energy wear coefficient (μm^3 / J) and b a residual value (μm^3).

The values of a, b and R^2 (linear interpolation coefficient) are given in Table 3, for both polished and shot peened specimens. These results confirm the slight influence of shot peening on the wear behaviour of Ti-6Al-4V under fretting conditions.

The fact that b is positive could be explained by the very rapid wear kinetic for very low number of cycles (corresponding to the establishment of a stabilised three bodies interface) but this explanation has to be confirmed by a study of wear focusing on the first thousands of cycles. Furthermore, at the end of 25000 and 50000 cycle tests, transfer is observed in the fretting scar and it is then more difficult to define and measure a wear volume.

TABLE 3. Parameters of the wear – energy relationship

	a (μm^3 / J)	b (10^6 μm^3)	R^2
Polished specimens	11994	60	0.98
Shot peened specimens	9717	80	0.98

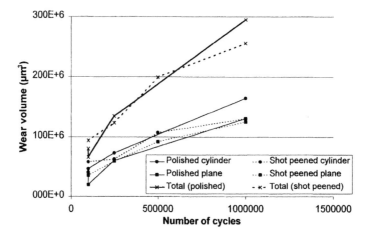

Figure 9. Evolution of the wear volumes with the number of cycles ($\delta^* = +/- 25 \ \mu$m).

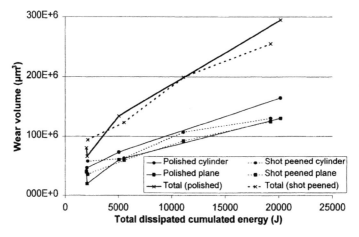

Figure 10. Relationship between the wear volumes and the dissipated energy
(function of the number of cycles) for $\delta^* = +/- 25 \ \mu$m.

6.   Conclusions and perspectives

The wear behaviour under fretting conditions of Ti-6Al-4V has been studied : the effect
of displacement amplitude and number of cycles have been investigated for polished
and shot peened specimens in the cylinder on plane contact configuration.

It appears that, with shot peened specimens, the initial friction coefficient is lower
than with polished specimens, but rapidly (as soon as the surface roughness induced by
shot peening disappears), the friction coefficient reaches a constant value of 1.0 in both
cases.

Displacement amplitude has a particular impact on the wear process in that there is
a non-linear relationship between the wear volume and the total dissipated cumulated

energy (as a function of the displacement amplitude). Further investigations (concerning the debris ejection and the effect of speed, for instance) are needed to understand this phenomenon.

Moreover, it has been demonstrated that shot peening has a very slight effect on the long term wear behaviour and the wear kinetic for high numbers of cycles. This also confirms that residual stresses do not have a key role in the tribological behaviour and the wear process.

In the future, the effect of shot peening on crack nucleation and propagation under fretting conditions will be studied, notably through the analysis of the stress relaxation induced by contact solicitations.

7. Acknowledgements

This study is financially supported by SNECMA Moteurs. The authors wish to thank V. Gros, Ph. Perruchaut and B. Brethes (from SNECMA Moteurs, Centre de Villaroche) for helpful discussion and R. Vargiolu and H. Zahouani for their technical assistance in performing and analysing the profilometry measurements.

8. References

Blanchard, P. (1991), "Usure Induite En Petits Débattements: TTS d' Alliages De Titane", *Ph. D. thesis*, Ecole Centrale de Lyon, n° 91-32.

Blanchard, P., Colombie, C., Pellerin, V., Fayeulle, S. and Vincent, L. (1991), "Material Effects In Fretting Wear: Application To Iron, Titanium, And Aluminium Alloys", *Metallurgical Transaction A* **22A**, 1535-1544.

Bowden, F.P. and Tabor, D. (1968), in *The friction and lubrication of solids: Part II*, Oxford University Press, London, England.

Fouvry, S., Kapsa, P. and Vincent, L. (1996), "Quantification Of Fretting Damage", *Wear* **200**, 186-205.

Fouvry, S., Kapsa, P., Zahouani, H. and Vincent, L. (1997), "Wear Analysis In Fretting Of Hard Coatings Through A Dissipated Energy Concept", *Wear* **203-204**, 393-403.

Mindlin, R.D. (1949), "Compliance Of Elastic Bodies In Contact", *Trans. ASME, Series E, Journal of Applied Mechanics* **16**, 259-268.

Rabinowicz, E. (1992), "Friction Fluctuations", in *Fundamentals of friction: macroscopic and microscopic processes*, Eds. Singer, I.L. and Pollock, H.M., NATO ASI Series; series E: Applied Sciences, **220**, 25-34, Kluwer Academic Publishers, Dordrecht, The Netherlands.

Voisin, J.M. (1992), "Méthodologie Pour l'Etude De l'Endommagement d'Un Contact Tube-Grille", *Ph. D. thesis*, Ecole Centrale de Lyon, n° 92-49.

Waterhouse, R.B. (1972), in *Fretting corrosion*, Pergamon Press, New York, USA.

WEAR MEASUREMENTS AND MONITORING AT MACRO- AND MICROLEVEL

N. K. MYSHKIN, M. I. PETROKOVETS and S. A. CHIZHIK
*V.A.Belyi Metal-Polymer Research Institute of Belarus National Academy
of Sciences, Kirov st. 32a, Gomel, 246050, Belarus*

Abstract. Wear as a surface loss of substance resulted from friction is considered in connection with various factors affecting tribosystem. These factors, first of all deformation and adhesion, are shown to be dependent on a scale factor as a result of combined effect of surface topography and material properties. The main trends in development of wear testing techniques, data collection, processing and presentation are discussed. Problems of test data reproducibility and comparability are shown to be very important and the attempts to solve these problems are examined. Scale factor in wear measurements is discussed. Concept of wear monitoring and its main purposes are considered in the context of providing non-failure, long-term operation at optimum friction performance of a tribosystem. Wear monitoring tools based on analysis of debris accumulation in a lubricated machine are reviewed.

1. Introduction

Tribology research is developing from the macroscopic models to the current attempts of understanding micro- and nanoscale processes of friction and wear. This development gives a new insight on the basic problems, first of all a relation of deformation and adhesion at friction. Both the components of friction force play a crucial role in the wear process (Bowden and Tabor, 1964; Kragelskii, 1982). We know a great diversity of wear modes, but their satisfactory classification is still under discussion. Bearing the friction dualism in mind, the mechanical wear modes may be ranked according to the deformation-to-adhesion relationship (Fig. 1). Given this, the adhesive and fatigue wear modes are in the extreme positions while the fatigue is mainly governed by deformation, whereas adhesion is dominant in adhesive wear. Such an approach to wear processes is quite simplified. Friction always occurs in certain environment, which produces a significant influence on the tribological processes through the chemical reactions. Although the reactions change the strain rate and interfacial junction strength, the dominant role of deformation and adhesion in wear processes remains intact. Because of this, we consider the mechanical wear only.

One particular point, which should be mentioned, is the great difference in wear rates covering ten orders of magnitude (the table). It is much wider difference than variation in friction coefficient basically varying in most practical cases between 0.1 and 0.5. The reason is that wear is much more complicated response of tribosystem than

B. Bhushan (ed.),
Fundamentals of Tribology and Bridging the Gap between the Macro- and Micro/Nanoscales, 423–438.
© 2001 *Kluwer Academic Publishers.*

friction (Czichos, 1978; Garbar, 2000; Kato,1997; Rigney; 1992) resulted from synergism of processes leading to accumulation of damage in material and its failure.

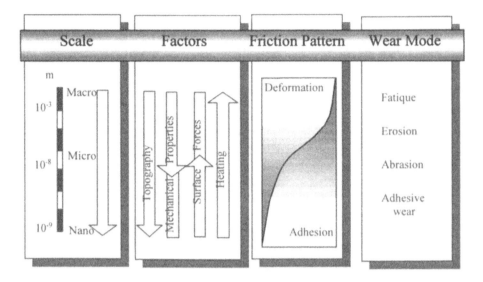

Fig. 1. Combination of factors effecting wear

TABLE. Wear resistance of common tribosystems

Friction unit	Worn part, material	Wear conditions		Wear rate *	Wear resistance grade
		Counterbody	Loading parameters, environment		
Cylinder-piston assembly of car engine	Piston ring, grey cast iron	Cylinder sleeve, grey cast iron	v = 7-20 m/s Oil M10Г1	10^{-12}-10^{-11}	12-11
Block brake	Block, friction plastic	Drum, special cast iron	q = 1 MPa, v = 10 m/s	10^{-7}-10^{-6}	7-6
Disc brake	Braking element, friction plastic	Braking disc, alloyed cast iron	q = 2.5 MPa, v = 25 m/s	10^{-10}-10^{-6}	10-6
Sliding bearing	Shaft, steel with solid lubricant coating	Bushing, carbon steel	$q \cong 20$ MPa, v = 0.5 m/s, T = 373 K	10^{-8}-10^{-7}	8-7
Lathe tool	Hard alloy	Workpiece, carbon steel	q = 400 MPa, v = 2 m/s, coolant	$5 \cdot 10^{-8}$	8

* worn layer thickness related to sliding distance

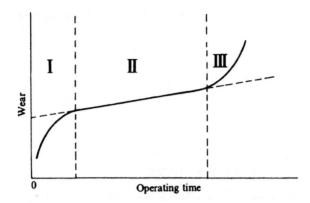

Fig. 2. Time-dependent wear of tribosystem: I – running-in; II – stationary wear; III – severe wear

Typical curve of wear in a tribosystem vs. time is presented in Fig. 2, it can be divided in three stages – running-in, normal wear, and severe wear. Third stage results from accumulation of damage or lubricant degradation and we need to escape or at least postpone it in a real machine operation. Presense of different stages in wear process makes wear simulation and prediction very difficult and we need to take account of this factor.

2. Basic Concepts

A real surface is composed of several roughness scales, which are superimposed on each other. Four scale levels are distinguished: atomic and molecular-scale one comprising the intrinsic features of material resulted its surface texture (subroughness), conventional roughness formed in machining or processing as a result of interaction of tool with material, waviness (mostly result of periodical effects of tool), and macrodeviations or errors in form of a given part (Fig. 3). We consider below the most important practical levels, roughness and subroughness. (Myshkin et al., 1998)

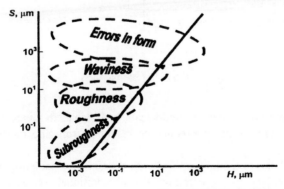

Fig. 3. Multi-scale roughness

The real contact area (RCA) is one of the basic notions adopted in tribology. It owes existence to the fact that a contact of real surfaces involves two characteristic scales, the characteristic size of nominal area and the asperity size. As is shown above, the latter in turn has a multi-scale character. This fact is in wide use when the real contact is simulated because of possibility to combine different mechanical models and different modes of material deformation. The whole set of contact models can be illustrated by the general scheme given in Fig. 4.

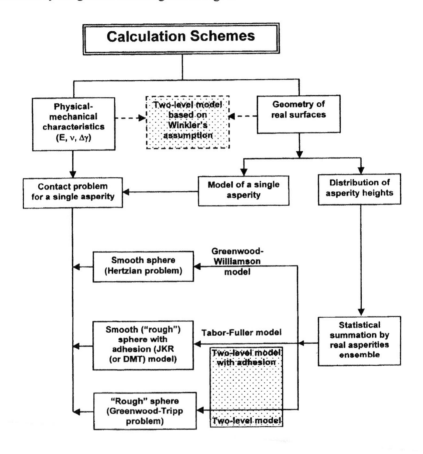

Fig. 4. Calculation scheme of contact with multi-scale roughness

We have concentrated on development of multi-scale ones (shaded squares in Fig.4). Without going into details, some consequences resulting from the models developed are given:

From the Greenwood-Williamson model modified with the aim of considering the surface (molecular) and/or capillary forces, it follows that an additional contact area between the surfaces appears. The area governs the adhesion component of friction, and its magnitude can be related to the adhesion parameter (Myshkin et al., 1999):

$$\Delta_c = \frac{1}{3\sigma}\left(\frac{9\pi}{8}\frac{\beta^2\Delta\gamma}{K}\right)^{2/3},$$

where σ is the mean-root-square of asperity height; β is the curvature radius of asperity summit; $\Delta\gamma$ is the interface energy; K is the reduced stiffness of contacting materials. With increasing the parameter Δ_c, the adhesion and RCA increase, the effect of surface force being essential at $\Delta_c > 0.2$. Of interest is the fact that the strength of adhesion is dependent on the combined effect of contact geometry (σ,β), mechanical parameter (K) and surface physics $(\Delta\gamma)$.

It is well known that the real contact area (RCA) is a set of separated contact spots. Each of them consists of a set of smaller spots, the total area of which was conditionally named "physical contact area". This area is formed by contact of nanometer-scale asperities (subroughness). The physical contact area is less than the RCA by one-two orders of magnitude. It has a clear meaning when comparing contact areas bearing the load and those conducting the electric current. Figure 5 can be an illustration of this concept presenting the AFM data processing with extraction of subroughness by median filtration of a digital image.

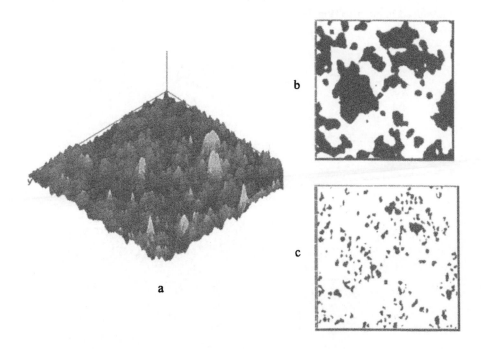

Fig. 5. Visualization of contact spots at various scales: a – AFM-image of an analyzed area (scan 15×15 µm); b – actual contact spots at microlevel; c – actual contact spots at submicrolevel (physical contact area)

428

A lot of methods are used in calculation of stresses and strain in the contact of rough solids and these methods are in continuous development (Williams, 1999), so we can conclude that the computation of wear as a summation of microscale damage is possible in principle. But taking in mind the complexity of wear as a phenomena developing simultaneously in different scales of time and space it seems unrealistic to rely on such computation in practice.

3. Wear Measurements

According to ASTM Standard G48-83, wear is damage to a solid surface, generally involving progressive loss of material, due to relative motion between that surface and a contacting substance or substances (Blau and Budinski, 1999). Wear measurements provide data on service life of materials, lubricants and machine components. Tests can be classified as laboratory, bench tests under practice-oriented conditions and field tests under operation conditions (Alliston-Grenier, 1997). First and second types of tests usually are accelerated in order to save time in materials and lubricants screening. The successful simulation requires the similarity between the real and test systems so the mating materials, lubricant and operating conditions should be the same. At the same time in order to accelerate the test, some of the parameters should be changed, most often pressure, velocity or temperature. A variety of wear-testing machines were introduced in practice (Fig.6).

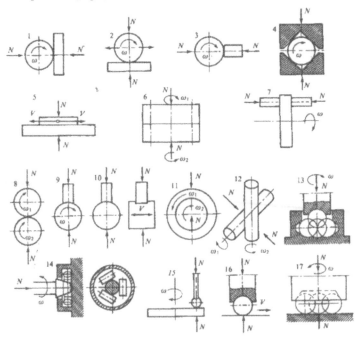

Fig. 6. Various schemes of wear tests: 1-4 - cylinder-to-plane contact; 5-7 – plane-to-plane contact; 8-12 – cylinder-to-cylinder contact; 13 – ball-to-ball contact; 14 – cone-to-cylinder contact; 15 – ball-to-plane contact

But there are three basic groups of machines considering the geometry of initial friction contact: point contact, line contact and conformal contact (can be plane or curvilinear). These three types can give a numerous variety of modifications. A number of them are manufactured as commercial machines available in the market. Even the larger number is used in laboratories all over the world by researchers who have created them for particular studies.

There are the following basic factors in providing the efficient wear test except of geometry: type of motion; type of loading; type of lubrication; environment control; specimen preparation. Each of these factors should be considered carefully in order to make test results reliable, reproducible and adequate to the system simulated. So, to be reliable in data, the wear-testing machine should have a control of ambient humidity and composition.

The accuracy and reliability of data are very important in measurements of testing parameters. Wear can be measured by measuring mass loss, change in linear dimensions, surface profile measurements, measurement of indentations size on worn surface, radioactive tracers etc. Concentration of debris in lubricant and their size, changes in the composition of surface layers and lubricant, surface morphology (presence of transfer films, cracks, pores etc.) are the other parameters which are usually controlled. This means that even a simple macroscopic wear test needs a lot of equipment integrated with a clear idea of finding the most important trend in tribosystem behavior for a given application.

4. Microwear Measurements

Fast development of precise measuring tools followed the invention of scanning tunnel microscope has given a new driving force in wear simulation (Bhushan, 1996; Schiffman, 1998;).

We have tried to simulate microscopic wear by scratching a smooth surface by the tip of contact AFM, which has modeled a single asperity (Fig. 7, Chizhik et al, 2000).

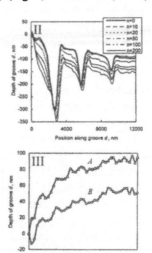

Fig. 7. Multiple scratching of silicon wafer by the AFM tip: (I) AFM image of the test area (scan size 14×14 micron, height amplitude 412 nm); a, b, c are reference grooves formed at one-pass scratching under load 0.5, 0.2 and 0.1 mN; grooves d, e, f, g formed after 200 cycles of reciprocal tip movement under the load 0.5 mN (grooves d and e) and 0.2 mN (grooves f and g); (II) profile of groove d after different number of scratching cycles; (III) change in groove depth at points A and B (a) vs. number of scratching cycles

Since asperities of different scale exhibit different behavior in frictional contact, it is reasonable that tribologists simulate the behavior of the smallest asperity using the nanoinstruments (AFM, LFM, and so on). This reduction in scale level of experiment reveals that mechanical properties of contacting materials become scale-dependent, and parameters such as Young's modulus and hardness differ not only in magnitude but also in their physical meaning. The physical interpretation of experimental data and their self-affinity in changing the scale factor is a basic challenge.

Figure 7 shows the results of wear test of silicon wafer. The reference grooves a, b and c were made at one-pass scratching under load of 0.5, 0.2 and 0.1 mN, respectively. The grooves d, e, f and g were formed after 200 reciprocating passes in the transverse direction. The load was 0.5 mN in case of scratches d and e and 0.2 mN for the grooves f and g. Figure 7 also shows the groove depth at points A and B depending on the number of cycles (time). Comparison of Figs.7 and 2 shows that patterns of nano- and macrowear are similar, but the nanowear curve is more complicated owing to small oscillations. They count in favor of the friction transfer and presence of ultrafine debris in the contact at nanolevel.

Experiments with microscratching simulate the friction and wear of a single asperity, but we should have in mind at least two important factors: (i) shape of AFM tip is not similar to the shape of separate asperities of a rough surface which have a variety of shapes; (ii) real asperities are distributed in height, they support different loads and can be deformed in different modes. Henceforth, a very difficult problem arises how to transfer microscratching data to real rough contact. Probably this can be made in simple cases, e.g. abrasive polishing by free abrasive particles of the same size and shape.

5. Presentation of Wear Data and Databases

Wear is expressed in specified units (length, volume or mass). The wear process is often described by wear rate. There is no single standard way to express wear rate. The units depend on the type of wear and the nature of the tribosystem in which wear occurs (Hutchings, 1992). Wear rate can be expressed, for example, as (1) volume of material removed per unit time, per unit sliding distance, per revolution of a component, or per oscillation of a body; (2) volume loss per unit normal force at unit sliding distance $(mm^3/(N\,m))$ which is sometimes called the wear factor or dimensional Archard coefficient; (3) mass loss per unit time; (4) change in a certain dimension per unit time; (5) relative change in dimension or volume with respect to the same changes in another (reference).

The reciprocal of wear rate is known as wear resistance which is a measure of the resistance of a body to removal of material by wear process. Relative wear resistance is sometimes considered using arbitrary standards.

There is no standard format in which the wear test data are reported. Attempts to standardize the wear data have been made by Czichos (Appendix A to DIN Standard 50320) and Blau (ASTM Committee G2). Nowadays it is common to present data on tribological behavior in the form of "map" where not only quantitative data on testing are given but also the modes of wear are related to the testing conditions of a given material.

One of the earlier examples of such presentation were "PV" maps used to summarize the information on wear of plastics under given combination of contact nominal pressure and sliding velocity. Such map presents two-dimensional diagram where the area of certain wears (e.g. linear wear of 25 micrometers or limit admissible wear) as a function of test parameters combination (Fig. 8). This type of data presentation is convenient for design engineer providing necessary range of data to fix the limit load and velocity in a given friction unit.

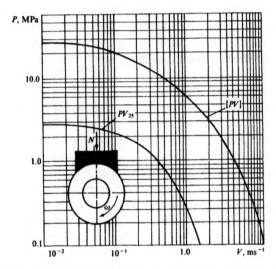

Fig. 8. Plot of PV data for polyamide-based composite when rubbing against steel: $[PV]$ - permissible linear wear; PV_{25} – data for wear equal 25 μm

Lim and Ashby (1987) have proposed further development of this approach in the form of wear maps. This format has used normalized axes parameters where wear, pressure and velocity are given related to nominal contact area, hardness, sliding distance, heat flow rate and thermal diffusivity. Areas of the map present various modes of wear and critical parameters for transition from one mode to the other. Recently an approach of triboscopy has been proposed which can provide three-dimensional wear data presentation in their space-time evolution (Belin, 1993).

A development of computerized databases providing data on friction unit performance under the given test or operation conditions is the important step in generalization of tribological data (ASM Handbook, 1992). But there is a serious obstacle in the basis of this job because the test data presented in database should be standardized and not dependent on the place where they have been collected. The data obtained in international project VAMAS (Versailles Project on Advanced Materials and Standards) can be an illustration to this matter. These data were collected from 31 institutions in seven countries. Test specimen kits have been distributed with instructions how to run the tests and present friction and wear data. Under special requirement of VAMAS to testing technique the spread of the wear data was in the range of 30%. So, in order to solve problem of reliable and comparable data collection, we need to standardize testing procedure, data processing and presentation.

One attempt to of this kind was made in Metal-Polymer Research Institute (Myshkin et al.,1997a). It has started from search for standard methods of testing. The result of this search was formulated as a standard procedure of rubbing the polymer conformable blocks against steel rings. This scheme has been approved by the International Standard Organization for simulation of the polymer plain bearings (ISO/TR 8285: 1992). Data including contact pressure, speed and real time values of friction coefficient, wear factor and temperature, are collected via the interface board to PC. The shell of the database includes the root part which contains test conditions, design features of friction units, materials and environment, mean values of triboparameters, critical points and extensions (data files) with a more detailed information on the materials behavior under specified conditions.. The main file contains the results of tribotests of polymer materials. Each experiment is described by five coordinates within the space of measured values p, v, T, K, f. Two more variables are added, vis. the test number and the experimental point number in the test.

Two modes can be entered to operate the database, one is passive (or searching) for the user to retrieve information serving for selection of materials for a given tribosystem, the other is active when the user can enter new standard reference data. The searching mode allows one to receive answers to requests, which can be combined into the following main groups. Direct requests: a) to list all materials in the database; b) to list the materials in a specific class; c) properties of a specific material (document search) under a given number or name. Inverse requests: a) to find material(s) with a specified property within a given range of parameters; b) to find material(s) according to a combination of properties, i.e. the material(s) satisfying several requirements at once. The data are tabulated or plotted graphically (Fig. 9).

Velocity: V = 0.1 m/s	Parameters		
	K	f	T, K
Pressure 0.1	0.54	0.24	293
0.25	1.72	0.27	294
0.5	4.59	0.19	295
1.25	10.10	0.24	301

Selection of P (MPa)

0.10	0.25 *	0.50	1.25	2.50	5.00	10.0

Fig. 9. Displayed graph of wear coefficient vs sliding velocity

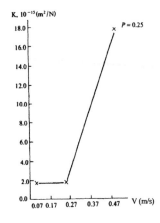

Creation of systems combining databases with information processing, and, if possible, with expert systems is useful in presenting and processing of wear data. An attempt of such creation was made when designing the Tribology Information System (ACTIS) in National Institute for Standards and Technology of USA. Approaches similar to ACTIS have been developed in other countries. Most of them were limited by certain databases for a given area of tribology. It has also been supposed that there will be expert systems created on the basis of data collection.

Discussing the topic of wear measurements and data processing we need to emphasize the urgent necessity of development international efforts in wear testing equipment and procedure standardization as the most efficient way to make wear data reliable and useful for a mechanical engineer.

The national standards on wear testing (ASTM, DIN, GOST, etc.) can be listed in several pages. The brief analysis of this long list shows that these standards don't cover many possible combinations of basic materials as well as types of wear because they have been introduced mostly for evaluation of particular products. At current stage the necessity of more general and internationally approved standardization is clear. This standardization can be made in the frame of the International Standard Organization and national standard committees.

6. Wear Monitoring

Considering wear as a response of tribosystem, we are interested in many cases not in measurements only but in monitoring the system state and prediction of its life. Basic purpose of wear monitoring is the possibility of wear control in a given unit and taking necessary maintenance measures in order to provide non-failure long-term operation at optimum friction performance of the unit (Myshkin et al., 1997a). This purpose can be achieved by using the monitoring tools. The other important purpose is to provide recommendations for improving the unit design and to select better materials, coatings and lubricants for the unit.

Initially wear monitoring techniques were developed based on non-direct evaluation of severity of tribosystem operation. Such most common non-direct methods include temperature and vibration control. Temperature and vibration sensors are easy to install in the most critical tribosystems of the machine, their readings can be collected and processed from the main control point. The drawbacks of non-direct methods are originated mostly in delayed response to wear mode and severity changes, effects of external factors, etc.

Direct methods of wear monitoring can be realized by continuous measurements of wear rates of certain critical components. But sensors measuring a change in dimensions or mass loss in most cases can't be installed in real machine units, so the other techniques of material loss control were developed including radioactive tracers or measuring the special grooves cut in certain places of worn surface. Measurement of the quantity and size of wear debris in lubricated systems can be a natural way to control and monitor wear (Hunt, 1993).

Analysis of debris accumulation in lubricant can be made by various methods. Its general scheme is given in Fig. 10.

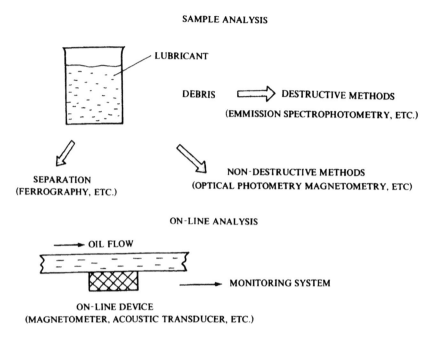

Fig. 10. Scheme of debris analysis in oil

The development of real-time monitoring methods, i.e., the use of on-line devices that provide continuous tracking of wear rate and severity in machine units, is a very promising direction in wear debris diagnostics. Such devices make it possible to evaluate the condition of machines and predict their wear on the basis of characteristics such as the wear particle concentration and the particle size distribution.

On the other hand, direct observation of the wear particles yields information on the type of wear. Shape, texture and color of debris can be used for identification of their source, wearing component (Myshkin et al., 1997b, Stachowiak and Podsiadlo, 1999).

In the cases when complete information on the wear particles is not required and it is sufficient to determine their formation rate, the simpler and less expensive field diagnosis methods are suitable. An optical-magnetic detector (OMD) used for wear monitoring of heavy machinery in industry can be an example of such monitoring techniques (Myshkin et al., 2000).

Variation in optical density of used and clean oil and their differences under the effect of magnetic field were measured by OMD. Optical density can characterize total contamination of the oil by oxidation and aging products, contaminating dust and wear debris. It can be a measure of contamination of the used oil compare to clean one defined as a total contamination index.

Variation in optical density of the used oil under effect of magnetic field is proportional to ferrous wear debris concentration in the oil. It is explained by the effect of the field on the particles in the optical cell, which results in the particle removal from the optical axis to periphery of the cell. Change in optical density can be related to concentration of ferrous debris in the oil. Rolling bearings of the turbine and gearbox, with higher rate of debris generation in the latter, were potential sources of wear debris in the real air compressor. As these components were made of steels the oil contamination by ferrous debris can be a measure of wear of the components defined as a wear rate index.

OMD has been applied in wear monitoring of air-compressor system at metallurgy plant of Pohan Iron and Steel Co. (South Korea) for two years together with vibration and temperature control detectors placed in several places (Fig.11).

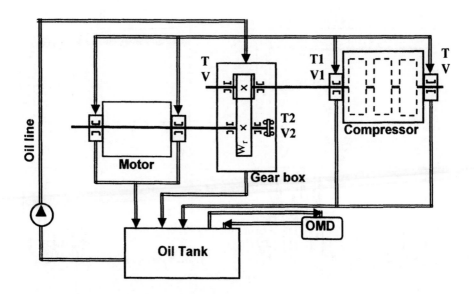

Fig. 11. Schematic view of air-compressor system: V, T – vibration and temperature sensors; OMD – optical-magnetic detector

436

Figure 12 presents the data on monitoring the compressor during 30 days before a serious failure occurred due to breakdown of the steel collar preventing the extraordinary slippage of pinion in the gearbox in the axial direction. Examination of Fig.12a shows that the signal related to gearbox shaft displacement and acceleration has became to grow only in a few days before the breakdown. Temperature control of both bearings of the shaft hasn't given any warning data as is seen from Fig.12.b. At the same time oil contamination and wear rate indices (Fig.12.c) have given early warning three weeks in advance of failure and emergency warning at least ten days before the failure caused the system stop and significant economic loss.

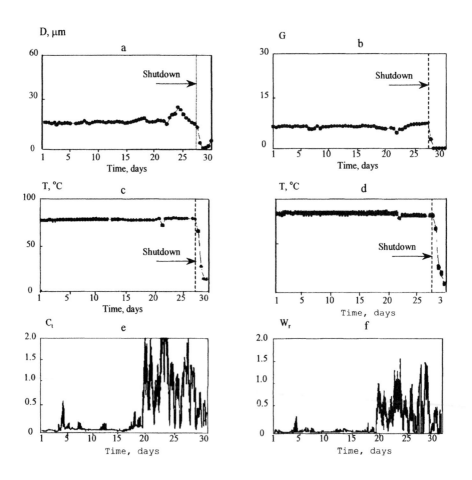

Fig. 12. Monitoring in 30 days of real time: displacement of compressor shaft (a) and acceleration at gear box unit (b), temperature of compressor shaft bearing (c) and gear box bearing (d), OMD readings in total contamination index (e) and wear rate index (f).

The given example of wear debris case study can illustrate the efficiency of integrated approach to the machine maintenance in industrial conditions.

Condition monitoring (predictive maintenance is another notion used often in USA) is considered to be an efficient tool in providing operation efficiency of machinery, reducing the cost of maintenance, saving materials, energy and labor. The whole concept is based on combination of diagnostic devices, on-line monitoring systems, tracking the data on machine operation and making recommendations based on the knowledge accumulated in database. A monitoring system combined with the expert system providing necessary corrections in operation conditions or other maintenance measures is a final goal of the concept. The structure of possible laboratory service for condition monitoring in industry is shown in Fig. 13.

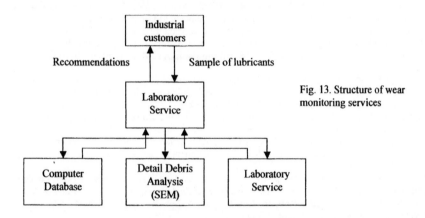

Fig. 13. Structure of wear monitoring services

It includes data collection from the industrial customers at certain time intervals established for each type of machines (gear transmissions, engines etc.); supply of samples to laboratory; preliminary screening of samples and comparison of the data to computer track data on this particular equipment; additional analysis in case of sharp deviations of current data from normal trend, and expert evaluation. This evaluation should be supported by databases and in ideal case include the automated expert system.

7. Conclusions

In order to overcome the current difficulties in wear simulation, measurements and control we need to use reasonable levels of sophistication. In many cases a simple wear test can efficiently replace the complicated calculation model but we need to arrange the test to be simple and clear in realization and explanation of its data. Computation of wear as a summation of microscale damage is possible in principle. But it makes sense only in simple cases when one wear mode is dominant and the tribosystem operates in stationary wear stage.

The necessity of more general and internationally approved wear testing standardization is clear. This standardization can be made in the frame of the ISO and national standard organisations.

438

Condition monitoring can be an efficient tool in controlling wear as a tribosystem response to operation, thus providing service efficiency of machinery and reducing the cost of maintenance. It should combine diagnostic devices, on-line monitoring systems, tracking the data on machine operation and expert knowledge accumulated in database.

Acknowledgement

The work reported was partially funded under INTAS grant 99-0671.

References

Alliston-Greiner A.F. (1997), Test Methods in Tribology, in *New Direction in Tribology* (I.M. Hutchings, ed.), pp. 85-93, Mechanical Engineering Publications Limited, London.

Belin M. (1993), Triboscopy: A New Quantative Tool for Microtribology, *Wear* **168**, 7-12.

Bhushan B. (1996), *Tribology and Mechanics of Magnetic Storage Devices*, Springer-Verlag, N.Y.

Blau P.J. and Budinski K.G. (1999), Development and Use of ASTM Standards for Wear Testing, *Wear* **225-229**, 1159-1170.

Bowden F.P. and Tabor D. (1964), *The Friction and Lubrication of Solids*, Clarendon Press, Oxford.

Chizhik S.A., Goldade A.V., Korotkevich S.V., and Dubravin A.M. (2000b), Friction of Smooth Surfaces with Ultrafine Particles in the Clearance, *Wear* **238**, 25-33.

Czichos H. (1978), *Tribology: A System Approach to the Science and Technology of Friction, Lubrication and Wear*, Elsevier, Amsterdam.

Friction, Lubrication, and Wear Technology (1992), ASM Handbook, vol. 18, ASM International, USA.

Garbar I.I. (2000), Critical Structures of Metal Destruction under the Process of Wear, *ASME Journal of Tribology* **122**, 361-366.

Hunt T.M. (1993), *Handbook of Wear Debris Analysis and Detection in Liquids*, Elsevier, London.

Hutchings I.M. (1992), *Tribology: Friction and Wear of Engineering Materials*, Arnold, London.

Kato K. (1997), Wear Mechanisms, in *New Direction in Tribology* (I.M. Hutchings, ed.), pp. 39-56, Mechanical Engineering Publications Limited, London.

Kragelskii I.V. (1982), *Friction and Wear*, Pergamon Press, Elmsford.

Lim S.C. and Ashby M.F. (1987), Wear-Mechanisms Maps, *Acta Metall.* **35** (1), 1-24.

Myshkin N.K., Petrokovets M.I., and Chizhik S.A.(1996), Tribology: A Bridge from Macro to Nano, in *Micro/Nanotribology and Its Applications* (B. Bhushan, ed.), pp. 385-390, Kluwer Academic Publishers.

Myshkin N.K, Chang Kyun Kim, and Petrokovets M.I.(1997a), *Introduction to Tribology*, Cheong Moon Gak, Seoul.

Myshkin N.K., Kwon O.K., Grigoriev A.Ya., Ahn H.-S., and Kong H. (1997b), Classification of Wear Debris Using a Neural Network, *Wear* **203-204**, 658-662.

Myshkin N.K., Petrokovets M.I., and Chizhik S.A. (1998), Simulation of Real Contact in Tribology, *Tribology International* **31**, 79-86.

Myshkin N.K., Petrokovets M.I., and Chizhik S.A. (1999), Basic Problems in Contact Characterization at Nanolevel, *Tribology International* **32**, 379-385.

Myshkin N.K., Markova L.V., Semenyuk M.S., Kwon O.K. and Kong H. (2000), On-Line Opto-Magnetic Detector and Its Application for Air Compressor System Monitoring, in *Proceeding of 12th International Colloquium "Tribology 2000-Plus"*, vol. III, pp. 1571-76, Esslingen, Germany.

Rigney D.A. (1992), Some Thoughts on Sliding Wear, *Wear* **152**, 187-192.

Schiffman K.I. (1998), Microfriction and Microwear Experiments on Metal Containing Amorphous Hydrocarbon Hard Coatings Using an Atomic Force Microscope, in *Tribology Issues and Opportunities in MEMS* (B. Bhushan, ed.), pp. 539-558, Cluwer Academic Publishers, Netherlands.

Stachowiak G.W. and Podsiadlo P. (1999), Surface Characterization of Wear Particles, *Wear* **225-229**, 1171-1185.

Williams J.A. (1999), Wear Modelling: Analytical, Computational and Mapping: A Continuum Mechanics Approach, *Wear* **225-229**, 1-17.

SLURRY EROSION: MACRO- AND MICRO- ASPECTS

HECTOR McI. CLARK
Metallurgist
Maros u. 25, 1 em. 2
H-1122 Budapest
HUNGARY.

Abstract

Erosion by slurries – the removal of material from a container wall by the contact with it of solid particles suspended in liquid - is described and the principal factors affecting the rate of material removal given. The complexities of the slurry flow and material wear phenomena require that test methods to measure rates of material loss by erosion employ known, controlled and reproducible conditions. Common test methods and the methods of assessing wear are noted. Continuing problems are the achievement of comparability of test conditions and the measurement of small amounts of erosion wear. Comments are offered on the limiting particle-wall collision event, the range of wear rates typically measured and uncertainties in slurry erosion.

1. Introduction

Slurry erosion is the process of wear caused by the collision of moving solid particles (erodent) suspended in a liquid with the container wall. Erosion may be caused by the impact of isolated particles or by the motion of a bed or layer of particles across the surface. These wear processes may occur together. Industrially, such items as centrifugal pumps, valves and pipe-work in mines or the petroleum industry are susceptible to rapid damage and failure and wear depths of several millimeters may be produced within days or hours of service. Frequent replacement of parts involves downtime and production loss, apart from the inherent replacement cost of the equipment itself.

Scientific studies of erosion (whether gas-solid or liquid-solid) have been conducted for more than 50 years (Wahl and Hartstein 1946). Several factors were seen as important in erosion, but in particular, the collision of particles with larger mass or higher collision speed with the container surface produced more serious damage, (van Riemsdijk and Bitter 1959). The complexities of the flow and impact processes have raised problems in formulating test methods that reflect industrial conditions because of the difficulty of isolating single variables. In a given apparatus, change in one variable, say flow speed, leads to variation also of other variables such as particle impact angle on the test surface. In the following discussion, the effect of corrosion on material removal is ignored.

B. Bhushan (ed.),
Fundamentals of Tribology and Bridging the Gap between the Macro- and Micro/Nanoscales, 439–444.
© 2001 *Kluwer Academic Publishers.*

2. Factors Affecting Erosion Rates

- Concentration of Particles in the liquid. Higher concentrations produce greater erosion wastage per unit of time, but the effect is not linear. When measured per unit of erodent mass, higher concentration reduces erosion damage, (Addie et al., 1996).
- Slurry Flow Speed and Particle Impact Speed. Higher speeds dramatically increase erosion wastage, with exponents reported between 2 and 4.
- Impact Angle of Particles on the wearing surface. Ceramics show maximum wear rates for 90° (normal) impact, while most metals show a well-defined maximum for oblique impact angles of about 30°, (Bitter, 1963).
- Particle Size. The impact of larger particles produces more damage than the impact of the same mass of smaller particles, erosion being proportional to particle diameter.
- Particle Shape. Sharp, jagged particles are more effective in producing wastage than rounded particles.
- Particle Density. Particles of higher density give rise to greater energy dissipation at impact and give rise to more damage per particle impact.
- Particle Hardness or friability. Particles that are hard and do not shatter at impact are much more effective in producing erosion than softer, more friable particles.
- Nature of the suspending liquid. Those of low density and low viscosity are associated with higher erosion rates.
- Nature of the container (target) material. Erosion resistance is imparted by high hardness with some associated toughness – hence the success of metal carbides in applications requiring erosion resistance.

3. Test Methods

It is not possible to give here an exhaustive review of experimental methods for slurry erosion testing. Common to all is the use of erodent (silicon carbide, alumina, quartz sand or ores) with particle sizes typically less than 1 mm diameter suspended either in water or, less usually, oil. Several test methods and examples of apparatus can be found in the literature (e.g., Roco et al., 1987). Three methods have been frequently used.

3.1 SLURRY JET

Pre-mixed fresh (not previously used) slurry is pumped through a small-bore ceramic tube, giving rise to a jet (of known velocity) which is directed, usually at 90°, against the target specimen (e.g., Zu et al., 1990). A circularly-symmetrical depression is worn in the target, the amount of material loss being determined either by weight loss or by the use of a surface profile measuring device, although data from the latter may be compromised by the absence of an unworn surface reference datum. Flow and particle impact analysis of the slurry jet has been attempted. Other problems exist, e.g., spreading of the jet at the target surface, interference of particles with each other before and after impact and a spectrum of particle velocities in the jet. These problems – and their analysis - are exacerbated by directing the jet at angles other than 90° to the target. The method does not reflect common industrial conditions in which low-angle particle impact prevails.

3.2 SLURRY POT

Two small-diameter cylindrical specimens, placed parallel to and on either side of a central shaft, are rotated through a bath of slurry so that the liquid flow is directed normal to the cylinder axes. Such an arrangement is straightforward to model so that the frequency of impact, position, and mean normal and tangential velocities of impacting particles can be calculated, allowing the energetics of particle impact to be compared with wastage from the cylindrical specimen (Wong and Clark, 1993). Arrangements using more complex specimen shapes – flat paddles etc. – do not allow of flow analysis because of greater flow complexity and turbulence. For the slurry pot test times must be short since slurry particles themselves wear producing less erosion as a function of time.

3.3 CORIOLIS EROSION TESTER

Originally proposed in 1984, (Tuzson, 1984), this apparatus has recently been further developed (Clark et al., 1999). Fresh slurry is delivered to the central chamber of a rapidly rotating fixture, which contains two diametrally-opposed 1 mm wide x 6 mm deep passages formed in tungsten carbide specimen holders. As the slurry is flung outwards by centrifugal force, particles settle onto the specimen surface as a result of the Coriolis force during their passage, producing erosion at low impact angles. Simple plate specimens are used at rotation speeds of up to 7000 rpm. Wear scars 1 mm wide of increasing depth with distance from the rotation centre are formed and are assessed using a cross-section profile measuring device. The test method is very rapid and has been used for both bulk specimens as well as coatings. Particle trajectories and impact conditions in this fixture have been modelled by Xie et al. (1999). The method has been shown to discriminate clearly between the erosion resistance of different materials including steels of varying hardness and closely to simulate the slurry flow and low impact angle damage processes in industrial equipment. A machine is currently in operation at the Budapest University of Technology and Economics in the laboratory of Dr. Eleöd András.

4. Measuring Erosion Wear

Slurry erosion wear is often measured as mass loss alone but this can provide only a measure of mean wear rates. Techniques using profile-measuring devices can provide much more information, but only in tests where the flow and particle impact conditions are known (Wong and Clark, 1993). Surface profile data measured about the circumference of cylindrical specimens from a pot tester before and after testing give wastage as a function of angular location. Typically, wear measurement systems store data as a file that can be analyzed to yield wear rates as a function of particle impact angle. LVDT (linear variable differential transformer) probes used for flat or cylindrical specimens can read to 0.1 µm, so that the factor limiting the precision of wastage measurements is often the roughness of the unworn surface.

Analysis of cylindrical test specimens is facilitated by the occurrence of damage at all impact angles on a single specimen. Normal (90°) particle impact occurs at the stagnation line of the cylindrical section giving rise to what has been termed "deformation wear"(Bitter, 1963). Subtracting this component from the total wear at each angle yields the so-called "cutting wear", each value being associated with a characteristic, or "specific" energy for the removal of unit volume of material. These values can be used to describe quantitatively the material response to erosion.

5. Particle-Wall Interaction

Particle-wall interactions are complex (Brach, 1991). On impact, particles may imbed, rebound or become entrapped, plough, slide or roll, depending on the properties of the particles, the impact conditions and the wall material. Typical eroded surfaces of ductile metals, seen in the SEM, are highly disturbed, but once developed, remain essentially unchanged. Some information can be obtained by examination of single impacts particularly with respect to the length-of-contact between particle and eroding surface, (Xie et al., 1999). Disturbance of the wall surface is caused by energy transfer from particle to wall, some of which energy may be returned to the particle as a rebound velocity. In slurries, liquid has also to be squeezed out from beneath the incoming particle for impact to occur. The effect of this "squeeze film" leads to cushioning of the impact through a reduction in particle approach velocity (Clark and Burmeister, 1992). Indeed, if the particle has a low kinetic energy (small size or low speed), impact may be prevented and the particle becomes trapped at the wall surface since its rebound velocity is zero. Under these circumstances the nature of the wear process changes as wastage occurs by particles abrading the surface under conditions of constrained low-angle impact rather than by unconstrained single impacts. However, both processes may be present at the same time (Clark, 1993).

Knowledge of the impact velocity and frequency of particles in erosion allows the calculation of a mean value of the damage (volume of material removed) by a single impact. The specific energy for deformation wear of mild steel is about 100 Jmm^{-3}. The kinetic energy of impact of a 200 μm diameter quartz sand particle colliding with the target surface at 10 ms^{-2} is about 0.5 μJ. If all this energy is consumed in material removal (i.e., no rebound), the wear volume corresponds with the removal of about 5 x 10^{11} atoms/impact. Under these circumstances even single impacts are properly termed 'micro-', rather than 'nano-', events. Such analysis is deceptive, however, since the removal of a single debris particle is often the result of multiple particle impacts as the surface material is deformed and cracks propagate.

6. Material Response in Erosion

Ductile metals show damage by slurry erosion at all flow speeds and particles sizes. Deformation wear rates are typically, two or three times lower than those for particle impact angles of about 30°, at about which angle the cutting wear maximum occurs (van Riemsdijk and Bitter, 1959). Material removal under normal impact has been shown to proceed by a process of platelet formation under repeated impact and consequent crack growth. For this reason calculations of the mean wastage/particle impact are of limited value.

Ceramics characteristically show peak wastage in slurry erosion under normal impact, (Clark and Wong, 1995). The rate of wastage under oblique impact can be satisfactorily expressed as a function of the normal component of particle impact energy. Material loss is presumed to take place by a process of crack initiation and slow growth under the influence of repeated impacts. Cylinder profile measurements of slurry erosion damage on ceramics (Pyrex glass and alumina) has shown that wastage effectively ceases at some small value of the impact angle, indicating the presence of a threshold for damage. Experimental uncertainties do not permit the quantitative evaluation of this threshold in energy terms.

Polymers have been shown to exhibit a distribution of wear with particle impact angle, under more severe conditions of erosion, closely similar to that of ductile metals, but with the additional presence of a damage threshold at low impact angles. The damage threshold may be a reflection of the resilience and ability to dissipate elastic energy that polymers show. Under less severe conditions UHMWPE has been found to be strikingly resistant to damage (Clark, 1996). The origin of this behaviour remains unexplained.

7. Material Wear Rates

An acceptable level of wear must be defined in terms of the context of the application. A user of slurry pumps might be delighted to see wear rates of 1 mm/year on pump components, a level unacceptable in electronic equipment. In general, laboratory test methods have focussed on accelerated wear, with the inherent assumption that results can be extrapolated directly to low wear rates. The effect of particle-particle interactions in even nominally dilute slurries makes such extrapolation of uncertain reliability. A wear depth of, say, 3 μm after 20 minutes of erosion testing on a carbide specimen – at about the limit for reliable measurement – still implies a wear depth of 1.5 mm/week. Wear rates on less resistant materials may easily be 100 times greater.

8. Uncertainty in Erosion Measurements

Unfortunately, slurry erosion does not allow of great analytical precision since the phenomenon is beset by uncertainties, including the size and shape of particles in the erodent sample. Unless spherical particles are employed, non-uniformity in the shape of particles is inevitable. Discussions continue on the best, most useful, way to quantify particle shape. The quantitative connection between erodent particle impact on the one hand and erosion wear on the other remains unclear. Further, liquid flow is subject to turbulence and vortex formation, which give rise to a range of particle impact velocities even for nominally identical particles. Vortical flow in pumps is well known as responsible for high local wear rates, often described as "gouging". In view of these experimental difficulties, inherent in even the most careful of laboratory experiments, it can be seen that the reliable prediction of wear rates in industrial equipment is a major challenge. The problem has recently been treated, with some success, by incorporating experimental wear data into a computer programme to analyse the complex flow and particle impact dynamics in oilfield control valves, (Forder et al., 1998).

9. Concluding Remarks

Experimental test methods for slurry erosion remain essentially empirically-based and tend to be principally concerned with accelerated wear testing corresponding with wear rates that would be unacceptably high in almost any industrial context. Increasingly, however, flow modelling has been used to define particle trajectories and impact dynamics to give a quantitative approach to wear rate prediction, even though such models generally refer to wear by dilute slurries. Further work on the nature, energetics and effect of single particle impacts, the role of the coefficient of restitution and the relation to wear rates is warranted in the context of the nature and origin of very low erosion wear rates. Clarification of the effect of particles agglomerated into a moving layer on wear rates would be useful. Efforts are also necessary to elucidate the origin, and the means of control, of those flow conditions that give rise to the highest wear rates in industrial equipment. Work will certainly continue on the development of commercial materials with high damage thresholds, high hardness, toughness and erosion resistance.

10. References

Addie, G., Pagalthivarthi, K. V. and Visintainer, R. (1996), "Slurry pump wear factors", *Pumps and Systems Magazine*, October, 18-27.

Bitter, J. G. A. (1963), "A study of erosion phenomena, Part I", *Wear* 6, 5-21.

Brach, R. M. (1991), *Mechanical Impact Dynamics*, John Wiley, New York.

Clark, H. McI. (1993), "The surface squeeze film in slurry erosion", *Proc. 6th Internat. Congress on Tribology, EUROTRIB 93*, (M. Kozma, ed.), Hungarian Scientific Society of Mechanical Engineers 5, pp. 134-139.

Clark, H. McI. (1996), "The slurry erosion wear of UHMWPE under conditions of particle impact and abrasion-erosion", *Proc. Internat. Tribology Conf. Yokohama 1995*, Japanese Society of Tribologists, Tokyo 1, 229-234.

Clark, H. McI. and Burmeister, L. C. (1992), "The influence of the squeeze film on particle impact velocities in erosion", *Int. J. Impact Engng.* 12, 415-426.

Clark, H. McI., Hawthorne, H. M. and Xie, Y. (1999), "Wear rates and specific energies of some ceramic, cermet and metallic coatings determined in the Coriolis erosion tester", *Wear* 233-235, 319-327.

Clark, H. McI. and Wong, K. K. (1995), "Impact angle, particle energy and mass loss in erosion by dilute slurries", *Wear* 186-187, 454-464.

Forder, A., Thew, M. and Harrison, D. (1998), "A numerical investigation of solid particle erosion experienced within oilfield control valves", *Wear* 216, 184-193.

Van Riemsdijk, A. J. and Bitter, J. G. A. (1959), "Erosion in gas-solid systems", in *Proc. Fifth World Petroleum Congress, Section VII, Engineering, Equipment and Materials*, pp. 43-58, Fifth World Petroleum Congress, Inc., New York.

Roco, M.C., Nair, P. and Addie, G. R. (1987), "Test Approach for dense slurry erosion", in *Slurry Erosion: Uses, Applications and Test Methods, ASTM STP 940*, (J. E. Miller and F. E. Schmidt, Jr., eds) American Society for Testing and Materials, Philadelphia, 1987 pp. 185-210.

Tuzson, J. (1984), "Laboratory slurry erosion tests and pump wear rate calculations", *ASME J. Fluids Eng.* 106, 135-140.

Wahl, H. and Hartstein, F. (1946), *Stahlverschleiss*, Franck'sche Verlagshandlung, Stuttgart.

Wong, K. K. and Clark, H. McI. (1993), "A model of particle velocities and trajectories in a slurry erosion pot tester", *Wear* 160, 95-104.

Xie, Y., Clark, H. McI. and Hawthorne, H. M. (1999), "Modelling slurry particle dynamics in the Coriolis erosion tester", *Wear* 225-229, 405-416.

Zu, J. B., Hutchings, I. M. and Burstein, G. T. (1990), "Design of a slurry erosion test rig", *Wear* 140, 331-344.

MACRO- AND MICRO KELVIN PROBE IN TRIBOLOGICAL STUDIES

A.L.ZHARIN
Belarussian State Research & Production Powder Metallurgy Concern
41 Platonov Street
Minsk 220071
Belarus

1. Introduction

The Kelvin method of electronic work function (EWF) measurement of contact potential difference (CPD) technique is an excellent non-destructive monitoring technique. Lord Kelvin offered the CPD in 1898. The CPD technique was developed considerably parallel with quantum theory of solids. As researchers were trying to correlate experimental data with theory, the EWF was explained according to the fundamental quantum mechanical parameters of solids. However, a strong influence of surface conditions on the experimental results was found and the technique was practically forgotten. Later, problems in the measurement of surface conditions have gained a special importance with the development of solid-state electronics. However, systems of surface analysis began to appear during the same years. Such systems were complicated devices attached to ultrahigh vacuum systems. These systems have overshadowed the CPD technique. An analysis of published papers has shown that surface analysis systems yield interesting results when conducting fundamental experiments with pure model surfaces. Results are not reliable for most of engineering surfaces. It is explained that surface analysis systems, in most cases, do not analyse the surface, but instead analyse artefacts on the surface. According to our experience, CPD does give reliable information about the surface.

In the present paper, the results of more than 20 years worth of the author's works with CPD and its application for tribological studies are described.

2. Techniques of an Electron Work Function Measurement

The EWF can be measured by direct and indirect techniques. Direct techniques induce electron emission, and their accurate sensing requires operation in a vacuum. The indirect techniques do not require high vacuum conditions. The techniques of EWF

B. Bhushan (ed.),
Fundamentals of Tribology and Bridging the Gap between the Macro- and Micro/Nanoscales, 445–466.
© 2001 *Kluwer Academic Publishers.*

determination on CPD are most convenient from the point of view of rubbing surface monitoring. Therefore, we shall consider it in detail.

Let us consider the contact phenomena in the case of two metals with a thin vacuum gap (Figure 1(a)) (Chalmers, 1942). The EWF (φ, Figure 1) is the difference between the Fermi level (E_F) of the metal and the surface potential barrier. In Figure 1(a), the surface potential barrier is shown by a dotted line, while the solid horizontal lines show the Fermi level. This figure corresponds to the initial time, when the metals separated by a distance d_0, at which an effective exchange of electrons stipulated by a thermionic emission is possible. The regularities of the thermionic emission are described by the Richardson - Dushman equation: $j = AT^2 \exp(-F/kT)$, where T is the temperature, k is a Boltzmann constant and A is a Richardson constant.

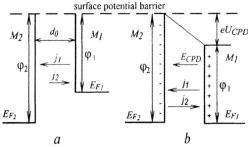

Figure 1. The contact phenomena in the case of two metals with a thin vacuum gap

If $\varphi_2 > \varphi_1$, then $j_1 > j_2$ and there will be an electron transfer from the right to the left (as it is shown in a Figure 1(a)). The first metal will be charged positively and the second metal will be charged negatively. The electric field E_{CPD} and, accordingly, the potential difference U_{CPD} will appear in a gap. An equilibrium condition implies an equality of Fermi levels. The thermoemission currents are aligned, i.e. $j_1 = j_2$. This equality means that the potential barrier $\varphi_1 + eU_{CPD}$ should be equal to the potential barrier φ_2. In other words: $\varphi_1 + eU_{CPD} = \varphi_2$. Thus, after equilibrium establishment the CPD becomes: $U_{CPD} = (\varphi_2 - \varphi_1)/e$. The case above is only a model. In practice, equilibrium (equality of Fermi levels) is attained by an external electric circuit.

The condenser technique is the most widely applied CPD measurement. The metal plates M_1 and M_2 form a parallel-plate capacitor of capacitance C (Figure 2(a)). The stipulated CPD charge Q on the capacitor will be: $Q = CU_{CPD}$. The gap between the capacitor plates periodically changes due to the vibration of one of the plates. Therefore, the capacitance periodically varies, an alternating current varies with the frequency of plate vibration, and the alternating voltage appears on the high-Ohmic resistor. A source of compensation voltage U_{comp} is included, the output of which can be adjusted so that it will compensate the CPD. In this case:

$$Q = C(U_{CPD} + U_{comp}) = 0 \qquad (1)$$

The alternating voltage from the vibrating-reed capacitor applies to the input of an electrometric preamplifier (an amplifier with a high input resistance) and to the indicator of a zero signal. The compensation voltage is equal to the CPD when Q is set to zero (Eq. 1), from which the CPD is determined. This is the basis of the large number of devices for the CPD measurement.

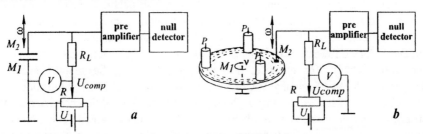

Figure 2. Functional diagram of the generalized Kelvin technique (*a*) and its modification for friction testing (*b*)

Another step in the development of the CPD measurement technique was the application of automatic compensation of a measurable value (Palevsky et al. 1947; Simon, 1959). After amplification, the output signal of the vibrating-reed capacitor is rectified by a phase-sensitive detector and fed back to the vibrating-reed capacitor for CPD compensation. An automatic record of the EWF changes is thus performed.

For a topology study on a surface without determination of the EWF value a technique using nonvibrating capacitor was offered, i.e. a kind of Kelvin technique that allows measurement of EWF changes without vibrating the capacitor plates (Danyluk and Zharin, 1999). The basis of such an approach is as follows. The generalized equation for the current in a circuit containing the Kelvin capacitor is:

$$i = U_{CPD}\frac{dC}{dt} + C\frac{dU_{CPD}}{dt} \qquad (2)$$

In the classic Kelvin technique, the second term of Equation (2) is equal to zero, i.e. the current in the circuit is provided by the change in capacitance only. If one scans a surface with a non-vibrating reference electrode, a current in the circuit will appear according to the second term of Equation (2), owing to potential spots on the sample surface. Albeit simple, technique of a non-vibrating capacitor has been shown to be reliable (Zanoria, 1995; 1996; 1997). One only needs a high-Ohmic preamplifier. Therefore, it can be used as a built-in sensor for surface monitoring. The non-vibrating probe does not provide the EWF value, but its derivative. It can be used for monitoring of local spots on a surface and for surface topology studies, but not for long term changes in integrated EWF of the surface.

Another modification of the condenser technique, which is promising for tribological applications, is the ionisation technique. In this technique, the air gap between the capacitor plates is ionised by a source of radioactive particles. The ions are moved in the electric field to the opposite charged plates, i.e. there is an ionic current between the capacitor plates. The potential that arises is a function of the ionic current that compensates the CPD. Thus, the potential difference between plates will be equal to the CPD and can be measured with an electrometric voltmeter. Currently the first steps in its usage for tribology are underway at the Georgia Institute of Technology (Danyluk and Zharin).

Since the invention of noncontact AFM, the electrostatic force has been used to measure electrical properties of various samples without contact. The applied voltage between the conductive tip and the sample generates the mechanical deflection of the cantilever. Many different instruments are included under the heading Electrostatic Force Microscope (EFM). For instance, Scanning Potentiometry, Scanning Kelvin Probe Microscope (SKPM), Scanning Maxwell Stress Microscope (SMM) are included in the EFM category (See, for example, Hong J., 1999). The difference among these instruments relates to the way in which the principle is implemented. These techniques seem to be very attractive for tribology studies. Currently it is used for studies related to microtribology at the Ohio State University (DeVecchio and Bhushan, 1998)

3. Background of Kelvin Technique Application for Tribological Studies

The forces of interaction with a crystal lattice do not allow electrons to escape from the metal, i.e. there is a potential barrier on a metal surface. Only those electrons with sufficient energy to overcome these forces can escape from the metal (Friedel, 1976). Two terms: the dipole and the quantum-mechanical exchange-correlation, contribute to the surface potential barrier (Mahan et al. 1974). A classic analogue of the exchange-correlation term is an image force. The dipole term arises because the wave function of an electron has a "tail" at the metal surface. The density of electrons near the metal surface is not equal to zero and is not compensated by a positively charged ionic skeleton of the metal lattice. Therefore, there is a dipole layer on the metal surface.

The electrons in the metal have energy even at absolute zero. This energy is termed the Fermi level. Therefore, for the removal of an electron from the metal, it should receive an energy equal to the difference between the surface potential barrier and the Fermi level. This energy is termed as the EWF.

The EWF is one of the fundamental characteristics of the condensed (solid) metallic state. It correlates with some physicochemical and mechanical characteristics of metals. This characteristic applies to the study of the phenomena of deformation and destruction of metals, and to the study of adsorption and desorption.

The surface changes considerably under friction. Many of the parameters vary, among these are roughness, temperature, chemical and phase composition, defect density and others. In addition, a variety of physical and chemical processes are in force. The formation and destruction of oxide and lubricating films, adsorption and desorption of lubricants and gases take place at the rubbing surface. Practically all these processes will influence the EWF, directly or indirectly.

The chemical composition of alloys influences the position of the Fermi level considerably and, accordingly, affect the surface EWF. This issue has been well-described in the literature (e.g., Yamamoto et al. 1974). Generally, the chemical composition of the bulk does not vary under friction. However, considerable changes in the chemical composition at the rubbing surface can occur. Therefore, the Kelvin technique in general allows for the study of tribochemical processes.

The Kelvin technique's applicability for monitoring a rubbing surface that has a lubricant may be a problem. Apparently, the following processes will contribute to the EWF change with a deposition of lubricant: the adsorption of lubricant molecules, the chemical interaction of lubricant with oxides and the material, the formation of a dipole layer on a metal - lubricant boundary and others. A screening property of the lubricant layer can contribute to measurement results also. Pekar et al. (1947) have found that the EWF of the metal surface covered with an oxide or semiconductor layer should essentially not vary if the layer thickness is much less than the Debye length of screening. The Debye length of screening depends on the density of free charge carriers (electrons, ions and holes) in the substance. Metal oxides and the majority of lubricants are dielectrics or semiconductors with a low density of free charge carriers. In these cases the Debye length of screening has a high value (sometimes centimetres) (Zharin, 1996). As it is known, there is a high density of free electrons in metals. Therefore, for metals and alloys the thickness of a layer contributing to the EWF measurement is equal to several interatomic distances. It follows, then, that when studying surfaces covered by dielectric film (lubricant, oxide etc.) one will only study the metal behaviour through the dielectric film. In the case of lubricated friction, the influence of the change of the lubricant film thickness to the EWF can be neglected.

The adsorption processes on the metallic surface strongly change its EWF. It is governed by the redistribution of electrons at the formation of an adsorptive bond between a free atom or molecule and the surface dipole layer (Cherepnin, 1973). The EWF change of the metallic surface by adsorbed atoms or molecules is directly proportional to their effective electrical dipole moment and the degree of surface occupation. The high sensitivity of the EWF to the adsorption has led to wide use of this parameter for the study of adsorption processes of organic and inorganic molecules and atoms, and for a study of the oxidation. The adsorption processes must be taken into account for studies of lubricated friction. The adsorption of molecules creates a double electric layer on the "metal - lubricant" boundary. Certainly, taking into consideration the contributions of all processes happening on a rubbing surface simultaneously, is rather complicated. Therefore, it is convenient to consider the contribution of the double electric layer as constant. In practice, one can create a condition, trough the control of lubrication (for example, flowing), where this contribution will be a constant.

Mechanical perturbations can influence the EWF through the Fermi level. The compactness of the metal lattice varies under mechanical compression or tension. Thus, the position of Fermi level varies and creates an EWF change (Andreev et al. 1963). The EWF will grow with an increase in compression for the majority of metals. The typical

EWF change is about several μeV/bar (Craig et al. 1969). Thus, the EWF varies insignificantly with changes in lattice compactness in the elastic regime.

Crystalline defects in metals influence the EWF to a greater degree. The greatest influence on EWF is rendered by linear defects (boundary and screw dislocations). The atoms in the neighbourhood of a dislocation are under a considerable hydrostatic pressure created by its stress field (Gutman, 1974). Latishev et al. (1975) estimated the local EWF change in a dislocation core on the surface theoretically. They found that at a dislocation core the EWF is reduced by approximately 0.3 eV. Therefore, the integrated value of EWF can vary considerably with a change in the defect density. Mints et al. (1975) have found theoretically that the dislocations can result in a decrease of an integrated surface EWF by 10^{-2} - 10^{-1} eV.

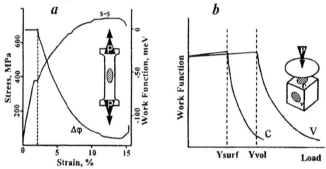

Figure 3. Results of simultaneous measurements of the regular stress - strain diagram and the work function for medium carbon steel (*a*) and schematic representation of experiments on contact deformation (*b*).

We carried out additional experiments on the influence of deformation the EWF for simple cases of loading. The test machine "Instron" was used. The Kelvin probe was mounted on a lateral surface of the sample. The sample elongation and EWF were recorded simultaneously. Typical results for the EWF change under a simple monoaxial tension for a case of soft steel are shown in a Figure 3(*a*). It follows from the figure, that up to the yield point the EWF varies insignificantly. However, there is a sharp EWF decrease. The nucleation and movement of dislocations in surface layers can explain such EWF decrease. This is consistent with Vishniakov's (1975) conclusion that the EWF integrated value can vary considerably with a change in defect density. In our case, it varies with changes in dislocation density on the surface. It is known (Vishniakov, 1975), that the dislocation density in the material can be increased up to some critical value only. After this, the equilibrium between the nucleation and coalescence of dislocations is achieved. To our minds, it can explain the saturation of the EWF under the further deformation in a plastic area. It is necessary to point out, that similar EWF behaviour at deformation takes place also for materials not having a yield point. This fact is promising for the use of the Kelvin technique for general mechanical testing.

Generally, the behaviour of EWF dependence on deformation (both tensile and compressive) is identical for the majority of metals. There is a small increase in EWF in the elastic regime caused by the change in the Fermi level due to the crystal lattice dilatation. It decreases quickly at the hardening stage. The speed of the EWF change decreases as dynamic recovery is approached (Vishniakov, 1975). Thus, the EWF can characterize the mechanical processes on a metal surface both in the elastic and plastic regimes.

Problems of deformation processes in thin surface layers are of particular interest in tribology. For the majority of metals and alloys the thickness of the metal layer, which contributes to the EWF, has a value approximately equal to the interatomic distance. The other feature of the technique is that the measurements are produced integrally on a rather large segment of the surface equal to the dimension of the reference electrode (in our case, ~ 5 mm^2).

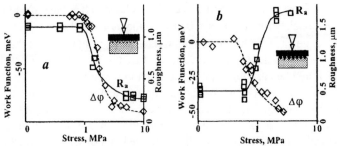

Figure 4. Results of simultaneous measurements of the surface roughness and the work function depending on load for cases of "flat on rough" (*a*) and "rough on flat" (*b*) loading.

A sample was mounted in the test machine and was compressed through a glass plate within 30 seconds. Then it was removed from the test machine and the EWF of the lateral and contact surfaces were measured. The roughness parameter (*Ra*) of the contact surface was also measured. This procedure was carried out with incremented load and dependencies of the EWF of lateral and contact surfaces and the roughness of the contact surface on the applied load were found. In Figure 3(*b*), such dependences and the method of loading are shown schematically. In this case, EWF measurements of the lateral surface are indicative of the behaviour of the bulk of the material.

In Figure 3(*b*) it is shown, that the character of the dependency of EWF on the contact (*c*) and lateral (*v*) surfaces, on the applied load, is similar. In both cases, a sharp decrease in the EWF is observed at some load value. The decrease of the curve for the contact surface (*Ysurf*) is observed under a smaller load than for the lateral one (*Yvol*). It is possible to explain it by the deformation of the contact surface asperities and by the beginning of plastic processes (reaching yield point) under smaller loadings than for the bulk of material. The surface roughness does not vary until some critical value under small loads (Figure 4). The roughness begins to decrease under a further load increase for the case in Figure 4(*a*). The contact surface roughness does not decrease, but increases under the sample loading with rough plates (Figure 4(*b*)). From Figure 4 it is clear that the EWF changes occur a little bit earlier than roughness changes begin.

Hence, the beginning of the EWF decrease in this case signals a yield point in the surface layer even before the plastic deformations are measured by traditional methods.

Thus, the Kelvin technique can be applied for a study of deformation of thin surface layers. In particular, it can be used for the determination of a contact deformation type and yield points at contact deformation. There is an opportunity of experimental verification and improvements of the existing contact theories of rough surfaces.

In the Kelvin technique, the average distance between plates of a vibrating-reed capacitor greatly exceeds the dimensions of the asperities of a surface and the variations of the roughness values do not affect the results generally (Dydko, 1961). However, the dependence of the surface EWF on roughness should be considered in connection with a physical condition of the surface. The presence of surface defects, their character and density will influence the EWF. Nazarov et al. (1990) have applied the CPD technique in the monitoring of processing of mirror metallic surfaces. They have mentioned that, by using a conventional technique, it is possible to monitor roughness up to 0.1 μm only. The CPD technique can only be applied for surfaces with roughness less than 0.1 μm.

From the above description, it follows, that the basic contribution to the EWF under friction would be introduced by the deformation. More exactly EWF changes with: the density change of dislocation cores on a surface, density change of point defects on the surface, and change of atomic roughness. The next most significant is the contribution from a change in chemical composition of the surface. It is possible to interpret the results from friction, basically, from the point of view of defect dynamics.

The adsorption of both lubricant molecules and molecules of ambient gases also contribute to the EWF considerably. However, the influence of adsorption processes can be minimized by creating constant conditions of lubrication and ambient gases when conducting long-time experiments.

4. Application of the Kelvin Technique for *in-situ* Monitoring of Rubbing Surfaces

Figure 2(*b*) shows a generalized modification of the Kelvin technique suitable for friction testing (Zharin et al. 1978). The device for the CPD measurement in the dynamics of the friction should meet high requirements concerning a noise reduction and stability of operation since high level of vibrations, acoustic and electromagnetic noise originate from the friction test rig. It is necessary to provide an effective guard from mechanical, electrical and magnetic disturbances. The basic requirement for friction testing machines is the presence of a reliable grounding for the rotating sample and the presence of a segment of friction track not covered by a counter sample.

The essential point is the relationship between the reference electrode dimensions and the friction track width. At the large reference electrode the device will "feel" a material located near a friction track that can give an error in measurement. In order to eliminate this, reference electrode should be selected to have a narrower friction track (2 and 7 mm accordingly for results described below).

The generalized circuit of the Kelvin probe was developed with the above requirements in mind (Figure 5(*a*)) (Zharin et al. 1979; 1996; 1998). In Figure 5(*a*) the reference electrode 3 is vibrated by a modulator 4 and is located at a distance less then 0.5 mm from the rubbing surface of the rotating disc 2. Together they form a capacitor. The alternating current induced by the periodic change in the capacitance is converted to a voltage signal by a high Ohmic resistor 9 and is amplified by a preamplifier 5. From the output 5, the signal comes to the phase detector together with the reference signal from the modulator 4. Upon detection, the signal goes to the integrator 8 and then via the high Ohmic resistor 9, to the reference electrode to compensate its EWF. The probe maintains an output voltage equal to the CPD and a recorder 10 (chart recorder, data acquisition etc) records it. When necessary, an integrator with a time constant much exceeding the period of revolution of the specimen can be incorporated between the recorder input and the probe output to record the integrated EWF of the rubbing surface. The Figure 5(*b*) shows a photograph of CPD probe mounted on the tree pin on disk friction testing machine.

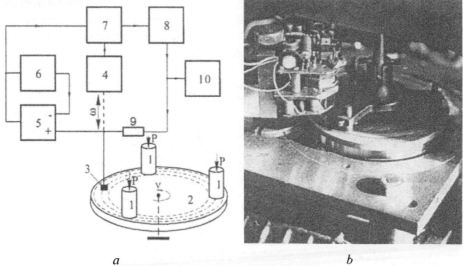

a   *b*

Figure 5. Schematic diagram of measurement system (*a*) and photograph of Kelvin probe mounted on the tree pin on disk friction tester (*b*). 1 - pins; 2 - a rotating grounded disk; 3 - a reference electrode (nickel or gold); 4 – an electromechanical modulator; 5 - a differential amplifier with high-Ohmic inputs; 6 - a feedback circuit; 7 - a phase-sensitive detector; 8 - an integrator; 9 - a high-Ohmic resistor; 10 – a recorder.

The experience of designing the Kelvin probes has shown that the modern electronics allows one to make them as a small device entirely located near the rubbing pair. Practically any friction machine may be equipped with the Kelvin probe when the surfaces do not fully overlap. This includes reciprocating machines also (Zharin et al. 1990). Some technical problems of the CPD probe have been considered in (Zharin et al. 1993; 1996; 1998).

5. Experimental Examples of Kelvin Technique Application

The Kelvin technique can be used in two experimental modifications: (a) monitoring of the EWF with a small time constant of the signal integration and its synchronization with the measurement of sample position, allowing the investigation of the topology of the EWF; (b) monitoring the EWF with the time constant of the signal integration exceeding the period of revolution of the sample, allowing investigation of integral EWF changes over the entire rubbing surface. In the former case, one can study the rubbing surface topology with sliding time, as shown in Figure 6(a). However, in most of the experiments conducted we used the second approach, because, in our opinion, the properties of the entire rubbing surface are responsible for the tribological behaviour of materials.

Figure 6(b) shows the variation with time of the rubbing surface EWF for some materials. At the initial friction contact, a sharp decrease is followed by a sharp increase of the EWF relative to the initial value. Our interpretation is that the initial transients correspond to changes in the original oxide or other surface layers and running-in of the sliding surfaces. After a certain time, which depends on material properties, the friction regime stabilizes. From Figure 6(b), one can find running-in finishing and the beginning of steady state friction conditions (Zharin et al. 1989).

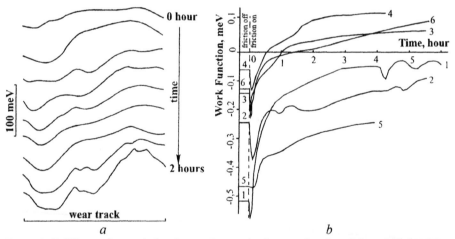

Figure 6. Rubbing surface work function topology evaluation as a function of time of friction (a) and variation of the rubbing surface work function with time for some materials (b). 1 - brass; 2 - bronze (Al); 3 - bronze (Sn); 4 - commercial copper; 5 - stainless steel; 6 - medium carbon steel.

Examples of rubbing surface EWF responses on some perturbations are shown in Figure 7. Figure 7(a) shows the EWF responses of different materials with the same lubricant (the initial EWF values of materials are equalized). The speed of the lubricant layer formation can be determined from such results. It is possible that this could be a fruitful way to study lubricant properties. This technique is applicable for sliding electric contact investigation as well. Figure 7(b) shows the responses of the rubbing surface EWF when the electric current passes through the friction contact.

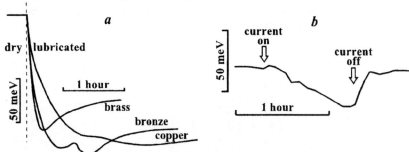

Figure 7. Rubbing surface work function responses using the same lubricant for different materials (a) and rubbing surface work function response when applying an electric current through the friction contact (b).

At present the technique is used for two major purposes: (1) to determine the critical points with respect to changes in normal load, with relevance to the materials selection and optimisation, and (2) to determine the kinetics of friction processes, including periodic changes which may be related to those in fatigue.

6. Critical Points with Respect to Changes in Normal Load

The studies of a wide range of materials have shown a similar qualitative dependence of the rubbing surface EWF on the normal load (Figure 8). Generally, there are three specific zones (Zharin et al. 1995). There is an increase in the EWF with load in zone I. At a higher load, there is a qualitative change of the curve. In zone II, EWF decreases. With further loading in zone III there is very little change until the beginning of scoring, which is detected by increases in friction force and bulk temperature of the sample. During scoring, the value of the EWF decreases sharply. Hence, according to the character of the EWF changes, one can record changes in the material surface layers that do not manifest themselves in the external friction parameters (for example, friction force and bulk temperature).

It has been found that during the long-run trials the wear rate is very low under loads corresponding to zones I and II. For zone III, the damaged spots on the surface and high wear rate are observed. The additional investigations of specimens via independent methods as well as studies concerning changes of the EWF during simple uniaxial deformation (Figure 3) allow us to interpret the results obtained on the base of dislocation interactions. We suggest that the first zone corresponds to mainly elastic and early stages of plastic deformation without a significant increase in dislocation concentration near the surface. In the second zone, plastic deformation dominates, but the density of dislocations increases with load (in this zone the EWF decreases with increasing load, as shown in Figure 3). In the third zone, the plastic processes also dominate, but the dislocation concentration do not changes significantly; i.e., there is a dynamic process involving the generation and annihilation of dislocations and the creation of micropores and microcracks.

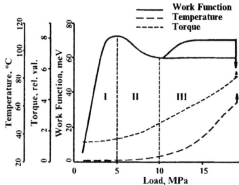

Figure 8. Bronze rubbing surface electron work function, torque and surface temperature vs. normal load curves.

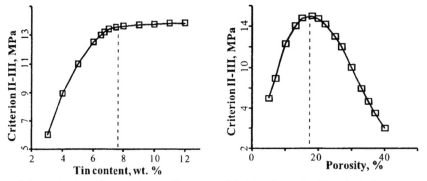

Figure 9. Examples involving powder metallurgy materials (*a*) and porosity (*b*) optimisation.

Studies of a wide range of materials have shown that the transition load-separating zone II and zone III could serve as an objective experimental criterion to estimate the serviceability of the tribological materials. This transition is probably related to the critical transition between mild and severe wear for a given material. The criterion developed is fruitful for developing new materials for tribological applications (Zharin, 1996). Examples involving materials and porosity optimisation are provided in Figure 9.

7. Periodic Changes of Rubbing Surface Electron Work Function Integral Value

Within the third zone described above, the kinetics of steady state friction is characterized by regular periodic changes of the rubbing surface EWF integral value (Zharin et at. 1979). The period, amplitude and harmonic contents of such changes are determined by the properties of materials and testing conditions (Zharin et al. 1981). The existence of the EWF periodic changes was confirmed independently at the Ohio State University (Kasai et al. 1998; 1999).

The EWF periodic changes were observed most clearly the in case of friction of a 63-37 brass (disk) (Figure 10). The amplitude of EWF periodic changes exceeds 100 meV, and the period is over 30 minutes. Several thousand passes of the pin on the sample surface occurred over one period. The shape of the EWF periodic changes is close to sinusoidal. The periodic changes were observed in the cases of reciprocated and lubricated friction also.

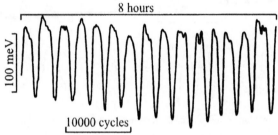

Figure 10. Record of 63-37 brass rubbing surface electron work function periodic changes over 8 hours. (Pins were made of phosphor bronze. The contact pressure is 0.12 MPa and the period of disk rotation is 0.42 s^{-1}.)

The studies of a wide range of materials have shown that the form of rubbing surface EWF changes can be divided into the following groups:

(a)   EWF does not vary with friction within the limit of measurement errors. This group is represented by materials with low strength properties working under dry friction and most materials working at lubricated friction under low loading. (Kasai et al. (1998) have found that periodic changes disappear under decreased atmospheric pressure, i.e. in vacuum condition.)

(b)   The rubbing surface EWF changes have a periodic character with a shape close to sinusoidal. This group is represented by single-phase solid solutions on a copper base at dry friction.

(c)   The rubbing surface EWF changes with a destroyed periodicity, i.e. superposition of periodic EWF changes with different periods and amplitudes. This group is represented by multiphase copper-based solid solutions working under dry friction.

(d)   The rubbing surface EWF changes have a periodic character with a "saw tooth" shape, i.e. the EWF rather slowly increases and quickly decreases. This group is represented by the most cases of steels at dry friction and most of materials under boundary lubrication. A "spike-like" behaviour (Kasai et al. 1999) should be included in this group.

Thus, the period, amplitude and harmonic contents of the EWF changes depend on chemical and phase composition of materials, their mechanical characteristics, and testing conditions. The experiments were carried out on devices that have an integration time that exceeds the time of one revolution of the disk, i.e. the EWF value was integrated over the rubbing surface. Therefore, periodic changes of EWF reflect physical processes happening on the whole rubbing surface.

The question is whether the periodic changes of the rubbing surface EWF integral value are damping in time. The statistical analyses have been carried out to examine this and related points.

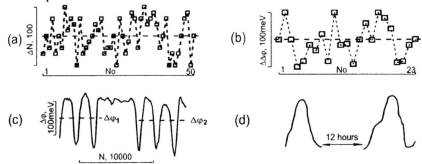

Figure 11. Statistical regularities of rubbing surface work function periodic changes

Data for the statistical analyses were collected during long duration tests (over 50 days) of brass (disc) - phosphoric bronze (pins) at a normal pressure 0.15 MPa under the same conditions. The records of the periodic changes of the rubbing surface EWF integral value were statistically processed and then the results were validated using statistical criteria. From the statistical analyses it follows that the periodic changes of the rubbing surface EWF integral value have the following properties: (*a*) they are identical i.e. they have the same statistical distribution during repeated tests under similar testing conditions (Figure 11(*a*)), (*b*) they are non-damping (Figure 11(*b*)), (*c*) they are self-restorable (Figure 11(*c*)) and (*d*) they conserve the same phase of period after the interruption and resumption of sliding (Figure 11(*d*)).

It follows from statistical studies, that the rubbing surface integral value of the EWF periodic changes are stable and non-damping. On the other hand, it implies that there is a synchronization of the periodic changes over the rubbing surface.

The property (*d*) means that the surface maintains its conditions at the conclusion of sliding. It allows the investigation of surface and subsurface layers in the particular points of the rubbing surface EWF period by means of regular techniques.

Studies of surface and subsurface conditions at four characteristic points during period of EWF change were carried out. These characteristic points are selected: positive and negative peaks and the midpoints of the EWF period increase and decrease. The disks were 4 mm thick and had a diameter 100 mm. They are made of brass 37-63. The pins were made of phosphor bronze. The samples were mounted in the friction-testing machine equipped with the Kelvin probe and tested before they reached a required point of EWF change, and then they were prepared for further studies.

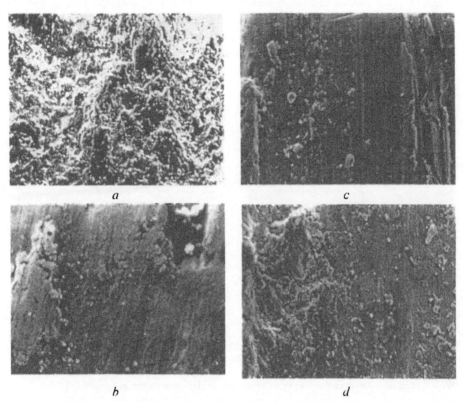

Figure 12. Microphotographs (SEM) of brass worn surfaces at four characteristic points based on the period of the work function periodic changes. All have the same magnification (x400). (*a*) - the negative peak, (*b*) - the midpoint of the work function period increase, (*c*) - the positive peak, (*d*) - the midpoint of the work function period decrease.

Figure 12 shows scanning electron micrographs of the worn surfaces of brass 37 - 63 at these points. The worn surface has the highest visible "defectness" at the point of the negative peak. Upon further sliding, the worn surface smoothes out. At the midpoint of the EWF increase (Figure 12(*b*)) there are only individual areas containing visible "defectness" similar to those shown in Figure 12(*a*). The surface was smoothest at the positive peak (Figure 12(*c*)). Then at the midpoint of the EWF decrease, the separate damaged areas appeared on the worn surface, as can be seen in Figure 12(*d*).

Figure 13 shows the microhardness distribution below the worn surfaces at the above mentioned four points. The minimum subsurface workhardening is observed at the point of negative peak (Figure 13(*a*)). As a result of further sliding, work hardening

460

increases, reaching the values shown in Figure 13(b) and 13(c). In these three cases, the shape of the microhardness distribution remains unchanged and is approximated by a power-law curve. The shape of the curve of microhardness distribution changes radically for the midpoint case of decreasing period (Figure 13(d)). At a depth of more than ~40 μm the microhardness distribution is described by the power-law curve, i.e., the form of the distribution remains the same as for the cases described above. At depths of less than ~40 μm the microhardness distribution is linear.

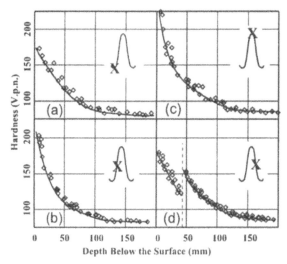

Figure 13. Microhardness distribution under the brass worn surface in four characteristic points based on the period of the work function periodic changes. (a) - negative peak, (b) - midpoint of the work function period increase, (c) - positive peak; (d) - midpoint of the work function period decrease. The microhardness was determined at crack-free region in the case (d)

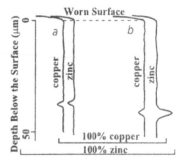

Figure 14. Distribution of brass 37-63 components below the worn surface. (a) - at the location of subsurface crack, (b) - at the crack-free region and (c) - SEM Zn Kα image of the worn surface cross-section at the midpoints of the work function period decrease.

The alloy components distribution below the worn surface was studied by looking at cross-sections (normal to the worn surface) of the specimens corresponding to the same four points. The concentration nonhomogeneities of the alloy components were detected on the specimens corresponding to the midpoint of the decreasing EWF only and at the same depth below the surface where cracks were observed. Figure 14 shows distributions of the brass 37 - 63 components below the worn surface: (*a*) - at the location of the subsurface crack, (*b*) - at the crack-free region. The crack was identified by a simultaneous reduction of the concentration of both alloy components. In the region free of the subsurface cracks and at the same depth where they were observed (approximately 40 μm), there was a redistribution of alloy components, i.e. an increase in zinc and a decrease in copper concentrations. The increase in the zinc concentration was significant, reaching ~8%. The above observations are additionally illustrated by an SEM *Zn Kα* image of the worn surface cross-section (Figure 14(*c*)), were the near crack region looks lighter due to the higher zinc concentration.

The friction tests were carried out using relatively low sliding speed. An increase in zinc concentration was observed significantly below the worn surface. That is why it is reasonable to suppose that the redistribution of alloy components is not related to thermal effects in the subsurface layers. The behaviour can be explained by the transfer of zinc atoms through the crystal defects into the defect accumulation zone.

Similar studies were carried out with specially prepared model copper alloys: *Cu + 6%Sn* and *Cu + 8%Sn*. It should be noted that the first alloy belongs to α -solid solutions and the second one contains δ - phase inclusions. Similar to the case of brass, these materials showed cracks parallel to the worn surface at the midpoint of the EWF decrease, but the subsurface cracks in bronze, *Cu + 6%Sn*, were ~35 - 40 μm deep and in *Cu + 8%Sn* were ~70 - 75 μm deep. Figure 15 shows panoramic optical micrographs (separate 200x micrographs are combined) of the cross-sections made normal to the worn surface and perpendicular to the sliding direction (here and after, (*a*) is *Cu + 6%Sn* and (*b*) is *Cu + 8%Sn*). The length of the subsurface cracks is considerable (the width of the micrographs 15(*a*) and (*b*) are approximately 2 mm). In some cases, very long cracks were observed; they are extended practically to the full width of the friction track (~8 mm). The same cross-sections are shown in detail in the Figure 16.

The micrographs shown in Figures 15 and 16 demonstrate that there is no significant plastic flow in the near surface. This was usually observed in numerous subsurface damage studies (see, for example, Alpas, 1993). Another feature of the micrographs is that the damaged zone below the worn surface looks like extended system of voids but not like a crack. The metallographic structure of the material band between the surface and damaged zone was not revealed by the metallographic etching; the bulk of material under the band shows a nice metallographic structure. The latter can be explained by a subgrain structure formation in this band. The modification of the microhardness distribution law as it is shown in Figure 13(*d*) could be an indirect confirmation of the above observation.

Samuels et al. (1980) have found a generally similar structural arrangement. Their schematic representation of an abraded brass surface includes the subgrain band on the surface. It is reasonable to extend their schematic representation by adding voids under the subgrain band, as shown in Figure 18. In this case, stresses appearing on the rubbing surface are transferred to deeper layers by the subgrain band, without dislocation nucleation and accumulation within the individual subgrain. The main processes governing subsurface fatigue are the nucleation of voids and their growth, while crack propagation processes play an insignificant role.

a

b

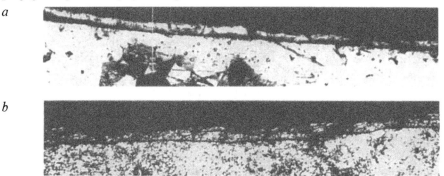

Figure 15. Panoramic optical micrographs (the width corresponds to ~ 2 mm) of the cross-sections made normal to the worn surface and perpendicular to the sliding direction at the prefracture stage. (*a*) - *Cu* + *6%Sn*, (*b*) - *Cu* + *8%Sn*).

a   *b*

Figure 16. Detailed micrographs of the same cross-sections as in Figure 15. (*a*) - *Cu* + *6%Sn*, (*b*) - *Cu* + *8%Sn*.

Two things are responsible for the nucleation of voids and their growth. The first one is the appearance of considerable shear stresses on the boundary between the subgrain band and the bulk material, as, in general, the reordering of near surface dislocations into subgrain boundaries leads to changing of the near surface band volume. The second one is void nucleation and growth. To our minds, the second case is more reasonable as it is indirectly confirmed by the alloy components redistribution (Figure 14) that could be explained by a movement of a number of the dislocations, together with their impurity atoms (zinc), to the accumulation zone.

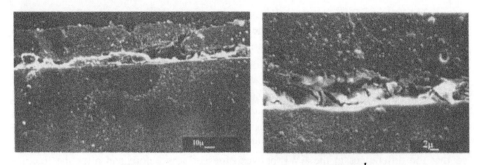

a   *b*

Figure 17. Micrographs of the subsurface features in details. (*Cu* + 6%*Sn*, unetched)

Let us consider the kinetics of processes in the rubbing surface from the point of view of the subsurface fatigue processes described above. Initially the pins smoothes the outermost surface layers and creates a surface with certain physical properties and with a certain EWF. Simultaneously, the dislocation density in the subsurface layers grows due to repeated stresses. When the dislocation density reaches a critical value, reordering of the near surface dislocation dense into subgrain boundaries will occur. The reordering process can start when both conditions (certain critical dislocation density and certain critical stress level) are reached simultaneously. That is why, below the subgrain band, where stresses are not sufficient, dislocation density is continuing to increase without reordering. Further dislocation density increase leads to void nucleation and growth under the subgrain band according to the known mechanisms. For example, in materials with low stacking fault energy like the tested brass, voids can be nucleated from the flat dislocation pile-ups and the dislocation movement can grow the voids (Hirt et al. 1972). The continuation of the process described above can lead to unstable conditions when a relatively thick surface band is almost separated from the bulk material, but still remains under friction because the mechanical properties of the subgrain band are high enough to keep its integrity. The density of defects on the outermost surface layers may remain relatively low due to, for example, dislocation annihilation on the free surface.

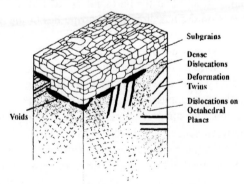

Figure 18. Worn surface schematic representation for the prefracture stage of subsurface fatigue.

464

The failure of one or several weak spots of the subgrain band will increase the load on the remaining segments of the band. Subsequently, further destruction will behave like an avalanche i.e. the destruction process will be synchronized over a considerable segment of the surface.

After subgrain band destruction and removal, one of the walls of the subsurface voids will become a rubbing surface. The smoothed outermost surface will be replaced with a highly defective surface. This new surface is smoothing by the pins again and its EWF is increasing. In parallel, conditions for the next cycle of the subsurface fatigue are accumulating. By the next cycle, physical properties of the surface change significantly, so that the differences in the EWF between the surfaces before and after the fracture become significant. From the point of view of the outermost worn surface, the processes described above can be illustrated by the microphotographs in Figure 12. Figure 12(a) shows the surface just after subgrain band removal. After that, the surface layer is smoothed down (Figure 12(b) and (c)). Figure 12(d) shows a prefracture stage when spots of destruction appear.

The above model is consistent with Kasai's et al. (1998, 1999) observations of metallic and oxidized surfaces at different points of the EWF period.

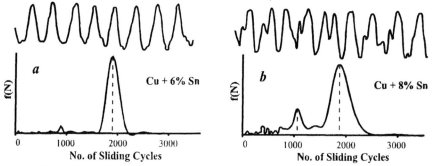

Figure 19. The rubbing surface work function periodic changes and their spectrums. (*a*) - single-phase bronze, (*b*) double-phase bronze.

It has been demonstrated that the harmonic content of the rubbing surface EWF periodic changes depends on the metallophysical properties of the material (Zharin et al. 1993). In Figure 19, the rubbing surface EWF periodic changes and their spectrums are shown for the case of the single-phase (*a*) and double-phase (b) bronzes, described above. The single-phase material clearly exhibits strictly regular periodic EWF changes and its spectrum has only one maximum. The periodic changes contain additional harmonic and an additional maximum in the spectrum in the case of a double-phase material (*b*). This additional maximum is apparently due to the difference in the fatigue fracture mechanism between single- and double-phase materials. Double-phase materials have additional obstacles, as secondary phase inclusions keep defects (dislocations) from movement. It can be assumed that the additional process of the fatigue damages accumulation (in secondary phases, solid inclusions, etc.) causes additional harmonics in the spectrums of the EWF periodic changes.

The description above can be confirmed also by comparing the microphotographs in Figure 15 and 16. Case (a) is related to the single-phase bronze and case (b) is related to the double-phase bronzes. The band includes some additional minor damage (case (b)), which do not appear in case (a). These results may be considered an experimental confirmation of the fact that several fatigue mechanisms may operate simultaneously in subsurface layers of multiphase materials during sliding.

Based on the fatigue-like nature of the rubbing surface EWF periodic changes integral value, one expects a correlation with known characteristics of bulk fatigue fracture. Curves similar to bulk fatigue curves (Weller's curves) were plotted using the work function periodicity data (Zharin et al. 1993, 1996, 1998). Since it is possible to plot curves similar to those of Weller using data on the rubbing surface periodic EWF changes, approaches used for the conventional fatigue processes can be applied to evaluate fatigue fracture parameters for the rubbing surfaces.

The last question is why the subsurface voids shown above were not observed earlier. They appear to exist for only a relatively shot time and it is necessary to stop the test and observe the rubbing surface when surface conditions change. Such surface condition changes must be monitored using a very sensitive technique, such as that used in this paper.

8. Conclusion

The described technique is applicable for investigations and analysis of a wide range of processes involving rubbing surfaces, including running-in processes, formation and depletion of lubricating layers, responses of a surface to variations of friction conditions, fatigue processes, etc.

The Kelvin technique is practically unique in its sensitivity, versatility and simplicity for practical implementation.

9. References

Alpas A., Hu H. and Zhang J. (1993), "Plastic Deformation and Damage Accumulation Below the Worn Surface", *Wear* **162** 188-195.

Andreev A.A. and Polige Ia. (1963), "Work Function Change Under Cold Deformation of Molybdenum and Tungsten in Ultrahigh Vacuum", *Proc. USSR Academy of Science* **152** (5) 1986 – 1088 (in Russian)

Chalmers J.A. (1942), "Contact Potentials", *Phil. Mag.* **33** 399 - 430.

Cherepnin N.V. (1973), *Sorption Phenomenon in Vacuum Technique*, Soviet Radio, Moscow (in Russian).

Craig P. (1969), "Direct observation of stress - induced shifts in contact potentials", *Phys. Rev. Let.* **22** (14) 700.

Danyluk S., Zharin A.L. et al. (1999), *The non-vibrating capacitance probe*, US patent 5,974,869.

DeVecchio D. and Bhushan B. (1998), "Use of a Nanoscale Kelvin Probe for Detecting Wear Precursors", *Rev. Sci. Instrum.* **69** 3618 - 3624.

Dydko G.V. (1961), "On Contact Potential Difference Measurements by Condenser Technique", *USSR J. of Exp. Tech.* **5**, 128 - 130.

Hirt J. and Lote I. (1972), *Theory of Dislocations*, Atomizdat, Moscow (in Russian).

Friedel J. (1976), "The Physics of Clean Metal Surfaces", *Ann. phys.* **1**, No. 6, 257 - 307.

Gutman E.M. (1974), *Mechanochemistry of Metals and Corrosion Protection*, Metallurgy, Moscow (in Russian)

Hong J. (1999), *Electrostatic Force Microscopy in the Noncontact Mode and Its Applications*, PhD Thesis, PSIA Corp., Seoul, Korea.

Kasai T., Rigney D. and Zharin A. (1998), "Changes Detected by a Non-Contacting Probe during Sliding", *Scipta Mater.* **39**, 561-567.

Kasai T., Fu X., Rigney D. and Zharin A. (1999), "Application of a Non-contacting Kelvin Probe During Sliding", *Wear* **225-229**,1186.

Lord Kelvin (1898), *Phil. Mag.* **46**, 82.

Latishev A.N., Molotski M.I., Chibisov K.V. (1975), "An Interaction of the Chemisorbed Particles with Dislocations", *Proc. USSR Academy of Science*, **224** (4) 880 – 882 (in Russian).

Mahan G., Schaich W.L. (1974), "Comment of the Theory of Work Function", *Phys. Rev.* **B10** (6), 2647 - 2654.

Mints R.I., Melekchin V.P. and Partenski M.B. (1975), "Exoelectrons Emission Relation with Work Function in Metals", *USSR J. of Metals Physics*, 40, No. 4, 886-889 (in Russian).

Nazarov U.V., Postagonov B., Geigopov G.I. and Domashka N.V. (1990) "The Basis of Nanotechnology", *Russian Proc. Mashinconstruction*, **1**, 29 - 31. (in Russian).

Palevsky H., Swank R.K. and Grenchik R. (1947), "Design of Dynamic Condenser Electrometer", *Rev. Sci. Instrum.* **18**, 297 - 314.

Pekar S.I. and Tomasevich O.F. (1947) "Thermionic Emission from Metals, Covered by Thick Semiconductor Layer", *USSR J. of Technical Physics* 17 (12), 1339 – 1342.

Rivere H., (1969) "Work Function. Measurements and Results", in *Solid State Surface Science* **1**, Dekker, NY.

Samuels L. E., Doyle E. D. and Turley D. M., (1980), *Fundamentals of Friction and Wear of Materials*, ASM Materials Science Seminar, 13.

Simon R. (1959), "Work function of iron surfaces produced by cleavage in vacuum". *Phys.Rev.* **116** (3) 613-617.

Vishniakov Ia. D. (1975), *Modern technique for investigation of deformed crystal structure*, Metallurgy Press, Moscow, (in Russian).

Yamamoto S., Susa K., Kawabe U. (1974). "Work function of binary compounds". *Japan J. Appl. Phis.* **2** 209.

Zanoria E., Hamall K., Danyluk S. and Zharin A.L., (1995) "Surface Wear Monitoring with a Non-vibrating Capacitance Probe", *Journal of KSTLE* **11**, 40 – 46.

Zanoria E., Hamall K., Danyluk S. and Zharin A.L., (1997) "The Non-Vibrating Kelvin Probe and Its Application for Monitoring Surface Wear", *Journal of Testing and Evaluation, JTEVA* **25**, No. 2, 233 – 238

Zanoria E.S., Danyluk S., Bhatia C.S. and Zharin A.L., (1996) "Kelvin probe measurements of wear of a magnetic hard disk", *Advances in Information Storage Systems* 7 181 – 191.

Zharin A.L. and Shpenkov G.P. (1978), *Device for friction pair monitoring*, USSR Patent no. 615379.

Zharin A.L. and Shpenkov G.P. (1979), "Macroscopic effects of delamination wear", *Wear* 56 309 – 313

Zharin A. and Guenkin V. (1981), "On rubbing surface electron work function periodicity", *Soviet J. Frict. and Wear* **2**(1) 91-95.

Zharin A.L., Guenkin V.A and Roman O.V., (1986), "Connection of Periodic Changes of Electron Work Function of a Rubbing Surface with Fatigue Damage", *Soviet J. Friction and Wear* 7(2) 112-120.

Zharin A.L., Genkin V.A., Fishbein E.I., Shipitsa N.A. and Terekhov A.L. (1989), "Method for Run-in of Friction Assembly Materials", *Soviet J. Friction and Wear* 10 530 - 534.

Zharin A.L., Genkin V.A., Fishbein E.I., Shipitsa N.A., Terekhov A.L. and Barkun E. (1990) "Determination of Contact Deformation Mode from the Electron Work Function", *Soviet J. Friction and Wear* 11, 144 - 146.

Zharin A. and Guenkin V. (1990), "Study of Friction Processes with Reciprocating Risplacement", *Soviet J. Frict. and Wear* 11 128-131.

Zharin A.L., (1993), "Techniques of Friction Monitoring", *Soviet J. Friction and Wear* 14(3) 111-120.

Zharin A.L., Shipitsa N. and Fishbein E. (1993), "Some Features of Fatigue at Sliding Friction", *Soviet J. Frict. & Wear* 14(4) 13-22.

Zharin A.L., Fishbein E.I., and Shipitsa N.A., (1995), "Effect of Contact Deformation upon Surface Electron Work Function", *Soviet J. Friction and Wear* 16(3) 66-78.

Zharin A. (1996), *Contact Potential Difference Technique and Its Application in Tribology*, Minsk (in Russian)

Zharin A.L. and Rigney D. (1998), "Application of the Contact Potential Difference Technique for On-Line Rubbing Surface Monitoring (Review)", *Tribology Letters* 4 205 - 213.

THERMOMECHANICS OF SLIDING CONTACT

When Micro Meets Macro

A. SOOM, C.I. SERPE and G.F. DARGUSH
State University of New York at Buffalo
Buffalo, NY 14260 USA

Abstract

When relatively hard surfaces that make asperity contact slide against one another at speeds on the order of a m/s or more very high local temperatures can be generated. If the sliding conditions are of sufficient severity and duration, thermal distortions at local or component levels can occur. Overall thermal deformations are primarily determined by thermal gradients. We have found that during the early parts of a sliding interval, all heat input is confined to small volumes at individual asperities which form "hot mounds" surrounded by much larger cool regions where there is little or no temperature change. Thermal distortions are essentially non-existent and overall component level interactions, including thermal softening, are not much different from isothermal. Eventually, more of the surface and near surface regions are heated and the possibility of component level thermal deformation increases.

The modeling of such problems with reasonable fidelity using modern numerical approaches (*e.g.*, finite or boundary element methods) remains a challenge. Asperity level up to component level analysis must be performed simultaneously. When the problem is isothermal, the macro level interactions can be handled by replacing the rough surface by a distributed contact compliance while retaining smooth surface geometry in the analyses. However, with thermomechanical problems, the assumption of smooth surfaces, however implemented, leads to thermomechanical deformations and macroscopic contact pressure distributions that are not found in practice.

In this paper, we present some examples of these phenomena both via experimental observations, including shapes of wear tracks, and coupled thermomechanical finite element analyses of smooth and rough surface contacts. Neither view provides a complete picture of the interactions. We describe the current state of the art as we understand it and discuss approaches being taken to model the full problem.

Keywords

sliding contacts, thermomechanical modeling, contact stiffness, finite element analysis, rough surfaces

1. Introduction

Over the past sixty years, significant advances have been achieved in our general understanding of contact mechanics and friction, *e.g.*, Johnson (1985), Bhushan (1999), Kragelskii (1982), Williams (1994). However, the analysis and simulation of the behavior of even simple sliding systems is still limited by a lack of detailed physical understanding

B. Bhushan (ed.),
Fundamentals of Tribology and Bridging the Gap between the Macro- and Micro/Nanoscales, 467–485.
© 2001 *Kluwer Academic Publishers.*

of particular situations and the large computational burden associated with most practical sliding contact problems. This is especially true when heat is generated at high speed sliding contacts, *e.g.*, at velocities above one meter per second at metallic interfaces. Complicated surface textures result in highly localized, often brief (lasting a few tens of microseconds), dynamic contacts that interact in ways that are only partly understood. Many applications are of a transient nature from both mechanical and thermal points of view. Material properties and constitutive relations for near surface layers, different from the bulk, messy surface chemistry, and combined elastic and plastic deformations, complicate matters in ways that challenge experimental, analytical or computational attempts to penetrate these problems. Clutches, brakes, rubbing seals and numerous other sliding contacts in machines and manufacturing processes are designed daily, based much more on evolutionary practice and poorly understood tests than on analysis and fundamental understanding. It has become clear that surface roughness down to the micro level and the three-dimensional nature of the contact must often be included to describe the macroscopic behavior of sliding components. Interfaces represent one of the last frontiers of mechanical design.

Existing computational mechanics software is not well suited for these problems. Although several finite element codes include capability for coupled thermomechanical analysis, severe computational demands occur due to bandwidth requirements associated with three-dimensional problems, particularly when finite sliding contact is involved. The computational demands of these problems are substantial, requiring high performance computing to solve engineering problems. No studies, covering micro to macro levels have been undertaken to date.

With the complexity and uncertainty that is inherent in these problems, it is clear that a combined experimental and computational approach is needed. Computational results must be validated experimentally. Material and computational parameters need to be tuned within realistic ranges to match experiments. A number of approaches are possible. The most promising ones include building from simple models with known geometry, material properties and operating conditions. We believe that the time is approaching when anticipated computational capabilities can be employed to solve practical thermomechanical sliding problems. In the following sections, we describe the physical problem in more detail and present several computational models that we have used in an attempt to develop a better understanding.

2. The Physical Problem

2.1 ISOTHERMAL STATIC AND SLIDING CONTACTS

Modern understanding of contact behavior and friction of can be traced to the pioneering work of Bowden and Tabor (1950, 1964). They forcefully articulated the concept, still useful today, that friction forces are generated by the rupture or shearing of the real area of contact between two bodies. The real area of contact is that which is needed to carry the normal (and tangential) loads and is composed of a number (tens, hundreds, or thousands) of asperity contacts. These localized microcontacts account for only a small fraction, usually less than one per cent, of the nominal or apparent contact area. Archard (1953) took a deterministic approach while Greenwood and Williamson (1966) took a statistical approach to show that collections of hemispherical asperities of various heights which deform elastically according Hertz Theory result in a real area of contact that is proportional to the normal load. By setting the friction force equal to the real area of contact times a shear strength of the asperity contacts, the Amontons-Coulomb friction relation can be explained. Work has continued to the present day to extend and refine rough surface contact models by including plastic deformations (*e.g.*, Chang *et al.*, 1987, Johnson, 1980, Kapoor *et al.*, 1994, Lee, 1996, Oden and Martins, 1985, Martins *et al.*, 1990, Tworzydlo *et al.*, 1993, Majumdar and Bhushan, 1990, 1991). Finite element analysis of rough contacts has been presented by Webster and Sayles (1986), Bailey and

Sayles (1991), Sayles (1996), Lubrecht and Iaonnides (1991), Kovmopoulos and Choi (1992), Lee (1996) and Polonsky and Keer (1999). All these formulations are quasistatic, and do not explicitly include sliding motion. The information that is obtained from these models and computations usually include the real area of contact, typical contact size, the number of contacts and some indication of the degree of plastic deformation at the contact. Most dry engineering interfaces contain both plastic and elastic asperity contacts.

Also useful is the work of McCool (1986, 1987) that relates surface profile measurements to the parameters (average asperity tip radii, asperity height standard deviation and asperity density) that appear in some of the theoretical models, *e.g.*, Greenwood and Williamson (1966) and Chang *et al.* (1987).

It is worth noting that the presence of surface roughness renders the interface region considerably more compliant than if the surfaces were smooth and flat. Many of the models cited above allow contact stiffness to be estimated directly or inferred from the force-deflection characteristic. This contact compliance can cause pressure and stress distributions to differ from the smooth surfaces that are assumed to envelop elastic bodies in most analytical and computational formulations. In some finite element programs the contact compliance can be included via gap elements, obviating the need to include detailed surface roughness in many types of analyses of statically and dynamically interconnected mechanical or structural components.

More than two hundred papers have been written on this and closely related topics over the past fifty years. Detailed reviews have been prepared by Oden and Martins (1985), Martins *et al.* (1990), McCool (1986) and Bhushan (1996, 1998), among others. This extensive literature provides an excellent, if not total, understanding of the isothermal contact of rough surfaces. However, thermomechanical problems remain much more challenging.

2.2 THERMOMECHANICAL PROBLEMS

The research on thermomechanical effects at rough sliding contacts has followed different paths·from the isothermal contact analyses discussed above. There are essentially two thrusts in the literature on sliding interface behavior with frictional heat generation. One focuses on so-called flash temperatures at localized asperity contacts. The second deals with various aspects of smooth surface modeling of extended contacts including temperature fields at the interface or at the component level, thermoelastic instabilities and the development of hot spots.

When contact between surfaces is highly localized at a number of asperity contacts, very high temperature rises, (hundreds of degrees Celsius and higher), can be reached when sliding speeds exceed a few meters per second, even at moderate applied pressures. The average temperature rise, over the whole surface, may only be a few degrees. The flash temperature is the maximum temperature occurring at the surface during a single asperity to asperity or asperity to surface interaction between two bodies. Since there may be many asperity contacts of different sizes interacting at once, there will be an average flash temperature and a distribution of flash temperatures. The flash temperature concept is generally attributed to Blok (1937). Archard (1953, 1959), Carslaw and Jaeger (1959) and others developed solutions for stationary (slow) and moving (fast) regimes based on the Peclet number, va/K, where v is the velocity, a the contact radius, and K the thermal diffusivity. Since flash temperature analyses deal only with the temperature rise at and around points of direct contact, the temperature rise of the remainder of the surface and subsurface has been treated in an ad hoc manner. The maximum "total" temperature, T, at a microcontact is composed of a bulk temperature, T_B, of the body in the absence of frictional heat generation, an average surface temperature due to accumulated heat from other asperity contacts, T_A, and the flash temperature rise at the contact in question, T_F, so that

$$T = T_B + T_A + T_F$$

Lim and Ashby (1987), Tian and Kennedy (1993) and Wang and Kovmopoulos (1994a,b) take different approaches to estimating T_A. Wang and Kovmopoulos use the surface temperature that one would find if the heat were applied uniformly over the nominal contact area. Lim and Ashby consider an equivalent large asperity and take T_A to be a temperature within a layer of unspecified depth. Tian and Kennedy consider the thermal resistance of the contacts as determining the overall heat flow. This points to a limitation with all flash temperature approaches. By focusing on the surface, taking ad hoc modeling of the contributions contacts other than the one in question, and ignoring thermal expansion, only temperatures can be estimated. Full field information and thermal distortions, which depend on temperature gradients, are not captured. Convective and radiative losses are usually ignored without significant loss of accuracy.

One area where flash temperatures can give useful insight is in explaining changes in friction with sliding speed. For example the coefficient of kinetic friction of steel against steel surfaces can decrease by a factor of three or four as the sliding speed increases from less than one to 20 or 30 meters per second. Lim et al. (1989) have compiled data from a variety of sources. Only recently have Molinari et al. (1999) shown quantitatively that this is almost certainly due to thermal softening associated with increasing flash temperatures with speed. This heating only needs to occur in the outermost few microns, or even less, to affect friction. The real area of contact changes rather slowly unless there are vibrations in a direction normal to sliding. Some type of friction-speed or friction-temperature relation must be incorporated implicitly or explicitly into any thermomechanical modeling.

Thermal expansion due to frictional heating does, of course, occur. A comprehensive review of all aspects of thermomechanical contact behavior was given by Kennedy (1984). Burton (1980) gives a very good overview of thermoelastic problems from a physical point of view. He discusses the concept of a "thermal asperity" which can emerge from a smooth or rough surface to become higher than neighboring regions of the surface. In some cases waves or patterns of "hot spots" develop. Thermal instabilities were first analyzed by Barber (1969). He and coworkers have continued to work on these problems (e.g., Azarkin and Barber, 1987, Yeo and Barber, 1994). Dow and Burton (1973) and Dow (1979) have developed criteria for critical velocities above which thermoelastic instabilities occur for ring and blade geometries. Kennedy and Ling (1974) found moving hot regions in their finite element analysis of high energy aircraft brakes. Experimental observations often agree with these predictions although full field transient solutions are not available.

More recently, Varadi et al. (1996), Neder et al. (1998), Polonsky and Keer (1999), Liu and Wang (1999) and Liu et al. (2000) have begun to develop computational approaches suitable for three-dimensional thermomechanical analysis. However, these ideas are at an early stage of development, and have not yet been incorporated in commercial finite element codes. Consequently only very limited studies on the thermomechanics of sliding contact can be performed. Several simplified approaches that we have used are presented in Sections 4-6, but first some of our observations from physical experiments are discussed in the following section.

3. Experimental Observations

Our interest in these problems arose from studies of the friction, wear and engagement modeling of electromagnetic clutches. A typical armature and pulley (rotor) configuration are shown in Fig. 1. Both components are of mild steel. During an engagement process the armature is pressed against the pulley plate, which is spinning at constant speed. The relative sliding speed varies from the nominal pulley speed to zero in a few tens of milliseconds. A nominal contact pressure of 1 MPa is typical. Experimental data was determined for plates with circular slots forming three rings (armature) and four rings (pulley) that are connected together. The contact takes places at six circular tracks, 1.5 to 2.2 mm wide. The mean radius, used to establish a nominal sliding speed in m/s, of the pulley/armature plates is 0.05 m.

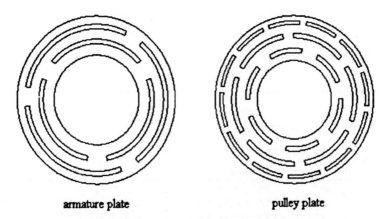

armature plate   pulley plate

Figure 1. Electromagnetic clutch plates: armature and rotor/pulley

A comprehensive characterization of the sliding process requires investigation of surface characteristics and their evolution during operation, of friction and friction mechanisms, as well as of wear and wear mechanisms. Table 1 summarizes the four types of tests that were performed. Much more extensive detail on all points addressed can be found in Serpe (1999).

Test A was designed for friction studies. Short, continuous sliding tests, typically lasting 150 milliseconds were run at 5000 rpm and up to 4000 operations. At five hundred engagement intervals, and more frequently at the beginning of the test, a series of engagements was also run at different speeds and friction measurements were recorded. For this test, a reduced plate geometry was used, made up of the four innermost contact circles from the original six contact tracks plate. Tests B, C and D were conducted on full plate geometry.

Test B represents engagements at 5000 rpm of relatively long duration. Parts were run up to 1000 and 5000 operations respectively and then were used for surface characterization and contact stiffness studies. Test C was designed for surface characterization, friction studies and wear studies. Data was also used for contact stiffness studies. The parts were run at the same speed for a long number of operations, under short, realistic engagement times. Profile measurements were recorded every few thousand operations. Test D represent parts run on vehicles in the field. Consequently, they were tested at a distribution of engagement speeds and large number of operations. The components were used in surface characterization and contact stiffness studies.

We have examined the evolution of the surface characteristics, by means of scanning electron microscopy, optical techniques, chemical analysis and profilometry. Corresponding measurements of friction, wear and contact stiffness have been made.

The effect of sliding action on the rubbing surfaces is shown in the scanning electron micrographs of Fig. 2 (plan view) and Fig. 3 (section in the direction of sliding). The appearance and nature of the contacting surfaces change over time due to surface tractions and wear processes. The initial machining marks (grinding of the armature and turning of the pulley) are removed. Abrasion and delamination scars, material transfer layers, glazed patches and wear debris of various sizes and shapes, and black oxides are found on the worn surfaces. Severe plastic strains are observed between the new and worn pulley or armature. The deformed layer can reach a thickness of 30 to 40 μm for parts that have a large number of operations.

Table 1. Summary of tests and test conditions

TEST	Sample	Speed at engagement rpm/, (m/s)	Average sliding speed rpm, (m/s)	Typical engagement time ms	Total number of operations	Total sliding distance, km
A	• short, constant speed sliding test • series of 100 - 500 operations, sliding at 5000 rpm • 20 operations at various other speeds between series • test used in friction studies					
	A1	2000 rpm (10.5 m/s) 4000 rpm (21.0 m/s)		150	4000	15.8
	A2	5000 rpm (26.3 m/s) 7000 rpm (36.8 m/s)				
B	• engagements at 5000 rpm • test used in surface characterization and contact stiffness studies					
	B1	5000 rpm	2500 rpm	250	1000	2.8
	B2	(26.3 m/s)	(13.15m/s)		5000	14
C	• engagements at same speed • test used in surface characterization, contact stiffness, friction and wear studies					
	C1	800 rpm (4.2 m/s)	400 rpm (2.1 m/s)	8.5	248000	4.4
	C2	1400 rpm (7.4 m/s)	700 rpm (3.7 m/s)	13.8	157000	8.0
	C3	2000 rpm (10.5 m/s)	1000 rpm (5.25 m/s)	18.6	57600	5.7
	C2	5000 rpm (26.3 m/s)	2500 rpm (13.15m/s)	44.3	20000	11.6
D	• engagements at various speeds • parts tested in real operating conditions (field) • test used in surface characterization, contact stiffness and wear studies					
	D1	various	1005 rpm	20.0	200000	40
	D2		(5.3 m/s)		250000	50

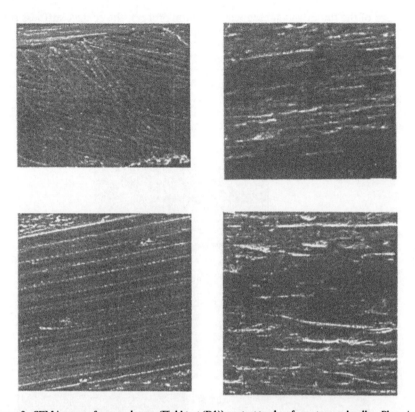

Figure 2. SEM images of new and worn (Field test (D1)) contact tracks of armature and pulley. Plan view

Friction and wear depend on the nature of the real contact area between the surfaces, which is dependent upon the distribution, size and shapes of asperities. The quantitative assessment of the topographical features of the surfaces is of interest for modeling of contact stiffness and understanding running in processes. Profile measurements were recorded from test C components. Pairs of radial and tangential tracks at the same location were analyzed separately, and combined in a mean square sense to determine equivalent values for the three-dimensional isotropic surface. The signals have been digitally filtered, so that only the important features that affect the interaction between the contacting surfaces are analyzed. Regions of the surface with deep wear marks were avoided since their presence significantly distorted the data and the load was carried by the tops of the "asperities." A eighth order Butterworth band-pass filter 25.4 mm - 800 mm has been used (tip radius of the stylus is 10 mm). The equivalent arithmetic (Ra) values for armature and pulley are shown over time in Fig. 4. The armature roughness increases over the first km of sliding after which it approaches a steady value. There is also an initial increase in the pulley roughness which also decrease to a value somewhat rougher than the armature. There is some variability in the data due, in significant part, to the limited number of samples taken.

An important effect of roughness on the mechanical behavior is on the contact stiffness. At a given nominal pressure, the contact stiffness in the elastic range is nominally proportional to the inverse of the surface roughness. Using a vibration resonance technique, we have measured the contact stiffness of segments of the wear tracks at different stages of the run in process for the test C components. The results of these measurements for the C3 tests (2000 rpm engagements) are shown in Fig. 5. At each stage of the run in and subsequent operation process, the contact stiffness per unit area is measured at a number of nominal applied pressures. In each case the stiffness increases roughly in proportion to the applied pressure, reflecting of an expotential force-deflection characteristic suggesting that an expotential distribution of asperity heights can approximate the surface. Of more immediate interest is the change of contact stiffness with time. When the clutch runs in, the contact stiffness decreases as the composite roughness increases. As a steady state sliding condition is approached, the contact stiffness increases again as the surfaces become smoother but still remain rough. These parameter changes can, among other things, affect the elastic stability of the sliding system.

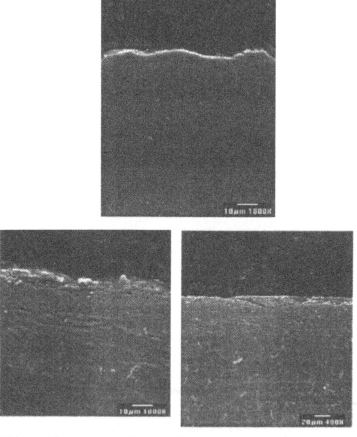

Figure 3. SEM images of new and worn (Field test (D1)) contact tracks of armature and pulley. Section showing sliding direction.

Our wear studies, based on single speed engagements, can be summarized by the wear coefficients shown in Fig. 6. The wear coefficient starts quite high, with higher initial values associated with lower engagement speed tests. The wear coefficient decreases quite quickly to a steady state value at the higher engagement speeds. At the lower speeds (800 and 1400 rpm) the approach to steady state is much slower. We ascribe this to the much slower development of oxide layers which moderate wear. Since the sliding speed varies throughout an engagement a number of wear regimes may be traversed. We characterize the wear as delamination modified by oxidation, although the processes

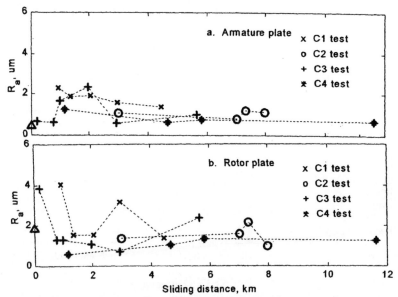

Figure 4. Evolution of surface roughness (Ra) of the armature and pulley during clutch engagements on test stand at various speeds. C1-800 rpm; C2-1400 rpm; C3-2000 rpm and C4-5000rpm

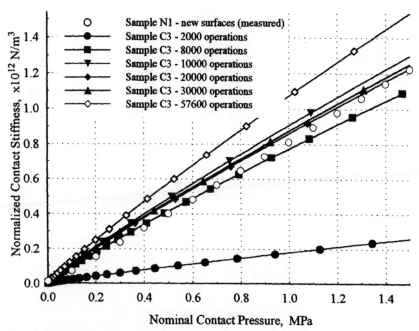

Figure 5. Evolution of computed (from profilometry data) contact stiffness during running in and operation of an electromagnetic clutch. Test condition C3-2000 rpm engagements.

476

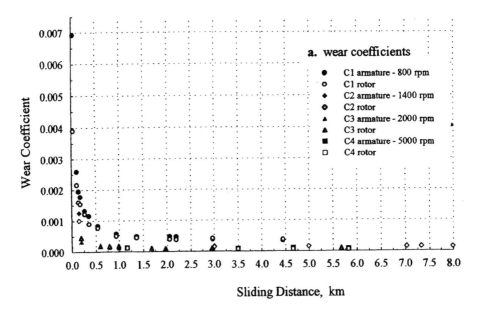

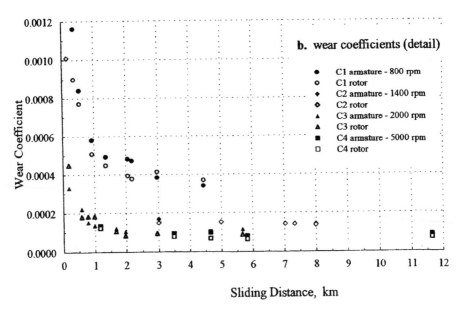

Figure 6. Measured wear coefficients during running in and operation of an electromagnetic clutch. k = 3HV/PL. . C1-800 rpm; C2-1400 rpm; C3-2000 rpm and C4-5000rpm

involved are really more complex. Steady state wear coefficients of around 0.0001 can be viewed as on the high side of "mild wear." If we examine the wear shape of individual tracks, scanned radially with a profilometer, we observe that the wear tracks on the armature and pulley conform. This is shown in Fig. 7. The horizontal dashed line is at the level of the original unworn surface. It is worth noting, that a localized Archard type wear law (local wear depth proportional to normal pressure) should lead to wear tracks on the two surfaces that are mirror images of each other. Finite element wear modeling, assuming isothermal contact conditions, and employing a modified Archard type model (Serpe, 1999) results in the predicted shape shown by the curved dashed lines in Fig. 7. The agreement is quite good. Also, Fig. 7 indicates that the wear depth is approximately 1 nm per engagement. Clearly this wear does not take places uniformly. It is a complex process of particle formation, break up, agglomeration, transfer, and removal.

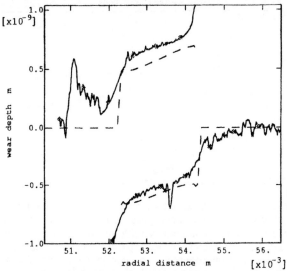

Figure. 7. Wear track profiles after 57600 engagements (test C3-2000 rpm). Dashed lines indicate predicted profile using modified Archard model and 2-D axisymmetric isothermal FE analysis. Upper: armature. Lower: rotor

Friction tests were performed under test A conditions with focus on friction-speed characteristics. The armature and pulley plates have been run at 5000 rpm (26.3m/s) for 4000 operations, with short, constant speed sliding tests. Measurements of speed, normal load and friction torque were recorded at few hundred engagements interval at several sliding speeds (10.5, 21.0, 26.3 and 36.8 m/s). The friction coefficient was calculated from the friction torque and applied normal load.

The mean friction coefficient out of the four sets of data for each speed was calculated. The results are presented in Fig. 8. There is a run-in period which extends for 500 to 1000 engagements. During the transition period the friction coefficients show quite large fluctuations. After run-in the friction coefficients decrease and tend to reach more steady characteristics, although some fluctuations are also present. From this we inferred the steady state friction-speed characteristic also shown in Fig. 8. The decrease of friction with speed is due to flash temperature heating at the asperity contacts which softens (decreases the shear strength) the real area of contact, whatever the material (steel, oxide or some mixture) is present within a particular microcontact. We estimate that the flash temperatures can reach 500 to 1000 degrees Celsius. The friction coefficient is highly speed-dependent which must be taken into account in any realistic modeling.

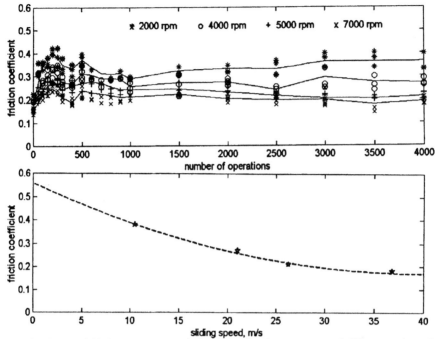

Figure. 8. Test stand friction measurements during 150ms duration constant speed sliding operations. Test conditions A1 and A2.

While the experimental observations we have shown are for a particular mechanical component, that compared to some, undergoes fairly extensive wear, it includes many of the complexities that one encounters in high speed thermomechanical contact. In attempting to model these problems, for example in a finite element context, one must be very careful to now push the model beyond reasonable limits. We believe that three dimensional coupled thermomechanical analyses with full surface roughness must be performed to fully model many high speed sliding problems.

4. Smooth Surface Models

We now present results from some finite element models of frictional sliding. First the frictional sliding of two concentric rings is considered as depicted in Fig. 9a. The surface roughness is represented by a set of nonlinear springs with an exponential load-deflection characteristic. An axisymmetric coupled thermomechanical analysis is performed using the ABAQUS (1998) finite element package. The model is shown in Fig. 9b. A uniform pressure loading is applied to the top of the upper ring and ramped to the full value of 1 MPa over the first 1 ms of the sliding process. The sliding time is 30 ms during which the sliding speed decreases linearly from 5000 rpm to zero. Once engagement is completed the analysis is continued for an additional 70 ms. The bottom surface of the lower ring is constrained in the axial direction. A user defined subroutine specifies the friction versus speed model and calculates the amount of frictional heat generated locally at each contact point. A linear friction-velocity model is employed at the local level. The coefficient of friction increases from 0.2 at the beginning of sliding to 0.4 at rest. All lateral surfaces are assumed to be insulated. The contact between the surfaces may be modeled as either hard or soft. The hard contact represents contact between smooth and flat surfaces. The present thermomechanical analysis considers a hard contact (H1) and two softened contact cases (Sf2 and Sf3). The Sf2 model is representative of worn steel surfaces with an r.m.s. surface roughness on the order of 1 μm, while the Sf3 model represents a softer surface than one would find with steel surfaces. The contact stiffness values reach 0.2 N/μm/mm2 for Sf2 and .002 N/μm/mm2 for Sf3 at 1 MPa contact pressure. All components respond elastically, following the standard thermoelastic constitutive relationships and material properties.

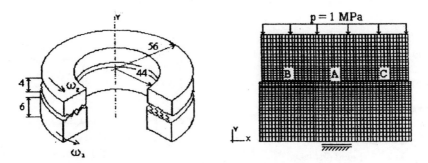

Figure 9. The Model Problem: a) Sliding Rings; b) Finite Element Model

Figure 10a shows the contact pressure distributions for the hard contact case at the end of the load step (t=1 ms), at maximum temperature time (t=12 ms) and at the end of analysis (t=100 ms). As a result of thermal expansion and distortion of the ring, the contact pressure increases at the outer and inner contact points at the beginning of load application. When the temperature reaches its maximum value, the contact is concentrated near points B and C. The contact pressure distributions for the softened case Sf2, presented in Fig. 10b, show that the contact is nearly uniform over the interface at the end of the load step. With further sliding the pressure distribution and the thermal deformations cause the localization of the contact along a line in the middle of the ring (near point A). The contact pressure in the soft contact Sf3 case, not shown here, has a uniform distribution throughout the analysis because the very compliant characteristics of the soft interface prevents the localization of the contact and therefore significant distortions of the rings. The maximum contact pressure and temperature values for Sf2 case are considerably higher than for the hard contact since all the load is carried over a single area. Accordingly, the maximum temperatures reached at the corresponding points are 320°C for soft contact Sf2 case and 190°C for the hard contact case. The very soft case Sf3 results in a much lower temperature rise due to the uniformity of the contact pressure distribution. Throughout the process the maximum temperature reached is 24°C, in agreement with a one dimensional analytical solution assuming uniform pressure over the interface.

The localization of the contact over one or two small areas, as in H1 and Sf2 cases, appears to be strongly dependent on the relationship between temperature gradient, thermal expansion and the compliance of the interface. The axisymmetric nature of the model also accentuates the contact localization on the same spot(s) over the entire analysis. In practical applications the contacts move around over the nominal contact area, as asperity contacts are continuously formed and broken. As the asperities heat up and cool down many times during relatively short periods, the increase in surface temperature as well as contact pressure are more uniform than the one predicted by the above model. In order to account for this, two more situations of the realistic Sf2 soft contact case were investigated by modifying the thermal properties of the near surface layer. The specific heat within the first element layer, of 50 μm thickness, has been increased by 5 times and

480

100 times the initial value (cases Sf2x5 and Sf2x100, respectively).

The contact pressure for Sf2x5 case is shown in Fig. 11a. The pressure distribution at the end of load application is uniform. As sliding continues and heat is generated the contact tends again to localize in the middle of the rings, although over a larger contact area surrounding point A. When the specific heat is increased 100 times within the first layer, case Sf2x100, the contact pressures retains a more uniform characteristic, as shown in Fig. 11b. The maximum surface temperatures are reduced due to increased heat capacity of the near surface layer and more heat conducted into the bodies. The maximum temperature reached at the contact is 67°C in case Sf2x5 and 3.1°C for Sf2x100 case.

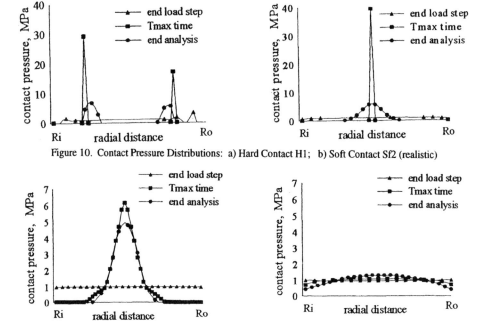

Figure 10. Contact Pressure Distributions: a) Hard Contact H1; b) Soft Contact Sf2 (realistic)

Figure 11. Contact Pressure Distributions: a) Sf2x5 Case (5 times specific heat); b) Sf2x100 Case (100 times specific heat)

5. Single Asperity Sliding

In order to examine localized thermomechanical behavior, the frictional sliding of a single spherical asperity over a flat rigid surface is now considered. The three-dimensional asperity model consists of a disk of outer radius r_o and thickness h, with a central asperity having a radius of curvature R, as shown in Fig. 12. The disk is first pressed against a rigid half-space and then slid with an appropriate velocity for 10 ms. This asperity is imagined to be one of many in a regular array. Consequently, thermal and mechanical symmetry boundary conditions are applied on the outer edges of the disk, that is, the surfaces at $r = r_o$ are insulated and constrained in the radial direction. The back surface of the disk, opposite the asperity tip, is loaded with a uniform pressure p and maintained at an ambient temperature T_a. Coulomb friction with a velocity-dependent coefficient μ is utilized between the asperity tip and the rigid half-space. The value of μ is assumed to vary from 0.4 at 4.2m/s to 0.2 at 26.3m/s. Frictional heating is introduced at the interface with 50% of the total heat entering the asperity. Heat generation due to inelastic deformation is also included.

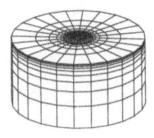

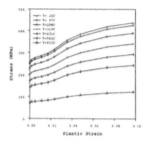

Figure 12. Asperity Model   Figure 13. Stress-Plastic Strain Model

The geometric parameters required for the model are estimated by using the elastoplastic asperity model of Chang *et al.* (1987), along with data from profilometry studies of steel surfaces under moderate contact pressures on the order of 1MPa. For the baseline results presented here, the asperity tip radius of curvature is assumed as $R = 500\mu$m, with an approximate contact spacing of 500μm. Thus, the outer disk radius is selected as $r_o = 250\mu$m. Furthermore, the disk thickness is specified as $h = 250\mu$m in order to maintain a sufficient distance from the asperity for the imposition of the back surface boundary conditions. A nominal contact pressure $p = 0.64$MPa is assumed in the baseline analysis.

The constitutive behavior of mild steel in the asperity is represented by a thermally-sensitive elastoplastic model with a von Mises yield surface and isotropic hardening. The elastic modulus $E = 207$GPa and Poisson ratio $\nu = 0.3$ are specified. Figure 13 provides the stress-plastic strain curves that were utilized. These curves are based upon data in Simmons and Cross (1955) and Boyer (1987). The temperature dependence of E is ignored, along with any similar dependencies in the thermal properties. Thus, the conductivity, density, specific heat and coefficient of thermal expansion are taken as 50W/m°C, 7800kg/m^3, 460J/kg°C and 10.8×10^{-6}/°C, respectively.

Baseline results obtained using the ABAQUS (1998) finite element code for a sliding velocity $v = 10.5$m/s are presented in Fig. 14. In all of these plots, the asperity velocity is directed from left to right. Due to the very small scale of the asperity, most of the surface temperature rise is achieved during the initial 1 ms. While the transient solution may be of some interest, only the response at the end of 10 ms of constant velocity sliding is presented here.

The contour plots provided in Figs. 14a-d isolate a cross-section through the asperity. Figure 14a displays the temperature, Fig. 14b shows the von Mises equivalent stress, Fig. 14c presents the normal stress σ_{33}, and Fig. 14d depicts the equivalent plastic strain. (The units reported in the Fig. 14 stress plots are 10^{12}Pa.) Starting from the surface, the layers of solid elements represent the behavior 4, 9, 17 and 29μm beneath the surface. The contact pressures can be estimated from Fig. 14c. This indicates a maximum level of approximately 600MPa, which is reasonably consistent with the values obtained from corresponding contact mechanics calculations. Figure 14d illustrates that the depth of the plastic zone for this asperity is approximately 15-20μm. This is also reasonably consistent with the depth of the plastically deformed layer that is observable from SEM images of steel components subjected to similar sliding conditions, shown for example in Fig. 3.

482

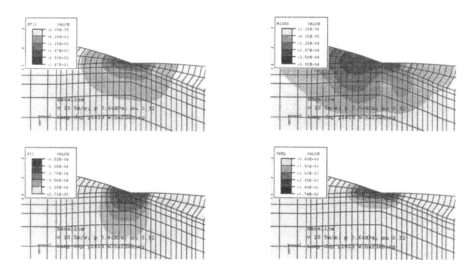

Figure 14. Single Asperity Results on Cross-Section: a) Temperature; b) Von Mises Equivalent Stress; c) Normal Stress; d) Equivalent Plastic Strain

6. Two-dimensional and Three-dimensional Sliding

While the single asperity model provides some useful information concerning sliding contacts, analyses that consider the multiscale nature of the thermomechanical processes are needed. Here we consider a two-dimensional finite element idealization of sliding contact between rough surfaces. The rough surface profiles were developed from profilometry measurements of clutch plates obtained during the experimental program described in Section 3. The ABAQUS code was employed for the analysis. Figure 15 shows a series of temperature contours for the two sliding bodies. The finite element analysis considers the two bodies sliding at 26.3m/s under 1MPa nominal contact pressure. During the simulation the upper body is stationary while the lower one is moving at the given speed for $3\mu s$. As the contact takes place at the tip of the contacting asperities, local temperatures reach significant values even in this very small time interval. The maximum temperature reached at single asperity contacts corresponds to the flash temperature. Since there may be many asperity contacts of different sizes interacting at once, there will be a distribution of flash temperatures. From the analysis, we find that high flash temperatures occur even while the overall temperature rise of the surface may be orders lower.

Although these results have some characteristics that are consistent with physical experiments, stress levels resulting from this analysis are well below yield. Consequently, we find that two-dimensional models are not adequate for simulating the behavior in the near-surface layer. The process is inherently three-dimensional. Unfortunately, existing engineering analysis software cannot cope with the computational demands associated with the full three-dimensional thermomechanical sliding contact problem.

However, several research efforts are now underway. Varadi *et al.* (1996) and Neder *et al.* (1998) have performed three dimensional thermomechanical analyses of very small plates with realistic surface textures sliding for a few microseconds, nominally bridging the micro/macro gap. The results look promising but components of any size that slide for any appreciable length of time would probably require computational capabilities many orders of magnitude greater than is available in most engineering offices or research laboratories. The work of Wang, Keer, and their co-workers also points in the direction that will lead to a better solution of these problems (*e.g.*, Polonsky and Keer, 1999, Liu and Wang, 1999, Liu *et al.*, 2000). Additionally, in order to address the sliding contact problem, the present authors are developing a multi-level boundary element method based upon an unsteady three-dimensional thermomechanical formulation that includes near-surface nonlinearities.

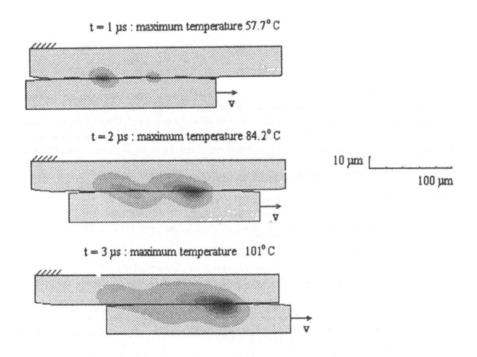

Figure 15. Heat Generation and Temperatures in Two-dimensional Analysis of Rough Surface Sliding Contact

7. Conclusions

We have shown that one can gain some insight into the short time thermomechanical behavior of sliding systems with significant interfacial heat input by assuming isothermal conditions. This is possible since the heat enters the bodies at discrete asperity contacts while the remainder of the bodies remain quite cool and exhibit almost no thermal gradients. Smooth surface models, with contact stiffness, can provide a reasonable model of macroscopic pressure distributions. However, once thermal distortion begins to occur at the interface or at the component level, one needs to accurately model the problem at the asperity level to capture even macro level behavior. Based on the calculated thermomechanical behavior of a single model asperity, it seems that meaningful results, representing the essential physics of the problem, could be obtained. At present, the solution of such problems using commerical codes requires impractically large computation times. We expect that specialized approaches, employing multiscale techniques, Fourier methods, wavelets or hybrid analytical and computational methods will be developed to perform fully three-dimensional concurrent computer simulation of asperity level and component level thermomechanical response of sliding systems.

484

References

Archard, J.F. (1953), "Contact and Rubbing of Flat Surfaces," *Journal of Applied Physics*, 24, 981-988.

Archard, J.F. (1959), "The Temperature of Rubbing Surfaces," *Wear*, 2, 438-455.

Azarkin, A. and Barber, J.R. (1987), "Transient Contact of Two Sliding Half-Planes With Wear," *Journal of Tribology*, 109, 598-603..

Bailey, D.M. and Sayles, R.S. (1991), "Effect of Roughness and Sliding Friction on Contact Stresses," *Journal of Tribology*, 113, 729-738.

Barber, J.R. (1967), "The Influence of Thermal Expansion on the Friction and Wear Process," *Wear*, 10, 155-159.

Barber, J.R. (1969), "Thermoelastic Instabilities in the Sliding of Conforming Solids," *Proc. Roy. Soc. London*, A312, 381-394.

Blok, H. (1937), "Theoretical Study of Temperature Rises of Actual Contact under Oiliness Lubricating Conditions," *Proc. Gen. Disc. on Lubrication*, Instn. Mech. Engrs., London, 2, 222-235.

Bowden, F.P. and Tabor, D. (1950), *The Friction and Lubrication of Solids*, Part 1, Oxford University Press.

Bowden, F.P. and Tabor, D. (1964), *The Friction and Lubrication of Solids*, Part 2, Oxford University Press.

Bhushan, B (1996), "Contact Mechanics of Rough Surfaces in Tribology: Single asperity contact," *Applied mechanics reviews*, 49, 5, 275-298.

Bhushan, B. (1998), "Contact mechanics of Rough Surfaces in Tribology: Multiple Asperity Contact," *Tribology Letters*, 4, 1-35.

Bhushan, B. (1999), *Principles and Applications of Tribology*, John Wiley and Sons, Inc., New York, NY.

Boyer, H.E. (1987), *Atlas of stress-strain curves*, ASM International, Metals Park, Ohio.

Brandt, A. and Lubrecht, A.A. (1990), "Multilevel Matrix Multiplication and Fast Solution of Integral Equations," *J. Computational Physics*, 90, 348-370.

Burton, R.A. (1980), "Thermal Deformation of a Frictionally Heated Contacts," *Wear*, 59, 1-20.

Carslaw, H.S. and Jaeger, J.C. (1959), *Conduction of Heat in Solids*, Oxford University Press.

Chang, W.R. Etsion, I. and Bogy, D.B. (1987), "An Elastic-Plastic Model for the Contact of Rough Surfaces," *Journal of Tribology*, 109, 257-263.

Cho, S.-S. and Komvopoulos, K. (1997), "Thermoelastic Finite Element Analysis of Subsurface Cracking Due to Sliding Surface Traction," *Journal of Engineering Materials and Technology*, 119, 71-78.

Day, A.J., Harding, P.R.J. and Newcomb, T.P. (1984), "Combined Thermal and Mechanical Analysis of Drum Brakes," *Proc. Instn. Mech. Engrs.*, 198D, 287-294.

Day, A.J. (1988), "An Analysis of Speed, Temperature and Performance Characteristics of Automotive Drum Brakes," *Journal of Tribology*, 110, 298-305.

Day, A.J., Tirovic, M. and Newcomb, T.P. (1991), "Thermal Effects and Pressure Distributions in Brakes," *Proc. Instn. Mech. Engrns.*, 205, 199-205.

Dow, T.A. and Burton, R.A. (1973), "The Role of Wear in the Initiation of Thermoelastic Instabilities of Rubbing Contact," *Journal of Lubrication Technology*, 71-75.

Dow, T.A. (1979), "Thermoelastic Effects in a Thin Sliding Seal - A Review," *Wear*, 59, 31-52.

Floquet, A. and Dubourg, M.C. (1994), "Nonaxisymmetric Effects for Three-Dimensional Analysis of a Brake," *Journal of Tribology*, 116, 401-408.

Greenwood, J.A. and Williamson, J.B.P. (1966), "Contact of Nominally Flat Surfaces," *Proceedings of the Royal Society London*, A295, 300-330.

Hutchings, I.M. (1992), *Tribology, Friction and Wear of Engineering Materials*, CRC Press, Boca Raton, FL.

Johnson, K.L. (1980), "Aspects of Friction, in Friction and Traction," *Proceedings of the 7th Leeds-Lyon Symposium on Tribology*, 3-10.

Johnson, K.L. (1985), *Contact Mechanics*, Cambridge University Press.

Kapoor, A., Williams, J.A. and Johnson, K.L. (1994), "The Steady State Sliding of Rough Surfaces," *Wear*, 175, 81-92.

Kennedy, F.E.Jr. and Ling, F.F. (1974), "A Thermal, Thermoelastic, and Wear Simulation of a High-Energy Sliding Contact Problem," *Journal of Lubrication Technology*, 497-507.

Kennedy, F.E.Jr. (1984), "Thermal and Thermomechanical Effects in Dry Sliding," *Wear*, 100, 453-476.

Komvopoulos, K. and Choi, D.-H. (1992), "Elastic Finite Element Analysis of Multi-Asperity Contacts," *Journal of Tribology*, 114, 823-831.

Kragelskii, V. (1982), *Friction and Wear*, Pergamon Press, London.

Lee, K and Barber, J.R. (1994), "An Experimental Investigation of Frictionally-Excites Thermoelastic Instability in Automotive Disk Brakes Under a Drag Brake Application," *Journal of Tribology*, 116, 409-414.

Lee, S.C. (1996), "Behavior of Elasto-Plastic Rough Surface Contacts as Affected by Surface Topography, Load and Material Hardness," *Tribology Transactions*, **1**, 67-74.

Lim,S.C and Ashby, M.F. (1987), "Wear Mechanism Maps," *Acta Metallica*, **35**, 1, 1-24.

Lim,S.C, Ashby, M.F. and Brunton, J.H. (1989), "The Effect of Sliding Conditions on the Dry Friction of Metals," *Acta Metallurgica*, **137**, 767-772.

Liu, G. and Wang, Q. (1999), "Thermoelastic Asperity Contacts, Frictional Shear, and Parameter Correlations," Paper 99-TRIB-45, *STLE/ASME Tribology Conference*, Orlando, FL, Oct 1999.

Liu, S., Rodgers, M., Wang, Q. and Keer, L. (2000), "A Fast and Effective Method for Transient Thermoelastic Displacement Analyses," *Wear*, to appear.

Lubrecht, A.A. and Iaonnides, E. (1991), "A Fast Solution of the Dry Contact Problem and Associated Sub-surface Stress Field Using Multi-level Techniques," *Journal of Tribology*, **113**, 128-133.

Majmudar, A. and Bhushan, B. (1990), "Role of Fractal Geometry in Roughness Characteristics and Contact Mechanics of Rough Surfaces," *Journal of Tribology*, **112**, 205-216.

Majmudar, A. and Bhushan, B. (1991), "Fractal Model of Elastic-Plastic Contact Betweeen Rough Surfaces," *Journal of Tribology*, **113**, 1-11.

Martins, J.A.C., Oden, J.T. and Simoes, F.M.F. (1990), "A Study of Static and Kinetic Friction," *Intl. J. Engrg. Sci.*, **28**, 29-92.

McCool, J.I. (1986), "Comparison of Models for the Contact of Rough Surfaces," *Wear*, **107**, 37-60.

McCool, J.I. (1987), "Relating Profile Instrument Measurements to the Functional Performance of Rough Surfaces," *Journal of Tribology*, **109**, 264-270.

Molinari, A., Estrin, Y and Mercis, S (1999), "Dependence of the Coefficient of Friction on the Sliding Conditions in the High Frequency Range," *Journal of Tribology*, **109**, 264-270.

Neder, Z., Varadi, K., Man,.L. and K. Friedrich (1998), "Numerical and Finite Element Contact Temperature Analysis of Steel-Bronze Real Surfaces in Dry Sliding Contact," Paper 98-TC-5D-1, *ASME STLE Tribology Conference*, Toronto, Oct 1998.

Oden, J.T. and Martins, J.A.C. (1985), "Models and Computational Methods for Dynamic Friction Phenomena," *Comp. Meth. Appl. Mech. Engrg.*, **52**, 527-634.

Polonsky, I.A. and Keer, L.M. (1999), "Fast Methods for Solving Rough Contact Problems: A Comparative Study," Paper 99-TRIB-4, *STLE/ASME Tribology Conference*, Orlando, FL, Oct 1999.

Sayles, R.S. (1996), "Basic Principles of Rough Surface Contact Analysis Using Numerical Methods," *Tribology International*, **29**, 8, 639-649.

Serpe, C.I., Dargush, G.F. and Soom, A. (1998), "Contact Stiffness and the Thermomechanical Response of Sliding Rings," *International Symposium on Impact and Friction of Solids, Structures, and Machines*, Ottawa, Canada, June.

Serpe, C.I. (1999), *The Role of Contact Compliance in the Deformation, Wear and Elastic Stability of Metallic Sliding Rings: Experiments and Computational Analysis*, Ph.D. Dissertation, State University of New York at Buffalo.

Simmons, W.F. and Cross, H.C. (1955), *Elevated-temperature Properties of Carbon Steels*, American Society for Testing Materials, Philadelphia.

Tian, X. and Kennedy, F.E.Jr. (1993), "Contact Surface Temperature Models for Finite Bodies in Dry and Boundary Lubricated Sliding," *Journal of Tribology*, **115**, 411-418.

Twordzydlo, W.W., et al. (1993), "New Asperity-Based Models of Contact and Friction," *Contact Problems and Surface Interaction in Manufacturing and Tribological Systems*, ASME, PED 67 /TRIB **4**, 87-104.

Varadi, K., Neder, Z. and K. Friedrich (1996), "Evaluation of Real Contact Areas, Pressure Distributions, and Contact Temperatures During Sliding Contact Between Real Metal Surfaces," *Wear*, **200**, 55-62.

Wang, S. and Komvopoulos, K. (1994a), "A Fractal Theory of the Interfacial Temperature Distribution in the Slow Sliding Regime: Part I - Elastic Contact and Heat Transfer Analysis," *Journal of Tribology*, **116**, 812-823.

Wang, S. and Komvopoulos, K. (1994b), "A Fractal Theory of the Interfacial Temperature Distribution in the Slow Sliding Regime: Part II - Multiple Domains, Elastoplastic Contacts and Applications," *Journal of Tribology*, **116**, 812-823.

Webster, M.N. and Sayles, R.S. (1986), "A Numerical Model for the Elastic Frictionless Contact of Real Rough Surfaces," *Journal of Tribology*, **108**, 314-320.

Williams, J.A. (1994), *Engineering Tribology*, Oxford University Press Inc. New York, NY.

Yeo, T. and Barber, J.R. (1994), "Finite Element Analysis of Thermoelastic Contact Stability," *Journal of Applied Mechanics*, **61**, 919-922.

NANOSTRUCTURING OF CALCITE SURFACES BY TRIBOMECHANICAL ETCHING WITH THE TIP OF AN ATOMIC FORCE MICROSCOPE

M. MÜLLER, TH. FIEDLER AND TH. SCHIMMEL
Institute of Applied Physics and Institute of Nanotechnology,
University of Karlsruhe, D-76128 Karlsruhe, Germany

1. Introduction

Due to their sharp tips, scanning probe microscopes open the possibility for surface modification and surface manipulation on the nanometer scale and even on the atomic scale. The manipulation and arrangement of atoms adsorbed on a surface was demonstrated with the scanning tunneling microscope under UHV and at low temperatures [Eigler et al. (1990), Crommie et al. (1993)]. It was also demonstrated that it is not only possible to manipulate atoms adsorbed on a solid surface, but also to modify the lattice of a solid itself. Reversible switching of atomic-scale structures was demonstrated even under environmental conditions, i.e. at room temperature and in ambient air [Schimmel et al. (1991), Fuchs et al. (1991)].

On the other hand, it is not only interesting to find techniques for atomic-scale switching and atom-by-atom manipulation. It is also desirable to develop nanomachining techniques which allow for the controlled abrasive structuring of solids on the nanometer-scale. First experiments of this kind were already performed, using the atomic force microscope (AFM) either in the tapping mode or in the contact mode to modify surfaces e.g. by ploughing or scratching with the AFM tip on different semiconductors [Irmer et al. (1998, 1999), Schumacher et al. (1999)] and on layered dichalcogenides [Schimmel et al. (1995)]. Still, processes are of great interest which allow the nanostructuring of solids with precision on the atomic scale. Here, we demonstrate a new technique which allows to modify surfaces with atomic-scale precision using frictional forces between an AFM tip and a solid surface, and we show nanostructuring experiments on calcite surfaces applying this technique.

2. Experimental

For the experiments, a home-built scanning force microscope was used. Combined normal and lateral force detection was made with a beam deflection detection system using a four quadrant photodiode [Meyer et al. (1990), Marti et al. (1990)]. V-shaped silicon nitride cantilevers with a length of 140 µm and a nominal bending force constant of 0.1 N/m were used.

B. Bhushan (ed.),
Fundamentals of Tribology and Bridging the Gap between the Macro- and Micro/Nanoscales, 487–494.
© 2001 *Kluwer Academic Publishers.*

488

The scanning direction in the images is from left to right in the forward scan and from right to left in the backward scan. In the lateral force images, low values (dark areas in the images) in the forward scan (and bright areas in the backward scan, respectively) correspond to a high value of the frictional force between tip and sample. The topographic images shown in this article were taken in the deflection mode of the AFM. All images are presented in a linear grey-scale and represent the original raw data. All experiments were performed at room temperature and in ambient air.

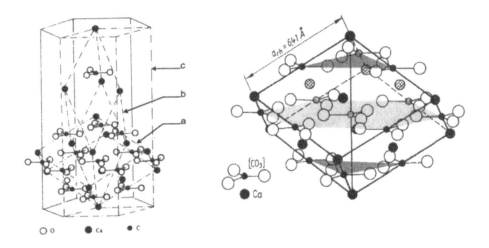

Figure 1. Crystal structure of calcite. On the left side the hexagonal unit cell is shown. Within this cell as well as on the right side, the cleavage rhomboeder is displayed, whose surfaces correspond to the (10$\bar{1}$1) cleavage planes.

Calcite exhibits a hexagonal cubic crystal structure (see Fig. 1) [Schröke et al. (1991), Rösler (1988)]. Single crystals of calcite can easily be cleaved parallel to the (10$\bar{1}$1) plane. On calcite (10$\bar{1}$1) surfaces lattice resolution imaging has been achieved by AFM. [Ohnesorge et al. (1993)], showing atomic resolution at very low forces in a liquid environment. Processes of crystal decomposition and growth on such surfaces were also investigated by AFM [Hillner et al. (1992)]. Stipp et al. (1996) studied calcite cleavage planes by AFM in air, showing that after cleavage, changes in the surface structure occurred due to the interaction with the ambient environment.

For the experiments presented here, naturally grown calcite crystals were cleaved parallel to the (10$\bar{1}$1) direction using a razor blade. As surface changes are observed immediately after cleavage in ambient air [Stipp et al. (1996)], nanostructuring experiments were only performed after a stable surface morphology had developed, i.e. after approximately 60 minutes.

3. Results and Discussion

3.1. TRIBOMECHANICAL ETCHING ON CALCITE

Calcite (10$\bar{1}$1) surfaces were imaged by AFM in the contact mode immediately after cleavage. They show atomically flat terraces separated by monolayer and multilayer steps. As already reported by Stipp et al. (1996), changes in the surface structure were observed due to the interaction with the humidity of the air, which leads to both local dissolution processes and at the same time to a partial coverage of the surface with thin layers most likely consisting of polycrystalline calcite. While these layers exhibit an almost constant thickness of the order of several nanometers over areas of the order of several micrometers, they are no longer atomically flat. After waiting for one hour after cleavage, no further changes were observed by AFM even after two days at ambient conditions.

Surface modification was performed on these layers in the contact mode of the AFM. Typical load forces of the order of 10 – 100 nN were applied. Scanning across step edges leads to mechanical wear due to the interaction with the AFM tip. To increase the rate of this tribomechanical etching process, the vertical position of the sample was periodically modulated by adding a sinusoidal voltage to the z-electrode of the scanning piezo of the AFM (see Fig. 2). The modulation leads to a periodic bending of the cantilever. As the length of the cantilever beam is constant, the periodic cantilever bending also leads to a periodic lateral movement of the AFM tip on the sample surface parallel to the direction of the cantilever axis. The basic idea for applying this technique

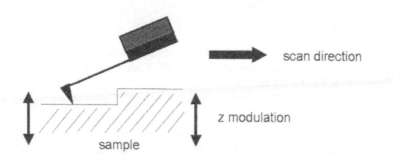

Figure 2. Schematic illustration of the modulation technique used for tribomechanical etching.

is the following: when the tip meets a step edge or an atomic-scale defect, material can be removed each time the AFM tip crosses the step edge or defect during its periodic lateral movement. If the modulation frequency of the z-position of the sample coincides with a mechanical resonance frequency of the cantilever in contact with the sample, structuring is found to be significantly more efficient than at other frequencies. This can be understood easily, as for a fixed excitation amplitude the amplitude of the resulting oscillation is higher than in the case of non-resonant frequencies. In this way, rates of tribo-mechanical etching are achieved which are up to four orders of magnitude higher than those observed by scan-induced etching without applying a z-modulation.

3.2. CONTROLLED WRITING OF NANOSTRUCTURES

This technique of tribomechanical etching described above can be applied to generate well-defined nanostructures with the AFM tip. Fig. 3 shows an example of a nanostructure in the shape of the number "8" written on the calcite surface layer (line width approx. 25 nm). The image was taken in the contact mode of the AFM with the same tip used for writing the structure. In order to generate this structure, the tip was

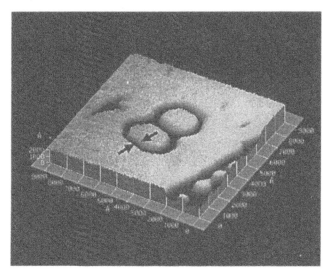

Figure 3. Topographic AFM image of a tribomechanically etched nanostructure on calcite. Line width approx. 25 nm (between the arrows). Scan size: 0.9 µm x 0.9 µm.

moved on the sample along a computer-controlled trace in the shape of an "8". This structuring cycle was repeated 100 times. During the structuring, the z-position of the sample was modulated sinusoidally at a frequency of 40.2 kHz, corresponding to the first bending resonance frequency of the cantilever used in the experiment in contact with the sample surface. The structuring was done at an average force load of the order of 10 nN. The rate of the computer-controlled tip-advancement along the structure was 2 µm per second.

Fig. 4 is a 2.5 µm x 2.3 µm topographic image of the same structure. While the bright parts of the image correspond to the layer formed after cleavage due to the influence of humid air, the darker parts correspond to the original calcite single crystal surface. The structure written into the surface layer exhibits the same height value as the calcite single crystal, indicating that the structuring process for the parameters applied occurred whithin the surface layer, but did not affect the intact single crystal. This result can be

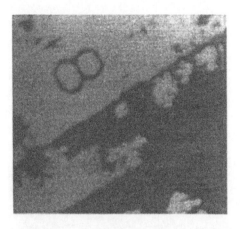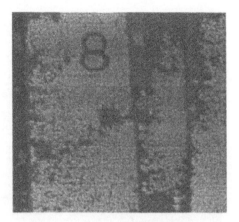

Figure 4. Topographic AFM images of the artificial structure and the calcite layer on top of the calcite cleavage plane (dark). Scan sizes: 2.2 μm x 2.0 μm in a) and 2.5 μm x 2.3 μm in b).

easily understood, as detaching an atom from a largely disordered film or from a step edge is expected to cost less energy and especially needs a lower activation energy than detaching an atom from an intact surface plane of a single crystal.

3.3. CONTRAST BY LATERAL FORCE IMAGING

Fig. 5 gives the corresponding lateral force microscopy (LFM) images of the structure shown in Fig. 4. The scan size is 0.9 μm x 0.9 μm and the scanning area corresponds to that of Fig. 3. A clear contrast inversion is observed between the forward scan image of Fig. 5a (scanning direction from left to right) and the corresponding backward scan image of Fig. 5b. This indicates that the lateral force contrast is due to differences in the coefficient of tip-sample friction and not due to topographic artefacts. The area of the single crystal calcite (10$\bar{1}$1) surface (upper right corner of the image) appears dark in the forward scan and bright in the backward scan, while for the presumably polycrystalline layer, the contrast is just the other way round. It is clearly seen in the images that the value of the lateral force signal of the artificially generated nanostructure (the "8") exactly corresponds to that observed for the intact single crystal surface (see upper right corner of the same image). This is in agreement with the interpretation suggested above that the structuring was performed within the surface layer that formed on top of the calcite single crystal. It also indicates that lateral force microscopy reveals a frictional contrast between single crystal calcite (10$\bar{1}$1) cleavage planes and more disordered or polycrystalline films. This is very plausible as the coefficient of microscopic friction does not only depend on the material, but also on parameters like the microscopic order or the roughness on the nanometer-scale.

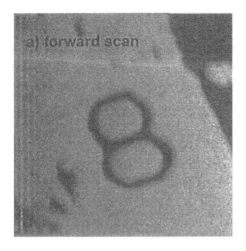 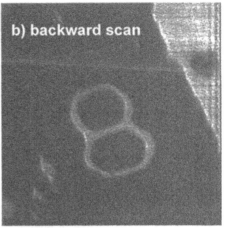

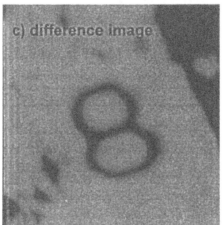

Figure 5. Lateral force images scanned from left to right (a) and from right to left (b). Image c) gives the difference between image a) and image b) representing the friction force between tip and sample during the scan. Scan size: 0.9 μm x 0.9 μm.

3.4. TIME-STABILITY OF THE STRUCTURES

To investigate the time stability of the generated structures, the surface was investigated for more than two days following the structuring process.

Fig. 6 shows a topographic AFM image of the structure shown in Fig. 3 after 48 hours at room temperature and in ambient air (scan size: 1.3 μm x 1.3 μm). Both on the

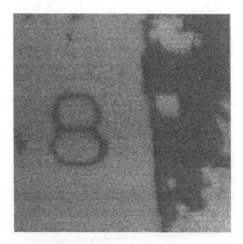

Figure 6. Topographical AFM image of the structure of Fig. 3 48 hours after structuring. Scan size: 1.3 μm x 1.3 μm.

surface and on the nanostructure itself, no measurable changes were found. Even the value of the line-width of the structure did not change its value.

4. Conclusions

In summary, a new technique was demonstrated which allows to generate defined nanostructures with the tip of an atomic force microscope using frictional forces on the atomic scale. By inducing atomic-scale wear processes locally in a controlled manner, material is detached with an etching rate of the order of one lattice constant per z-modulation cycle, thus allowing ultra-precision machining with the AFM tip on an oscillating sample. The process was demonstrated on calcite under ambient conditions, and the resulting structures were subsequently imaged with the same AFM tip which was used for writing the structures.

Acknowledgements

This work was supported by the "Deutsche Forschungsgemeinschaft" within "SFB 195" (Project B11).

494

5. References

Crommie, M. F., Lutz, C. P. and Eigler D.M. (1993), "Confinement of Electrons to Quantum Corrals on a Metal Surface", *Science* **262**, 218-220.

Eigler, D.M., Schweizer, E. K. (1990), "Positioning Single Atoms with a Scanning Tunneling Microscope", *Nature* **344**, 524-526.

Fuchs, H. and Schimmel, Th. (1991) "Atomic Sites of a Bare Surface Modified with the Tunneling Microscope", *Adv. Mater.* **3**, 112-113.

Hillner, P. E., Gratz, A. J., Manne, S. and Hansma, P. K. (1992), "Atomic Scale Imaging of Calcite Growth and Dissolution in Real Time", *Geology* **20**, 359-362.

Irmer, B., Blick, R. H., Simmel, F., Gödel, W., Lorenz, H. and Kotthaus, J. P. (1998) *Appl. Phys. Lett.* **73**, 2051-2053

Irmer, B., Simmel, F., Blick, R. H., Lorenz, H., Kotthaus, J. P., Bichler, M. and Wegscheider,W. (1999), *Superlattices and Microstructures*, **25**, 785-795

Marti, O., Colchero, J. and Mlyneck, J. (1990), "Combined Scanning Force and Friction Force Microscopy of Mica", *Nanotechnology* **1**, 141-144.

Meyer, G. and Amer, N. M. (1990), "Simultaneous Measurement of Lateral and Normal Forces with an Optical Beam-Deflection Atomic Force Microscope", *Appl. Phys. Lett.* **57**, 2089-2091.

Ohnesorge, F. and Binnig, G. (1993), "True Atomic-Resolution by Atomic Force Microscopy" *Science* **260**, 1451-1456.

Rösler, H. J. (1988), *Lehrbuch für Mineralogie*, Deutscher Verlag für Grundstoffindustrie, Leipzig, Germany.

Schimmel, Th., Fuchs, H., Akari, S. and Dransfeld, K. (1991), "Nanometer-Size Surface Modifications with Preserved Atomic Order Generated by Voltage Pulsing", *Appl. Phys. Lett.* **58**, 1039-1045

Schimmel, Th., Kemnitzer, R., Küppers, J., Kloc, Ch. and Lux-Steiner, M., (1995), „Nanometer-Scale Machining of Covalent Monolayers Investigated by Combined AFM/LFM", in: Güntherodt, H. J., Anselmetti, D. and Meyer, E. (eds.), *Forces in Scanning Probe Methods, NATO ASI-Series*, Kluwer, Dordrecht, pp. 519-524.

Schröke, H. and Weiner, K.-L. (1991), *Mineralogie*, Walter de Gruyter, Berlin, Germany.

Schumacher, H. W., Keyser, U. F., Zeitler, U., Haug, R. J. and Eberl, K. (1999), „Nanomachining of Mesoscopic Electronic Devices Using an Atomic Force Microscope", *Appl. Phys. Lett.* **75**, 1107-1112

Stipp, S., Gutmannsbauer, W. and Lehmann, T. (1996), "The Dynamic nature of Calcite Surfaces in Air", *American Mineralogist* **81**, 1-8.

ATOMIC-SCALE PROCESSES OF TRIBOMECHANICAL ETCHING STUDIED BY ATOMIC FORCE MICROSCOPY ON THE LAYERED MATERIAL NbSe2

R. KEMNITZER[1], TH. KOCH[1,2], J. KÜPPERS[1], M. LUX-STEINER[3] AND TH. SCHIMMEL[2,4]
[1]Experimentalphysik III, Universität Bayreuth, D-95440 Bayreuth, Germany,
[2]Institut für Nanotechnologie, Forschungszentrum Karlsruhe,
D-76021 Karlsruhe, Germany,
[3]Hahn-Meitner-Institut, D-14109 Berlin, Germany,
[4]Institut für Angewandte Physik, Universität Karlsruhe,
D-76128 Karlsruhe, Germany

ABSTRACT. Layered materials play an important role as solid state lubricants. At the same time, they are ideal model systems for the study of microscopic processes which lead to tribochemical wear. Here, we report on a microscopic study of wear induced by the scanning tip of an atomic force microscope (AFM). For this purpose, freshly cleaved surfaces of the layered dichalcogenide NbSe2 were scanned with the tip of an AFM at force loads of the order of 100 nN. Due to frictional forces between tip and sample, processes of tip-induced wear could be observed. The tribomechanical etching of the sample was studied and analysed on the nanometer scale. Three different microscopic processes were identified which contribute to friction-induced wear and layer decomposition: i) tribomechanical etching and wear on the atomic scale due to lateral forces at defects and step edges; ii) lateral force induced cutting of islands of the topmost NbSe2 layer and iii) delamination of smaller islands (diameters < 20 nm) due to lateral forces between the AFM tip and the island.

1. Introduction

Friction and wear are often closely related to tribochemical processes, i.e. chemical processes which are activated as a consequence of frictional forces. Such processes do not only influence the composition and properties of lubricant layers. They also provide a mechanism of surface wear. A detailed understanding of processes leading to tribochemical reactions therefore is of great technological as well as scientific interest. But although processes of tribochemical wear play such an important role, not very much is known about the atomic-scale processes leading to the tribomechanical activation and breaking of chemical bonds. Yet, such an understanding is the key for a basic understanding of tribochemical processes.

B. Bhushan (ed.),
Fundamentals of Tribology and Bridging the Gap between the Macro- and Micro/Nanoscales, 495–502.
© 2001 Kluwer Academic Publishers.

Here, we use the tip of an atomic force microscope (AFM) as a tool which allows to induce the rupture of chemical bonds locally on the atomic scale. Using an AFM also allows to image the resulting surface changes due to the atomic-scale wear processes in situ and in real time.

For a detailed study of such processes, layered materials are of special interest for several reasons. On the one hand, layered materials like graphite or molybdenum disulfide (MoS_2) play an important role both as solid state lubricants and as additives for liquid lubricants. On the other hand, they provide an ideal model system for such studies: macroscopic crystals of these substances are readily available. Freshly prepared, well-defined surfaces can be easily obtained by cleaving. And finally, these surfaces exhibit atomically flat terraces which can be easily imaged by scanning probe techniques. For the experiments shown in this article, the layered compound niobium diselenide ($NbSe_2$) was used. As this compound exhibits a higher reactivity at ambient conditions as compared to graphite or MoS_2 tribochemical reactions could easier be observed, while at the same time $NbSe_2$ exhibits the same crystal structure as MoS_2.

2. Experimental

2H-$NbSe_2$ crystals grown by the vapor phase transport technique [Lux-Steiner (1991), Späh (1986)] were used for the experiments. Freshly prepared (0001) surfaces obtained by cleaving and as-grown (0001) crystal surfaces were used. A commercial AFM control unit (Park Scientific Instruments) was applied in connection with a home-built AFM equipped with a laser deflection detection system and a four quadrant photodiode. Two different commercially available V-shaped Si_3N_4 cantilevers with pyramidal tips and nominal bending force constants of 0.21 N/m and 0.37 N/m, respectively, were used. For torsion, a lower boundary of the force constant of 25 N/m and 50 N/m, respectively, is estimated. The experiments were performed at room temperature and in air. All images were taken by contact-mode AFM in the constant force mode. Both forward and backward scan were performed with the same scanning velocity. The images shown in this article were taken in the forward scanning direction, i.e. from left to right. They are given in a linear grey-scale, increasing brightness corresponding to increasing height. All images represent unfiltered raw data.

3. Results and Discussion

3.1 IN SITU OBSERVATION OF ATOMIC-SCALE WEAR BY AFM

Fig. 1 shows the crystal structure of niobium diselenide. Similar to MoS_2, $NbSe_2$ has a layered structure with covalent bonds within the Se-Nb-Se sandwich layer and van der Waals interaction between the layers [Gavarri et al. (1989)]. On the as-grown crystal surfaces, atomically flat terraces are found, which have been investigated with AFM by different groups [Bando et al. (1987), Dahn et al. (1988), Parkinson (1990), Schimmel et al. (1995)]. True atomic resolution was demonstrated by AFM at monolayer step

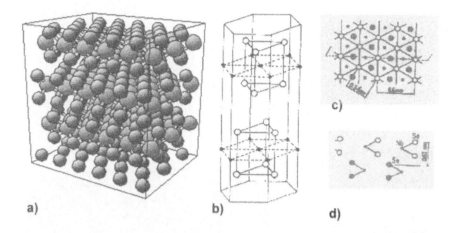

Figure 1. Crystal structure of 2H-NbSe$_2$. The crystal consists of Se-Nb-Se sandwich layers interacting with each other by van der Waals forces. a) and b): three-dimensional illustrations of the layered structure, c) top view and d) side view of the layers.

edges and kinks [Schimmel et al. (1999)]. The terraces are separated by steps with a height of 0.63 nm, which corresponds to the height of one single sandwich layer, or by multiples of this value.

To investigate tribomechanical wear on NbSe$_2$ surfaces on the atomic scale, NbSe$_2$ (0001) cleavage planes were imaged by contact mode AFM and the change of the surface structure from image to image due to the scanning tip was studied. Fig. 2 shows selected images out of a sequence of more than 80 successively taken AFM images. The scan size in the images is 400 nm x 400 nm. The total normal force of the tip on the sample was 70 ± 20 nN (including capillary forces between tip and sample). Each image scan consists of 256 line scans, and the scan rate was 5 lines per second.

The images of Fig. 2 show atomically flat terraces separated by monolayer steps. The brighter areas in the images represent the surface of the topmost Se-Nb-Se sandwich layer. Due to previous scanning, the topmost layer already has numerous holes with a depth of approx. 0.63 nm, corresponding to the monolayer thickness. The images of Fig. 2 represent the same scanning area as shown in the first image of this figure (top left) after 10, 20, 30, 40, 50, 57, 60, 63, 67, 70 and 73 further image scans, respectively. It is clearly seen that the area covered by the topmost layer is continuously decreasing until finally after more than 100 image scans (not shown in Fig. 2), it is completely removed from the scanning field of the AFM tip.

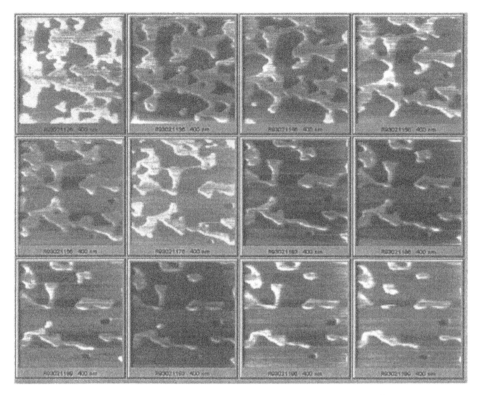

Figure 2. Twelve AFM images of the same 400 nm x 400 nm scanning area on a NbSe$_2$ (0001) surface. Due to the mechanical interaction with the scanning AFM tip, tribomechanical etching of the topmost NbSe$_2$ layer is observed.

Analysing the observed wear process more closely, three different mechanisms of microscopic wear can be distinguished, which will be discussed in more detail below:

1) The tribomechanical etching process: at step edges, the interaction with the AFM tip leads to wear on the atomic scale.

2) The cutting process: narrow necks within the topmost layer are separated by cutting with the AFM tip within one scan-line.

3) The delamination process: NbSe$_2$ monolayer islands which are below some critical diameter are detached from the sample due to interaction with the AFM tip and pushed outside the scanning field by the AFM tip.

3.2 THE ATOMIC-SCALE MACHINING PROCESS

When the tip is crossing a monolayer step edge in the direction from the lower towards the higher terrace, increased lateral forces between tip and sample are observed. Frequently, stick-slip processes are found at upward step edges on NbSe$_2$ (0001) surfaces. At the same time, it is found that material is detached from the step edge at a

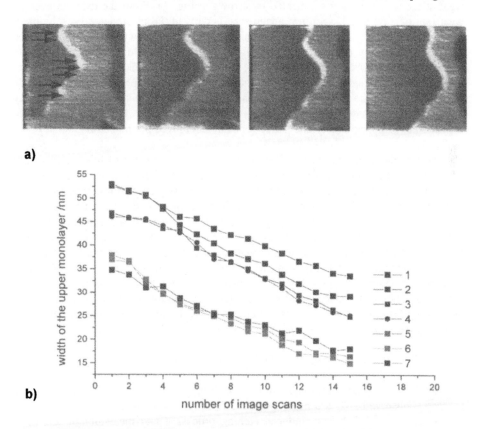

Figure 3. a) Four AFM images of the same 100 nm x 100 nm scanning area out of a series of 15 consecutive image scans, showing tribomechanical etching due to tip-sample interaction at the edges of the terraces. b) The width of the upper monolayer at the seven different positions indicated by the arrows as a function of the number of image scans. The topmost arrow in Fig. 3a corresponds to curve 1 in Fig. 3b, the lowest arrow to curve 7, respectively.

rate which is of the order of one atomic lattice constant per scan line. This is clearly seen in Fig. 3. The four images of Fig. 3a are AFM topography images of the same area of a NbSe$_2$ (0001) cleavage plane. The images represent from left to right the 1st, 4th, 8th and 12th image, respectively, out of a series of 15 images taken at a total force load of 70 nN. The scan size is 100 nm x 100 nm, and the scanning area consists of 64 scan lines per image scan, corresponding to a distance between neighbouring scan lines of 1.56 nm. The images show a topmost NbSe$_2$ layer which is partially covering the image field. The area and width of this layer is decreasing during the scanning process due to tip-induced wear at the edges. This is illustrated in the diagram of Fig. 3b, which gives the lateral width of the topmost layer as a function of the number of scans from the 1st to the 15th image scan. Curves 1 to 7 correspond to the widths of this layer from left to right at the seven positions indicated by arrows in Fig. 3a. From the data, an average wear rate of (1.5 ± 0.17) nm per image scan is derived. From the lateral resolution at the upward step edges, the contour length of the tip-sample contact is estimated to be also approx. 1.5 nm, a value which is roughly a factor of five higher than the in-plane lattice constant of NbSe$_2$ within the (0001) surface, which is 0.33 nm. This would mean that at the step edge, the tip is in direct lateral contact with approx. 5 unit cells of the step edge at the same time. The fact that the observed wear rate corresponds to roughly five lattice constants per image scan means that within the accuracy of the measurement, approximately one lattice constant is removed per line scan.

This also explains that no wear debris is found as a consequence of this mechanically induced atomic milling process. Even when zooming out of the scanning field in which the wear occurred, no wear particles are found as a consequence of this process. As the material is detached from the step edges lattice constant by lattice constant, only small molecular fragments or even atoms are detached at a time. As the experiments were performed under ambient conditions, the surface is known to be covered by a contaminant layer mainly consisting of water. Radicals and ions formed in the atomic milling process at the step edges in this way could react with molecules, radicals and ions of this layer. The mobility of the products of the tribomechanically induced reactions is obviously sufficiently high so that they cannot be observed in the AFM images. The atomic milling process at step edges thus represents a mechanism of wear without generation of wear particles.

3.3 THE TRIBOMECHANICAL CUTTING PROCESS

In addition to the tribomechanically induced etching process described above, a cutting process is observed. When a neck is formed within the topmost layer of NbSe$_2$ – e.g. due to the tribomechanically induced etching process – and the width of this neck parallel to the scanning direction is below a certain critical value, the scanning tip makes a cut through this neck, thus separating it into two parts.

The four consecutive images of Fig. 4a-d show an example of such a cutting process (scan size: 220 nm x 220 nm). The black arrow in the image of Fig. 4a points at a narrow neck in the topmost NbSe$_2$ layer. In the image of Fig. 4b, the neck is still narrower, while during the scan of the image given in Fig. 4c, the tip cuts the remaining neck within one scan line. The two separated parts of the upper layer are seen in the subsequent image of Fig. 4d. Immediately before the cutting process occurred, the neck at its narrowest point still had a width of more than 5 nm. This means that chemical bonds have been broken mechanically with the tip of an AFM, the cut involving at least 15 lattice constants, which were cut within one scan line.

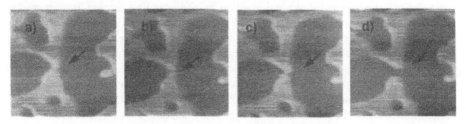

Figure 4. Cutting of a neck in the topmost NbSe₂ monolayer with the scanning AFM tip (see arrows). Image size: 220 nm x 220 nm.

Again, this is a process which involves the rupture of chemical bonds due to lateral force interaction with the AFM tip. The AFM experiments show that this process does not lead to the generation of wear particles or debris. The radical sites and/or ionic sites created on the surface during this process most probably react with species within the water layer on the surface.

3.4 THE DELAMINATION PROCESS

When islands on the surface have become sufficiently small, the lateral forces between tip and sample are sufficient for detaching the islands from the surface by disrupting van der Waals bonds between adjacent layers. From the lateral force and the island area, information on adhesion forces of the island on the surface can be derived.

An example of such a delamination process is given in Fig. 5, which shows three consecutive image scans of the same scanning area before, during and after delamination of a small island of the topmost NbSe₂ layer. The island is indicated with an arrow in Fig. 5a. In Fig. 5b, a blurred trace is seen instead of the island. The trace is

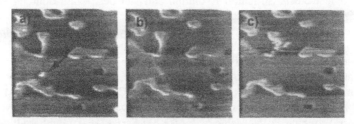

Figure 5. Three consecutive AFM images of the same 400 nm x 400 nm scanning area of a NbSe₂ (0001) surface, showing the detachment of an island of the topmost layer during AFM imaging (see arrow).

stretching from the position of the original island upward and indicates that the island has been removed from its original position by the scanning tip and is then moved upward within the imaging field by the AFM tip. The upward movement within the scanning field can be explained easily, as scanning was performed line by line, beginning with the lower end of the image. In Fig. 5c, finally, the island is completely removed from the scanning area, a process which involves debris particles with diameters on the nanometer scale, in contrast to the other two processes of microscopic wear, which only lead to the generation of small molecular fragments on the atomic scale. In the case of the island shown in Fig. 5, the island diameter was approx. 20 nm, which means that the island consisted of approx. 8500 atoms.

4. Conclusions

Using an atomic force microscope in the contact mode, atomic-scale processes of tribomechanical wear and etching could be investigated in situ and in real time under ambient conditions. Three processes of nanoscale wear could be identified, two of them involving the rupture of covalent bonds resulting in tribomechanical etching processes on the atomic scale. The experiments show at the same time that the atomic force microscope is an ideal instrument for the in situ investigation of nanoscale mechanisms of tribomechanical etching and wear.

5. References

Bando, H., Tokumoto, H., Mizutani, W., Watanabe, K., Okano, M., Ono, M., Murakami, H., Okayama, S., Ono, Y., Wakiyama, S., Sakai, F., Endo, K. and Kajimura, K. (1987), " Effects of Atomic Force on the Surface Corrugation of 2H-NbSe$_2$ Observed by Scanning Tunneling Microscopy", *Jpn. J. Appl. Phys.* **26**, L41-53.

Dahn, D.C., Watanabe, M. O., Blackford, B. L. and Jericho, M. H. (1988), "Tunneling Microscopy of 2H-NbSe$_2$ in Air", *J. Appl. Phys.* **63**, 315-322.

Gavarri, J. R., Mokrani, R., Vacquier, G. and Boulesteix, (1989), "Relations Between Anisotropic Defects Structural Evolution and van der Waals Bonding in 2H-NbSe$_2$" *Phys. Stat. Sol.* **10a** , 445-449.

Lux-Steiner, M. Ch. (1991), *Synthesis, Optoelectronic Properties and Applications of new Semiconductor Crystals*, Springer Verlag, Heidelberg, Germany.

Parkinson, B. (1990), "Layer-by-Layer Nanometer Scale Etching of Two-Dimensional Substrates Using the STM", *J. Am. Chem. Soc.* **112**, 7498-7502.

Schimmel, Th., Kemnitzer, R., Küppers, J., Kloc, Ch. and Lux-Steiner, M., Ch. (1995), "Giant Atomic Corrugations on Layered Dichalcogenides Investigated by AFM/LFM", *Forces in Scanning Probe Methods, NATO ASI-Series* (H.-J. Güntherodt, D. Anselmetti and E. Meyer, eds.), pp. 513-518, Kluwer, Dordrecht, Netherlands.

Schimmel, Th., Koch, Th., Küppers, J., Lux-Steiner, M. (1999), "True Atomic Resolution under Ambient Conditions Obtained by Atomic Force Microscopy in the Contact Mode", *Appl. Phys. A* **68**, 399-402.

Späh, R. (1986), Ph.D. Thesis, *Konstanzer Dissertationen Vol. 130*, Hartung-Gorre Verlag, Konstanz, Germany.

DETERMINING THE NANOSCALE FRICTION AND WEAR BEHAVIOR OF Si, SiC AND DIAMOND BY MICROSCALE ENVIRONMENTAL TRIBOMETRY

M. N. GARDOS
Raytheon Electronic Systems, Engineering Services Center,
El Segundo, CA 90245 USA

Abstract

This paper offers a review of the author's decade-long attempts to examine the tribochemical changes that occur with various crystallinities of (a) polished silicon (b) unpolished and polished polycrystalline diamond films, and (c) a commercially available version of polished polycrystalline α-SiC. Theory-based model experiments were performed in $\sim 1.33 \times 10^{-3}$ Pa $\cong 1 \times 10^{-5}$ Torr vacuum ($\sim 93\%$ of the residual gases is water vapor) and some in low partial pressures of hydrogen test atmospheres, at temperatures ranging from lab-ambient to 950°C. The apparatus used was a unique pin-on-oscillating-flat-type scanning electron microscope (SEM) tribometer specially built to fill the gap between an atomic force microscope and a conventional, bench-top friction and wear tester. Its primary purpose has been to examine the changes in the tribological behavior of a variety of bearing materials and solid lubricants, under realistic engineering (Hertzian) contact stresses in the GPa to MPa (from many thousands to hundreds of psi) range, as influenced by elevated temperatures in moderate vacuum and in low partial pressures of inert or reactive gases. The coefficient of friction and wear measurements were occasionally complemented by surface analyses to decipher the footprints of atomic-level surface interactions by the tribological behavior of essentially microscopic (~ 50 to 500 μm diameter) Hertzian contacts. All the friction *trends* indicate that the changes in *adhesion* (and thus the coefficient of adhesive friction) can be explained by the number of dangling (high-friction), reconstructed (reduced-friction) or adsorbate-passivated (low-friction) surface bonds developing on the counterfaces as a function of temperature and atmospheric environment.

1. Introduction

Microelectromechanical systems (MEMS) are gaining increasing importance and utility in miniaturized instrumentation for space, airborne, terrestrial and undersea applications. The powerful combination of significant weight savings combined with an otherwise unattainable redundancy of operation by a large number of tiny devices render MEMS as a critical technology for both military and civilian use.

The most often employed MEMS material of construction is single-crystal (XTL) or poly-XTL Si. With methods developed for highly integrated microelectronics, mechanical Si MEMS moving mechanical assembly (MMA) components are similarly co-fabricated on planar wafers and subsequently etched free for six-degrees-of freedom mechanical movements. While the manufacturing technology to fabricate them is well

B. Bhushan (ed.),
Fundamentals of Tribology and Bridging the Gap between the Macro- and Micro/Nanoscales, 503–523.
© 2001 *Kluwer Academic Publishers.*

established, the reliability of Si MEMS-MMAs remains unresolved. Just to cite one example, a single device such as a rotary microgyro in a nanosatellite (Hilton, 1998) could replace accelerometer-type gyros based on bending Si AFM beam/tip combinations with less thermal sensitivity, more accuracy and simpler electronics. Despite this fact, Si MEMS-MMAs have not seen any practical use anywhere because of the high friction and wear of unlubricated XTL and poly-XTL Si (Gardos, 1998).

The MEMS community is beginning to realize the shortcomings of Si for MEMS-MMAs (especially for extreme environment applications) and has begun to gradually shift to coating the micro-components with the more wear-resistant CVD SiC (Yasseen et al., 1999). Diamondlike-carbon, silicon nitride or thermally grown SiO_2 coatings did not perform nearly as well. The use of SiC, both as a coating on Si and as a self-standing alternative for MEMS-MMAs, was also proposed (Rajan et al., 1999, Yasseen et al., 1999). A recently developed TiC coating process might also lend itself to treating fully assembled Si MEMS-MMAs with a thin, wear-resistant film (Radhakrishnan et al., 2000), provided the coating constituent species reach the insides of narrow and tortuous bearing clearances. However, notwithstanding the possible success of these and similar methodologies and the coatings' wear resistance, both SiC and TiC are still high friction materials without additional lubrication (Tkachenko et al., 1979; Erdemir et al., 1996).

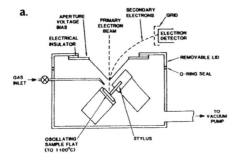

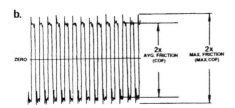

Figure 1. Schematics of (a) the SEM tribometer, and (b) the oscillatory friction traces.

In search of ideal MEMS-MMA tribomaterials and lubrication systems, a summary of a decade's worth of special friction and wear tests are described, performed with an SEM tribometer (Figure 1), see (Gardos 1994, 1996, 1996a, 1998, 1998a, 1999, 2000; Gardos et al., 1989, 1994, 1995, 1996, 1996a, 1999, 1999a, 1999b). The model experiments were guided by a friction hypothesis based on the state of the surface bonds. It will be shown here that the friction trends and the degree of wear are caused by the characteristic cohesive energy density differences and surface chemistry of silicon, α-SiC and polycrystalline diamond (PCD). The results shown in this paper indicate that neither Si nor α-SiC performs better than PCD films as a MEMS-MMA bearing material.

In particular, it is demonstrated that the 1.8-times strength of the C-C bond in PCD as opposed to the Si-Si bond in bulk silicon (Table 1) translates into more than *10,000 times* lower PCD wear rates under thermal ramping to and from 950°C, in moderate vacuum and in low partial pressures of hydrogen, under realistic engineering loads. The somewhat stronger Si-C bond leads to only a 10-times increase in wear rate measured with PCD under similar experimental conditions.

TABLE 1. Selected properties of cubic silicon, silicon carbide
and diamond.

Property	Silicon	SiC	Diamond
Atomic Distance in (111), nm	0.384	0.307	0.251
Diatomic Bond Energy (eV)	2.83	4.28	5.26
Young's Modulus (GPa)	188	380	1210

This ordering in tribological performance is heavily influenced by the interaction of the counterface surface bonds of Si, α-SiC and PCD as a function of the thermal-atmospheric environment. The respective material removal rates are controlled by shear-induced surface cracking caused by the adhesive interaction via the incipient linkage of dangling bonds between the sliding counterfaces. The unsaturated surface bonds are generated by (a) wear, and (b) heating in vacuum or low partial pressures of gasses above the particular desorption temperatures of physisorbed and chemisorbed adsorbates. The magnitude of the COF is material-specific, but can be defined by the number of dangling (high-friction), reconstructed (reduced-friction) or adsorbate-passivated (low-friction) surface bonds in the real area of contacts. This action is always perturbed by the presence of some wear debris or other third body with their own characteristic surface chemistries as a function of material, load and the thermal-atmospheric environment. Based on experiments designed by a theoretical model of atomic-level behavior, a connection is made between *nanometric* surface phenomena and the *micro/macroscopic* friction and wear performance of Si, SiC and PCD.

2. Theory of Interface Bonding

2.1. INTERACTION AT THE ATOMIC SCALE

The crystal structures of silicon (cubic) and diamond (cubic or hexagonal) are based on the tetrahedral coordination of C and Si in the lattice. The atomic-level architecture of the cubic β-SiC can be viewed as a version of the cubic silicon and diamond structure, in which every other C is replaced by a Si. The hexagonal α-SiC is derived from the closely related lonsdaleite lattice (the hexagonal allotrope of diamond). Due to fundamental bonding considerations, Si cannot form a hexagonal structure.

The electronically most significant XTL facets of Si are the (100) and (111). The (100) and (111) wafers, along with poly-XTL Si, are the most often used base materials for MEMS-MMA fabrication. Coincidentally, the most prevalent crystal textures for PCD films are also the pyramidal (111) and the tile-shaped (100) microstructure, see Figure 2.

As to α-SiC, it has several polymorphs that are closely related structurally, normally designated as *polytypes*. The hexagonal (2H, 4H, 6H) α-SiC polytypes differ only in the crosslinking of the identically puckered-corrugated layers. However, in contrast with silicon and diamond, α-SiC crystallizes in a polar form, because the basic building block is a carbon atom tetrahedrally bonded to four silicon atoms. These tetrahedra are arranged so that all atoms lie in parallel planes: one plane composed entirely of Si atoms alternating with a plane of C atoms. The distance from the C plane to the Si plane above is 3-times the distance from the C plane to the Si plane below.

506

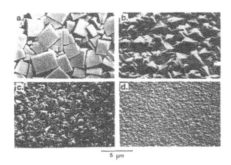

Figure 2. SEM photomicrographs of various PCD film textures: (a) [100] growth, (b) pyramidal [111] orientation, (c) micropyramidal, and (d) cauliflowered texture.

This asymmetry results in a polar crystal, whose opposing faces exhibit different reactivity, ergo different tribochemical behavior. Since both Si and C are Group IV elements, they tend to make three sp^2 (planar) bonds and leave an unpaired electron in the p_z orbital, when bonding in a threefold coordination on the surface. Since the degree of sp^2 hybridization depends largely on the separation in energy of the s and p valence levels (for C, it is ~8.4 eV; for Si ~7.1 eV), the sp^2 hybridization of the Si atom will be more complete than that of the C atom (Ueno et al., 1998). Therefore, the C-plane is more electron-rich.

This difference in surface electron density affects both the bonding and the reactivity of these chemically different SiC surfaces (Lee et al., 1982; Muelhoff et al., 1986; Didziulis et al., 1991; Ueno et al., 1998).

As schematically shown by the simple 2-D cross-section of hydrogen-terminated (111) and (100) of Si or cubic diamond in Figure 3, the desorption of the σ-bonded surface hydrogens upon sufficient energy input leaves different dangling bond configurations behind on the respective habit planes (Gardos, 1996b). On unreconstructed (111) surfaces there is only one hydrogen per surface atom, therefore only one dangling bond per atom extends normal to the surface plane, whereas two hydrogens (dangling bonds) per surface atom extend from the (100) plane. Notwithstanding the more pronounced surface corrugation of α-SiC due to the atomic size and bonding mismatch, the schematic in Figure 3 (and the associated argument) also apply to the α-SiC(0001) hexagonal basal planes, which are essentially identical to the β-SiC(111) cubic plane.

If an entire plane were removed by cleavage from the respective cubic structures, the lowest critical resolved shear strength (111) would be the preferred shear plane, because only single backbonds are broken between planes. Cleaving (100) off its mating subplane requires breaking two backbonds per surface atom. For example, the surface energy (γ) of a freshly-cleaved Si(111) is always lower (1.83×10^{15} cm^{-2}) than that of (100) (3.12×10^{15} cm^{-2}) in the absence of adsorbates, because of the number of dangling bonds per unit area generated at the parted (111) versus (111) interfaces is less. The $\gamma_{Si(100)}$, even after the (2x1) reconstruction of the fractured surfaces is 1.36 J·m^{-2}, while the $\gamma_{Si(111)}$ represented by Figure 4 (a pseudo-3-D version of the (111) facet in Figure 3), is only 1.23 J·m^{-2}. In the same vein, the calculated $\gamma_{C(100)}$ of freshly cleaved (unreconstructed) diamond is some 80% higher than the $\gamma_{C(111)}$ (Gardos, 1996b). The cross-sectional schematic of the high surface energy, dangling bonds-containing Si(111) or C(111) sliding against each other is depicted in Figure 5. Depending on their register at lattice-matched (identical Si, SiC or diamond facets paired with themselves or lattice mismatched (i.e., Si versus diamond, Si versus SiC or diamond versus SiC) interfaces (Figure 6), the unpaired electrons of each dangling bond will try to spin-pair with one from a similar, singly-occupied orbital protruding from the mating counterface. The resulting attempt at incipient chemical bonding leads to progressively greater atomic-level adhesion. This, in turn, translates into higher adhesive friction.

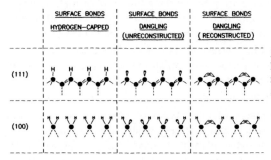

Figure 3. Cross-sectional schematics of hydrogen-terminated, dehydrogenated and reconstructed (111) and (100) surfaces of Si, β-SiC and diamond.

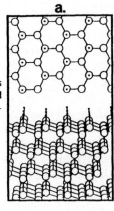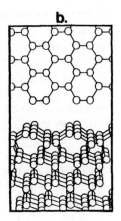

Figure 4. Top and oblique side-view schematics of diamond: (a) the "ideal", hydrogen terminated (111), and (b) the (2x1)-reconstructed (111) - - note missing hydrogens.

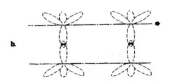

Figure 5. Cross-sectional schematic of dangling bonds-terminated counterface atoms of Si or diamond (111) in a self-mated tribocontact.

Figure 6. Cross-sectional schematic of dangling bonds-terminated counterface atoms in a lattice-mismatched interface (e.g., Si versus diamond).

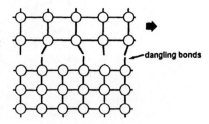

508

The characteristic architecture of the various crystal facets leads to the types of reconstruction structures that follow hydrogen desorption on some sufficiently energetic input (e.g., heating). Reconstruction is favored in covalent semiconductors, because the energy gain derived from pairing of the dangling bond orbitals is larger than the bond distortion energy on going from an unreconstructed to a reconstructed configuration. As depicted in Figures 3 and 4, the hydrogen-denuded (111) completely reconstructs to (2 x 1) to lower the high surface energy of the facets terminated with dangling bonds. The (100), on the other hand, will initially release only one hydrogen (i.e., at a lower temperature) than the H-release off (111), because the strength of the silicon-hydrogen and carbon-hydrogen bonds is lower in the SiH_2 and CH_2 configurations. Regardless of the development of other types of possible reconstruction structures, e.g., the highly stable Si(7x7) and a variety of different complex reconstruction mosaics forming on high-index planes of wear- (or polishing)-miscut crystallites of the poly-XTL materials (Homma et al., 1993; Gardos, 1998, 1998a, 1999), the linking of dimers into these π-bonded chains leads to a reduction in the surface energy. It is expected that this reduction translate into lower adhesion and adhesive friction.

According to this model, one way to induce high friction is by generating dangling bonds via heating at temperatures high enough and at atmospheric pressures low enough to drive off the surface adsorbates. Continued heating to even higher temperatures provides the activation energy necessary for reconstruction. This radical change in the surface state should result in some significant reduction in friction. Then, cooling the surfaces is attendant with losing the energy needed to maintain reconstruction. The ensuing *deconstruction* (i.e., breaking of the dimers and the regeneration of dangling bonds) causes the friction to increases to a higher level once more. On continued cooling, however, the surface temperature becomes sufficiently low for dissociative chemisorption of gasses present in the test environment. The dangling bonds become passivated, reattaining the low friction originally observed during room temperature (RT) sliding. Barring any tribochemical generation of a third body in the interface with surface behavior incompatible with this friction model, the complete cycle should be repeatable in the same wear track.

2.2. INTERACTION ON THE GRAIN SIZE SCALE

At the asperity level, engineering bearing surfaces interact at isolated real areas of contact (A_r), see Figure 7.

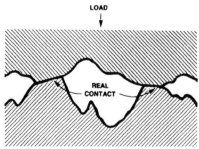

Figure 7. Cross-sectional schematic of the real area of asperity contact (A_r).

To simplify the understanding of the friction and wear process on the microscopic scale, the progressive, time-dependent wear of cauliflowered (Figure 2d) and micropyramidal (Figure 2b) PCD films may be modeled by wearing them against an atomically smooth diamond counterface (Figure 8).

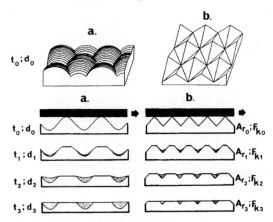

Figure 8. Top: schematic representation of (a) cauliflowered, and (b) pyramidal diamond film textures, before sliding (t_0 = zero time; d_0 = zero distance). *Bottom:* schematic representation of the surfaces depicted on top, progressively worn by an atomically smooth diamond counterface as a function of increasing sliding time and distance. The equally increasing A_r is commensurate with a progressively greater (i) amount of debris trapped in the valleys, and (ii) friction force F_k (F_k = shear strength of the interface x A_r).

Ideally, adhesive friction in the A_r can be reliably measured only in the absence of any third-body in the contact, unless the surface chemical behavior of its constituents is identical to that of the mother substrate. If the original surfaces are sufficiently rough to trap all the debris in the depressions ("valleys"), this state is automatically attained. The SEM photomicrographs of cauliflowered (Figure 9) and micropyramidal (Figure 8) surfaces of PCD films polished by wear indicate the realistic nature of this model, so long as the amount of wear debris generated does not exceed the volume of the available valleys (sinks) between asperities. Where it does, maximum articulation of the predicted friction *trends* depends on (a) the physico-chemical state of the high surface-to-volume ratio debris particles or other tribocatalytically generated surface layers, and (b) an interfacial layer sufficiently incomplete to allow substantial counterface asperity contact in the A_r. Unfortunately, tribometric generation, trapping and plowing of some third body material is inevitable. Therefore, it is more realistic to look for qualitative manifestation of the predicted friction *trends* as a function of heating and atmospheric test environment than to provide highly accurate *values* of the tribosystem-dependent *adhesive* friction.

The wear-polished polycrystalline surfaces in Figures 9, 10 and 11 always exhibit some crystallographic disorder. Therefore, a variety of reconstruction cells can be formed in the A_r. However, the Hertzian areas in engineering concentrated contacts are still very large relative to the size of the π-chains comprised by the self-bonded dimers. It follows that a global reduction in COF may be explained by a local decrease in the surface energy within each mosaic piece of a reconstructed nanoregion, regardless of the directionality of the chains. Therefore, extending the *gedanken* friction model from the atomic level to the grain size regime is reasonable.

510

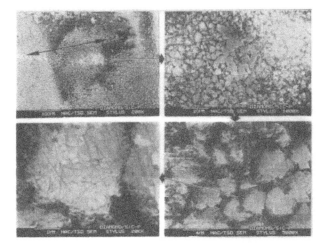

Figure 9. SEM photo-
micrographs of the cauli-
flowered PCD film on
the SEM tribometer pin
tip after repeated, thermally ramped friction and
wear tests in vacuum.
The arrow indicates the
direction of oscillation;
from (Gardos 1999).

Figure 10. SEM photomicrographs of variously textured PCD
films subjected to SEM tribometry: a C(111)-micropyramidal
film deposited on a α-SiC pin,
and the mainly (100)-oriented
film a on a Si(100) flat before
and after mechanical polishing;
from (Gardos 1999).

3.   Test Equipment and Methodology

The SEM tribometer described schematically in Figure 1 and the associated test
procedures designed for Si, SiC and PCD were established to induce the conditions
defined by the *gedanken* model described above.

3.1.   TEST EQUIPMENT, METHODOLOGY AND SPECIMENS

3.1.1. Test Procedures
The SEM tribometer, as well as the procedures for preparing and testing silicon and
PCD test specimens, are given in (Gardos 1994, 1996, 1996a, 1998, 1998a, 1999, 2000;
Gardos et al., 1989, 1994, 1995, 1996, 1996a, 1999, 1999a, 1999b). The particular
methodologies pertaining to Si (Gardos 1996, 1996a, 1999, 1999b) and PCD (Gardos
1994, 1996, 1999; Gardos et al., 1989, 1990, 1994, 1995, 1996a, 1998a, 1999b) have
been described in the indicated references and the related papers cited therein. The
general thermal ramping test procedure was slightly modified for the present work to
accommodate testing of α-SiC.

Essentially, a small flat is oscillated against a hemispherically tipped pin dead-weight-loaded to a level suitable for a given material combination. Testing is done either in the vacuum of the SEM column ($\sim 1 \times 10^{-5}$ Torr; over 90% water vapor in the residual gas atmosphere) or in a lidded Knudsen cell-like subchamber backfilled and purged continuously with low partial pressures of selected gases such as air, H_2, He, N_2 and O_2. The friction forces are archived, averaged and converted in real time by a desktop computer using commercial data logging and analysis software.

A constantly exchanged gas environment offers several benefits. If the wear rate of a particular materials combination is found sufficiently low, several thermally ramped experiments can be repeated without breaking either the SEM column vacuum (or the particular test atmosphere in the subchamber), because low wear precludes the need to photograph the wear scars after each test. The normal load, the number of tests that can be run with the repeatedly reused pin in the same or different wear tracks on the flat and the frequency of SEM photomicrography of the pin tip between tests are material-dependent. These factors are selected to optimally determine the time-dependent evolution of the pin tip wear scar size (used for wear rate measurements) with minimal downtime of the apparatus. An additional advantage is the higher thermal conductivity of the various partial pressures of gasses flowing through the lidded and preferentially pumped tribometer chamber. It allows cooling of the friction force strain gages to near RT in a reasonably short period of time between tests. The length of the standard 2000-cycle test thermally-ramped to temperatures as high as 1000°C was determined during preliminary experiments in vacuum, where the friction force transducers are unavoidably heated by residual thermal transfer from the hot stage. The water-cooled nature of the stage and strategic use of thermal insulation could only mitigate eventual heating of the transducers to the permitted upper limit of $\sim 45°C$.

Where specimen cleanliness must be maximized for after-test AES/XPS examination, high purity argon gas is admitted in the SEM column (and subchamber) to bring the pressure to atmospheric. Note that no *in situ* video or SEM photomicro-graphy of the e-beam-exposed flat tracks is done during any of the P_{gas} experiments to (a) eliminate any electron-enhanced desorption of the friction-controlling surface adsorbates, and (b) prevent e-beam-induced surface-chemical reactions.

The stage housing the flat and its heater assembly is oscillated at 0.5 Hz, with a small average sliding speed of 3 to 4 mm/s. The exact length of the friction stroke depends on the magnitude of the COF. Higher friction is commensurate with a somewhat shorter wear track on the flat and *vice versa*. The actual length of a track on the flat, therefore the total length of sliding calculated by the number of oscillatory cycles), are measured from low magnification SEM photomicrographs taken after completion of each experiment. After-test surface analyses were performed not only by SEM examination of the worn pin tip and the flat scar sites of interest, but also by occasional probing of selected wear sites with an AES/XPS spectrometer.

The temperature of the SEM tribometer hot-stage is ramped to a predetermined temperature high enough to remove the chemisorbed adsorbates from the initially cleaned specimen surfaces in the tribocontact. Then, cooling back to RT enables chemisorption of the gasses present in the SEM column or the lidded subchamber. Thermal ramping is accomplished at the standard heating and cooling rate of 35 C/min selected empirically to permit at least a 2000-cycle test in vacuum without overheating the force transducers.

The SEM-tribometric behavior was interpreted by (a) the temperature- and atmosphere-dependent COF trends and inflections as a function of the counterface materials and test environment, (b) surface analyses of the worn specimens (where warranted), and (c) surface-physical and -chemical literature data gleaned from the literature on single-crystal and polycrystalline diamond, Si and α-SiC normally examined in the static (non-tribological) mode. Since the surface temperature of the heated flat (measured by a thermocouple embedded below the Pt heater strip) is not the true temperature of the sliding interface, it was estimated by observing the initial COF increase as a function of thermal ramping. Specifically, the temperature of the sudden rise in friction was correlated with literature data on the temperatures needed to thermally generate the dangling bonds on Si and diamond during temperature-programmed desorption. This approximation was used to estimate the temperature of the sliding interface not otherwise measurable in the SEM tribometer (Gardos, 1998, 1999; Gardos et al., 1994).

The data on poly-XTL α-SiC presented in this paper are new and previously unpublished. The particular ceramic employed was a commercially available, pressure-densified version with a unique microstructure highly suitable for tribological and ceramic armor applications (Figure 11).

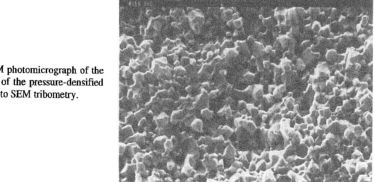

Figure 11. SEM photomicrograph of the fracture surface of the pressure-densified α-SiC subjected to SEM tribometry.

The vacuum test procedures originally designed for Si and PCD were slightly altered to suit the tribological behavior of α-SiC. The new procedure consisted of the following steps: (a) running two experiments in the same wear track, (b) breaking vacuum to remove the pin still in its holder for SEM photomicrography, (c) reassembling the pin/holder assembly into the apparatus, thus maintaining the plan-parallel nature of the already formed wear scar on the pin tip and the yet-undisturbed flat, (d) move the flat sideways under the pin with the special electronic stage installed for that purpose to expose the used (but cleaned) tip to a previously heated but yet-unworn part of the flat, and (e) running two additional test in the second wear track. The results, along with post-test (*ex situ*) AES/XPS analyses of the pin tip scar, provided some preliminary information on the appearance and chemistry of the statically or tribochemically generated surface layers in vacuum. Note that surface chemical analyses of vacuum- and hydrogen-tested Si and PCD are problematic (Gardos 1999). Both the literature and the preliminary AES/XPS of the worn α-SiC surfaces indicated that their analyses are more manageable and meaningful.

3.1.2. Specimen Cleaning

Early experimentation with the fine-cauliflowered PCD described in Figure 9 was carried out by cleaning the specimens with hot n-hexane before each test. The polished PCD pin/flat combination shown in Figure 10 was cleaned by washing with hot hexane *and* hot chromic acid between experimental rounds. Note that single-crystal diamond and PCD films are routinely cleansed by various acid-type treatments to remove any sp^2-bonded residue and other contaminants from the surface. Therefore, the solvent-cleaned specimens were boiled briefly in a hot chromic acid solution, followed by several successive rinses with hot, triple-distilled water. The acid-cleaned diamond is free of organic and inorganic contaminants, but it becomes oxidized in the process. Annealing at high temperatures in vacuum and subsequent surface analysis had shown that the oxides are tenacious. The concentration of the oxygenated surface moieties does not diminish until ~1000°C is reached (Gardos 2000).

All the polished Si and α-SiC samples were cleaned by a hexane wash followed by a HF + water rinse. This procedure, which assures complete cleanliness *and* hydrogenation of all the surface bonds of Si (Gardos 1996,1996a, Gardos et al., 1994) is also commonly used to clean SiC by removing adventitious carbon and the native oxide (Rahaman et al., 1986; Singh et al., 1998; Shin et al., 1998; Vathulya et al., 1998). However, the dangling bonds on the surface of SiC are not passivated as homogeneously and effectively as on the various Si crystallinities (Achtzinger et al., 1998; Ueno et al., 1998). For that, annealing SiC in H_2 at 1000°C, hydrogen plasma treatment or ion-implantation with hydrogen is needed. Not having access to this process, it was presumed that upon HF-washing, rinsing and drying in air, the bulk surface oxide was removed, leaving the surface bonds terminated with -OH and other oxygenated moieties.

4.   Data and Discussion

4.1.   EARLY VACUUM EXPERIMENTS WITH PCD AND Si

The earliest wide temperature range PCD experiments (Figure 12) were performed with the ~1.5 μm thick, self-mated fine-cauliflowered films shown in Figure 9, sliding first in vacuum and then in 13 Pa (0.1 Torr) lab air (Gardos et al., 1990). Their reduced surface roughness (0.05 to 0.1 μm AA/CLA) came with a somewhat lower purity than that of the textured films described in Figure 10, because the methane pressure (and thus the sp^2-bonded grain boundary phase of the fine-cauliflowered layers) had to be increased in the reaction chamber to enhance the nucleation rate (Gardos 1999).

The "rabbit ear"-like stepfunction-with-trough COF signatures in Figure 12 were the first circumstantial evidence of the *gedanken* friction hypothesis. Accordingly, incipient linking of the sliding counterfaces by unsaturated bonds on heating and their passivation by benign adsorbates at sufficiently low temperatures were presumed to be responsible for the radically increased and reduced adhesion and friction. Continued heating of the progressively degassed and worn surfaces causes the dangling bonds to reconstruct, thus reducing the surface energy and the COF. However, once the heating stops and the rubbed surfaces cool below a certain temperature, the activation energy needed to keep the bonds reconstructed is lost. There is a significant rise in friction from the deepest part of the COF trough due to regeneration of the dangling bonds

514

ance of another COF peak. As the contact is cooled further to the temperature where the adsorbates (mainly the residual H_2O in the SEM column) chemisorb and passivate the deconstructed surface bonds. As a result, the lowest possible surface energy and a large reduction in COF are attained.

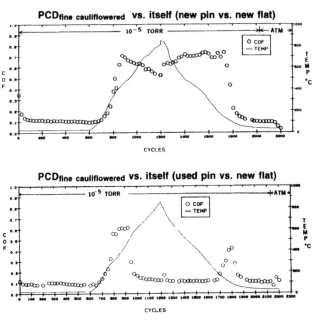

Figure 12. The coefficient of friction (COF) of the PCD film depicted in Figure 9 as a function of temperature, tested in the SEM tribometer against itself in ~1×10^{-5} Torr vacuum; from (Gardos et al., 1990).

At that time, however, the trough-like "bathtub" (dip-and-rise) curve in the COF at the highest temperatures in vacuum could be attributed to surface re(de)construction only if this reduction in friction was not caused by shear-generated diamond-to-graphite transformation during continued heating, and then a graphite removal rate higher than its formation rate on cooling. Inasmuch as diamond is less stable than graphite by 0.45 kcal/mol at 1 atm. (Gardos 1999), some shear-induced phase transformation in the A_r was a possibility. The probability of variable graphitization notwithstanding, the worn pin tip in Figure 9 (also representative of the films deposited on the flat) indicated several advantages to using cauliflowered films as proof-of-concept specimens:

1. The real areas of contact were essentially free of wear debris. The small amount that did form became displaced into the depressions between the asperity plateaus.
2. The pin tip is worn at all times during a test. The appearance of this highly stressed (and possibly partially graphitized) site indicated that the prevalent mode of surface damage (incipient wear) was tensile stress-induced microcracking under the high shear stress (high COF) conditions between the worn micro-mounds. Cracks in the A_r formed mainly normal to the direction of sliding, followed by the development of a webbed crack network.

The validity of the friction hypothesis predicting the "rabbit ear"-type COF signature was further strengthened by obtaining a similar vacuum friction trend provided by a polished PCD-coated poly-α-SiC pin (Figure 10) mated against a polished, Si(100) flat, see Figure 13 (Gardos et al., 1994). Since Si cannot graphitize, the probability of PCD graphitization in the A_r was significantly reduced.

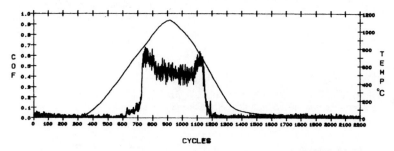

Figure 13. The coefficient of friction (COF) of the polished PCD film on a pin depicted in Figure 10 as a function of temperature, tested in the SEM tribometer against a polished Si(100) flat in ~1x10^{-5} Torr vacuum; from (Gardos et al., 1994).

4.2 FRICTION OF PCD AND POLY-Si IN VACUUM AND PARTIAL PRESSURES OF HYDROGEN

To further clarify the effect of PCD graphitization in the A_r on the characteristic COF signature, the same PCD pin tip used against the Si(100) flat was repolished, acid-cleaned and retested against an unpolished (rough) PCD flat (both depicted in Figure 10). The typical COF signature was temperature-independent, consistently remaining at a value of 0.3 to 0.4 (Figure 14). This unusual, graphite-like behavior was observed before by SEM tribometry (Gardos et al., 1997, 1997a). Auger analysis of the unused film on the flat and the debris piles collected at the end of the strokes (Figure 15) offered the explanation for this shear induced phase transformation in the A_r. The C_{KLL} peaks in the debris spectrum indicated the presence of amorphous carbon and possibly graphite, with the new film displaying a spectrum typical of diamond (Gardos 1999; Gardos et al, 1995).

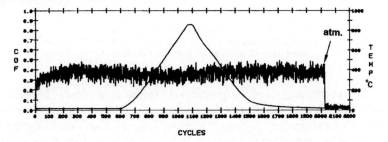

Figure 14. The coefficient of friction (COF) of the polished PCD film on a pin tested against an unpolished PCD flat (both depicted in Figure 10) by SEM tribometry as a function of temperature, in ~1x10^{-5} Torr vacuum; from (Gardos et al., 1995).

516

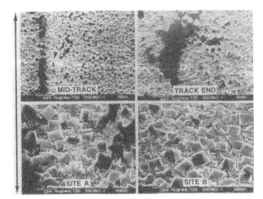

Figure 15. SEM photomicrographs of the wear track on the used, unpolished PCD flat (depicted in Figure 10 in the as-deposited form) after SEM tribometry as a function of temperature, in ~1x10⁻⁵ Torr vacuum, (see Figure 14). Double-headed arrow indicates the direction of oscillatory sliding; from (Gardos et al., 1995).

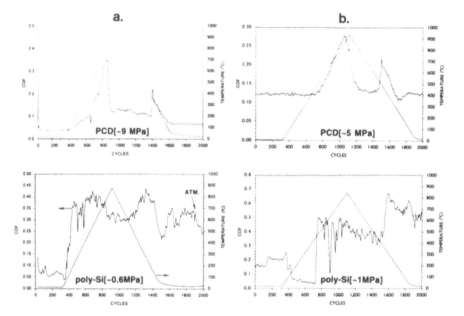

Figure 16. Selected representative SEM-tribometric COF curves of the polished PCD pin versus polished PCD flat depicted in Figure 10, complemented by those of polycrystalline Si couples tested (a) in vacuum, and (b) in 0.2 Torr H₂; from (Gardos, 1999).

The flat with the carbonized/graphitized wear tracks in Figure 15 was then polished and cleaned by hot hexane and hot chromic acid. It was made to slide against the slightly worn and similarly re-cleaned (polished) pin tip. Due most likely to the drastic reduction in the micro-Hertzian stresses in the real areas of contact, the graphitization effect disappeared and the typical stepfunction-with-trough COF signatures were regained both during testing in vacuum and in partial pressures of hydrogen (Figure 16), (Gardos 1999). The accuracy of the friction hypothesis was further reinforced by (a) a large number of PCD tests in hydrogen, where the characteristic friction trend was duplicated during repeated sliding in the same tracks (Gardos 1998a, 1999), (b) standard thermally-ramped Si experiments also shown in Figure 16, and (c) slow thermally-ramped SEM tests of both the various Si crystallinities and the polished PCD specimens in vacuum (Gardos 1998, 1998a; Gardos et al., 1999, 1999a and b).

The following arguments explain the similarities and differences in the Si and PCD COF signatures in Figure 16:

1. The ΔT between the heater thermocouple and the true surface temperature of the flat was estimated to be less in hydrogen than in vacuum (Gardos 1998; Gardos 1999). A better heat transfer to the flat notwithstanding, the presence of the gas *raised* the apparent adsorbate desorption temperatures of both Si and PCD. This increase is attributed to the Le Châtelier Rule: if a reaction (the desorption of an adsorbate) results in a gas (e.g., H_2) as a product, increasing the partial pressure (activity) of this gas in the tribometer subchamber will retard the reaction, i.e., desorption will occur at higher temperatures.

2. The substantial reduction in the COF of PCD on cooling in vacuum was interpreted as the footprint of tribocatalytically enhanced chemisorption of *molecular* water (i.e., the residual atmospheric moisture in the SEM column) or molecular hydrogen purposely introduced into the subchamber.

3. There was also some reduction in the COF of the various Si crystallinities in vacuum, on continued cooling to near RT, again attributed to the chemisorption of residual water vapor in the SEM column. However, the final RT friction values were not as low as the starting COF observed at the same temperature. This was explained by the large increase in the real and apparent areas of contact resulting from the high wear rate of Si (Figure 17).

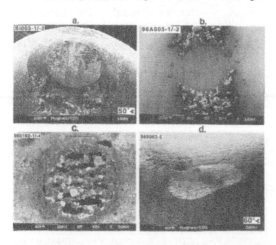

Figure 17. SEM photomicrographs of worn pin tips after SEM tribometry: (a) poly-Si in vacuum (15 g load, 2000 cycles), (b) poly-Si in 0.2 Torr dry H_2, (15 g load, 2000 cycles), (c) poly-α-SiC in vacuum (28 g load, 8000 cycles), and (d) PCD in vacuum and low partial pressures of H_2 (28 g load, 63,200 cycles); from (Gardos 1998, 1998a, present work).

4. In contrast, the final COF_{Si} in hydrogen at RT was higher than in vacuum. Although this trend is consistent with the lack of activation energy for the chemisorption of H_2 on cooling [2], it is difficult to believe that surface bonds remain unsaturated at or near RT in the presence of hydrogen. If this phenomenon is not caused by unpassivated free radicals but by the interaction of variously reconstructed surfaces with energies changing as a function of temperature in the presence of adsorbates, the SEM tribometer is not able to separate these variables. Work is ongoing to help clarify this anomaly.

5.   The general level of COF$_{PCD}$ is significantly lower than COF$_{Si}$ in partial pressures of H$_2$, at any temperature. Molecular hydrogen appears to be a better gas-phase "lubricant" for PCD. Since the worn flat was not imaged during any of the experiments in partial pressures of gasses, the lubricative surface reactions were not e-beam activated.

4.3. FRICTION OF α-SiC IN VACUUM

The friction data generated to date with the hexane+HF-cleaned specimens in Figure 18 indicate that the "rabbit ear"-type COF signatures are similar to those previously observed with all the Si crystallinities and PCD. The superimposed friction curves of PCD, poly-Si and poly-SiC in Figure 19 emphasize this remarkable trend. Its repeated manifestation is the strongest circumstantial evidence to date for the friction hypothesis based on the interaction of dangling, reconstructed and passivated surface bonds.

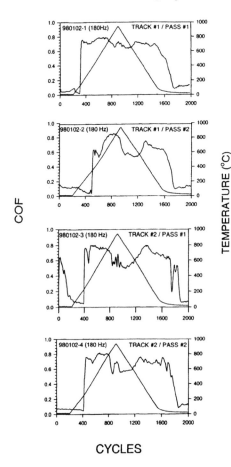

CYCLES

Figure 18. SEM-tribometric COF curves of hexane+HF-cleaned α-SiC (Figures 11 and 17c) friction couples tested repeatedly in vacuum.

The causes of different COF magnitudes as a function of material can only be speculated upon. Inasmuch as both the Si and α-SiC interfaces contained some residual debris during the friction tests (Figure 17), the behavior of the surface moieties and bonds on the trapped microparticles did nor seem to erase the characteristic stepfunction-with-trough COF signatures. Discounting any plowing effects, the friction hypothesis predicts that the magnitude of adhesive friction should depend on the *number* of dangling, reconstructed and passivated surface bonds present at any one time in the A$_r$. As to signature similarities of the cubic and hexagonal structures of the respective chemical homologs, the cubic and hexagonal (lonsdaleite) crystal structures of diamond and the equivalent (111) surfaces of the β (cubic) and the (0001) basal planes of the α-(hexagonal)-SiC versions are essentially identical. The randomly oriented and wear-miscut XTL grains on the polycrystalline surface of α-SiC (Figures 11 and 17) should behave globally the same way as the miscut crystallites of PCD (Figure 10).

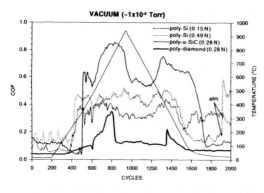

Figure 19. Normalized superposition of SEM-tribometric COF curves on poly-Si, poly-α-SiC and PCD friction couples, from (Gardos 1998, 1998a, 1999, present work); also see Figures 16 and 18.

Preliminary XPS analyses of the pin tip scar and the high-resolution spectra of carbon, silicon, and oxygen present a consistent picture of the dominant surface species on the scar depicted in Figure 20.

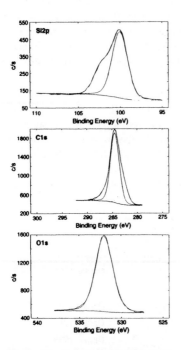

Figure 20. Deconvoluted XPS peaks of the α-SiC wear scar in Figure 17c.

In the case of C1s emission, carbon was shown to be in three binding energy states, averaging 282.7 eV, 284.6 eV, and 286.7 eV. The lowest value corresponds with reference values for carbidic carbon (280.7 to 283 eV), while the middle value is in the range of graphitic carbon (284.2 to 285.0 eV). Neither of these regions overlaps with any other. The high-energy value best matches with the expected energies of carbon atoms singly or doubly bonded to oxygen (286.0 to 288.0 eV). This would include alcohol, ether, ketone, and aldehyde functionalities, but exclude carboxyls and carbonates.

The generally weak Si2p emissions manifested themselves in two dominant components: a low energy peak at 100.0 eV and a higher peak at 102.2 eV. The low energy peak falls at the low end of the reported values for carbides (99.9 to 100.9 eV), while the higher peak overlaps best with silicates (102.0-103.0 eV). The reported range for SiO_2 does not begin until binding energies in excess of 103.2 eV are reached, so the surface oxides do not seem to be stoichiometric silica.

The O1s emissions appeared only as single peaks in all the measured spectra. These peaks were centered around 532.0 eV, which is lower than the values typically reported for SiO_2 (532.5 to 534.3 eV), but is similar to values seen for organic carbonyls such as *p*-benzoquinone and benzamide.

In general, carbon appeared in chemical states appropriate for carbide, carbon-graphite and partially oxidized compounds. The carbidic carbon clearly matched the low energy carbidic silicon peak, and the oxygen present was in a binding energy range appropriate to partially oxidized carbon compounds. The high binding energy Si peak appeared to come from a non-silica source that produced a shift for oxygen, which cannot be resolved from the carbon-shifted oxygen. This means that the presence of some silicon oxycarbide species in the third body (generated by some gas-phase reactions with the residual water vapor in the SEM column) cannot be discounted.

4.4. WEAR OF Si, α-SiC AND PCD

The wear rate of polished PCD samples (characteristic of the standard thermal ramping procedure) was found to be in the 10^{-16} m^3/N·m range , with a measurable difference in the wear of the polished pin tip sliding against either the unpolished or the polished flat (Table 2). The 3 to 4×10^{-16} m^3/N·m values for the polished specimen couple, as measured by the final scars on the pin tip after all the standard thermal ramp experiments in vacuum and hydrogen, are in good agreement with the wear rate previously found for the fine-cauliflowered PCD tested in vacuum. The remaining depressions in the mechanically polished surfaces (clearly visible in Figure 10) acted as debris traps for particle-free measurement of adhesive friction (Gardos 1998a, 1999; Gardos et al., 1999c).

The SEM-tribometric wear of PCD appears to be the same in vacuum and hydrogen (Gardos 1998a; Gardos et al., 1999b). Silicon behaved the same way (Gardos 1998; Gardos et al., 1999a). Due to its lower friction and wear rate measured under higher Hertzian stresses, polished PCD appears to be a better bearing material than Si or α-SiC (at least in vacuum) for extreme environment MEMS-MMA applications.

TABLE 2. Wide temperature range SEM tribometric wear rate of poly-Si, poly-α-SiC and PCD in vacuum and hydrogen.

Pin	Flat	Atmosphere	Normal Load (g)	Pin Wear Rate (m^3/N·m)	Ref.
poly-Si	poly-Si	vac.[1]	15	1.6×10^{-12}	Gardos 1998
poly-Si	poly-Si	H$_2$[2]	15	1.3×10^{-12}	Gardos 1998
α-SiC[3]	α-SiC[3]	vac.	28	1.5×10^{-15}	present work
PCD[4]	PCD[4]	vac.	50	4×10^{-16}	Gardos 1999
PCD[5]	PCD[6]	vac.	28	8.4×10^{-16}	Gardos 1999
PCD[5]	PCD[5]	vac.+H$_2$[7]	28	3.9×10^{-16}	Gardos 1999
PCD[5]	PCD[5]	vac.+H$_2$[7]	28	2.6×10^{-16}	Gardos 1999

[1] vacuum, 2000-cycle test.

[2] 0.2 Torr, 2000-cycle test.

[3] microcrystalline, see Figures 11 and 17; 8000-cycle tests.

[4] fine-cauliflowered growth morphology, see Figure 9; ~8000-cycle tests.

[5] polished, see Figure 10.

[6] unpolished, see Figure 10.

[7] worn for 10,000 cycles in vacuum and 4,000 cycles in 0.2 Torr hydrogen.

[8] worn for 10,000 cycles in vacuum and 52,300 cycles in 0.2 Torr hydrogen.

5.   Conclusions

SEM-tribometric tests of poly-Si, poly-α-SiC and PCD in ~1.33×10^{-3} Pa = 1×10^{-5} Torr vacuum and 26 Pa (0.2 Torr) H_2, from RT to 950°C repeatedly confirmed a characteristic trend for the COF as a function of temperature. This trend has been interpreted by a hypothesis connecting the wear-torn interfaces and the thermal desorption of adsorbates in vacuum with the generation of dangling σ bonds on the surfaces. In spite of the fact that the SEM tribometer has no in-situ surface analytical capability, the extensive Si, α-SiC and PCD work completed to date has allowed the interpretation of the friction trends as a function of temperature and atmosphere in terms of atomic-level surface behavior.

Incipient linking of the sliding counterfaces by the unsaturated bonds on heating, and their passivation by certain adsorbates on cooling, are suggested as the main causes of radically increased and reduced adhesion and friction, respectively. Strong circumstantial evidence is given for the trough-like "bathtub" curve dip in the COF at the highest temperatures, attributed to surface re(de)construction. Continued heating of the progressively degased and worn Si and PCD surfaces in vacuum appears to dimerize the dangling bonds to lower the surface energy, unhindered by the ongoing tribological action. However, once the heating stops and the rubbed surfaces cool below a certain temperature, the loss of activation energy needed to keep the bonds reconstructed causes a significant rise from the deepest part of the COF trough due to deconstruction. This rise culminates in another COF peak, where surface adsorbates begin to rapidly passivate the regenerated dangling bonds to lower the COF once more on further cooling to RT. To date, the friction and surface-analytical results, combined with the related literature data, helped explain (a) the high torque and wear rate of miniaturized tribomechanical parts made from Si, (b) why molecular hydrogen can act as an atomic level lubricant for both Si and PCD, and (c) why PCD might be a better bearing material for MEMS-MMA systems than Si or α-SiC.

The results of the *microscopic* model experiments performed with the SEM tribometer are offered as preliminary proof of the friction hypothesis. The work needs to be duplicated and refined by wide temperature range surface analytical tribometry techniques not yet developed. The new apparatus and methodology must be able to show, with improved definition, the dynamic changes in the state of the surface bonds and the associated tribochemical changes.

6.   Acknowledgements

Messrs. Lindon Melton, Steve Gabelich, Bruce Buller and Drs. Dan Demeo and Kibbey Stovall of the Raytheon Electronic Systems are recognized for their able assistance in running the experiments and reducing the data. Portions of this work have been supported by (a) DARPA during the 1985-1993 "Tribological Fundamentals of Solid Lubricated Ceramics" program (Part I), and (b) AFOSR under past and current Contract Numbers F49620-95-C-0002 and FA9620-98-C-0009 (Maj. Hugh De Long and Maj. Paul C. Trulove serving as the AFOSR/NL Program Managers). The author is equally grateful to his management for their ongoing encouragement and continued appreciation of tribology in an electronic sensors-oriented company.

522

7.   References

Achtzinger, N., Grillenberger, J., Whitthuhn, W., Linnarsson, M.K., Janson, M. and Svensson, B.G. (1998), "Hydrogen Passivation of Silicon Carbide by Low-energy Ion Implantation," *Appl. Phys. Lett.*, 73, 945-947.

Didziulis, S.V. Lince, J.R., Fleischauer, P.D. and Yarmoff J.A, (1991), "Photoelectron Spectroscopic Studies of the Electronic Structure of α-SiC," *Inorg. Chem.*, 30, 672-678.

Erdemir, A. and Fenske, G.R. (1996), "Tribological Performance of Diamond and Diamondlike Carbon Films at Elevated Temperatures, *Tribol. Trans.*, 39, 787-794.

Gardos, M.N. (1994), "Tribology and Wear Behavior of Diamond," Chapter 12 in *Synthetic Diamond: Emerging CVD Science and Technology* , (K.E. Spear, and J.P. Dismukes, eds.), Electrochem. Soc. Monograph, John Wiley and Sons, New York, NY, 533-580.

Gardos, M.N. (1996), "Surface Chemistry-Controlled Tribological Behavior of Si and Diamond," *Tribol. Lett.*, 2, 173-187.

Gardos, M.N. (1996a), "Tribological Behavior of Polycrystalline and Single-Crystal Silicon," *Tribol. Lett.*, 2, 355-373.

Gardos, M.N. (1998), "Advantages and Limitations of Silicon as a Bearing Material for MEMS Applications," in *Tribology Issues and Opportunities in MEMS,* (B. Bhushan, ed.) Kluwer Academic Publishers, 341-365.

Gardos, M.N. (1998a), "Re(de)construction-induced Friction Signatures of Polished Polycrystalline Diamond Films in Vacuum and Hydrogen," *Tribol. Lett.*, 4, 175-188.

Gardos, M.N. (1999), "Tribological Fundamentals of Polycrystalline Diamond Films," *Surf, Coat. Technol.*, 113, 183-200.

Gardos, M.N. (2000), "Tribo-oxidative Degradation of Polished Polycrystalline Diamond Films in 0.2 Torr Partial Pressure of Oxygen," paper presented at the 26th Leeds-Lyon Symposium on Tribology, 14-17 Sept. 1999, U. of Leeds, UK, (in press).

Gardos, M.N. and Ravi, K.V. (1989), "Tribological Behavior of CVD Diamond Films," *Electrochem. Soc. Proc.* Vol. 89-12, 475-493.

Gardos, M.N. and Soriano, B.L. (1990), "The Effect of Environment on the Tribological Properties of Polycrystalline Diamond Films," *J. Mater. Res.*, 5, 2599-2609.

Gardos, M.N. and Ravi, K.V. (1994), "Surface-Chemistry-Controlled Friction and Wear Behavior of Si(100) vs. Textured Polycrystalline Diamond Film Tribocontacts," *Dia. Films & Technol.*, 4, 139-165.

Gardos, M.N. and Ravi, K.V. (1995), "Carbon-Graphite-Like Friction Behavior of Polycrystalline Diamond Sliding Against Itself in Vacuum," *The Electrochem. Soc. Proc.* Vol. 95-4, 415-424.

Gardos, M.N., Adams, P.M., Barrie, J.D. and Hilton, M.R. (1997), "Crystal-Structure-Controlled Tribological Behavior of Carbon-Graphite Seal Materials in Partial Pressures of Helium and Hydrogen. I. Specimen Characterization and Fundamental Considerations," *Tribol. Lett.*, 3, 175-184.

Gardos, M.N., Davis, P.S. and Meldrum, G.R. (1997a), "Crystal-Structure-Controlled Tribological Behavior of Carbon-Graphite Seal Materials in Partial Pressures of Helium and Hydrogen. II. SEM Tribometry," *Tribol. Lett.*, 3, 185-198.

Gardos, M.N. and Gabelich, S.A. (1999), "Atmospheric Effects of Friction, Friction Noise and Wear with Silicon and Diamond. Part I. Test Methodology," *Tribol. Lett.*, 6, 79-86.

Gardos, M.N. and Gabelich, S.A. (1999a), "Atmospheric Effects of Friction, Friction Noise and Wear with Silicon and Diamond. Part II. SEM Tribometry of Silicon in Vacuum and Hydrogen," *Tribol. Lett.*, 6, 87-102.

Gardos, M.N. and Gabelich, S.A. (1999b), "Atmospheric Effects of Friction, Friction Noise and Wear with Silicon and Diamond. Part III. SEM Tribometry of Polycrystalline Diamond in Vacuum and Hydrogen," *Tribol. Lett.*, 6, 103-112.

Hilton, M.R. (1993), "Tribological Approaches to Micromachine Design and Fabrication for Space Applications," in *Micro- and Nanotechnology for Space Systems: An Initial Evaluation,* (eds. H. Helvejian and E.Y. Robinson), Aerospace Report No. ATR-93(8349)-1, The Aerospace Corp., El Segundo, CA, 127-135.

Homma, Y., Suzuki, M., and Tomita, M. (1993), "Atomic Configuration Dependent Secondary Electron Emission from Reconstructed Silicon Surfaces," *Appl. Phys. Lett.*, 62, pp. 3276-3278.

Lee, D.H. and Joannopoulos, J.D. (1982), "Ideal and Relaxed Surfaces of SiC," *J. Vac. Sci. Technol.*, 21, 351-357.

Muelhoff, L., Choyke, W.J., Bozack, M.J. and Yates, J.T. Jr., (1986), "Comparative Electron Spectroscopic Studies of Surface Segregation on SiC(0001) and SiC(0001)," *J. Appl. Phys.*, 60, 2842-2853.

Radhakrishnan, G., Adams, P.M., Robertson, R. and Cole, R. (2000), "Integration of Wear-Resistant Titanium Carbide Coatings into MEMS Fabrication Processes," *Tribol. Lett.* and references therein (in press).

Rahaman, M.N., Boiteux, Y. and De Jonghe, L.C. (1986), "Surface Characterization of Silicon Nitride and Silicon Carbide Powders," *Am. Ceram. Soc. Bull.*, 65, 1171-1176.

Rajan, N., Mehregany, M., Zorman, C.A. and Stefanescu, S. (1999), "Fabrication and Testing of Micromachined Silicon Carbibe and Nickel Fuel Atomizers for Gas Turbine Engines," *J. MEMS*, 8, 251-257.

Shin, W., Hikosaka, T., Seo, W.-S., Ahn H.S., Sawaki, N. and Koumoto, K. (1998), "Fibrous and Porous Microstructure Formation in 6H-SiC by Anodization in HF Solution," *J. Electrochem. Soc.*, 145, 2456-2460.

Singh, N.N. and Rys, A. (1998), "Electrical Characterization of 6H-SiC Metal Oxide Semi-conductor Structures at High Temperature," *J. Electrochem. Soc.*, 145, 299-302.

Tkachenko, Yu. G., Pilyankevich, A.N., Britun, V.F., Bazilevich, V.D., Opanashchuk, N.F., Dyban', Yu.P., Yurchenko, D.Z. and Yulyugin, V.K. (1979), "Frictional Characteristics and Contact-Zone Deformation Behavior of TiC in its Homogeneity Range," *Soviet Powder Metallurgy and Metal Ceramics*, 18(6), 45-51.

Ueno, K., Asai, R. and Tsui, T. (1998), "4H-SiC MOSFET's Utilizing the H_2 Surface Cleaning Technique," *IEEE Electr. Dev. Lett.*, 19(7), 244-246.

Vathulya, V.R., Wang, D.N. and White, M.H. (1998), "On the Correlation between the Carbon Content and the Electrical Quality of Thermally Grown Oxides on p-type 6H-SiC," *Appl. Phys. Lett.*, 73, 2161-2163.

Yasseen, A.A., Zorman, C.A. and Mehregany, M. (1999), "Surface Micromachining of Polycrystalline SiC Films using Microfabricated Molds of SiO_2 and Polysilicon," *J. MEMS*, 8, 237-242.

ON SOME SIMILARITIES OF STRUCTURAL MODIFICATION IN WEAR AND FATIGUE

L.PALAGHIAN, S.CIORTAN, M.RIPA
Mechanical Engineering Department, "Dunarea de Jos" University
Str. Domneasca, No.47, 6200, Galati, Romania

1. Introduction

The analysis of fatigue and wear damage in air and corrosive environment has to use different methods of investigation depending on the degradation phase characterizing the surface tested.

Taking into account that both damaging processes take place in the superficial layer and in the adjacent one, the different damaging stages may be studied by the methods of solid state physics, statistics and mechanics of solids, etc. All these methods have to point out the modifications at a microscopic level characterizing the kinetics of plastic deformation.

2. Physical pattern of damage by fatigue and wear

The structure of the superficial layer is essentially different from that of the base material. It is characterized by a high level of dislocations' density and by the presence of the inter- and intra- crystalline micro-cracks having a great influence on the fatigue resistance characteristics involved in friction processes.

The presence of stress concentrators leads to the appearance of high traction stress on the surface and thus the ultimate tensile strength is substantially diminished. The internal plastic deformations on the surfaces lead to a higher gradient for the dislocations. If a critical level is reached, the dislocations within the superficial layer tend to unite the existing micro-pors and the micro-cracks, leading to a decreasing in the hardness of this layer. Furthermore, at the boundary between layers, some of them being rich in dislocations others having micro-pors and micro-cracks, the macroscopic crack develops.

This periodicity in accumulating the crystalline lattice defects followed by a damage of the superficial layer (Suh, 1973; Zharin et al. 1986) is characteristic for the wear processes and obeys the general laws of damaging by fatigue.

In the case of fatigue, for a certain level of the applied cyclic stress, the damaging process has several characteristic phases. These phases are: a) incubation period, b) initiation and development of fatigue micro-crack, c) initiation and development of the macroscopic crack, d) final damage. After Forsyth (1962) the phases a) and b) form the

B. Bhushan (ed.),
Fundamentals of Tribology and Bridging the Gap between the Macro- and Micro/Nanoscales, 525–528.
© 2001 *Kluwer Academic Publishers.*

526

first stage of the fatigue damaging process in the same time. Among other theories, the dislocations' theory allows one to study the different phases of dislocations mutual interaction and with unstable structures of crystalline lattice. Within the superficial layer these may get on the free surface making a specific profile of the surface. After that, the number of the crystalline grains involved in the process of plastic micro-deformation increases influencing the modification of the inelastic characteristics of the material. Hence, we can find similarity in surface layer modifications in wear and in fatigue.

3. Methodology of structural changes estimation

The structural modifications in the superficial layer during fatigue and wear in air and in corrosive environment may be studied by physical and physic-chemical methods. Among physical methods, X-ray diffractometry can be used.

One of the physical-chemical methods that can give additional information is the electrode potential measurement.

For wear intensity evaluation we propose to use the electrode potential measurement for a roller-block couple tribomodel. The roller was made of Fc 200 cast iron and the block was made of OL 32 steel, test environment was a 3% NaCl – water solution.

The fatigue tests were runned on OL 32 steel flat samples, in symmetrical bending conditions. The sample was put into the same corrosive environment as in wear (3% NaCl – water solution). The electrode potential measurements were done with an electronic potential-meter both in wear tests and in fatigue test, according to the schemes presented in Figure 1.

The X-ray diffractometry allowed one establishing the structure modifications located in superficial layers (Palaghian et al., 1994; Palaghian, 1997). We measure dislocation density evolution φ obtained by I_{fon}/I_{max} ratio (I_{max} = maximum intensity of the diffraction line) and micro-stresses of second order, evaluated by determining the width of diffraction line β.

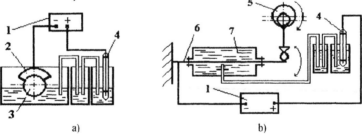

Figure 1. The electrode potential measurement methods for wear (a) and fatigue test (b)
1 - electronic potential-meter; 2 - Fc 200 block; 3 - OL 52 roller; 4 - HgCl electrode;
5 - eccentric bending system; 6 - sample; 7 - corrosive environment chamber.

4. Experimental results

The wear tests have been done with a steel-steel couple, at boundary lubrication.

The fatigue tests were done with plane bending of samples made of the same steel (OL 32).

The evolution of the value I_{fon}/I_{max} for the tests is presented in Figure 2.

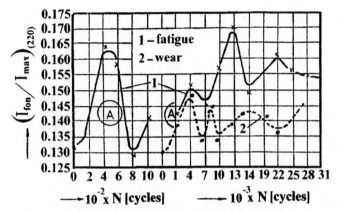

Figure 2. The evolution of the dislocations' density during fatigue and wear processes

For corrosive environment data are presented in Figure 3 and Figure 4 including the electrode potential in wear and fatigue tests, dislocation density evolution and internal second order micro-stresses.

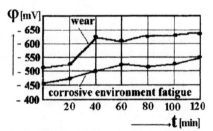

Figure 3. The electrode potential evolutions for wear and fatigue tests

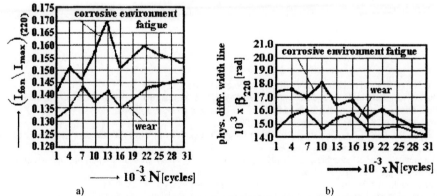

a)   b)

Figure 4. The evolution of the internal second order micro-stresses (a) and the evolution of the dislocation density (b)

528

All measurement shows that both wear and fatigue damages have an accumulating character (Manson et al., 1981). The structural modifications of superficial layer have stages characterized by successive stresses and releases, dislocation density accumulation and relaxation.

At macrostructure level, the corrosion on the walls of the crack is anodic and in their tip catodic. It goes in three stages. First stage is the initial jump of the electrode potential, due to the metal surface's oxide films damage (T_1 period). Second stage is the electrode potential slow evolution (T_2 period). Third stage is the final growth of the electrode potential, in the fatigue case due to the fatigue macroscopic cracks (T_3 period).

The experimental studies show that the damage in the superficial layers in wear and corrosion fatigue, results from microstructure modifications (dislocations evolution, internal micro-stresses) which leads, at macrostructure level, to the cracks propagation.

The analysis shows that both for wear and fatigue, structural modifications take place in several phases, the most intense changes being recorded for the first phase that occurs due to the external loading. The experimental measurement of the dislocation density confirms the theoretical consideration that fatigue as well as wear is periodic processes with general oscillating character. One may note that the highest intensity of the processes is found within the superficial layer.

The structural modifications demonstrate that the damaging processes for both fatigue and wear result from dislocation mobility in air and in corrosive environment.

5. Conclusions

Damage in the superficial layers, in wear or fatigue, takes place in levels of microstructure and macrostructure. For both levels there are similarities but their intensity is greater for fatigue.

At the microstructure level, both for fatigue and wear in air and in corrosive environment, the degradation has several specific stages the most intense modifications taking place in the first one, the dislocations mobility having an important contribution.

The experimental data confirm the fatigue processes cause the hypothesis that in the superficial layers in wear the damage and they may be described by their specific parameters.

6. References

Suh, N.P. (1973), "The Delamination Theory of Wear", Wear, 1, 111-124.
Zharin, A.L., Genkin, V.A., and Roman, O.V. (1986), "The Interrelation of Periodical Changes in the Electron Work Function of Friction Surface and Fatigue", Friction and Wear, 7, 330-341.
Mott, M.T. (1958), Acta Metallurgica, 6, 195-197.
Forsyth, P.J. (1962), "A Two Stage Process of Fatigue Crack Growth", in Proc. Crack Prop. Symp., pp. 94-96, The College of Aeronautics, Cranfield.
Palaghian, L. and Gheorghies, C. (1994), Proc. of the 10th International Conference on Experimental Mechanics, pp. 1191-1193, Lisbon.
Palaghian, L. (1997), Proc. of the First World Tribology Congress, pp. 765, London.
Manson, S.S. and Halford, G.R. (1981), International Journal of Fracture, pp. 169-192.

THE MESOSTRUCTURE OF SURFACE LAYERS OF METAL UNDER FRICTION WITH RELATIVELY HIGH CONTACT STRESS

I.I. GARBAR
Department of Mechanical Engineering
Ben-Gurion University of the Negev,
P.O.Box 653, Beer-Sheva, 84105, Israel

Abstract

Meso-level, which is on the scale of 0.1-1 μm, is known as the gap between nano- and macro-levels. From the structural point of view, it is the level of cells and fragments. These structures are formed in the surface layers of metal during friction, and they strongly affect tribological metal properties.

Attention is being given to the following aspects of mesostructures formed during friction: (a) evolution of dislocation density and distribution: (b) formation of the boundaries of the fragments; (c) space and time-dependent changes of mesostructures in surface layers; (d) the role of mesostructures in strengthening surface layers of metal and their wear resistance. The relation between wear mechanisms and mesostructures formed in the surface layers of metals, and the role of fragments in hardening, negative hardening and failure of material under plastic deformation during friction, are discussed.

The results of transmission electron microscopy (TEM) and X-ray diffraction (XRD) studies of mesostructures formed in the surface layers of metals are reported. Different levels of fragmentation have been observed, depending on both the kind of material, friction and wear conditions. Based on the properties of mesostructures formed during friction, some conclusions concerning the choice of optimal friction conditions are presented.

1. Introduction

In the late 60s and early 70s, Embury et al. (1966), Langford and Cohen (1969), Rack and Cohen (1970) and Langford et al. (1972) found that cellular structure is fully developed in drawn iron wires. Langford and Cohen (1975) showed the size, shapes and crystallographic orientation of the cells. The distribution of cellular disorientation as a function of wire-drawing strain was presented. It was shown that such a structure provides the strengthening effect in drawn iron wires. Analysis by Kuhlmann-Whilsdorf (1998) showed that the late III and IV stages of deformation are observed in these results.

B. Bhushan (ed.),
Fundamentals of Tribology and Bridging the Gap between the Macro- and Micro/Nanoscales, 529–535.
© 2001 *Kluwer Academic Publishers.*

Nakajima and Mizutani (1969) and Inman and Kohn (1971) studied the structural changes of metals after friction. They found that a similar dislocation structure is formed near the surface as in drown wire. In our early papers (Garbar and Skorinin, 1974; Garbar and Skorinin, 1978), and those of Van Dijk (1977), Ives (1979) and Dautzenberg (1980) as well as in some more recent publications (Rainforth et al.., 1992; Hughes et al., 1994;) it was shown that a mesostructure is developed in the surface layers of some BCC and FCC metals during friction. According to the results of these investigations, the dimensions of the surface layer fragments depend on the materials and the friction conditions and range from several nanometers to 0.3-0.7 μm.

2. Time-dependent changes in the structure of surface layers

TEM investigations show that the dislocation structure changes during the running-in period. The tests were carried out in a reciprocating friction machine. Frictional contact was established between two flat surfaces. The initial structure is characterized by single dislocations that are distributed statistically uniformly throughout the metal volume. At the beginning of running-in period, the structure is characterized by elevated dislocation density and dense dislocation networks that form wall embryos of the cellular structure. We also observed that cells orient along the direction of friction. The next stage of deformation is characterized by regions of fully formed cellular and fragmented[1] structure. The fragments are greatly reduced in size and show a relatively greater disorientation during the running-in period. Toward the end of this period, a totally fragmented structure is formed.

The evolution of the fragment structure in the surface layers of metal during sliding is shown schematically in Fig. 1.

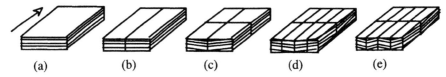

| (a) | (b) | (c) | (d) | (e) |

Figure 1. Scheme of formation of the fragmented structure during sliding; arrow shows the sliding direction.

The size of the fragments decreases, and their mutual disorientation increases during this period as indicated by the fine lines on the sides and ends of the fragments.

---

[1]The specific structure of metals formed under active plastic deformation, in particular in their surface layers during friction, has been defined in several papers as "fine-grained", "subgrained", "cellular" and so on. This often leads to misunderstandings in the interpretation of the results because each of these names corresponds to its own type of structure. Bay et al. (1992) gave a well justified subdivision of deformation microstructures. They show that "dense dislocation walls are formed as the grains are subdivided into cell blocks". Nevertheless, we use the term "fragmented" structure, because this term, which includes all microstructures formed under large plastic deformation, is more appropriate for cases where precision determination of each individual disorientation (by the XRD method, for example) is impossible.

The fragments orient along the sliding direction[2]. This structure reaches equilibrium. A further running produced no significant changes in the dimensions or misorientation of the fragments. The sequence of structural changes follows the progression of strengthening of metal surface layers under friction.

3. Space-dependent changes in the structure of surface layers

The investigations of the structure of surface layers of metal under friction show that the size of the fragments decreases near the surface and the mutual disorientation between the neighboring fragments increases. The scheme of such a structure is shown in Fig. 2.

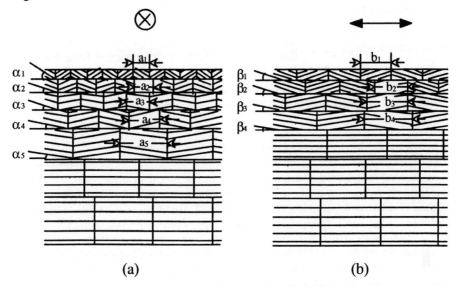

(a)   (b)

Figure 2. Scheme of the fragmented structure of surface layers. Perpendicular sectional view of the sliding surface: (a) normally to the sliding direction; (b) parallel to the sliding direction.

Small fragments extended in the sliding direction (see footnote [2]) are formed in the surface layer. There are no free dislocations within the fragments. Disorientation between the neighboring fragments is several degrees. On the individual boundaries disorientation can reach tens of degrees. As shown in Fig. 2 the size of the fragments increases with the distance from the surface:

$$a_i < a_{i+1}, \; b_i < b_{i+1}$$

---

[2]The study of the dislocation boundaries in cold deformed metals show that some of them lie on crystallographic slip planes while others are non-crystallographic. The letters have a macroscopic orientation with respect to sample axis (Winther et al., 1997). Liu et al. (1998) show that the angle between dislocation walls and rolling direction slightly decreases with increasing strain. Hughes and Nix (1989) show that the deformation structure formed in stages III and IV is characterized by laminar subgrains parallel to applied shear stress. This allows one to suggest that the prevailing orientation of the fragments almost parallel to the sliding direction, can be attributed to the high level of deformation of surface layers of metal under friction.

and their mutual disorientation decreases:

$$\alpha_i > \alpha_{i+1}, \ \beta_i > \beta_{i+1},$$

where, according to Fig. 2, a_i and b_i are the sizes of the fragments normal and parallel to the sliding direction, correspondingly, and α_i and β_i are the mutual disorientations between the fragments around the axes parallel and normal to the sliding direction, correspondingly.

As the distance from the surface increases, one can see the sequence of all stages of plastic deformation which correspond to its decrease down to the structure similar to the initial stage.

The sizes and disorientation of the fragments change in such a way that

$$a_i < b_i \text{ and } \alpha_i > \beta_i.$$

At the same time,

$$a_i/b_i < a_{i+1}/b_{i+1},$$

that is, the sizes of subsrtuctural elements in both the sliding direction and normal direction in the plane of friction become equal, as the distance from the surface increases. Fig. 2 shows alternating signs of crystallographic rotation of fragmented structure around both the parallel and the perpendicular axis. The define structure of surface layers is observed in both directions.

4. Work hardening of surface layers of metal during friction

Formation of fragmented structure in the surface layers is characteristic to metals and the level of fragmentation is associated with the strengthening of the metal during friction (Garbar and Skorinin, 1975; Garbar and Skorinin, 1978; Kuhlmann-Wilsdorf and Ives, 1983; Kato, 1993; Hughes et al., 1994; Perrin and Rainforth, 1997; Garbar, 1997). The disorientation between neighboring fragments ranges from several degrees to several tens of degrees. The walls of fragments restrict dislocation transfer. Therefore the fragment size indicates the level of work hardening, and can be used instead of the grain size in the Hall-Petch equation.

The generalized Hall-Petch equation is

$$\sigma_s = \sigma_0 + KG(b/d)^n$$

where σ_s is the flow stress; σ_0 is the friction stress; K is a constant; G is the shear modulus; b is the Burgers vector; d is the size of the microstructural elements.

These elements are cells and geometrically necessary boundaries (GNB) formed during deformation. In this case according to Hughes (1992), equation (1) can be written as:

$$\sigma_s = \sigma_0 + K_1 Gb/d_1 + K_2 G(b/d_2)^{0.5}$$

where $K_1 Gb/d_1$ and $K_2 G(b/d_2)^{0.5}$ are respectively the contributions of the cells and GNB to flow stress.

Therefore in all cases if

$$d \rightarrow \min, \ \sigma_s \rightarrow \max.$$

In other words, the metal flow limit significantly increases and the work hardening of the metal is maximal if the size of fragments reaches the minimum value inherent for this material, and the mutual disorientation between the neighboring fragments is also maximal. At this point σ_s and σ_{UTS} will have increased by as much as 3 or 4. When this occurs, the maximal ability of the metal to undergo wear is reached.

The level of work hardening depends on the material and on the magnitude of the friction stresses. In a recent paper of Garbar (2000), it was shown that the level of work hardening under friction also depends on the relation between the plastic deformation rate and the rate of relaxation processes. The maximal strengthening of metal during friction can be reached only when the plastic deformation rate is lower than the rate of relaxation processes. The minimal size of fragments and correspondingly the maximal level of work hardening can only be reached in this case.

5. Role of mesostructures in the wear process

Fragmented structures in the surface layers determine not only the level of work hardening, but also the processes and mechanisms of negative hardening and failure of metal during friction, that is the wear processes. Coincidentally, with fragmentation of surface layers their embrittlement takes place. As more developed fragmented structure is formed, there is less possibility of further structural changes. Therefore, the strengthening process is accompanied by lack of plasticity which leads to the destruction of surface layers as the friction forces increase.

It should be noted that the fragmented layer depth, which also depends on the materials being studied and the test regimes, is in all cases greater than or equal to the thickness of the particles formed during wear process. The particles themselves also have a fragmented structure (Sun, (1982); Garbar and Sher, 1999), and, obviously, their formation is due to fragmentation. Consequently, the metal damage that leads to the formation of such particles takes place within the fragmented layer. Therefore, its structure and properties predetermine the kinetics and mechanisms of wear processes during friction.

It is well known, that in the case of uniaxial tension, initial nucleation of the cracks takes place on the fragment walls oriented along the direction of deformation (Rybin, 1986). The mechanisms of crack nucleation in likely fragmented structures are similar under uniaxial tension and wear processes.

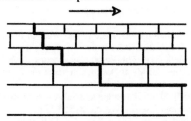

Figure 3. Scheme of crack propagation on the fragment walls and wear particle formation; arrow shows the sliding direction.

534

In the case of friction, such nucleation and the following propagation of the cracks initially should take place on the fragment walls which are parallel to the plane and direction of sliding (Fig. 3), where the processes of stress relaxation are hampered. On the contrary, these processes are facilitated on the fragment walls perpendicular to the sliding plane, where the surface outlet of dislocation is possible. Therefore, here the stress relaxation can take place and nucleation of the cracks is not likely to occurred. One result is the delamination mechanism of wear, proposed by Suh (1973).

Reinforth et al, (1992) show fine microcracks at crystallite boundaries under the friction surface. Garbar et al. (1976), Wert et al. (1989) and Hanlon et al. (1997) observed the mechanism of wear particle formation similar to those in Fig. 3 with different materials and under different friction conditions.

6. Summary

The mesostructures, namely fragmented structures, range from several nanometers to 0.3-0.7 μm, formed in the surface layers of metal as a result of plastic deformation under friction. In many aspects, they determine the tribological properties and behavior of material. Hardening, negative hardening and failure of metal during sliding depend on these structures. The detailed study of substructural mechanisms of wear particle formation can, in our opinion, open up the possibilities for the quantitative prediction of wear under different friction conditions.

7. References

Bay, B., Hansen, N., Haghes, D.A. and Kuhlmann-Wilsdorf, D. (1992), "Evolution of F.C.C. Deformation Structures in Polyslip", *Acta Metallurgica and Materialia*, **40**, 205-219.

Dautzenberg, J.H. (1980), "The Role of Dynamic Recrystallization in Dry Sliding Wear", *Wear*, **60**, 401-411.

Embury, J.D., Keh, A.S. and Fisher, R.M. (1966), "Substructural Strengthening in Materials Subject to Large Plastic Strains", *Transactions of the Metallurgical Society of AIME*, **236**, 1252-1260.

Garbar, I.I. and Skorinin, Yu. V. (1974), " Peculiarities of the Structural State of Deformed by Friction Layer of Low-Carbon Steel", *Mashinovedenie* **6**, 83-87.

Garbar, I.I. and Skorinin, Yu. V. (1975), "Investigation of the Surface Layer Structure Under Friction," *Mashinovedenie* **5**, 106-109.

Garbar, I.I., Severdenco, V.P., and Skorinin, J.V. (1976), "Formation of Wear Products in Sliding Friction", *Soviet Physics Doclady*, **20**, 778-780, (Russian original: Doklady Academii Nauk SSSR, **225** (1975), 546-548).

Garbar, I.I. and Skorinin, Yu.V. (1978), "Metal Surface Layer Structure Formation Under Sliding Friction", *Wear* **51**, 327-336.

Garbar, I.I. (1997), "The Effect of Load on the Structure and Wear of Friction Pair Materials (Example of Low-Carbon Steel and Copper)", *Wear*, **205**, 240-245.

Garbar, I.I. (1999), "Diffraction Methods of Tribosystem Diagnostics", *Tribotest Journal*, **6-1**, 79-93.

Garbar, I.I. (2000), "Critical Structures of Metal Destruction under the Process of Wear" , *Trans. ASME, J. of Tribology*, 122, (2000), 361-366, (also as 99-Trib-10).

Hughes, D.A. and Nix, W.D. (1989), Strain Hardening and Substructural Evolution in Ni-Co Solid Solutions at Large Strains, *Materials Science and Engineering*, **A122,** 153-172.

Hughes, D.A. (1992), "Microstructure and Flow Stress of Deformed Polycrystalline Metals", *Scripta Metallurgia et Materialia*, **27**, 969-974.

Hughes, D.A., Dawson, D.B., Korellis, J.S. and Weingarten, L.I. (1994), "Near Surface Microstructures Developing under Large Sliding Loads", *J. of Materials Engineering and Performance*, **3**(4), 459-475.

Hanlon, D.N., Reinforth, W.M. and Sellars, C.M. (1997), "The Effect of Processing Route, Composition and Hardness on the Wear Response of Chromium Bearing Steels in a Rolling-Sliding Configuration", *Wear*, **203-204**, 220-229.

Inman, M.C. and Kohn, E.M. (1971), "Transmission Electron Microscopy in Lubricant Evaluation", *Wear* **17**, 33-49.

Ives, L.K., 1979, "Microstructural changes in Copper due to Abrasive, Dry and Lubrication Wear," *Proc. Wear of Materials*, 1979, ASME, pp. 246-256.

Kato, K. (1993), "Friction and Wear", *Materials Science and Technology*, R.W. Cahn et al., ed., VCH, Vol. 6, pp. 635-680.

Kuhlmann-Wilsdorf, D. (1998), "Questions You Always Wanted (or Should have Wanted) to Ask about Workhardening", *Mat Res Innovat*, 265-297.

Kuhlmann-Wilsdorf D. and Ives L.K. (1983), "Sub-surface Hardening in Erosion Damaged Copper as Inferred from the Dislocation Cell Structure, and its Dependence on Particle Velocity and Angle of Impact, *Wear*, **85**, 359-371.

Langford, G. and Cohen, M. (1969), "Strain Hardening of Iron by Severe Plastic Deformation", *Transactions ASM*, **62**, 623-638.

Langford, G. and Cohen, M. (1975), "Microstructural Analysis by High-Voltage Electron Diffraction of Severely Drawn Iron Wires", *Metallurgical Transactions A*, **6A**, 901-910.

Langford, G., Nagata, P.K., Sober, R.J. and Leslie, W.C., (1972), "Plastic Flow in Binary Substitutional

Metallurgical Transactions, **3**, 1843-1849.

Liu, Q, Juul Jensen, D. and Hansen, N. (1998), "Effect of Grain Orientation on Deformation Structure in Cold-Rolled Polycrystalline Aluminium", *Acta Materilia*, **46**, 5819-5838.

Nakajima K. and Mizutani Y. (1969), "Structural Change of the Surface Layer of Low Carbon Steels due to Abrading", *Wear* **13**, 283-292..

Perrin, C. and Reinforth, W.M. (1997), "Work Hardening Behaviour at the Worn Surface of Al-Cu and Al-Si Alloys", *Wear* **203-204**, 171-179.

Rack, H.J. and Cohen, M. (1970,) "Strain Hardening of Iron-Titanium Alloys at Very Large Strains", *Materials Science and Engineering*, **6**, 320-326.

Rainforth, W.M., Stevens, R. and Natting, J. (1992), "Deformation Structures Induced by Sliding Contact", *Philosophical Magazine A*, **66**, 621-641.

Rybin, V.V., 1986, *Large Plastic Deformation and Failure of Metals* (in Russian), Metallurgia, Moscow.

Sun, T. C, (1982), "Technique for Preparation of Wear Debris Particles for Transmission Electron Microscopy", *Wear*, **79**, 385-388.

Suh, N.P. (1973), "The delamination theory of wear", *Wear*, 25, 111-123.

Van Dijck, .J.A.B. (1977), "The Direct Observation in the Transmission Electron Microscope of the Heavily Deformed Surface Layer of a Copper Pin After Dry Sliding Against a Steel Ring", *Wear*, **42** 109-117.

Wert , J.J., Srygley, F., Warren, C.D. and McReynolds, R.D. (1989), "Influence of Long-Range Order on Deformation Induced by Sliding Wear", *Wear*, **134**, 115-148.

Winther, G, Juul Jensen, D. and Hansen, N. (199 7), "Dense Dislocation Walls and Microbands Aligned *Acta Materialia*, **45**, 5059-5068.

THE EFFECT OF IMPACT ANGLE ON THE EROSION OF CERMETS

Irina HUSSAINOVA, *Dr.Sc.* and Jakob KUBARSEPP, *Prof.*

Department of Materials technology, Tallinn Technical University,
Ehitajate tee 5, Tallinn, 19806, Estonia

Abstract

The erosive wear resistance of ceramic metal composites with different composition, structure and properties has been investigated. It has been shown that the wear resistance of cermets cannot be estimated only by hardness. The differences in wear resistance between cermet materials with equal hardness levels can be attributed to differences in their resistance to fracture and modulus of elasticity. The present paper discusses some features of the material damage during the process of interaction of abrasive particle with the target surface. Solid particle erosion tests have been performed on three grades of cermets with silica erodents to study the effect of impact angle on erosion rate. The impact angles varied from 30 to 90^0.

1. Introduction

Widespread use of cermets as materials for working elements of various equipment, machine parts and cutting tools in advanced manufacturing processes may be attributed to their unique combination of desirable properties such as high hardness, strength, stiffness and wear resistance.

Because of a lack of common information about different types and grades of cermets, it is of interest and importance to test the materials and to identify their wear behaviour under different conditions to choose the optimum metal-matrix composite. WC-based cobalt-bonded hard metals are most widely used because of their excellent wear resistance-strength combination. The shortage of tungsten and cobalt and their poor corrosion resistance in corrosive mediums and elevated temperatures restricts application of these hard metals. For this reason, the so-called tungsten free hard metals have been developed and adopted in industry. Nickel and iron or their alloys are used as binders in these metal-matrix composites. The most widely known tungsten-free cermets are based on TiC and Cr_3C_2 cemented with Ni and Mo alloys. Steel-bonded cermets form a special group among hard metals. They are of interest because they are inexpensive and heat treatable.

B. Bhushan (ed.),
Fundamentals of Tribology and Bridging the Gap between the Macro- and Micro/Nanoscales, 537–542.
© 2001 *Kluwer Academic Publishers.*

2. Test materials and experimental details

The cobalt content of the WC-Co hard metals tested was 8 - 15 mass %. The content of TiC in TiC-based cermets investigated was 40 - 80 mass %. Four grades of those materials contained bonds with different composition and structure. All TiC-based cermets were sintered in vacuum. The average grain size of carbides was, depending on composition, 2.0 - 2.7 µm, porosity 0.1 - 0.2 vol.%. In the Cr_3C_2 - Ni composites, the content of metal binder was 15 - 30 mass %. Properties of the materials tested are shown in Table 1.

TABLE 1. Composition and properties of cermets investigated.

Grade	Carbide content, wt.(%)	Composition and structure of binder	Vickers hardness number HV	Density ρ, kg m^{-3}	Transverse rupture strength R_{TZ}, GPa	Modulus of elasticity E, GPa
BK8	92 WC	Co	1350	14500	2.3	650
BK15	85 WC	Co	1200	13900	2.35	560
TZC40	40 TiC	FeCr9Si1.5	1150	6100	1.83	300
TZC60	60 TiC	FeCr7Si1.5	1360	5800	2.0	380
TH20A	80 TiC	Ni13 Mo7	1378	5500	1.08	400
TH40A	60 TiC	Ni26 Mo14	1190	5770	1.32	380
K31	85 Cr_3C_2	Ni	1410	6970	0.9	340
KE3	70 Cr_3C_2	Ni	980	7190	1.4	320

The abrasive used in this work was silica, because it is the most common naturally occurring erodent. The particles were sieved into the required size fractions before erosion testing. The erodents were of size 0.1 - 0.3 mm. The mechanical properties of abrasive particles are as following: Hardness $HV = 1100$ and density $\rho = 2150$ kg m^{-3}.
An SEM micrograph of used erodents is shown in Fig.1. The silica particles are rather rounded.

Figure 1. SEM of used SiO_2 particles.

A centrifugal four - channel device was used to accelerate the abrasive particles for the studies of erosive wear resistance. Specimens with dimensions 20x12x5 mm were tested. The surface to be impacted was polished and then cleaned in acetone prior to impacting it with the projectiles. An accuracy of 0.1 µg could be obtained for the target mass loss measurements.

Investigations of the steady state erosion rate were made as a function of the impact velocity and impact angle. Wear conditions were: impact angle, deg - 30, 45, 60, 75, 90 and impact velocity, ms^{-1} – 31, 61. Four test samples were used from each cermet grade.

The erosion rate was determined as volume loss of the target sample per mass of erodent particles (mm^3 /kg) to facilitate the comparison of target materials with different densities.

3. Results

Figures 2 – 4 illustrate the erosion rates of three cermet grades plotted against the angle of impact at a particle velocity of 61 ms^{-1}.

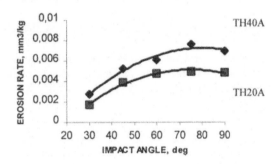

Figure 2. Effect of angle on the erosion rate of TiC-based cermets.

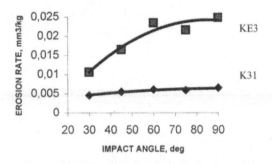

Figure 3. Effect of angle on the erosion rate of Cr$_2$C$_3$-based cermets.

As it can be seen, there are some differences in wear behaviour of materials tested. TiC- based cermets exhibit the maximum erosion rate at an impact angle of 75^0, but Cr$_3$C$_2$- based ones have poorer erosion resistance at 90^0. This is closer in behaviour to brittle ceramic materials. Erosion of brittle materials occurs by the propagation and intersection of cracks caused by impacting particles. At an impact angle of 90^0 the stress is remarkably higher than at an impact angle of 30^0 and, as a result, wear rates are higher. In comparison, ductile materials exhibit maximum erosion rates at impact angles between 30 and 60^0.

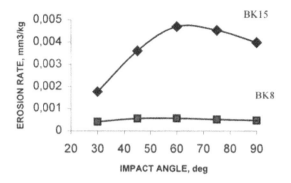

Figure 4. Effect of angle on the erosion rate of WC-Co cermets.

The two tungsten carbide - cobalt cermets, on the other hand, exhibit a maximum erosion rate at 60^0, that is closer to the behaviour of ductile materials. There are clearly two competing mechanisms of erosion, one responsible for the loss of the softer binder phase cobalt and the other leading to the loss of the brittle WC, which would explain the shift in maximum from 30 to 60^0.

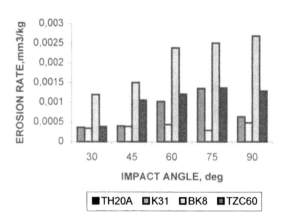

Figure 5. Comparative evaluation of the relative performance of materials tested.
Impact velocity – 31 ms^{-1}.

A comparative evaluation of the relative performance of materials tested (Figs.5 and 6) shows that erosion resistance of the relatively hard alloy K31 compare unfavourably with that of others grades. The intergranular cracking appearing in the abrasion tests of Cr_3C_2-base cermets (Pirso, 2000) may be explained by a weak bonding between carbide grains and a weak intercarbide boundaries.

Examination of wear debris and worn surfaces showed that the main mechanisms were spalling of large individual carbide grains and gradual wear of others on a very small scale. The mechanical and wearing properties of each individual carbide grain and the boundaries between adjacent grains and binder are therefore important for the overall performance of the cermet materials. As it was observed (Ball, 1999), the tungsten carbide based materials do not exhibit any transition into a high wear level that can be attributed to the tough binder phase and the ability of carbide grains to withstand contacts without microfracture.

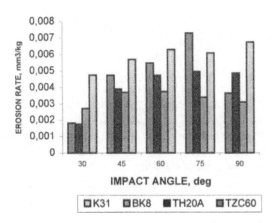

Figure 6. Comparative evaluation of the relative performance of materials tested.
Impact velocity – 61 ms^{-1}.

The relatively brittle TiC-based cermets may be an alternative to WC- based ones in the case of the shallow impact angles and the low kinetic energy of erodents. If the angle of impact becomes higher, the advantage of WC-based grades over others is increased.

Figure 7 shows steady state erosion rates of the tested materials impacted by silica particles travelling at a velocity 61 ms^{-1} under an initial angle of 75^0.

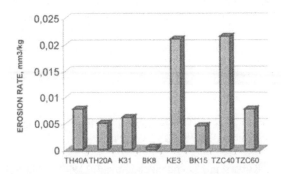

Figure 7. Steady state erosion rates of materials investigated.

In all tests, BK8 clearly outperforms the other materials. The durability of TiC and Cr$_3$C$_2$- based cermets compares unfavourably with that of WC- based ones and the wear resistance difference between different families of cermets at the same hardness reaches 400%. Remarkable differences between erosion rates could not be explained by differences in hardness or shear strength. Therefore for erosive wear resistance evaluation, hardness is only useful as a first approximation. It must be noted that the hardness values are based on low strain rate hardness measurements, but the hardness values actually relevant to the impact conditions occurring in erosion may be somewhat different.

During erosion processes the fracture of material starts locally and in most cases, in the binder phase (Kubarsepp, 1994). Carbide grains lose their protective binder and the eroded surface is almost entirely covered with the exposed carbides. If material hardness exceeds that of abrasive, the erodent particles cause only negligible plastic flow in target material. The degree of elastic penetration and therefore energy transmitted to a surface depends on the elastic modulus and if the latter is high, negligible elastic penetration occurs. Under these conditions the impact of abrasive particles may cause a small-cycle fatigue failure of the carbide skeleton and carbide grains (Hussainova, 1999). Thus, the elastic modulus is an even more important parameter for wear resistance than hardness. Among materials investigated, WC-8%Co alloy has the highest modulus of elasticity.

4. Conclusions

Based on the erosion results presented for three types of cermets, a maximum volume loss occurred for (1) WC-Co cermets at an impact angle of 60^0; (2) TiC-based cermets at an impact angle of 75^0; (3) Cr_3C_2-(Ni,Mo) cermets at an impact angle of 90^0. The maximum position depends on the material response to impact.

In the present study, the relative ranking of the materials investigated with respect to erosion could be explained by their different values of modulus of elasticity, whereas the hardness seems to be of minor importance.

The TiC-based cermets may be an alternative to WC- based ones in the case of low impact angles and low kinetic energy of erodents. The TH20A alloy may be an attractive candidate material for the erosive wear problem provided the hardness of the erodent is greater than that of the target at impact angles lower than 75^0.

5. References

Ball, A., and Feng, Z. (1999), "The erosion of four materials using seven erodents – toward an understanding." *Wear* **233** 233 - 245.

Hussainova, I., Kubarsepp, J. and Shcheglov, I. (1999), "Investigation of impact of solid particles against hardmetal and cermet targets." *Tribology Intern.* **32**, .337- 343.

Pirso, J., Viljus, M. and Kubarsepp, J. (2000) "Abrasive erosion of chromium carbide based cermets." *Proc. of 9th Symposium on Tribology NORDTRIB 2000*, Porvoo, **3**, 959 –964.

Reshetnyak, H. and Kubarsepp, J. (1994) "Mechanical Properties of Hard Metals and Their Erosive Wear Resistance." *Wear* **177**, 185 – 192.

WEAR MECHANISM OF CARBON MATERIAL – STEEL SLIDE BEARING IN POLLUTED ATMOSPHERE

A. POLAK
Cracow University of Technology
ul. Warszawska 24
PL-31-155 Krakow, POLAND

S. PYTKO
University of Mining and Metallurgy
Al. Mickiewicza 30
PL-30-059 Krakow, POLAND

1. Introduction

Automotive vehicles and other machines often operate in an atmosphere of increased dustiness. Normally, there are no conditions in which the air is free from dust. The highest dustiness is present on all kinds of building sites, in mines (shaft and strip ones), quarries and gravel pits. Off-road vehicles, tractors and farming machines, as well as military vehicles also operate in the conditions of increased dustiness. Industrial areas are characterised by a great deal of dustiness.

The mechanism of abrasive interaction in the metal-metal pair has been quite well known. This kind of interaction depends on hard asperities or hard particles being pushed into the surface of the co-operating elements, and thus during relative motion they cause its destruction. In the case of a metal-carbon friction pair a different mechanism can be expected. This results from a considerable difference in hardness between a metal and carbon material, and between carbon material and hard abrasive particle. This is also a result of occurrence of material transfer phenomenon (Langlade et al., 1993; Brandle et al., 1996; Dryzek et al., 1999; Polak, 1999).

2. Experiments and Results

The field investigations were carried out to verify the laboratory results (Polak and Pytko, 2000) in real conditions of operating friction pairs. When choosing the research objectives it was taken into consideration that the bearing should operate in a considerably dusted place and at variable loads.

An intermediate lever of a steering system for a passenger car was chosen because it operates in difficult conditions. It is situated close to the road surface and thus in the

B. Bhushan (ed.),
Fundamentals of Tribology and Bridging the Gap between the Macro- and Micro/Nanoscales, 543–548.
© 2001 *Kluwer Academic Publishers.*

atmosphere of considerable dustiness. Variable loads act on the lever, and it is subjected to rotary motion.

As shown in Figure 1, a new experimental version was used for these studies, which was slightly different than the manufactured version of the lever. The new experimental version was treated only as a model of journal bearing, on which laboratory studies results were being verified. The original metal-rubber bushes were replaced by slide bearings (clearance 0.07 mm), and under the bold head and under the nut a sliding ring was used. These parts were made of carbon material of 50% graphitisation grade, modified by bearing alloy L83. During laboratory studies this material proved to have a small value coefficient of friction and low wear. An original bracket and a bolt with the original diameter were used. The thread length on the bolt was shortened in such a way that the sliding part was longer than the distance between the external edges of the bearing bush.

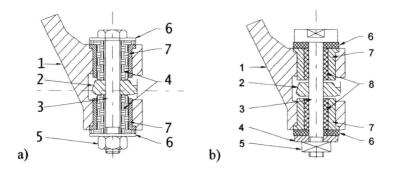

Figure 1. Construction changes in the bearing of intermediate lever of steering system for passenger car; a –manufactured version: 1 - housing, 2 – intermediate lever, 3 – bolt, 4 – metal-rubber bush, 5 – nut, 6 – washer, 7 – rubber-steel bush; b – experimental version: 1 – housing, 2 - intermediate lever, 3 – bolt, 4 – washer, 5 – nut, 6 – sliding ring, 7 – bush casing, 8 – bearing bush.

Field studies were carried out on a 15 000 km run during one year. It is an average run typical of a passenger car. It is important that the vehicle was utilized in difficult winter conditions (in temperatures below -10 °C, with road salinity and a large amount of quartz dust on it and in high temperatures in summer).

During the studies no irregularities occurred in the activity of the steering system, which could have been the result of inefficiency in the bearing of the intermediate lever, or excessive clearance in the intermediate lever bearings.

A phenomenon of bearing bush material transfer onto the steel surface was observed. A transfer film formed on the journal (Figure 2) on part A, mating with the bush and on part B, mating with the sliding ring. On the basis of observation at small magnification with an optical microscope, it can be seen that there are differences in the transfer film structure, resulting from different operating conditions (i.e. different values of pressure, speed and shape of the mating surfaces: the journal-bush pair and the plane-plane pair). A discontinuous transfer film was also formed on end faces of the: bolt head, nut washer and sliding ring (Figure 3, Figure 4). Similar laboratory results were presented by Polak and Pytko (2000).

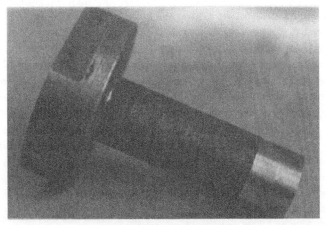

Figure 2. Journal sliding surface of the upper bearing of the intermediate lever steering system for a passenger car after field studies.

Figure 3. Sliding surface of the upper bearing of the intermediate lever steering system for a passenger car after field investigations – the friction pair: carbon washer and journal flange.

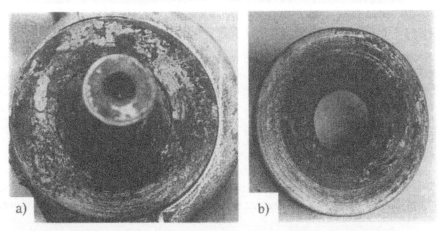

Figure 4. Slide surfaces of the lower bearing of the intermediate lever steering system for a passenger car after field investigations – the friction pair: a - – journal flange, b - carbon washer.

On the basis of microscope observations (at high magnification) and X-ray microanalysis, the presence of a discontinuous transfer film on the friction surfaces was found as well as the presence of iron oxides (Figure 6). The amount of quartz was low, and the dominant abrasive factor proved to be due to iron oxides.

Microscope observations allowed us to find out that the transfer film surface, both on the steel and the carbon material parts (Figure 5) has a similar structure to what was obtained in laboratory conditions. Therefore, the transfer film is discontinuous and heterogeneous. There are smoother areas (the right side of Figure 5) of increased thickness and areas built of loosely connected spherical wear particles (left side of Figure 5). On the smooth surface film cracks are visible, initiated most probably on the phase boundary. Lower thickness discontinuous film structures are also visible. Also, the distribution of abrasive particles is similar than that observed in the laboratory. These particles are not distributed uniformly on the whole surface but occur in groups (Figure 6). Therefore, it may be presumed that the mechanism of formaation of a transfer film under laboratory and field conditions was similar.

Figure 5. Slide surface of the upper bearing of the intermediate lever steering system for a passenger car made from carbon material after field investigations. SEI picture, enlargement 800 X.

Summing up the results of the research carried out in the presence of abrasive particles, it is possible to present the following wear mechanism for the steel-carbon material pair. The action of hard abrasive particles may cause micro-cutting, micro-plastic strains, or micro-cracking of the friction surface. Those processes occur as a result of interaction between particles fixed in counter-surface or loose particles relocating themselves (rotating or moving between mating surfaces), and surfaces of the journal and bush. The surface destruction caused by loose particles is lower than in the case of particles fixed in the counter-surface. This is because they relocate for only a short period, and for most of the time they just roll. In the mutual interaction between the abrasive particles and the friction surface adhesive phenomena and fatigue processes also occur.

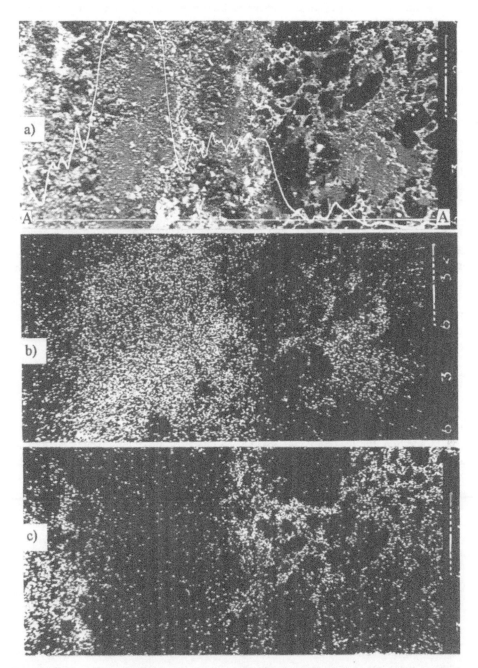

Figure 6. Sliding surface of the upper ring of the intermediate lever steering system for a passenger car made from carbon material after field investigations; a - SEI picture, enlargement 100 X, X-ray analysis of iron content on the surface along straight A-A; b – X-ray microanalysis of the surface - iron distributions; c- X-ray microanalysis of the surface – carbon distribution.

For the mechanism of loose abrasive particle interaction, the following factors are decisive: the difference between hardness of the particle and the material it mates, as well as its dimensions. The greater the hardness differences, the higher wear. However, there is a limit to the difference in hardness, above which the wear has a constant value. A similar dependence occurs when the dimensions of the abrasive particles are considered. Increasing the particle dimensions increased the wear, but only to a limit. Above it the wear remains constant.

In the case of carbon material - steel pair there is an additional factor influencing the destructive activity of the particle: a very large difference in hardness of the mating materials. This difference allows the abrasive particle to be driven into the carbon material surface more easily than the steel surface. In the case of carbon material – steel pairs this causes less intensive wear of the steel journal as compared with the metal – steel pairs.

Such a phenomenon is caused by the fact that the abrasive particle is weakly fixed in the carbon surface. Therefore its relocation (under the influence of tangent force acting on its destructive edge) is easy and occurs in such a way that this edge places itself in relation to the mating surface under such an angle at which no destruction occurs.

Hard abrasive particles, present in a dust atmosphere, do not cause processes preventing material transfer in the carbon material – steel tribological pair. In some conditions there is even a strengthening element of the transfer film structure formed on both mating surfaces, which protects and prevents these surfaces from destruction.

3. Conclusions

In summery it can be stated that field studies proved the possibility of using carbon material for sliding elements operating in difficult conditions.

In addition, the bearing could possibly be operated under unlubricated conditions. Although the adhesive and wear phenomena occur all the time in the bearing due to the destruction and rebuilding of the friction surfaces, the bearing as a whole continues to operate.

4. References

Brendle, M., Turgis P. and Lamouri S. (1996), "A general approach to discontinuous transfer films: The Respective Role of Mechanical and Phisiko-Chemical Interactions", *STLE Transactions*, **39**, pp. 157-165.

Dryzek, J., Polak, A. (1999), "Subsurface zone studied by positron lifetime measurements". *Tribology Letters*, **7**, pp 57-60.

Langlade, C., Fayeulle, S. and Oliver, R. (1993), "Characterisation of graphite superficial thin film achieved during friction", *Applied Surface Science*, **65/66**, pp. 83-89.

Polak, A. (1999), "Mechanism of transfer film formation on metal surface", *Applied Mechanics and Engineering*, **4**, pp. 241-245.

Polak, A., Pytko, S. (2000), "Wear Mechanism of Steel-Plastic Journal Bearing in Heavy Exploitation Conditions", in *Proceedings of the 48th National Conference on Fluid Power*, Chicago, USA, pp. 333-344, OMNIPRESS, Madison, Wisconsin, USA.

NANOMECHANICAL PROPERTIES OF BRITTLE MATTER

BODO WOLF and PETER PAUFLER
*TU Dresden, Institut für Kristallographie und Festkörperphysik
D-01062 Dresden, Germany*

Abstract. The paper focuses on inelastic deformation of brittle materials, quasicrystals in particular. It is shown, that quasicrystals exhibit the normal indentation size effect (ISE) at room temperature, and the inverse one at elevated temperatures. Load-depth-curves exhibit distinct discontinuities, and unloading hysteresis was observed. The concept of energetic hardness is introduced to the calculation of nanohardness in general, and energetic arguments are used to discuss the ISE. Phason defect facilitated dislocation glide is proposed as a possible material displacement process for the smallest indents, whereas fragmentation and grain boundary sliding may be responsible for deformation of quasicrystals in the microhardness range.

1.   Introduction

Nanoindentation testing was primarily developed to test nanoscale structures as thin layers on a substrate. This technique also proves useful for probing the mechanical properties of macroscopic brittle matter, particularly inelastic deformation (hardness testing). Quasicrystals are an interesting class of brittle materials which are very hard at room temperature due to missing translation symmetry, complex structure and covalent bonding (Urban et al., 1999). Macroscopic deformation tests such as uniaxial compression and tension fail at room temperature since quasicrystalline samples break before the onset of inelastic deformation.

Though many investigations were performed at high temperatures (Geyer et al., 2000; Feuerbacher et al., 1997) little is known concerning the room temperature deformation of quasicrystals (Wollgarten and Saka, 1999). We found that the brittleness is a length scale dependent property. The brittle-to-ductile transition temperature of most aluminium based quasicrystals is about or higher than 1000K, if the mentioned macroscopic tests are addressed (Yokoyama et al., 1993). In our own microhardness tests Al-based quasicrystals became ductile at about 750K, much below the macroscopic transition temperature (Wolf and Paufler, 1999a, b), and after nanoindentation at room temperature the AFM-inspection of the impressions revealed deformation without brittle destruction. Deformation based on spallation, fragmentation and microcracking requires a certain energy threshold to create the new fragment surfaces. Thus deformation on the smallest scales should be based on different processes.

In spite of the lack of translation symmetry dislocations can also exist in quasicrystals, and it is now well established that the high temperature quasicrystal deformation is based on dislocation glide (Guyot and Canova, 1999). The first evidence

B. Bhushan (ed.),
Fundamentals of Tribology and Bridging the Gap between the Macro- and Micro/Nanoscales, 549–556.
© 2001 *Kluwer Academic Publishers.*

for quasicrystal dislocation motion at high temperatures was obtained by Wollgarten et al. in 1993. Quasicrystals can be built up from at least two different unit cells according to certain matching rules to ensure local rotation symmetry. However, there is no overall translation symmetry. After having moved two quasicrystal lattice planes one over the other they will fit in many places, but not everywhere. Ensuring a best global fit requires localised atomic rearrangements that violate the mentioned matching rules. After sliding we thus find a network where atoms sit on wrong positions which is called a phason defect, and the introduced quasilattice distortion results in so-called phason strain. The formation of phason defects was found to contribute to macroscopic inelastic deformation at high temperatures which facilitate the atomic rearrangements by diffusion (Feuerbacher et al.,1997). For very small deformed volumes, these defects are also likely to occur at lower temperatures.

2.   Experimental

Nanoindentations were performed using a HYSITRON system (electrostatic transducer) attached to an Atomic Force Microscope (AFM) Nanoscope III of Digital Instruments. Some of the measurements were conducted using a small sample AFM (Multimode AFM) at Universität des Saarlandes, Saarbrücken; in other experiments the HYSITRON system was attached to a Dimension 3000 AFM at TU Dresden. A description of the nanoindentation system is given by Bhushan et al. (1996). High temperature microhardness tests took place at Ruhr-Universität Bochum, Institut für Werkstoffwissenschaften, using a self-constructed Vickers indenter under high vacuum (Wolf et al., 2000a). Experiments on quasicrystals were conducted on single-quasicrystalline specimens. Icosahedral $Al_{70}Pd_{21}Mn_9$ and decagonal $Al_{73}Ni_{14}Co_{13}$ were manufactured at FZ Jülich by Czochralski technique, icosahedral $Y_{10}Mg_{30}Zn_{60}$ and $Ho_{10}Mg_{30}Zn_{60}$ - prepared by top-seeded liquid encapsulated solution growth - were provided by Universität Frankfurt (Main). The samples were subject to mechanical grinding and polishing using diamond paste from 3 to 0.25µm grain size.

3.   Room temperature indentation of quasicrystals

Quasicrystals exhibit a distinct indentation size effect at room temperature, i. e. an increase of the hardness with decreasing load (Figure 1). Load-displacement curves of nanoindentations exhibit multiple discontinuities ("pop-in" effects, Figure 2). The number and the amplitude of the discontinuities are larger when using a corner-of-a-cube indenter, but they also occur in Berkovich indentations. An interesting observation is made when performing a loading-unloading-reloading cycling with increasing peak load (Fig. 3).

Normally the unloading is purely elastic, hence unloading and reloading curves coincide. In some cases a hysteresis loop is formed as with silicon (Figure 4) or with the III-V-semiconductor InSb (Figure 5). If the loop energy W_{loop} is drawn as a function of the load F, a relation $W_{loop} \sim F^{3/2}$ is found for silicon. From this we can conclude that the hysteresis loop formation is based on a volume effect since the characteristic length of the zone which is influenced by the indentation is proportional to $F^{1/2}$, thus the volume is proportional to $F^{3/2}$. The hysteresis is probably due to a pressure induced phase transformation as has been discussed for silicon in literature (Pharr et al., 1991) . InSb - under normal conditions existing in the sphalerite-structure – also has a hexagonal high pressure phase.

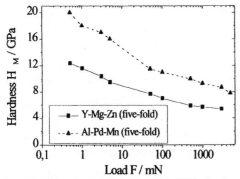

Fig. 1. Meyer hardness versus load (ISE) for i-$Al_{70}Pd_{21}Mn_9$ and i-$Y_{10}Mg_{30}Zn_{60}$ at 300K as derived from Vickers (F > 10 mN) and Berkovich tests (F < 10 mN). Surface perpendicular to the five-fold axis .

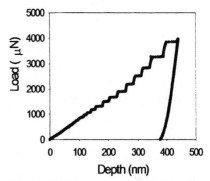

Fig. 2. Multiple pop-in events during nano-indentation of quasicrystalline icosahedral $Ho_{10}Mg_{30}Zn_{60}$.

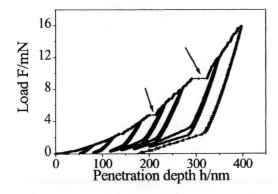

Fig. 3. Formation of "hockey-stick" like shaped unloading hysteresis loops after preceding pop-in events during loading (marked by two arrows) in quasicrystalline icosahedral $Y_{10}Mg_{30}Zn_{60}$.

In contrast to the semiconductors the unloading hysteresis in the quasicrystal is restricted to the lowest part of the unloading/reloading curve, giving it the shape of a hockey stick. Furthermore, the loop energy seems to be related to the occurring of discontinuities during loading. Every pop-in results in a considerable increase of the loop size. This phenomenon may be explained by the formation of lateral cracks, that open and close upon unloading and reloading, respectively. The introduction of phason strain into the sample could also explain this phenomenon.

The indentation size effect and the observation of pop-in phenomena may be explained by deformation of quasicrystals on the basis of fragmentation and grain boundary sliding as proposed by Wollgarten for Vickers microindents (Wollgarten and Saka, 1999). On the other hand the deformed volume in our experiments seems to be too small to explain extended cracking: there should be a load threshold for fragmentation. TEM-studies of very small indents in quasicrystalline Y-Mg-Zn did not show fragments, neither did they hint at phase transformations (Edagawa et al., 1997). It is therefore not unlikely that deformation on the smallest scale is based on quasicrystal dislocation formation and their motion, though this is unlikely for macroscopic deformation owing to the high energetic barriers for dislocation glide. On the other hand phason defects, intensely studied in atomistic simulations by the Stuttgart group of Prof. Trebin (Mikulla et al., 1995) are expected to facilitate dislocation motion. This together with the short motion paths in nanoindentation increases the probability of room temperature dislocation glide in small volumes. In chapter 5 the hardness is considered from the energetic point of view, which also delivers arguments in favour of this conclusion.

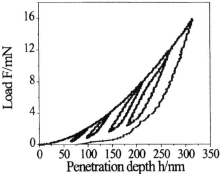

Fig. 4. Hysteresis loops during multiple cycling in single-crystalline silicon.

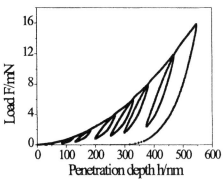

Fig. 5. Hysteresis loops during multiple cycling in single-crystalline InSb.

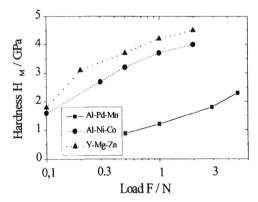

Fig. 6. Inverse indentation size effect of quasicrystals at elevated temperatures in microindentation experiments (T = 550°C for icosahedral $Al_{70}Pd_{21}Mn_9$ and decagonal $Al_{73}Ni_{14}Co_{13}$; and 320°C for icosahedral $Y_{10}Mg_{30}Zn_{60}$).

4.   Quasicrystal Hardness at Elevated Temperatures

At elevated temperatures quasicrystals exhibit an inverse indentation size effect : load increase results in a hardness increase (Figure 6). Unfortunately, high temperature measurements were restricted to microindents, since temperature drifts did not allow us to perform depth sensing nanoindentations at increased temperatures. Our model assumption is, that at elevated temperatures, but below the macroscopic brittle-to-ductile transition the deformation at intermediate length scales (microindentation) is based on fragmentation, but diffusional processes at the fragment interfaces facilitate the deformation similar to processes in superplastic materials. Since large indents result in large fragments, the diffusion is hindered by the long diffusion paths, in contrast to small indents with short lengths for material transportation. For a detailed discussion of quasicrystal hardness at elevated temperatures see (Wolf and Paufler, 1999a) and (Wolf et al., 2000a).

5. Energetical Aspects of hardness

5. 1. INFERRING THE HARDNESS FROM ENERGETIC DATA

The Meyer hardness H_M as load F over contact area A_C can also be understood as energy density - deformation energy dW_p per displaced volume dV_p :

$$H_M = F/A_C = Fdh_p/A_Cdh_p \qquad \text{(dh_p: displacement increment)} . \qquad (1)$$

For H_M = constant we can replace dW_p/dV_p by the quotient of differences:

$$H_e = H_M = W_p/V_p = (1/V_p)([\int Fdh]_{loading} - [\int Fdh]_{unloading}) , \qquad (2)$$

where W_p is the plastic energy: the area inside the loading/unloading cycle (Figure 7).

To determine V_p (irreversibly displaced volume) let us first consider the indenter volume V_1 "inserted" down to a depth h_c (contact depth under load) into the sample:

$$V_1 = A_Ch_c/3 \qquad \text{(ideally shaped pyramid or cone)} . \qquad (3)$$

This is larger than V_p owing to the elastic redeformation. The volume of the remaining depression is (h_r: depth of the remaining depression after unloading)

$$V_p = A_Ch_r/3 = (h_r/h_c)V_1 . \qquad (4)$$

The last relation can also be applied to an indenter of realistic shape. The indenter volume up to a height h from the very end is obtained by :

$$V_1 = \int A_C(h`)dh` . \qquad (5)$$

It has been proven useful to fit the area-height-function $A_C(h)$ by an expression

$$A_C(h) = \alpha_2h^2 + \alpha_1h^1 + \alpha_{1/2}h^{1/2} \qquad \text{(α fit parameter)} , \qquad (6)$$

resulting in

$$V_1(h) = (1/3)\alpha_2h^3 + (1/2)\alpha_1h^2 + (2/3)\alpha_{1/2}h^{3/2} \qquad (7)$$

that finally delivers

$$V_p(h_c) = (h_r/h_c)V_1(h_c) = [(1/3)\alpha_2h^2 + (1/2)\alpha_1h^1 + (2/3)\alpha_{1/2}h^{1/2}] h_r . \qquad (8)$$

5.2. APPLICATION OF THE ENERGETICAL HARDNESS CONCEPT TO INHOMOGENEOUS SAMPLES

Since depth sensing indentation offers easy access to energetic data, it is reasonable to apply energetic aspects to hardness determination. For samples of constant hardness the values H_M (equation 1) and H_e (equation 2) are equivalent. This does not apply to inhomogeneous samples. Owing to the energy integration, the energetic hardness H_e is influenced by the "penetration process history", i. e. by the depth dependent hardness. The influence of the depth dependent hardness on the load-depth curve varies with the shape of the indenter. A punch, for instance, deforms only directly beneath its base plane, whereas a cone deforms closer to the surface. This different surface sensitivity can be quantitatively expressed by calculating a weighted mean value of the indenter penetration which we call effective deformation depth $h_{d,eff}$. For weighing we use the volume element d^2V of material which is displaced in the depth h (Figure 8). If A_z (z) denotes the cross section area of the indenter in the height z, a depth increment $\Delta h = s$ displaces a total material volume dV which is on one hand $dV = A_C s$ (A_C : contact area), and on the other hand $dV = \int d^2V dz$, which results in

$$h_{d,eff} = \int h d^2V / \int d^2V = (1/A_C) \int h dA_z .$$ (9)

One obtains $h_{d,eff} = h$ for a punch (cylinder, prism of constant cross section), $h_{d,eff} = h/2$ for a sphere/paraboloid, and $h_{d,efff} = h/3$ for a cone/pyramid, i. e. an increase of the surface sensitivity in the sequence punch $\to$ sphere $\to$ pyramid. The different surface sensitivity of different impression bodies can now be used to better differentiate between layer and substrate hardness, for instance. A more detailed discussion is given by Wolf et al. (2000b).

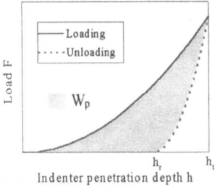

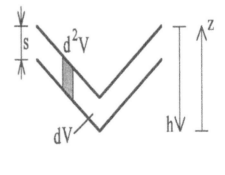

Fig. 7. Loading / unloading cycle with its irreversible deformation energy W_p (shadowed area) (h_r: residual depth after unloading, h_t: total penetration depth).

Fig. 8. Determination of $h_{d,eff}$ by weighing the penetration depth h or indenter height z with the (shadowed) volume $d^2V(z)$ of displaced matter.

5.3. INDENTATION SIZE EFFECT OF BRITTLE MATTER IN THE LIGHT OF ENERGETIC HARDNESS

Many materials exhibit a variation of the hardness with load and penetration depth, respectively. In most cases we have an increase of the hardness with decreasing load which is called normal indentation size effect (Lim and Chaudhri, 1999). Let L_c be a characteristic length of an indent (penetration depth, or diagonal length, e. g.). From the energetic point of view the normal indentation size effect can occur if deformation related structures are two-dimensional ($\sim L_c^2$), whereas the displaced volume is three-dimensional ($\sim L_c^3$). If fragmentation with subsequent grain boundary sliding is the basic deformation process, the deformation energy is in accordance with the above requirement. We can split the deformation energy W_p into the fragmentation energy W_{frag} and the energy W_{move} necessary to displace the fragments. W_{frag} is proportional to the newly generated fragment surface/interface area, whereas W_{move} - depending on the interfacial friction - proves also a two-dimensional quantity. The energetic hardness can be written as

$$H_e = W_p/V_p \sim L_c^2/L_c^3 \sim 1/L_c , \tag{10}$$

if the fragment size scales with the indent size. This will not exactly be the case, but it seems reasonable to assume, that larger indents result in larger fragment sizes, which gives rise to a hardness increase with an indentation size decrease.

We have already mentioned that the deformation could also be based on dislocation motion. In this case we have the formation of defect rich areas along the dislocation path (phason walls) which are two-dimensional structures, too. The formation of phason defects as an energy consuming process is thus also suited to explain the indentation size effect at room temperature as well as the occurrence of discontinuities in the load-depth curves of nanoindentations.

6.   Summary

Three different processes of material displacement during indentation of quasicrystals are proposed:
i) phason defect facilitated dislocation glide at room temperature for very small deformation volumes (nanoindentation),
ii) fragmentation and grain boundary sliding for micro- and macroindents, whereby diffusional processes at the fragment interfaces facilitate the deformation at elevated temperatures, and
iii) thermally activated dislocation motion at temperatures above the macroscopic brittle-to-ductile temperature.
The hardness can also be derived from plastic deformation energy W_p and irreversibly displaced material volume V_p. Particular attention must be paid to the correct determination of V_p because of considerable deviations from ideal indenter geometry in nanoindentation, and the influence of elastic sample redeformation. Measuring the energetic hardness of depth-inhomogeneous samples by indenters of different geometry is proposed to facilitate the determination of depth dependent hardness. Energetic aspects are useful to explain the indentation size effect, particularly with respect to different dimensionalities of hardness determining microstructures.

Acknowledgements

The authors are indebted to the teams of Prof. Urban (FZ Jülich) and Prof. Assmus (Universität Frankfurt/Main) for delivery of quasicrystals. Prof. Vehoff and co-workers

556

(Universität Saarbrücken), M. Kempf in particular, are thanked for valuable support in the field of nanoindentation. Furthermore, K.-O. Bambauer, Ruhr-Universität Bochum, is acknowledged for performing high temperature Vickers indentations. Financial aid by the Deutsche Forschungsgemeinschaft (DFG) and the Deutsche Akademie der Naturforscher Leopoldina, Halle (Saale), is gratefully acknowledged.

References

Bhushan, B., Kulkarni, A. V., Bonin, W. and Wyrobek, J. T. (1996), "Nanoindentation and picoindentation measurements using a capacitive transducer system in AFM", *Philos. Mag.* **A 74**, 1117 - 1128.

Edagawa, K., Suzuki, T. and Takeuchi, S. (1997), "Ultra-microindentation of a Mg-Zn-Y icosahedral quasicrystal", in *Proc. of the 6th International Conference on Quasicrystals*, Tokyo, 1997 (S. Takeuchi and T. Fujiwara eds.), World Scientific, Singapore.

Feuerbacher, M., Metzmacher, C., Wollgarten, M., Urban, K., Baufeld, B., Bartsch, M. and Messerschmidt, U. (1997), "Dislocations and plastic deformation of quasicrystals", *Mater. Sci. Engin.* A226 - 228, 943 - 949.

Geyer, B., Bartsch, M., Feuerbacher, M., Urban, K. and Messerschmidt, U. (2000),"Plastic deformation of icosahedral Al-Pd-Mn singlequasicrystals", *Philos. Mag. A* **80**, 1151 - 1163.

Guyot, P. and Canova, G. (1999), "The plasticity of icosahedral quasicrystals", *Philos. Mag.* **A 79**, 2815 - 2832.

Lim, Y. Y. and Chaudhri, M. M. (1999), "The effect of indenter load on the nanohardness of ductile metals", *Philos. Mag.* **A 79**, 2979 - 3000.

Mikulla, R., Roth, J.and Trebin, H.-R. (1995), "Simulation of shear stress in two-dimensional decagonal quasicrystals", *Philos. Mag.* **B 71**, 981 - 989.

Oliver, W. C. and Pharr, G. M. (1992), "An improved technique for determining hardness and elastic modulus using load and displacement sensing indentation experiments", *J. Mater. Res.* 7, 1564 - 1583.

Pharr, G. M., Oliver, W. C. and Harding, D. S. (1991), "New evidence for a pressure induced phase transformation during the indentation of silicon", *J. Mater. Res.* **6**, 1129 - 1130.

Urban, K., Feuerbacher, M., Wollgarten, M., Bartsch, M. and Messerschmidt, U. (1999),"Mechanical properties of quasicrystals", in *Physical Properties of Quasicrystals* (Z. M. Stadnik, ed.), pp. 361 - 401, Springer Verlag Berlin Heidelberg .

Wolf, B. and Paufler, P. (1999a), "Mechanical properties of icosahedral AlPdMn probed by indentation at variable temperatures", *phys. stat. sol.* **(a) 172**, 341 - 361.

Wolf, B. and Paufler, P. (1999b), "Mechanical properties of quasicrystals investigated by indentation and scanning probe microscopes", *Surf. Interface Anal.* 27, 592 - 599.

Wolf, B., Bambauer, K.-O. and Paufler, P. (2000a), " On the temperature dependence of the hardness of quasicrystals", accepted for publication, to appear in *Mater. Sci. Engin.* .

Wolf, B., Paufler, P. and Kempf, M. (2000b), "An energetic consideration of hardness", submitted to *Philos. Mag. A.* .

Wollgarten, M., Beyss, M., Urban, K., Liebertz, H. and Köster, U. (1993), "Direct evidence for plastic deformation of quasicrystals by means of a dislocation mechanism", *Phys. Rev. Lett.* **71**, 549 - 552.

Wollgarten, M. and Saka, H. (1999), "Microstructural investigations of the brittle-to-ductile transition in Al-Pd-Mn quasicrystals", *Philos. Mag.* A **79**, 2195 - 2208.

Yokoyama, Y., Inoue, A. and Masumoto, T. (1993), "Mechanical properties, fracture mode and deformation behaviour of $Al_{70}Pd_{20}Mn_{10}$ single-quasicrystal", *Materials Transactions, JIM* **34**, 135 - 145.

TESTING TRIBOLOGICAL BEHAVIOUR OF ION-BEAM MIXED SURFACE LAYERS

ZYGMUNT RYMUZA
MACIEJ MISIAK
Warsaw University of Technology
Institute of Micromechanics and Photonics
Chodkiewicza 8, 02-525 Warszawa, Poland
E-mail kup_ryz@mp.pw.edu.pl

ANNA PIĄTKOWSKA
JACEK JAGIELSKI
Institute of Electronic Materials Technology
Wólczyńska 133, 01-919 Warszawa, Poland

Abstract   Ion-beam mixed (IBM) ultrathin (about 50 nm thick) of Si, Mo and W sputtered films on high speed steel substrates have been studied to understand friction, wear and scratching behaviour using various tests. The scales of the tests were macro, meso and micro/nano to compare the effect of the method used on the obtained results. The distribution of the film material as a function of the depth after sputtering and after IBM was studied using RBS technique. The changes of the surface morphology and the surface composition within the wear and scratch traces have been analysed by means of the AFM and EDX techniques respectively. It was found that proper selection of method and the test conditions for scratching or wear testing of ion-implanted materials or materials with ion-mixed ultrathin films on the surface is very important.

1. Introduction

Ion implantation is an interesting method to control the surface properties of materials. Ion implantation is a doping process in which a beam of accelerated ions is directed towards the sample surface (see, for example Burakowski et al., 1999). The process is carried on at low temperatures, in general, from room temperature up to about 150 °C. The temperature is an independent parameter so the sample can be cooled down or additionally heated during the process. Subsequent implantation of different elements can be easily performed, in order to obtain a multi-elemental doping. The other process used to modify interface properties of materials in which a beam of accelerated ions is used, is Ion-Beam Mixing (IBM). In the IBM technique, a thin layer composed of the impurity material is first deposited on the sample surface. This layer is then bombarded with energetic inert ions. The incoming ions colliding with layer atoms transfer to them part of their kinetic energy ensuring the penetration of the layer atoms into the

B. Bhushan (ed.),
Fundamentals of Tribology and Bridging the Gap between the Macro- and Micro/Nanoscales, 557–564.
© 2001 *Kluwer Academic Publishers.*

underlaying substrate. Contrary to ion implantation, in the IBM process, the ion-beam only serves as the energy carrier. During the mixing process, the bombarded target is exposed to heavy radiation damage, leading to the formation of a layer characterised by a metastable, often even amorphous phase structure. The main advantages of IBM with respect to ion implantation are that in the latter method there is almost no impurity concentration limitations and simple ion implanters without mass separation could be used (Jagielski et al., 1999; Zhang et al., 1996). The IBM process is thus a less expensive and more reliable method for metallic impurity doping than classical ion implantation.

Various metals and alloys have been bombarded with different ions giving satisfactory results not only in the laboratory, but also in practical tests (Rymuza et al., 1996; Gerve, 1993). The process often provides the workpiece with multiple benefits, such as wear and corrosion protection, reduced friction and improved fatique resistance. Ion implantation or IBM processing offers a high degree of reliability and reproducibility. This feature of ion implantation or IBM process is crucial for a wide range of sophisticated high precision applications requiring the highest possible process yield. Ion implantation has been chosen for surface treatment of titanium-based artificial joints such as knees, hips, shoulders, and fingers by several major orthopedic manufactures (Sioshansi, 1989). Ion implantation has also been specified for treatment of bearings used in space mechanisms and also a large variety of tooling for stamping, shaping, cutting, and piercing applications are treated by the the ion beam process. As more consistent field results become available, it is anticipated that ion implantation and IBM will play a more important role in the processing of a larger variety of high precision, high technology, high value-added applications.

The modified layer of material after ion implantation or IMB processing is relatively thin (up to several hundreds nanometers). The method of testing to evaluate friction and wear behaviour of ion beam treated materials is very important. Typical methods (such as pin-on-disk test) used by many researchers in laboratory studies often lead to different results (Sioshansi, 1990; Chen et al., 1993; Wei et al., 1994; Fischer et al., 1991; Iwaki, 1987). We have tried therefore to use several methods to test tribological behaviour of IBM ultrathin films and to compare the results. The effect of geometry, the size of the tested tribological system and also the test conditions have been considered in the search of optimum test for IBM ultrathin films deposited on steel substrate.

2. Specimens

Disk specimens were made from high speed steel of composition 0.88%C, 4.2%Cr, 6.5%W, 5.0%Mo, 1.9%V. The diameter of the disks was 14 mm and thickness about 2 mm. All samples were thermally treated to ~64 HRC and polished to $R_a \sim 5$ nm. The 45-47 nm thick Al, Si, Mo and W layers were deposited on such substrates by the RF-sputtering technique. After layer deposition, samples were irradiated at room temperature with 100 keV (Al and Si) or 340 keV (Mo and W) Kr ions up to a dose of 5 x 10^{16} ions/cm^2. The energies of ions were chosen so that the penetration depths slightly exceeded the thickness of the deposited layer. Some samples used for comparison were 100 keV nitrogen ions (2x 10^{17} cm^{-2}) or 100 keV titanium ions + 50 keV carbon ions (3 x 10^{17} cm^{-2} + 1 x 10^{17} cm^{-2}) implanted.

3. Experimental

The layer thickness and depth profiles of the mixed atoms were measured by Rutherford Backscattering (RBS) technique. Scanning electron microscopy (SEM) and atomic force microscopy (AFM) permitted the visual inspection of the sample surfaces. An X-ray microprobe (EDX) of the worn samples was used for determination of composition changes in the wear tracks after tribological tests.

The wear and friction properties of mixed layers were analyzed in a ball-on-flat tribotester using 6.5 or 14.3 mm steel ball as a counterface. A constant force of 20 N was used (which corresponds to the maximum Hertz pressure of 0.17 GPa), the stroke was equal to 5 mm and the average sliding speed was 3.34 mm/s. The tests were carried out in air up to 100 oscillations. During the tests the tangential force was recorded, allowing the analysis of friction coefficient changes. The wear extent was measured via 3D profilometric analysis of the wear tracks. Numerical analyses of removed and built-up volumes were performed in ~1 mm long central track area.

Microscratch tests were performed using a diamond tip with two different radii of curvature : 30 and 2 μm. In the first case, the applied, constant load was 50 mN, the length of scratching was 3 mm and scratching speed was 1 mm/s. The multipass test was carried out. After performing the forward scratch the tip was raised over the surface and moved back to the initial position. The time of the stop on both start-end position was 1 s.

In the second microscratch test the radius of the diamond tip was 2 μm. The constant load of 3 mN was applied and the length of scratching was 200 μm. Also, the tests at linearly increasing load from 0 up to about 40 mN were performed on the distance about 1 mm. The scratching speed was 0.05 mm/s. The AFM on-line observation of the scratch trace was carried out.

4. Results and discussion

The experimentally determined impurity profiles performed by means of the RBS technique have shown that the depth of penetration (into the substrate material) of the material of the sputtered films of Al, Si, Mo and W depends on the dose of irradiation used during ion-mixing. The Al, Si, Mo and W atoms entered up to 50, 30, 60 and 50 nm, respectively, at the dose of 3 x 10^{16} Kr ions/cm^2 and up to 70, 45, 65 and 60 nm at the dose of 5 x 10^{16} Kr ions/cm^2.

The results of the tribological tests using oscillating steel ball are presented in Figs. 1 and 2.

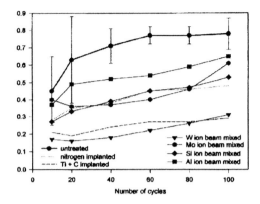

Figure 1. Friction coefficient for various samples used in experiments as a funktion of number of cycles. Oscillating ball 6.5 mm

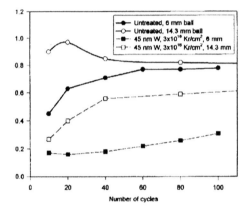

Figure 2. Friction coefficient for ion-mixed sample (45 nm thick W layer) and untreated steel (substrate) at two tests with oscillating ball of 6.5 mm or 14.5 mm diameter.

Three stages of the tribological processes can be recognized. The first stage is characterized by a rapid increase in the friction coefficient because of the removal of the contamination layer from the sample surface (mainly water molecules) acting as a lubricant. The second is the increase of the contact area between the counterface and the sample. The third stage, when the value of friction force almost saturates, corresponds to a three body wear process. The friction coefficient varies from 0.2-0.3 for Ti+C implanted and W-covered layers to 0.60-0.65 for Al layers, while 0.4-0.6 for Si and Mo layers as well as N-implanted layers. The average value for unlubricated steel-on-steel motion was about 0.8. One can note that all ion-beam treatments used led to a friction coefficient decrease. A significantly higher friction coefficient was found when the larger ball was used. The same effect was seen in the case of wear.

The topography of the scratch scars obtained using a scanning microscope with Roentgen microprobe (EDX) together with the profile cross-sections are presented in Fig. 3. The multipass of the diamond-tip effected in the raise and pile-up of the material. The differences were observed only in the width of the scratch scars. The more regular shapes of the boundaries of the scratch were observed in the case of the sample with tungsten ion-mixed film.

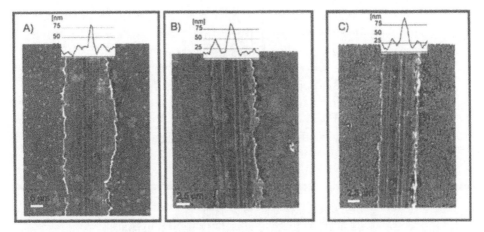

Figure 3. Scratch scars with diamond tip with 2 μm radius. A – untreated steel, B – tungsten layer, C- ion-mixed tungsten layer

The microscratch tests show similar results but in the case of a 30 μm diamond tip and multipass procedure the differences between tribological behaviour of the tested samples were more significant. The EDX analysis has shown the intensive oxidation of the worn surfaces (Fig.4).

The nanomechanical tests (nanohardness and modulus of elasticity) using a Hysitron Triboscope of the samples have shown that the tribological behaviour (wear resistance) of the investigated materials generally correlates with the demonstrated mechanical properties on a nanoscale (the maximum indentation depth was around the thickness of the layers deposited on the high speed steel). The higher dose of ions during ion-mixing effected in higher nanohardness of the layers.

All treatments used resulted in significant friction and wear reduction. The best results were obtained for tungsten layers. The decrease of the worn volume correlates with the reduction of the volume of wear debris accumulated on the track borders. It is very likely that this effect should be attributed to a lower adhesion of the wear debris to the metal surface in ion-beam treated samples.The effect is due to doping with the deposited atoms: the noble gas bombardment leads to a profile evolution similar to that observed for the untreated samples. The infuence of the chemical state modification of the surface is thus clearly seen. In many ion-beam treated samples, the formation of wear debris was not observed even when the worn depth greatly exceeded the thickness of the modified layer. This leads to the conclusion that the initiation of a favorable wear process is crucial for the tribological behaviour of the tested materials.

562

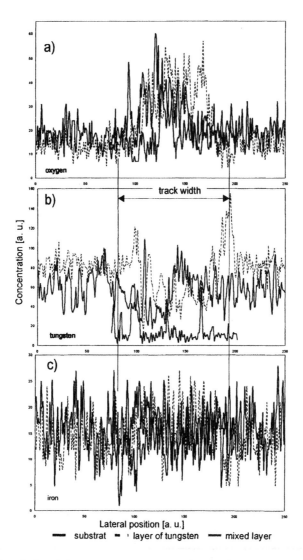

Figure 4. Variation of selected element concentration in scratch tracks for untreated steel (substrate), layer of tungsten and ion-mixed tungsten samples.

5. Conclusions

Ion-mixing is effective to improve the tribological behaviour of the tested system. The friction coefficient was lower compared to the friction coefficient of the untreated high speed steel and the wear rate was lower. The best material was found to be the tungsten ion-mixed layer. The tribological behaviour is a function of complex mechanical and chemical modification of rubbing surfaces and the tendency of wear debris agglomeration in the initial state of the friction process.

The ball-on-oscillating-plate tests helped to recognize the differences in tribological behaviour of the tested samples. It relates both to the friction and wear properties. The better tribological properties were found in the case of a smaller steel ball as the counterface. The qualitative differences in tribological behaviour of the tested samples in the case of the microscratch technique were similar to the case of the ball-on-oscillatory plate method. The ball-on plate method is time consuming; the microscratch tests give the possibility of the rapid investigation, but in this case expensive scratch test equipment is necessary.

Acknowledgements

The financial support from Polish State Committee for Scientific Research (grant no. 7T08C 03 013) is greatly acknowledged. We thank also to Prof. D. Treheux from Ecole Centrale de Lyon and Prof. J. von Stebut from Ecole des Mines Nancy, France for the possibility to use their test equipment.

References

Burakowski, T. and Wierzchon T. (1999), *Surface Engineering of Metals – Principles, Equipment, Technologies,* CRC Press, Boca Raton, Florida.

Chen A., Blanchard J., Conrad J.R., Fetherson P. and Qiu X. (1993), "A study of the relationship between wear rate and nitrogen concentration profile and application to plasma source ion implantation Ti-6Al4-4V alloy", *Wear* **165** , 97-101.

Ensinger W. (1998), "Modification of mechanical and chemical surface properties of metals by plasma immersion ion implantation", *Surface and Coatings Technology,* **100-101,** 341-352.

Fischer G., Welsch G.E., Kim M.Ch., and Schieman E.D. (1991), "Effects of nitrogen ion implantation on tribological properties of metallic surfaces", *Wear* **146,** 1-23.

Gerve A. (1993), "Improvement of tribological properties by ion implantation", *Surface and Coatings Technology,* **60,** 521-524.

"*Ion Beam Modification of Materials*" (1997), Conference Proceedings, last volume published in Nuclear Instruments and Methods in Physics Research **B127/128.**

Iwaki M. (1987), "Tribological properties of ion-implanted steels", *Materials Science and Engineering,* **90,** 263-271.

Jagielski J., Gawlik G., Turos A., Piatkowska A., Treheux D., Starczewski L. and Szudrowicz M. (1999), "Study of micromechanical properties of ion-beam mixed layers", *Nuclear Instruments and Methods in Physics Research* **B148,** 941-945.

Jagielski J., Piatkowska A., Rymuza Z., Kusznierewicz Z., Trehaux D., Boutard D., Thome L. and Gawlik G. (2000), "Micromechanical measurements of ion-beam treated steel", *Wear* **238,** 48-55.

Rymuza Z., Baszkiewicz J., Kusznierewicz Z., Kozubowski J., Krupa D., Barcz A., Gawlik G. and Jagielski J. (1996), "Wear of silicon-ion-implanted biomatarials", *Proceedings of the International Tribology Conference, Yokohama 1995, October 29-November 2, 1995,*Japanese Society of Tribologists, Tokyo, **3,** pp.1951-1956.

Sioshansi P. (1989), "Surface modification of industrial components by ion implantation", *Nuclear Instruments and Methods in Physics Research* **B37/38**, 667-671.

Sioshansi P. (1990), "Improving the properties of titanium alloys by ion implantation", *Journal of Materials,* **March**, 30-31.

Wei R., Shorgin B., Wilbur P.J., Ozturk O., Williamson D.L., Ivanov I.,and Metin E. (1994), "The effects of low-energy nitrogen-ion implantation on the tribological and microstructural characteristics of AISI 304 stainless steel *Journal of Tribology,* **116**, 870-876.

Zhang W., Zhang X., Yang D. and Xue Q. (1996), "Sliding wear study of ion-beam mixing Ni-Mo multilayer films on steel", *Wear* **197**, 228-232.

TRIBOLOGICAL STUDIES OF DLC FILMS CONTANING DIFFERENT AMOUNT OF SILICON COATED BY REACTIVE ION PLATING

Y. ÖZMEN[1], A. TANAKA[2], T. SUMIYA[3]
[1] Pamukkale University
Denizli Meslek Yüksekokulu, 20100 Denizli, TURKEY
[2] Mechanical Engineering Laboratory
Namiki 1-2, 305-8564 Tsukuba, JAPAN
[3] Nanotech Co.
Kashinoha, 277-0882 Kashiwa, JAPAN

Abstract

In order to synthesize DLC film on WC-Co alloy, C_6H_6 and $(CH_3)_6Si_2O$ mixture was used by employing ion-plating apparatus. The Si content in DLC film was varied arranging the mixture ratio of both gases. The film thickness of about 1 μm was obtained. Friction and wear experiments were conducted by using a reciprocating ball-on-plate system under relatively dry atmosphere (RH: 20% ±5%). Fairly small friction coefficients (0.06-0.15) and specific wear rates ($1-4 \times 10^{-7}$) were observed. When the Si content was increased above 5.9 at. % Si, tribological properties were deteriorated. The film containing about 1.6 at. % Si gave relatively good tribological results.

1. Introduction

Silicon doped diamond-like carbon (DLC) coatings hav been investigated by many researchers (Lee et al., (1997); Kim et al., (1999); Fountzoulas et al., (1998); Wu et al., (1999); Gilmore et al., (2000); Miyake et al., (1993); Gangopadhyay et al., (1997a) Oguri et al., (1992)) because of their promising results under humid environments. The excellent tribological properties of DLC films, mainly coated by PACVD method, have been verified in many tests in very dry air and dry nitrogen conditions (<5%) or in UHV (Donnet et al., (1994); Tanaka et al., (1998); Ogletree et al., (1998)). In addition, tribological results of DLC are very well known under high humid (>60%) environment. However, literature information on Si-DLC films synthesized by ion plating under moderately low humid ambient air (<25%) is scarce. The data show that

565

B. Bhushan (ed.),
Fundamentals of Tribology and Bridging the Gap between the Macro- and Micro/Nanoscales, 565–570.
© 2001 *Kluwer Academic Publishers*.

566

the large spread in the values of friction coefficient are caused by variations in the structure and composition of the films, Grill (1997). The majority of tribological studies of DLC films were conducted on the films deposited on silicon wafers. However, many engineering applications would require deposition of these films on metals and ceramics. Considering this point, the application of low friction coatings as a solid lubricants, wear life of DLC film appears to be dependent on not only the friction coefficient but also the hardness, toughness, adhesion, etc of the coating. In addition, the reported data do not permit one to confirm a correlation between the wear resistance and the friction coefficient.

The objective of the study is to (A) generate the tribological data of Si-DLC film coated on WC-Co alloy substrate, known mainly as a die or cutting toll material, by ion plating under moderately low humid ambient air (<25%), and to (B) try to understand the effect of Si and its concentration on its tribological properties.

2. Experimental Details

DLC films were synthesized on WC-Co alloy flats (20mmx20mmx5mm) by a mixture of C_6H_6 and $(CH_3)_6Si_2O$ for the tribological tests and also on Si wafer for film thickness measurement. Deposition of DLC film was done by a reactive ion plating method. A schematic diagram of the ion plating apparatus is shown in figure 1. Changing the ratio of both gases also changed the Si content in the films. The approximate total film thickness about 1-2 μm is composed of Si-DLC/DLC/Si on WC-Co substrate in all coatings except one, which is composed of pure DLC/Si for comparison. Si interlayer was chosen because DLC adheres well to silicon substrate (Gangopadhyay et al., (1997b), Özmen et al., (1999)). First, the substrate was ultrasonically cleaned in acetone and introduced into the chamber.

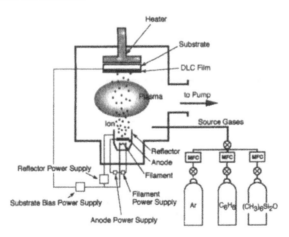

Figure 1. Schematic drawing of DLC coating process.

Then, it was preheated to 200 °C, and exposed to Ar ion bombardment for 40 min at a -2 kV bias voltage. Second, a Si interlayer was coated onto the test specimen by sputtering for 60 min to get 20-30 nm thickness approximately. Third, DLC film was deposited by letting only benzene vapor through the valves into the chamber for 30 min at -2 kV bias voltage. Then, $(CH_3)_6Si_2O$ gas was let into the chamber with desired ratio. The $C_6H_6/(CH_3)_6Si_2O$ ratios were selected as 1/0, 19/1, 9/1, 8/2, 6/4 and 4/6 resulting in Si contents (at. %) of 0, 1.6, 5.9, 8.9, 15.5 and 22.2 respectively, as measured by post-deposition XPS. The deposition rate was approximately 0.3-0.5 μm/h.

Tribological test were conducted by reciprocating ball-on-plate friction and wear test system in an airtight chamber. DLC-coated plate specimens sliding against stationary SiC ball (4.8mm in dia.) were examined under relatively low humidity condition (20% ±5%), which was obtained by introducing dry compressed air into the chamber. Sliding speeds were 25 cycle/min and 100 cycle/min, (5 mm stroke) for total duration of 2 hours. Specimens were cleaned in benzene and acetone solution ultrasonically, and dried in vacuum. Applied loads were selected as 1.06 N and 2.02 N.

3. Results and Discussions

Hardness of the films (figure 2) was measured by a nano indentation method. Scratch resistance (figure 3) was measured to clarify the adhesion of the films to the substrate by the apparatus described by Baba et al., (1986). Although the hardest (23.8 GPa) film is pure DLC, its adhesion is the poorest (about 320 mN). When a small amount of Si (1.6 at.%) is added to DLC film, adhesion is increased considerably (about 580 mN) without loosing much the film hardness value (22.4 GPa). Continuing to increase the Si content decreases the adhesion and hardness steadily, contrary to results of Miyake et al., (1992) and Lee et al., (1997).

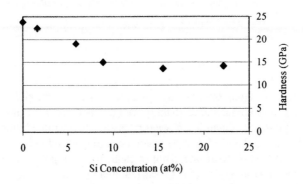

Figure 2. Nano indentation hardness of the DLC films vs. Si concentration.

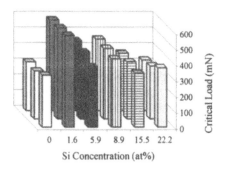

Figure 3. Critical loads vs. Si contents of DLC films. 133.57 [mN/mm] loading rate, 5 [deg] stage angle, 12.7 [mµ/s] stage speed.

The friction forces were recorded throughout the experiments and converted to friction coefficient. Average friction coefficients at the end of experiments are seen in figure 4. Each value corresponds to a new ball contact point and sliding track on the specimens. It is seen in figure 4 that increasing Si amount up to certain level (5.9 at.%) decreases the friction coefficient to 0.06-0.08. Then, it increases again above 0.1 with increasing Si. Gangopadhyay et al., (1997b) and Wu et al., (1999), state that most of the Si in these films is expected to be bonded to carbon. The higher coefficient of the films may be related to a more SiC-like surface. In addition, our ball specimen is made of SiC. It may increase the adhesion of two similar mating materials so that bearing higher friction coefficient. While, the friction coefficient trends are almost same versus Si concentration in all experimental conditions, DLC films containing Si 5.9, 8.9 and 22.2 at. % failed in an early stage under the experimental conditions 2 N normal load and 100 C/M velocity, with the exception of the film having Si 15.5 at. %. Generally speaking, Si amount after certain level (5.9 at.%) did not render the friction property of our DLC films synthesized by ion plating under low humid (20 % ± 5 %) ambient air, and there appears no clear advantage in terms of friction coefficient between Si-DLC/SiC pair. However, Si might be used up to this extend to tailor friction coefficient under relatively low humid ambient air.

Figure 5 gives the wear rate of these experiments versus the Si contents of the DLC films. It is clear that the effect of sliding speed is seen in the wear rate results given in figure 5. Total coating wear volumes were assessed by taking the profile of a wear track at three different points along the track by optic profilemeter at the end of each experiment. Wear rates are increased steadily with the increase of Si amount for the low speed 25 C/M experiments. In addition, wear rates are considerable higher for DLC films having Si 0, 1.6, 5.9 and 15.5 at. % when the sliding speed is 100 C/M. However, the effect of normal load on wear rate is not so clear. Increasing Si concentration after certtain level (at.% 5.9) adversely effect wear rate in all cases. As stated by some of the researchers (Grill (1997); Gilmore (2000)), wear and friction coefficient correlation is also difficult to establish for us. Like friction coefficient, wear is also affected badly from increasing amount of Si especially in high speed. Furthermore, our films also includes O coming from source

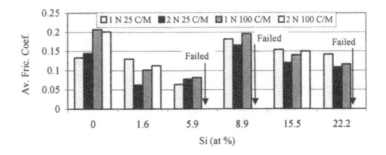

Figure 4. Average friction coefficient throughout the experiments

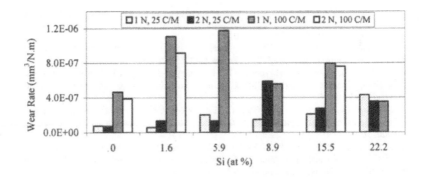

Figure 5. Wear rate at the end of test vs. Si contents.

gas that is expected to act either as lubricant or an agent for tribochemical wear depending on the relative humidity.

4. Conclusion

Small amount of Si (1.6 at.%) addition to DLC film, increased the adhesion of coating about twice (about 580 mN) without loosing so much from the hardness (22.4 GPa) of the film. Continuing to increase the Si content decreases the adhesion and hardness steadily. Increasing Si amount up to certain level (5.9 at.%) decreased the friction coefficient until 0.06-0.08. Then, it increases again above 0.1 with increasing Si more. Si amount after certain level (5.9 at.%) did not render the friction property of our DLC films synthesized by ion plating under low humid (20% ±5%) ambient air, and there appears no clear advantage in terms of friction coefficient between Si-DLC/SiC pair. However, Si might be used up to this extend to tailor friction coefficient under relatively low humid ambient air. Wear rates are considerable higher for almost all coatings when the sliding speed is 100 C/M. However, the effect of normal load on wear rate is not so clear. In addition, increasing Si amount seems adversely effect wear rate in all cases.

5. Acknowledgment

We, gratefully, acknowledge Mrs. A. Goto of Japan Fine Ceramic Center for XPS measurements.

6. References

Baba, S., Kikuchi, A., Kinbara, A. (1986), "A microtribometer for measurement of friction and adhesion of coatings", *J. Vac. Sci. Technol. A.*, **4** 6, 3015-3018.

Donnet, C., Belin, M., Augé, J.M., Grill, A., Patel, V. (1994), "Tribochemistry of diamond-like carbon coatings in various environments", *Surf. Coat. Technol.*, **68/69** 626-631.

Fountzoulas, C.G., Demaree, J.D., Hirvonen, J.K., Kleinmeyer, J.D. (1998), "Nitrogen ion implantation of silicon-containing diamond-like carbon (Si-DLC) coating synthesized by ion beam assisted deposition", *Surf. Coat. Technol.*, **103-104** 104-108.

Gangopadhyay, A.K., Willermert, P.A., Tamor, M.A., Vassel, W.C. (1997a), "Amorphous hydrogenated carbon films for tribological applications II. films deposited on aluminium alloys and steel", *Tribology International*, **30** 1, 19-31.

Gangopadhyay, A.K., Willermert, P.A., Vassel, W.C., Tamor, M.A. (1997b), "Amorphous hydrogenated carbon films for tribological applications I. Development of moisture insensitive films having reduced compressive stress", *Tribology International*, **30** 1, 9-18.

Gilmore, R., Hauert, R. (2000), "Comparative study of the tribological moisture sensitivity of Si-free and Si-containing diamondlike carbon films", *ICMCTF'2000*, San Diego.

Grill, A., Patel, V., Meyerson, B. (1991), "Tribological behavior of diamond-like carbon: effects of preparation methods and annealing", *Surf. Coat. Technol.*, **49** 530-536.

Grill, A. (1997),"Tribology of diamondlike carbon and related materials: an update review", *Surf. Coat. Technol.*, **94-95** 507-513

Lee, K. R., Kim, M. G, Cho, S. J., Eun, K.Y., Seong, T.Y. (1997), "Structural dependence of mechanical properties of Si incorporated diamond-like carbon films deposited by RF plasma-assisted chemical vapour deposition", *Thin Solid Films*, **308-309** 263-267.

Kim, M. G, Lee, K. R., Eun, K.Y. (1999), "Tribological behavior of silicon-incorporated diamond-like carbon films", *Surf. Coat. Technol.*, **112** 204-209.

Miyake, S., Kaneko, R., Miyamoto, T. (1992), "Micro- and macrotribological improvement of CVD carbon film by the inclusion silicon", *Diamond Films and Technology*, **1** 4, 205-217.

Miyake, S., Miyamoto, T., Kaneko, R. (1993), "Microtribological improvement of carbon film by silicon inclusion and fluorination", *Wear*, **168** 155-159.

Ogletree, M.P.D., Monteiro, O.R. (1998), "Wear behavior of diamond-like carbon/metal carbide multilayers", *Surf. Coat. Technol.*, **108-109** 484-488.

Oguri, K., Arai, T. (1992), "Two different low friction mechanisms of diamond-like carbon with silicon coatings formed by plasma-assisted chemical vapor deposition", *J. Mater. Res.*, **7** 6, 1313-1316.

Özmen, Y., Tanaka, A., Kumigai, T. (1999), "Friction and wear of diamond-like carbon films in dry and high humid air", in *Proc. of ADC/FCT' 99 Conference* (M. Yoshikawa, et al., eds.), pp. 289-294, Tsukuba, Japan.

Tanaka, A., Ko, M.W., Kim, S.Y., Lee, S.H., Kumugai, T. (1998), "Friction and wear of diamondlike carbon films deposited using different methods under different conditions", *Diamond Films and Technology*, **8** 1, 51-64.

Wu, W.J., Hon, M.H. (1999), "Thermal stability of diamond-like carbon films with added silicon", *Surf. Coat. Technol.*, **111** 134-140.

TRIBOLOGICAL ASPECTS OF WEAR OF LASER-SINTERED RAPID PROTOTYPE TOOLS

T. SEBESTYÉN[1,2] - J. TAKÁCS[2] - L. TÓTH[2] - F. FRANEK[3] -
A. PAUSCHITZ[3]

[1] BAYATI, Fehérvári út 130. H-1116 Budapest, Hungary
[2] Budapest University of Technology and Economics, Bertalan L. 2. 608.
   H-1111 Budapest, Hungary
[3] Vienna University of Technology, IFWT, Floragasse 7/2/E358 A-1040
   Wien, Austria

Abstract: The aim of this research is the investigation of surface properties, measurement of friction coefficient and wear rate, and determination of the reliability of selective laser-sintered (SLS) tool elements for use as prototype tools. The tests were carried out on a pin-on-ring machine. The pin was constructed of glass-fiber reinforced polyamide-imide, the ring of laser-sintered phosphorous bronze. During the tests the wear rate, the friction coefficient and the change in surface properties were measured at different temperature and loading conditions. The wear rate and the friction coefficient increase with normal force at temperatures up to 80 °C, but over this temperature the coefficient of friction decreases. No extreme destruction or failure occurred in the sintered structure during the tests.

1. Introduction

Technical prototypes and prototype tools can be made quickly from different materials saving cost and time, using the selective laser sintering (SLS) process. Even very complicated machine elements can be realized within a few days or hours since the geometric complexity does not influence the building time. The structure and mechanical properties of parts produced by SLS - in comparison with parts produced by conventional powder metallurgical technology - are more homogeneous. The density and strength might reach 75 - 98 % of the cast or forged structure, depending on the material used and the processing parameters. In the available technical literature there is a lot of data about thermal stability, mechanical and other properties (Eckstein, et al., 1997; van de Crommert et al., 1997; Coremans et al., 1997), little data can be found about tribological properties of SLS parts.

B. Bhushan (ed.),
Fundamentals of Tribology and Bridging the Gap between the Macro- and Micro/Nanoscales, 571–576.

2. Experimental

2. 1. METHOD OF MEASUREMENT

The aim of our investigations was the determination of tribological properties and reliability of SLS tool elements. Molding of plastics with reinforcing glass fibers results in severe, mainly abrasive, wear of dies. According to our earlier research results the plasma-sprayed oxide or carbide coatings increase the reliability of these dies (Csordás et al., 1997). During injection molding, a 100 N/mm^2 maximal tool pressure, 0.7 m/s injection speed and melting temperature (200-250°C or over) of polyamides are routinely encountered. In the tests we tried to exceed these limits for example by using room temperature. In reality plastic forms a film inside the tool, which makes flow of the plastic even easier. We chose the glass-fiber reinforced plastics because the ploughing, abrasive action of the hard fibers results in great wear. Furthermore we used three pins to increase the ploughing actions of fibers, directed perpendicularly to the sintered surface. The length of the channels in these tools were also taken into consideration, so duration of the tests was chosen to achieve a life ten times longer than that of a prototype we intend to create for a series of few hundreds of injection molded parts.

For this reason in the laboratory tests we tried to model the loading conditions of the injection molding tool with the following limits:
Normal force: 100...1500 N (Pressure: 2.5...37.5 N/mm^2)
Velocity: 0.016...0.048 m/s
Temperature interval: 23°C (room temperature)...180°C.
The duration of the tests with a certain set lasted 20...60 minutes. During the tests we did not use any lubricant. The ambient humidity was approximately 60 %.

2. 2. THE TEST RIG AND TESTING METHOD

2. 2. 1. *The Test Rig*

For the investigation of friction and wear properties of SLS tool parts we used the specialized high temperature pin-on-ring laboratory rig of TU-Wien, Institute for Precision Engineering, Tribology and Mechanical Engineering Section. The schematic and measurement method is shown on Fig. 1. (Barabolia et al., 1999).

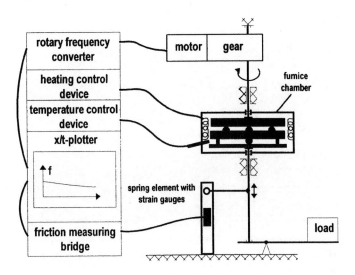

Figure 1. The schematic of high-temperature pin-on-ring machine and block diagram for measurement of wear, coefficient of friction and temperature.

2. 2. 2. *The SLS Specimen and the Counter Body*

For the measurements a special pin-holder apparatus was prepared (Fig. 2.).

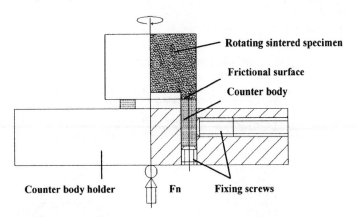

Figure 2. Arrangement of specimen and counter body during the measurements.

The specimens – the rings were sintered with a CO_2 laser from Ni-alloyed phosphorous bronze powder (EOSINT M Cu 3201). Five of ten specimens were filled with Bisphenol A/F epoxyresin. The surface structure of the laser-sintered parts is shown in Fig. 3.

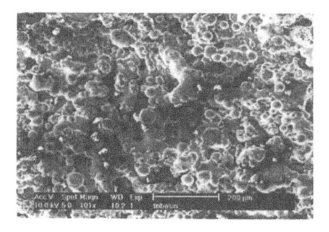

Figure 3. SEM image of the surface of the unfilled specimen.

The frictional surface of the specimen was prepared by lathe turning and smoothed with emery paper. The roughness (Ra) of the original laser treated surface was 10 – 15 μm and after grinding with emery paper it was 1 to 2 μm. The pin was glass-fibre-reinforced polyamide-imide and had a cylindrical shape with 5 mm diameter.

3. Results

3.1. WEAR MEASUREMENTS

Fig. 4. shows the wear value of sls ring and counter pin as the function of normal force.

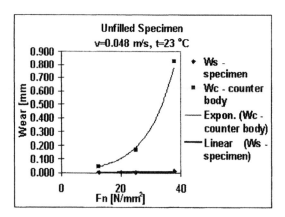

Fig. 4. Wear value of sintered body (Ws) and counter body (Wc) as a function of normal force (Fn) (v=0.048 m/s, t=23°C).

We measured the wear of the sintered specimen (Ws) and the counter body (Wc) in turn of normal forces, temperature and sliding speed.

3. 2. FRICTION COEFFICIENT

The friction coefficient was measured at different loading conditions in terms of normal force (Fig. 5.)

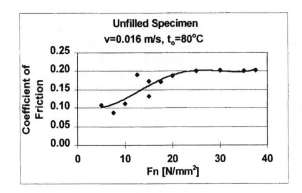

Fig. 5. The change of the frictional coefficient as a function of normal load at 0.016 m/s and on 80 °C.

4. Discussion

The wear rate of sls specimens under the selected conditions is small: it varied between 0.005 and 0.05 mm taking all our cases into consideration. The main wear mechanism is abrasive due to the ploughing action of glass fibers. The wear rate of plastic pins - in comparison with the laser-sintered rings - is high. The wear rates of filled and unfilled phosphorous bronze specimens were almost the same. This wear rate increases rapidly with the normal force, with the sliding speed and with temperature. We approached the measured data with exponential relationship between the F_n and Wc in the examined range of load. The rapid increase is a result of plastic deformation of the pin matrix. The change of speed did not have as strong effect on wear rate as normal force or temperature had.

The friction coefficient at low temperature up to 80°C (furnace temperature) slightly increases with normal force, but over 80°C the value decreases with normal force. At high temperature – around 150-180°C the measured values were lower (0.09-0.06) probably due to partial melting of the matrix.

On the worn surface we did not find cavities, pitting or parts broken away with oxidized surface. The surface roughness of the sintered specimens did not change significantly – it stayed between 1 and 2 μm - if they were prepared before application. Without preparation of the surfaces change of surface roughness was dramatic from 10 to 14 μm down below 10μm. As an additional problem, plastic stuck on the surface in the cases of unprepared specimens.

5. Conclusions

The sintered material resists the wearing effect of the glass-fiber-reinforced plastic surprisingly well. Although increases in temperature and pressure results in higher wear rate on both materials, the rate of increase on the sintered material is not as high as in the case of plastic. It is expectable that speed does not have a serious limiting effect during the application. The epoxy-infiltration of the sintered pieces did not have remarkable effect on wear resistance. Therefore, we found that the selected materials meet the requirements of most prototype tools.

Acknowledgments

The authors are grateful to the National Committee of Research and Development (OMFB) and the National Found of Basic Scientific Research (OTKA) for the support of the Research Theme No. T. 025929 (TeT A38).

References

Barabolia, A., Franek, F. and Pauschitz, A. (1999), Mechanisch-Dynamisch Prüfungen Tribotechnischer Funktionsschichten, *Einsatz und Charachterisierung Technischen Oberflächen für Tribologische Anwendungen*, Symposium 2. Dez. 1999., pp. 75-99.
Coremans, A., Groot, D. (1997), *Residual Stresses and Thermal Stability of Laser Beam Sintered Metal Parts, Proceedings of the LANE '97*, pp. 577-588, Meisenbach Bamberg.
van de Crommert, S., Esser, K. J., Seitz, S. (1997), *New Materials for New Applications in SLS® Selective Laser Sintering, Proceedings of the LANE '97*, pp. 623-628, Meisenbach Bamberg.
Csordás-Tóth, A., Gál, P., Kiss, Gy., Takács, J. and Tóth, L. (TU-Budapest), Franek, F., Pauschitz, A. (IFWT TU-Wien) (1997), *The Investigation of Friction and Wear Properties of Al₂O₃ Plasma-sprayed Coating at Room at Elevated Temperatures, WTC Abstract of Papers, 8 – 12 September 1997*, pp. 154.
Eckstein, M., Son, T., Wiesner, P. (1997), *Rapid Metal Prototyping Using Laser Radiation, Proceedings of the LANE '97*, pp. 555-559, Meisenbach Bamberg.

DISCUSSION FORUM REPORT: BRIDGING THE GAP BETWEEN MACRO- AND MICRO/NANOSCALE WEAR

Jorn Larsen-Basse
National Science Foundation
4201 Wilson Boulevard
Arlington, VA 22230
USA

1. Introduction

A forum was held to discuss issues and needs in bridging the gap between macro-, micro-, and nano-wear. As it turned out, the primary focus was on the macro-to-nano gap in tribology in general with some discussion of wear as well. The following summarizes the author's introduction and the audience input.

2. First item to consider: "Is there a gap?"

In addition to the gap in physical scales and our current inability to directly apply findings from nano-tribology to macro-tribological systems I believe there is a gap in understanding between the various branches of tribology. This has probably always been so, at least since serious application of tribological principles was first attempted. It is so because the field is highly multi-disciplinary and nobody can be a true expert in all of its aspects, from fluid dynamics to chemistry, from contact mechanics to system dynamics, from materials science to surface engineering, from molecular dynamics to finite element applications, from system design to materials selection, etc.

Thus, it is not unexpected that there should be a gap in understanding between the "traditional" tribologists and the newest branch of the field, nano-tribology. This new branch has evolved rapidly over the past decade for two primary reasons. One is the need to solve actual nano-tribology problems, such as those found in hard disk and other hardware systems of the age of information technology, and in processes to produce these systems, such as chemo-mechanical polishing of silicon wafers for IC chips. The other reason is the rather recent and sudden availability of a whole new set of instrumentation and modeling capabilities, which allow us to model friction interactions, conduct indentation and sliding experiments at the nano-scale, and image individual atoms and even manipulate them. This new instrumentation has made it appear possible to gain a much better knowledge of tribology basic, of what really happens at the surface during tribological events.

In the specific area of wear there may well be an even greater gap than for friction and lubrication. Wear at the macroscale has been the subject of much study and many of the general mechanisms are reasonably well understood. At the nano-scale this is not so. Wear occurs, but the mechanisms are largely unknown and not readily determined. They certainly do not produce wear particles of the size seen in common engineering systems, although they may well be operational there, as well. So far, there have been

B. Bhushan (ed.),
Fundamentals of Tribology and Bridging the Gap between the Macro- and Micro/Nanoscales, 577–581.
© 2001 *Kluwer Academic Publishers.*

too many other interesting phenomena to explore at the nano-scale but wear mechanism studies should come soon.

The new instrumentation has nurtured the perception that since we can now quite clearly see and model what happens in the top one or two layers of atoms then we should be able to extrapolate to a complete and full understanding of tribological events at the macro-scale. This perception is often promulgated by the enthusiasm of the many new practitioners of nano-surface science and nano-tribology, who for good reason tend to be untempered by exposure to what the traditional tribologists consider "real" tribology problems.

At the moment it often seems as if the various branches of the research community each approach the topic of tribology in total in the same manner as the group of blind men who each describe the whole elephant from the limited information they each get from the small part of the animal that they can touch. Also, it seems that each branch of tribology is convinced that only it knows best in which direction the field should move.

Is there a Gap?

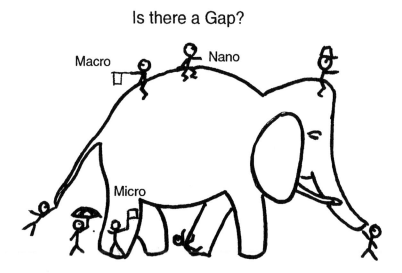

It is not new that the field of tribology has added a new community to its ranks. Since Amontons' paper in 1699, three hundred years ago, major changes in research emphasis have occurred a number of times. The main focus of studies from the time of Amontons' novel approach was the nature of friction and most of these studies were done by physicists. This persisted, at a modest rate, for about 170 years until oil was discovered. At that point, the need to lubricate the railroads and the availability of this new lubricant shifted the focus of tribologists to hydrodynamic lubrication and then specialists in fluid dynamics became the primary tribologists. About eighty years later the successes of the "scientific approach" in developing new technology for the war effort of WW2 spawned a new science age with science as "the eternal frontier". The feeling was then that a good understanding of fundamentals would almost automatically lead to a solution of all technological problems. The field of materials science was created at that time and newcomers from that field to tribology began to focus on wear, expecting to solve questions of wear mechanisms and design of wear resistant materials from basic materials science. That heady promise has now also bitten the dust, as Ken Ludema demonstrated early in this meeting.

Since about 1990 tribological investigations have had perhaps three foci –
1) engineering of surfaces for performance – coatings, surface treatments, ion implantation, etc., based on a general understanding of requirements and properties;
2) tribology for performance of micro- and nano-scale systems – hard disks, MEMS, etc., based also on a very general understanding of basics but constantly moving the performance limits into formerly unexpected size scales; and
3) the science of nano-tribology which uses the AFM and other highly advanced instrumentation as well as large-scale computer simulations to try to develop a basic understanding of all tribology events, from the nano-scale up.

Interestingly, this latter focus area has brought the physicists (and surface scientists) back into tribology, and studies of the "basic nature" of friction are again fashionable. The time periods of the major cycles outlined above have each been about one-half of the prior period – 170/80/40-45. If that trend continues one might expect some new emphasis to emerge around 2010-15.

Given the constant influx of new expertise and new brainpower to the tribology community it is to be expected that there will be gaps in knowledge and understanding between the various communities, occasionally resulting in the views or perceptions of one another being less than charitable, as caricatured below.

Mutual perceptions

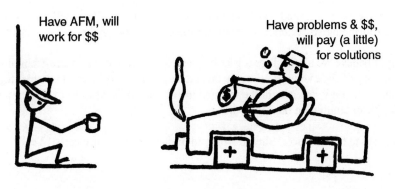

The answer to the question "Is there a gap" is then
"Yes, there are several gaps and with good reason".

3. Second item to consider: "Should we try to bridge the gap?"

No dynamic research community can be an island onto itself; it needs constant input from and interaction with the world around it. Thus, both nano-land and macro-land need connections and bridges to one another as well as to many other areas. Perhaps the bridge between the two need not be a ten-lane highway, which anchors them completely to one another, but at least it should be a footbridge for easy exchange of information, ideas and inspiration.

Bridges - do we need?

Nano is great - will it be big?

The macro-tribologists of the future will be called upon to push the performance envelope for tribological systems for greater efficiency, environmental acceptance, life-cycle performance and cost, etc. To be successful they will need as much fundamental understanding as at all possible, especially including atomic-level information and insight that the nano-tribologists can develop.

The nano-tribologists will be needed to enable operation of the many nano-technology systems on the horizon and also to develop much needed basic information about surfaces, their behavior and ways to modify and design them. They will need help from the macro-tribologists for realistic input parameters and reality check.

Finally, I believe we all need nano-tribology as part of the evolving nano-technology, in part because of the evolving utility of this area and in part because this topic, together with biotechnology, has caught the imagination of the public, as once exploration of the South Pole or landing on the Moon did.

The answer to the question "Should we try to bridge the gap?" is clearly "Yes".

4. Audience Input

A lively discussion ensued. Research needs presented by various members of the audience can be summarized in several subgroups as:

1. Materials-related tribological phenomena at the nano-scale: nano-thickness transfer films, substructure development during sliding, the greatly increased role of oxygen and atmospheric humidity in wear particle formation at the micro- and nano-levels, and the need to understand the behavior of monolayers, both polymeric or metallic. Also, we need reliable models, which span the size ranges, for example in contact mechanics.
2. Terminology – do the different communities really mean the same when they use terms such as friction, lubrication, and wear?
3. Testing – Is there reliable testing at the nano-tribology level, or even agreement on how to test? (It is difficult enough at the macro-level!). And, we do not even have a good mini-tribological tester.
4. Miscellaneous – there is a great need for integrated approaches, for using teams with expertise in various areas, and to learn from related fields, e.g., biomimetics, nano-biotechnology, etc. There will obviously be a need to provide tribology support in the many emerging new areas – bioengineering implants and tools, nano-manipulators for testing of biological material, etc.

Some additional, more specific areas of needed research:

a. Properties of surface layers and nano-structured material – fracture toughness, failure modes, effects of atmospheric humidity, thermal fatigue at stress singularities, wear, surface deformation, role of amorphous grain boundaries, effects of electrostatic phenomena, stiction;
b. Atomic level tribochemistry, atomic level electrochemistry – working in the double layer
c. Does AFM "wear" represent nano-wear in a real micro- or nano-system?
d. Long-term performance of moving MEMS and nano-components
e. Modeling of boundary lubricant films and "superslick" DLC coatings (and are they real?)
f. Nano-biotribology.

5. Summary

The discussion showed that there are gaps between the various communities. It was generally agreed, however, that the gaps are well worth bridging and that the various communities both need one another and have much to contribute in one another's fields. Meetings such as this serve well to bridge some of these gaps. The exciting research needs in the various branches of tribology, outlined in the discussions, will require more bridging both of attitudes and, for the particular case of wear, bridging between size scales of deformation and failure, surface interactions, environmental effects and of other phenomena normally not considered in wear studies, such as electrostatic fields.